TGF-β in Human Disease

Aristidis Moustakas • Keiji Miyazawa
Editors

TGF-β in Human Disease

Springer

Editors
Aristidis Moustakas
Department of Medical Biochemistry
and Microbiology
and
Ludwig Institute for Cancer Research Ltd
Uppsala University
Box 582, SE-751 23
Uppsala, Sweden
aris.moustakas@imbim.uu.se

Keiji Miyazawa
Department of Biochemistry
Interdisciplinary Graduate
School of Medicine and Engineering
University of Yamanashi
1110 Shimokato, Chuo
Yamanashi 409-3898, Japan
kmiyazawa@yamanashi.ac.jp

ISBN 978-4-431-54686-3 ISBN 978-4-431-54409-8 (eBook)
DOI 10.1007/978-4-431-54409-8
Springer Tokyo Heidelberg New York Dordrecht London

Printed on acid-free paper

Springer is part of Springer Science+Business Media (www.springer.com)

Preface

More than 30 years after the discovery and purification of transforming growth factor-β (TGF-β) as a secreted polypeptide from oncogenically transformed cells, this multifunctional cytokine has led major research efforts in diverse fields ranging from embryonic development to adult organ physiology and pathobiology of major diseases, including cancer and fibrotic, cardiovascular, and immunological diseases. Over the years, intense research efforts have resulted in a deep understanding of the signaling pathway engaged by TGF-β and many of the regulatory mechanisms of its signaling engine that allow us today to explain specific aspects of human disease progression based on TGF-β. Despite such major achievements, the field of TGF-β signaling remains one of the most highly evolving and rapidly changing fields of modern biomedical science, a fact that is confirmed by the high rate and impact of the publications produced in the field. This activity underscores the current importance of exploiting modern technological tools to expand and deepen the investigation of TGF-β pathway perturbations in human disease. This ultimate goal of the scientific field is feasible but far from being achieved.

The present book aims to provide a comprehensive coverage of major areas of human disease where the involvement of TGF-β and the mechanistic detail of understanding are firmly established. Simultaneously, the book aims to highlight major gaps of knowledge and future directions of research that can benefit human medical science. A good example is the strong potential of translational application of the TGF-β pathway in therapy for diverse human disorders. In reaching our goals, the book encompasses a broad series of topics with reference to specific human diseases. Owing to the vast number of human disorders where TGF-β is implicated, not all diseases are listed in the contents of this book. Furthermore, as TGF-β is a member of a large family of 33 cytokines that exhibit similarities but also unique features, the book often makes reference to some of the other growth factors of the TGF-β family but it does not comprehensively cover all of them. The core set of diseases where TGF-β action is well documented and are included in the book are cancer and cardiovascular and fibrotic disorders.

Our ambition is that this book will stimulate young scientists to enter the prolific TGF-β field and find new solutions to the many open problems remaining in this area of study. For this reason we have made a special effort to include authoritative educational chapters that provide a good introduction to the field for young doctoral students, postdocs, and clinical fellows. We believe that the book will also be a valuable reference for the pioneers and aficionados of the field, who can find accessible and well-illustrated material for their teaching, lecturing, and other activities, via which the importance of TGF-β biology needs to be disseminated to the world of science and to the public. To meet all these goals, we as editors recruited a select group of world leaders in specific subjects who have delivered outstanding examples of scientific literature. We are in debt to the meticulous work of all these authors and thank them for helping us realize our goal.

Uppsala, Sweden — Aristidis Moustakas
Yamanashi, Japan — Keiji Miyazawa

Contents

Part I
Introduction

Chapter 1
Transforming Growth Factor-β Signaling

Carl-Henrik Heldin

Abstract Members of the transforming growth factor β (TGF-β) family regulate cell proliferation, migration, and differentiation during embryonal development and in tissue homeostasis in the adult. They signal by inducing heteromeric complexes of type I and type II serine/threonine kinase receptors. Ligand binding activates the type I receptor kinase leading to phosphorylation of members of the Smad family, which after oligomerization are translocated to the nucleus where they together with other nuclear factors regulate the transcription of specific genes. TGF-β family members also signal via non-Smad pathways, including Erk, JNK, and p38 MAP-kinase pathways, the tyrosine kinase Src, the small GTPase Rho, and cleavage of the type I receptor whereby the intracellular domain is translocated to the nucleus where it drives an invasiveness program. The TGF-β signaling pathways are carefully regulated by posttranslational mechanisms, including phosphorylation, ubiquitination, acetylation, sumoylation, and PAR-ylation, as well as by positive and negative feedback mechanisms and cross talk with other signaling pathways.

Keywords Posttranslational modifications • Serine/Threonine kinase receptors • Signal transduction • Smad molecules • TGF-β

1.1 Introduction

The transforming growth factor-β (TGF-β) family of ligands has 33 members in humans, including TGF-β isoforms, activins, nodal, bone morphogenetic proteins (BMPs), and growth and differentiation factors (GDFs) (Derynck and Miyazono 2007). They have important roles as morphogens during embryonal development

C.-H. Heldin (✉)
Ludwig Institute for Cancer Research, Science for Life Laboratory,
Uppsala University, Box 595, SE-751 24 Uppsala, Sweden
e-mail: C-H.Heldin@LICR.uu.se

A. Moustakas and K. Miyazawa (eds.), *TGF-β in Human Disease*,
DOI 10.1007/978-4-431-54409-8_1, © Springer Japan 2013

and in the regulation of tissue homeostasis in the adult (Moustakas and Heldin 2009; Wu and Hill 2009). TGF-β family members are pluripotent and regulate cell growth, migration, death, and differentiation; aberrant signaling has been linked with various diseases, such as autoimmune diseases, cardiovascular diseases, and cancer. This review will focus on the mechanisms whereby TGF-β isoforms signal via Smad and non-Smad pathways. A remarkable aspect of TGF-β signaling, which will be discussed, is that the magnitude and duration of signaling is carefully controlled at many different levels, including the synthesis and activation of latent TGF-β isoforms, receptor activation and stability, and the activation and stability of Smad molecules and other downstream signaling molecules.

1.2 TGF-β Ligands

The three TGF-β isoforms are synthesized as precursor molecules that are secreted in latent forms and need to be activated before they can act on their target cells. The about 400 amino acid residue long precursors dimerize and are cleaved by furin-like proteases during secretion; the C-terminal TGF-β molecule thereafter remains bound to the N-terminal part of the precursor, the latency associated peptide (LAP). Within the LAP molecule, an α helix and a "latency lasso" trap TGF-β like a "strait-jacket" (Shi et al. 2011). The latent TGF-β complex often forms larger complexes with certain members of the latent TGF-β binding protein (LTBP) family of fibrillin-like molecules that confer interactions with components of the extracellular matrix (Hyytiäinen et al. 2004). TGF-β isoforms can be released from the latent complexes by exposure to low or high pH, or, more physiologically, by cleavage of LAP by certain proteases, by competition by certain matrix molecules, or by physical forces (Annes et al. 2003). Integrins have been shown to have a central role in TGF-β activation (Nishimura 2009); on the one hand integrins guide proteases to the latent complex, and on the other hand they transmit traction forces which lead to release of TGF-β from the latent complex (Buscemi et al. 2011).

The synthesis of TGF-β isoforms are controlled by external stimuli. Moreover, sortilin, which is structurally related to the yeast vacuolar protein sorting 10 (Vps10p) negatively regulates TGF-β signaling by diverting trafficking of precursor proteins to the lysosomes during transit through the biosynthetic pathway (Kwon and Christian 2011).

1.3 Signaling via TGF-β Receptors

TGF-β isoforms exert their effects on cells by binding to heterotetrameric complexes of two type I and two type II serine/threonine kinase receptors (Fig. 1.1). Altogether there are seven type I and five type II serine/threonine kinase receptors for TGF-β family ligands in humans. All three TGF-β isoforms bind to the TGF-β

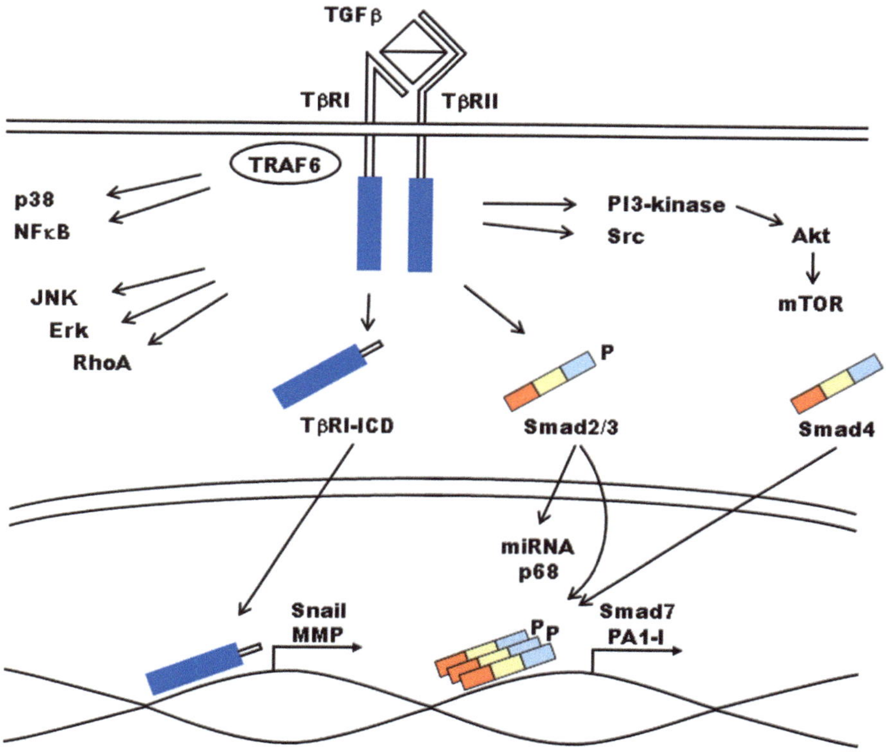

Fig. 1.1 Schematic illustration of major signaling pathways downstream of TGF-β receptors. For explanation, see the text

type II receptor (TβRII) and the TGF-β type I receptor (TβRI; ALK5) that are present on most cell types; in addition, they bind to another type I receptor (ALK1) that are preferentially present on endothelial cells (Goumans et al. 2003).

After TGF-β has induced a complex of TβRII and TβRI, TβRII phosphorylates TβRI in the GS domain located just upstream of the kinase domain (Wrana et al. 1994). Thereby an auto-inhibitory mechanism is perturbed and the TβRI kinase is activated and ready to phosphorylate its substrates, including members of the Smad family. In the ligand-receptor complex, TGF-β signaling is mediated by two autonomously functioning TβRI:TβRII pairs (Huang et al. 2011). Receptor activation by phosphorylation is counteracted by dephosphorylation by the PP2A phosphatase. This phosphatase contains a B regulatory subunit which occurs as two isoforms; the Bα subunit promotes and the Bδ subunit suppresses TGF-β signaling (Batut et al. 2008). Signaling via TβRI is moreover enhanced by sumoylation of the receptor, which possibly promotes the docking and phosphorylation of Smad molecules (Kang et al. 2008).

1.4 TGF-β Co-receptors

The interaction of TGF-β with TβRI and TβRII is enhanced by certain co-receptors, such as the proteoglycan TβRIII (also called betaglycan) and endoglin which is expressed preferentially on endothelial cells (Pardali et al. 2010). TβRIII undergoes ectodomain shedding whereby the extracellular domain is released by proteolysis and then act as an antagonist by scaffolding TGF-β and preventing it from binding to TβRI and TβRII (López-Casillas et al. 1994).

In keratinocytes, TGF-β signaling is negatively modulated by the glycosyl-phosphatidylinositol-anchored protein CD109, which is a member of the α2-macroglobulin family (Tam et al. 2003). The mechanism whereby CD109 exerts its negative effect on TGF-β signaling may involve promotion of TGF-β receptor localization in lipid rafts of the cell membrane and promotion of receptor degradation (Bizet et al. 2012).

The tetraspanin protein CD151/Tspan24, which interacts with integrins and many other receptor types, enhances TGF-β signaling; CD151 appears not to bind TβRI directly but may indirectly affect TβRI distribution (Sadej et al. 2010).

1.5 Functional Domains of Smad Molecules

Members of the Smad family of signal transducers are important substrates for serine/threonine kinase receptors (Fig. 1.2). They have conserved Mad homology (MH)1 and MH2 domains connected by a linker region. The N-terminal MH1 domain has a β-hairpin loop which can bind to DNA, and the C-terminal MH2 domain mediates interaction with receptors, other Smad isoforms and many other molecules (Moustakas and Heldin 2009). The linker region is subject to posttranslational modifications which affect interactions and the stability of Smad molecules.

The receptor-activated (R-)Smads are phosphorylated by the type I receptors in SXS motifs in their extreme C-terminals. The conventional TβRI and the type I activin receptor phosphorylate Smad2 and 3, whereas Smad1, 5, and 8 are phosphorylated by the receptors for most of the type I receptors for BMP and GDF isoforms, as well as ALK1. The activated R-Smads then form complexes with the common-mediator (Co-)Smad (Smad4), usually consisting of two R-Smad molecules and one Smad4 molecule; they are then translocated to the nucleus where they in cooperation with other nuclear factors regulate the transcription of certain genes (see below). Smad2 and Smad3 are 92 % identical in their sequences, but Smad2 has two extra sequences inserted in the MH1 domain which perturb DNA binding, thus giving the two molecules different functional effects.

Members of the subfamily of inhibitory (I-)Smads (Smad6 and 7) have the MH2 domain conserved but do not have any MH1 domain. They are induced after activation of serine/threonine kinase receptors and take part in a negative feedback

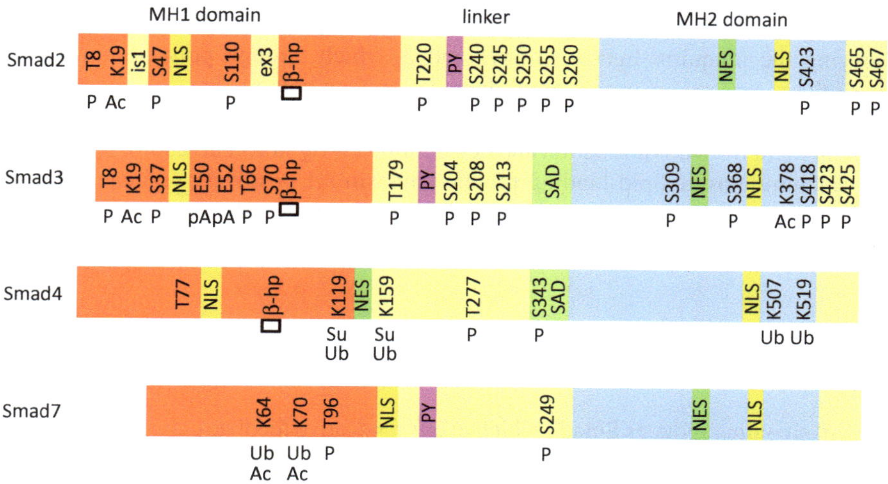

Fig. 1.2 Schematic illustration of the domain structures of the Smad molecules that are involved in TGF-β signaling. Amino acid residues undergoing posttranslationational modifications are indicated. *Is1* inserted sequence 1, *ex3* exon3, *NLS* nuclear localization sequence, *NES* nuclear export sequence, *PY* proline-tyrosine motif, *SAD* Smad activation domain, *β-hp* β-hairpin loop, *P* phosphorylation, *Ac* acetylation, *Ub* ubiquitination, *Su* sumoylation, *pA* poly-ADP-ribosylation

mechanism (see below). Whereas Smad7 interacts with both TGF-β receptors and BMP receptors, Smad6 is selective for BMP receptors.

1.6 Internalization of TGF-β Receptors

TGF-β receptor activation is followed by internalization via clathrin-coated pits into endosomes where the TGF-β receptor complex meets the R-Smad molecules which are delivered to them bound to SARA, a FYVE domain protein that resides in endosomes (Tsukazaki et al. 1998). Trafficking of TGF-β receptors to the endosomal compartment thus enables Smad activation (Hayes et al. 2002; Penheiter et al. 2002). The leucine-rich-repeat- and PDZ domain-containing protein ERBIN can compete with SARA for binding of Smad2 and 3 and thus modulates Smad activation (Sflomos et al. 2011).

Most of the endocytosed TGF-β receptor complexes are recycled back to the cell membrane to serve again in a process that is carefully regulated by RIN1, a Rab5 GTP-exchange factor (GEF) (Hu et al. 2008). This continuous endocytosis and recycling depletes ligand availability and is an important mechanism to fine-tune signaling (Clarke et al. 2009).

In addition to the clathrin-mediated internalization, TGF-β receptors can also be internalized via caveolae which leads to degradation of the receptor complex

(Di Guglielmo et al. 2003). Receptors entering this pathway interact with Smad7 carrying the ubiquitin ligases Smurf1 and 2, which through poly-ubiquitination mark the receptors for proteasomal degradation (see below). This process is controlled by the chaperone protein HSP90 which by binding to the receptors can prevent them from ubiquitination and degradation (Wrighton et al. 2008). Moreover, the calcium/phospholipid-binding protein Annexin A1 positively regulates TGF-β signaling possibly by interfering with TGF-β receptor endocytosis (de Graauw et al. 2010).

1.7 Control of TGF-β Receptor Signaling

TGF-β signaling induces Smad7 which exerts a negative feedback control on TGF-β signaling by competing with R-Smads for interaction with the receptors thereby lowering R-Smad phosphorylation (Hayashi et al. 1997; Kamiya et al. 2010; Nakao et al. 1997), by binding ubiquitin ligases of the Smurf family and bringing them to the receptors which thereby become polyubiquitinated and marked for proteasomal degradation (Kavsak et al. 2000; Zhu et al. 1999), and by binding the PP1C phosphatase and bringing it to the receptors thereby de-phosphorylating and de-activating the receptors (Shi et al. 2004). The negative feedback effect of Smad7 is balanced by TGF-β-induction of TGF-β-stimulated clone 22 (TSC22), which competes with Smad7 for binding to TβRI and thus promotes TGF-β signaling in a positive feedback mechanism (Yan et al. 2011).

The AMP-regulated kinase (AMPK) family member salt-inducible kinase (SIK), which is induced in parallel to Smad7 after TGF-β stimulation, binds to Smad7 and promotes receptor ubiquitination and degradation in cooperation with the ubiquitin ligase Smurf2 (Kowanetz et al. 2008; Lönn et al. 2012).

The ubiquitinated TGF-β receptors can be de-ubiquitinated by the de-ubiquitinase USP4 which binds directly to TβRI; the serine/threonine kinase Akt1 which is activated in a phosphatidylinositol 3′-kinase (PI3-kinase)-dependent manner by TGF-β and other growth factors, phosphorylates USP4 whereby it is re-located to the plasma membrane and promotes TβRI stability and TGF-β signaling (Zhang et al. 2012b). USP15, another de-ubiquitinase which interacts with the Smad7-Smurf2 complex, can also de-ubiquitinate TβRI and thus promote TGF-β signaling (Eichhorn et al. 2012). Interestingly, the USP15 gene has been found to be amplified in glioblastoma and the level of expression correlates with poor prognosis.

Smurf 1 and 2 bind to the PY motifs of I-Smads (and to R-Smads; see below); in addition to promoting ubiquitination and degradation of the TGF-β receptors, they promote ubiquitination and destabilization of Smad7 itself. In addition, the E3 ligases RNF12 (Zhang et al. 2012a) and Arkadia (Koinuma et al. 2003) bind to Smad7 and mediate its ubiquitination and destabilization; since these ubiquitin ligases do not ubiquitinate the receptors, they promote TGF-β signaling by weakening the negative feedback effect of Smad7. The E3 ligase Itch/AIP4 also binds to Smad7 and ubiquitinates it; however, it inhibits TGF-β signaling presumably by enhancing the interaction between Smad7 and TβRI (Lallemand et al. 2005).

1.8 Control of Smad Signaling by Posttranslational Modifications

The activity and stability of Smad proteins are carefully controlled by different mechanisms, including posttranslational modifications, e.g., phosphorylation, ubiquitination, sumoylation, acetylation, and poly-ADP-ribosylation (PAR-ylation) (Fig. 1.2).

1.8.1 Phosphorylation

In addition to the activating phosphorylation by type I receptors in the C-terminals, R-Smads have been shown to be C-terminally phosphorylated after hepatocyte growth factor stimulation, although the kinase involved has not been identified (de Caestecker et al. 1998). Moreover, Mps1, a kinase of the spindle checkpoint, has been shown to bind to Smad4, and activation of Smad2 in response to disruptions of the microtubule network by nocodazole (Dong et al. 2000) has been shown to be dependent on Mps1 (Zhu et al. 2007). Members of the WNK family of kinases have also been shown to phosphorylate R-Smads in their C-terminals, but they also phosphorylate other sites in Smads and thus have both positive and negative effects of the activity of Smads (Lee et al. 2007a).

R-Smads are also subject to negative regulatory phosphorylations; the Erk MAP-kinase phosphorylates them in the linker region, which prevents their accumulation in the nucleus (Kretzschmar et al. 1999). Other kinases that phosphorylate Smad2 or Smad3 in their linkers and suppress their activity include cyclin-dependent kinases (CDKs; Matsuura et al. 2004), p38 MAP-kinase (Kamaraju and Roberts 2005), JNK MAP-kinase (Mori et al. 2004), TGF-β-activated kinase 1 (TAK-1; Benus et al. 2005), MKP38 (Seong et al. 2010), and G-protein-coupled receptor kinase-2 (GRK-2; Ho et al. 2005).

Inhibitory phosphorylations occur also in the MH1 domain. Thus, phosphorylation of the MH1 domains of Smad2 or Smad3 by protein kinase C (PKC) interferes with DNA binding of Smad3 and thus inhibits Smad signaling (Yakymovych et al. 2001). Interestingly, the analogous amino acid residue in Smad4 can be phosphorylated by LKB1, which suppresses Smad4 activity (Morén et al. 2011). The calcium/calmodulin-dependent protein kinase II (CAMKII) phosphorylates Smad2 and Smad3, both in the MH1 domains and in the linkers, which inhibits Smad signaling (Wicks et al. 2000). In addition, glycogen synthase kinase 3β (GSK3β) phosphorylates the MH1 domain of Smad3, but not Smad2, which promotes its ubiquitination and proteasomal degradation (Guo et al. 2008). On the other hand, the Drosophila kinase Misshapen and its mammalian homologs, TNIK, MINK1, and MAPK4, phosphorylate all R-Smads, except Smad3, in their α-helix1 region and inhibit Smad signaling (Kaneko et al. 2011).

The activity of Smad4 is affected by phosphorylation by Erk MAP-kinase in a region in the linker referred to as the Smad activation domain which binds co-activators such as p300 and CBP; this phosphorylation enhances Smad signaling (Roelen et al. 2003). Interestingly, Smad7 can be phosphorylated in the corresponding region, whereas this phosphorylation does not affect the inhibitory effect of Smad7, it does influence its effect on transcription (Pulaski et al. 2001).

R-Smads are de-activated by de-phosphorylation of their C-terminal phosphorylations. Certain phosphatases dephosphorylate Smads in the nucleus, including PPM1A (also called PP2Cα) (Duan et al. 2006; Lin et al. 2006), which thus control the termination of Smad signaling. During hypoxia, PP2A selectively dephosphorylates Smad3 but not Smad2 (Heikkinen et al. 2010b). Other phosphatases, including the FYVE domain containing MTMR4 (Yu et al. 2010), reside in endosomes and thus titrate the magnitude of Smad activation. The inhibitory linker phosphorylation sites are dephosphorylated by small C-terminal domain phosphatases (SCP) 1, 2, and 3; these phosphates do not dephosphorylate the activating C-terminal phosphorylations (Sapkota et al. 2006; Wrighton et al. 2006).

1.8.2 Ubiquitination

The stability of R-Smads is affected by polyubiquitination by different E3 ligases. Some of these ubiquitin ligases act on non-activated Smads and thus titrate the steady state level of R-Smads, whereas others recognize activated Smads and thus contribute to termination of signaling.

Among the E3 ligases that act on nonactivated Smads are the HECT ubiquitin ligases Smurf1 and Smurf2 that bind to PY motifs in the linker regions of Smad1/5 and Smad1/2/3, respectively (Lin et al. 2000; Zhang et al. 2001; Zhu et al. 1999), Nedd4-2/Nedd4L (Kuratomi et al. 2005), Tiul/WWP1 (Komuro et al. 2004; Seo et al. 2004), WWP2 (Soond and Chantry 2011), SCF-ROC1 (Fukuchi et al. 2001) and CHIP (Xin et al. 2005).

Ubiquitin-mediated degradation of R-Smads is likely to be of major importance for termination of TGF-β signaling. TGF-β stimulation induces phosphorylations in the nucleus of the linker regions of Smad2 and Smad3 by CDK8 and CDK9 which serve as priming events for phosphorylation by GSK3 (Alarcón et al. 2009; Millet et al. 2009). The phosphorylations first enhance Smad transcriptional activity, then trigger ubiquitination of Smads by Nedd4-2/Nedd4L followed by proteasomal degradation (Aragon et al. 2011; Gao et al. 2009).

Recent studies have revealed that Smurf2 causes multiple mono-ubiquitination of Smad3 which prevents the formation of Smad complexes (Tang et al. 2011). Thus, ubiquitination can limit Smad signaling both by promoting Smad degradation and by inhibiting Smad interactions. The ubiquitinated R-Smads can be de-ubiquitinated by the de-ubiquitinase USP15 (Inui et al. 2011)

Smad4 is also subject to ubiquitination. The SCFβ^{TrCP} E3 ligase (Wan et al. 2004) and the E3 ligase JAB1/CSN5 (Wan et al. 2002) promote degradation of wild-type Smad4, whereas another member of the SCF family, SCF^{Skp2}, does not affect wild-type Smad4, but promotes degradation of mutated Smad4 found in cancers (Liang et al. 2004). Another ubiquitin ligase, ectodermin/TRIMM33/TIF1γ of the RING family, has been shown to be an E3 ligase for Smad4 (Dupont et al. 2005). Ubiquitination of Smad4 not only controls its stability but also affects its activity. Thus, mono-ubiquitination of Lys507 in Smad4 enhances its activity (Morén et al. 2003). Smad4 mono-ubiquitination can also occur on Lys519 by ectodermin/TRIMM33/TIF1γ which causes export of Smad4 from the nucleus and inhibition of signaling (Dupont et al. 2009). TGF1γ is also involved in epigenetic regulation of Smad signaling (see below). The ubiquitination of Smad4 is removed by the de-ubiquitinase FAM/USP9X (Dupont et al. 2009); other de-ubiquitinases have also been shown to act on ubiquitinated Smads, including UCH37/UCHL5 (Wicks et al. 2005).

1.8.3 Sumoylation

Smad4 has been shown to be sumoylated on Lys113 and Lys159 by the E3 ligase PIAS1. Sumoylation of these residues protects them from ubiquitination and thus prevents proteasomal degradation, thereby enhancing Smad4 signaling (Lee et al. 2003; Lin et al. 2003; Ohshima and Shimotohno 2003). However, there are also observations suggesting that sumoylation suppresses the effect of Smad4; thus, the adaptor protein Daxx represses Smad4 function by binding to sumoylated Lys159 (Chang et al. 2005). It is possible that the mechanism involves Daxx-dependent recruitment of histone deacetylases or silencing factors. Sumoylation has also been shown to promote nuclear export of Smad3, thus repressing signaling (Imoto et al. 2008).

1.8.4 Acetylation

The same lysine residues in Smad7 that can be ubiquitinated, i.e. Lys64 and Lys70, can alternatively be acetylated by the co-activator p300 (Grönroos et al. 2002). Thus, acetylation prevents ubiquitination and stabilizes Smad7. Thus, both sumoylation and acetylation can compete with ubiquitination and thereby fine-tune signaling. The acetyl groups on Smad7 can be removed by de-acetylases (Simonsson et al. 2005). Among the R-Smads, Smad2, but not Smad3, has been shown to be acetylated by p300 in the MH1 domain, in a manner that promotes TGF-β signaling (Simonsson et al. 2006; Tu and Luo 2007).

1.8.5 PAR-ylation

TGF-β stimulation promotes an interaction of Smad3 and Smad4 with poly-ADP-ribose-polymerase 1 (PARP1), whereby they are PAR-ylated. This decreases the binding of Smads to DNA and thus contributes to termination of signaling (Lönn et al. 2010).

1.9 Positive Control of Smad Signaling

In addition to C-terminal phosphorylation by TβRI, Smad signaling is enhanced by other mechanisms. After phosphorylation of Thr179 in the linker of Smad3, it binds the peptidyl-prolyl *cis/trans* isomerase PIN1. Whereas the knockdown of PIN1 had no effect on TGF-β-induced growth inhibition, it inhibited N-Cadherin expression, as well as migration and invasion of PC3 prostate cancer cells (Matsuura et al. 2010). In contrast, a suppressive role of PIN1 has also been observed; PIN1 was found to enhance the binding of Smurf2 to Smad2/3, resulting in enhanced ubiquitination and degradation (Nakano et al. 2009).

Smad3 also binds the pseudokinase Tribbles homolog 3 (TRB3) with its MH2 domain, thereby promoting nuclear localization of Smad3, possibly by preventing Smad3 from interacting with exportin 4 (Hua et al. 2011). Moreover, TBR3 binds to Smurf2 and promotes its degradation, thus enhancing Smad signaling further by limiting the ubiquitination and degradation of Smads.

1.10 Negative Control of Smad Signaling

In addition to posttranslational modifications, several other mechanisms control the magnitude and duration of Smad signaling. Thus, the transmembrane prostate androgen-induced protein (TMEPAI) interacts with R-Smads and prevents their binding to SARA, thereby suppressing Smad activation (Watanabe et al. 2010).

Smad signaling is also affected by mechanisms that change the amount of Smad molecules. For instance, the level of Smad3 is modulated by Ras activation and the levels of Smad3 are dramatically reduced in tumor cell lines with activated H-Ras (Daly et al. 2010). The mechanism involves effects on the mRNA level as well as stability of Smad3.

The zebrafish Piwi protein Zili suppresses Smad signaling by binding Smad4 and preventing complex formation with R-Smads (Sun et al. 2010); this control mechanism is important during early embryogenesis.

SnoN and Ski are negative regulators of TGF-β signaling which repress the transcriptional activities of Smad complexes by recruiting co-repressor complexes and blocking the interaction between Smads and co-activators (Akiyoshi et al. 1999; Luo et al. 1999; Stroschein et al. 1999; Wu et al. 2002). Moreover, Ski binds to TβRI and suppresses Smad activation (Ferrand et al. 2010).

1.11 Nucleocytoplasmic Shuttling of Smads

In order to perform their tasks as transcription factors, Smads need to be translocated to the nucleus. The nucleocytoplasmic shuttling of Smads is carefully controlled by different mechanisms for the various Smads molecules. Activation of R-Smads by receptor-mediated phosphorylation promotes nuclear accumulation, and it is likely that the time the Smad complexes spend in the nucleus determines the strength of the signaling (Inman et al. 2002; Nicolas et al. 2004; Schmierer and Hill 2005; Schmierer et al. 2008).

Among the TGF-β R-Smads, Smad3 has a putative nuclear localization signal (NLS) in the MH1 domain. Phosphorylated Smad3 interacts with importin-β1 of the nuclear pores and is taken into the nucleus by a mechanism that is dependent on the small GTPase Ran (Kurisaki et al. 2001; Xiao et al. 2000a; Xiao et al. 2000b). However, the NLS of Smad2 cannot interact with importin-β1 because of the inserted sequences in its MH1 domain (Kurisaki et al. 2001). Instead nuclear translocation is promoted by an epitope in the MH2 domain which can mediate interactions with FG-repeat-containing nucleoporins, such as CAN/Nup214 (Xu et al. 2000; Xu et al. 2002). Importin 7 and importin 8 have also been implicated in the nuclear import of Smad2, 3, and 4 (Xu et al. 2007; Yao et al. 2008). Moreover, Smad2 has been shown to be transported by kinesin-1 motors along microtubules through the cytoplasm; this transport mechanism is essential for Smad2 nuclear signaling (Batut et al. 2007). The export of Smad3 is dependent on a nuclear export sequence in the MH2 domain and on the Ran GTPase and exportin 4 (Kurisaki et al. 2006) or Ran binding protein 3 (RANBP3) (Dai et al. 2009); the mechanism of nuclear export of Smad2 is not known.

Several mechanisms have been proposed for the translocation of Smad4 into the nucleus. Thus, Smad4 can enter the nucleus in complex with R-Smads, by use of an NLS in the MH1 domain that interacts with importin-α (Pierreux et al. 2000; Xiao et al. 2003), or by interaction with CAN/Nup214 (Xu et al. 2003). The export of Smad4 is dependent on a nuclear export sequence in its linker region and involves binding to the exportin CRM1 (Pierreux et al. 2000; Watanabe et al. 2000).

There are also other mechanisms that regulate the nuclear residence of Smad complexes. Thus, the transcriptional co-activator with PDZ motif (TAZ) binds Smad complexes and anchors them at the chromatin by interaction with the anchor-recruited co-factor (ARC) protein ARC105 (Varelas et al. 2008).

In non-stimulated cells, Smad7, which has an NLS in its N-terminus, resides mainly in the nucleus. Upon TGF-β stimulation, Smad7 is exported from the nucleus (Itoh et al. 1998) in complex with Smurf1 (Ebisawa et al. 2001) or Smurf2 (Kavsak et al. 2000), which have NES epitopes and interact with CRM1 (Tajima et al. 2003).

1.12 MicroRNAs in Smad Signaling

Recent findings have shown that miRNAs have important roles in TGF-β signaling. Firstly, Smads have been shown to bind to the RNA helicase p68, which is a component of the Drosha complex that processes precursor miRNAs (Davis et al. 2008).

Moreover, Smads bind to SBE sequences in the stem structures of pri-miRNAs and thereby facilitates Drosha-mediated maturation of miRNAs (Davis et al. 2010).

Secondly, TGF-β induces several different miRNAs, including miR143/145 promoting smooth muscle cell differentiation via targeting of the transcription factor Klf4 (Davis-Dusenbery et al. 2011; Long and Miano 2011), miR-491-5p which targets the tight junction protein PAR-3 in proximal tubular epithelial cells thus disrupting cell junctions (Zhou et al. 2010), and miR-200 family members which regulate ZEB1 and ZEB2 that in turn have important roles in the regulation of EMT (Ahn et al. 2012).

Finally, Smad signaling is modulated by certain miRNAs. Thus, miR-155 targets Smad2 (Louafi et al. 2010) and miR-130a targets Smad4 (Häger et al. 2011), leading to decreased levels of these proteins and attenuated signaling.

1.13 Smads as Transcription Factors

Microarray analyses have revealed that TGF-β stimulation affects the transcription of several hundreds of genes. R-Smad/Smad4 complexes are of major importance as indicated by the dramatic effect of Smad4 knockdown on the transcriptional effects of TGF-β (Kowanetz et al. 2004).

1.13.1 DNA Binding of Smads

Smad3 and 4 bind DNA by an 11-amino-acid β-hairpin in their MH1 domains which contacts the major groove of DNA at the half-site 5′-GTCT-3′ and its reverse 5′-AGAC-3′ (Shi et al. 1998; Zawel et al. 1998). Whereas the most common form of Smad2 cannot bind DNA because of an inserted sequence immediately adjacent to the β-hairpin (Dennler et al. 1998; Shi et al. 1998), there is a splice variant of Smad2, lacking the inserted sequence, Smad2Δexon3, that does bind DNA (Yagi et al. 1999).

Since the DNA binding motif is short and thus common in the genome, and since DNA binding of Smads occurs at low affinity, the Smads are dependent on interactions with other transcription factors for their specificity.

1.13.2 Cooperation of Smads with Other Nuclear Factors

After Smad complexes and their transcriptional partners have bound to DNA, co-activators, such as the histone acetyltransferases p300 and P/CAF, are recruited, which facilitates initiation of transcription. Recent findings suggest that Smads bound to chromatin needs chromatin remodeling factors, such as Brahma-related gene 1 (BRG1) and ARC105 (Schmierer and Hill 2007; Xi et al. 2008).

Even if the most common configuration of Smad complexes is two R-Smad molecules and one Smad4 molecule, there are reports that the Smad4 molecule in the complexes may be replaced by ectodermin/TRIMM33/TGF1γ in the regulation of hematopoietic differentiation (He et al. 2006). In TIF1γ-depleted cells, Smad4 is more available for association with Smad2/3, leading to an enhanced TGF-β signaling which promotes EMT (Hesling et al. 2011). Moreover, in the epidermis of Smad4-null mice, the IκB kinase α (IKKα), which regulates the nuclear factor κB (NFκB) pathway, can form complexes with Smad2 and 3 and regulate keratinocyte differentiation by binding to the promoters of *Mad1* and *Ovol1* (Descargues et al. 2008).

By interacting with specific transcriptional co-factors, Smad complexes can induce groups of genes that in a coordinated manner regulate a specific response. Examples of such synexpression are the FoxO transcription factors that together with Smads regulate 11 genes that define the cytostatic, apoptotic, and adaptive response of keratinocytes (Gomis et al. 2006), the helix-loop-helix protein human homolog of Maid (HHM) which regulates a synexpression group of cell cycle and migration regulators in epithelial cells, but other responses in other cell types (Ikushima et al. 2008), and members of the Ets family of transcription factors and transcription factor activator enhancing-binding protein 2α (TFAP2a; Koinuma et al. 2009). Analysis of Smad2 binding sites in zebrafish early gastrulas furthermore unraveled cooperation with other transcription factors, such as FoxH1, Lef1/β-catenin, Oct1, and Gata6 (Liu et al. 2011). Moreover, comparison of Smad2/3 binding regions in HepG2 hepatoblastoma cells and HaCaT epidermal keratinocytes revealed that 81 % of the binding sites in HepG2 cells are not shared with those in HaCaT cells; however, 32.5 % of the Smad2/3 binding regions overlap with binding sites for hepatocyte nuclear factor 4α (HNF4α) (Mizutani et al. 2011). In addition, genome-wide mapping of Smad3 binding sites revealed that Smad3 cooperates with cell-type-specific master transcription factors, such as Oct4 in embryonic stem cells, Myod1 in myotubes, and PU.1 in pro-B cells (Mullen et al. 2011). Through these mechanisms, the cell-type-specific effects of TGF-β signaling are orchestrated.

Whereas Smad signaling is negatively controlled by ubiquitination and proteasomal degradation of Smad molecules themselves, nuclear Smad signaling can be enhanced by ubiquitination and degradation of transcriptional repressors. Thus, Smad complexes bind the ubiquitin ligase Arkadia which promotes ubiquitination and degradation of the interacting co-repressors Ski and SnoN (Le Scolan et al. 2008; Levy et al. 2007; Nagano et al. 2007), promoting transcription. Because of its ability to enhance the transcriptional activity of Smad, Arkadia acts as a tumor suppressor in colorectal cancer (Sharma et al. 2011). The E3 ligase activity of Arkadia is regulated by the RB1-inducible coiled-coil1 (RB1CC1) protein, which enhances TGF-β signaling by promoting ubiquitination of c-Ski (Koinuma et al. 2011). The level of SnoN is regulated by the anaphase-promoting complex (APC); in response to TGF-β stimulation, casein kinase (CK)II is activated leading to phosphorylation of Cdc27, a key component of APC, which targets SnoN for ubiquitin-mediated degradation (Zhang et al. 2011).

1.13.3 Nuclear Role of I-Smads

The transcriptional roles of R-Smad/Smad4 complexes are well established. However, there are indications that also Smad7 has a nuclear function. Thus, Smad7 interacts with the transcription factor MyoD and antagonizes the repressive effect of the MAP-kinase kinase MEK on MyoD function, thereby promoting myogenic differentiation (Miyake et al. 2010).

1.13.4 Epigenetic Regulation of Smad Signaling

Smad signaling is subject to epigenetic regulation. Thus, the co-repressor TGF-β-induced factor 1 (TGIF1) recruits histone deacetylase activity to Smad2 and thereby represses Smad signaling (Wotton et al. 2001; Wotton et al. 1999a, b). The ubiquitin ligase Fbxw7 targets TGIF1 for degradation and thus enhances the transcriptional activity of Smads (Bengoechea-Alonso and Ericsson 2010). Moreover, TGF-β suppression of the production of interleukin (IL)-2 by T cells involves Smad3-mediated recruitment of the histone H3 K9 methyl transferases Setdb1 and Suv39h1 to the proximal region of the IL-2 gene promoter (Wakabayashi et al. 2011). TIFγ/TRIMM33/ectodermin has a PHD finger-bromodomain which binds histone H3 that is unmethylated at K4 and R2, methylated at K9 and acetylated at K18 and K23; since its ubiquitin ligase activity is induced by histone binding, TIF1γ determines the time Smad complexes remain bound to their promoters by ubiquitination and inactivation of Smad4 (Agricola et al. 2011; Xi et al. 2011).

De-methylation of DNA has also been shown to be essential for regulation of a subset of TGF-β-dependent genes. Thus, TGF-β stimulates active de-methylation of the $p15^{ink4b}$ promoter in a process involving loss of the DNA methyltransferase DNMT3, allowing recruitment of Smad2/3, the CBP acetyltransferase and the DNA glycosylase TDG or the methyl CpG binding domain 4 (MBD4) protein to the same promoter region (reviewed by Thillainadesan et al. 2012).

1.13.5 Regulation of Growth Arrest by Smads

One of the characteristic effects of TGF-β is its ability to inhibit cell proliferation. Smad complexes have key roles in this process by regulating the transcription of genes coding for molecules involved in cell cycle control. Thus, TGF-β suppresses mitogenic signals, e.g. the Myc and Id transcription factors, and induces signals that inhibits the cell cycle, e.g. p15, p21, and p57 (reviewed by Massagué 2004).

1.13.6 Regulation of EMT by Smads

TGF-β is a potent inducer of EMT, i.e. a process during which epithelial cells lose their epithelial characters, such as polarity and cell–cell junctions, and acquire a more mesenchymal phenotype with increased production of matrix molecules, and cytokines and growth factors that stimulate cell migration. EMT is considered important for the invasiveness and metastasis of epithelial tumors. TGF-β regulates EMT by induction of a set of transcription factors, i.e. the basic helix-loop-helix proteins Twist and E47, the Zinc finger proteins Snail and Slug, the Zinc finger and homeodomain proteins ZEB1 and ZEB2, and FOXC2 (reviewed by Heldin et al. 2012). TGF-β induces ubiquitination and proteasomal degradation of the transcription factor Klf4 (Hu and Wan 2011), which is of key importance for the induction of Slug (Liu et al. 2012). Together, the action of these transcription factors leads to repression of E-Cadherin and other epithelial markers and enhancement of the expression of mesenchymal markers such as fibronectin (reviewed by Moustakas and Heldin 2012).

The induction of EMT by TGF-β is enhanced by ZEB1- and ZEB2-mediated downregulation of epithelial splicing regulatory proteins (ESRPs) (Horiguchi et al. 2012). This results in altered splicing of several proteins, including fibroblast growth factor (FGF) receptors 1, 2, and 3 which thereby are converted from the IIIb to the IIIc forms, causing a switch in ligand binding from FGF-7 and -10 to FGF-2 and -4 (Shirakihara et al. 2011; Warzecha et al. 2009).

Posttranscriptional mechanisms also affect TGF-β-induced EMT. Thus, the heterogenous nuclear ribonucleoprotein E1 (hnRNP E1) binds a 33 nucleotide element in the 3′ untranslated regions of the mRNAs of Dab2 and the cytokine interleukin-like EMT inducer (ILEI) and represses their translation. TGF-β induces phosphorylation of hnRNP E1 by Akt2, whereby it is released from the mRNAs which then can be translated, promoting EMT (Chaudhury et al. 2010).

1.14 Non-Smad Signaling

In addition to signaling via R-Smads, there are other signaling pathways activated in TGF-β stimulated cells (Fig. 1.1).

1.14.1 MAP-Kinase Pathways

It has been known for a long time that TGF-β activates the Erk, JNK, and p38 MAP-kinase pathways (Mulder 2000). These pathways have significant roles in TGF-β signaling, as is shown by the observation that TGF-β stimulation of Erk and JNK MAP-kinases drives the formation of aortic aneurysms in Marfan syndrome mice

(Holm et al. 2011). One mechanism of activation of Erk MAP-kinase is via docking of the adaptor protein Shc to the TβRI/TβRII complex, whereby it is phosphorylated on tyrosine residue(s) and become capable of binding the Grb2/Sos1 complex, thus activating the Ras/Erk MAP-kinase pathway (Lee et al. 2007b). On the other hand, TGF-β has been found to selectively activate Erk MAP-kinase in cells with high levels of TβRII, such as dermal cells, but not in cells with low levels of TβRII, such as epidermal cells in a manner which is not dependent on the kinase activity of TβRI (Bandyopadhyay et al. 2011).

Whereas activation of Erk may promote cell proliferation, activation of JNK and p38 MAP-kinases has been implicated in the apoptotic effect of TGF-β. JNK is activated via docking of the adaptor protein Daxx to TβRII (Perlman et al. 2001). Phosphorylation of Daxx by the homeodomain-interacting protein kinase 2 (HIPK2) activates the MAP-kinase kinases MKK4 and MKK7, which ultimately activates JNK (Hofmann et al. 2003). In prostate cancer cells, the p38 MAP-kinase is activated by the binding and activation of the E3 ligase TRAF6 to a motif in the juxtamembrane part of TβRI. TβRI also binds the TGF-β activated kinase (TAK)1 in a Smad7-dependent manner. TRAF6 then performs K63 poly-ubiquitination of TAK1, which thereby is activated, leading to activation of the downstream MKK3 or MKK6 and, finally, p38 (Edlund et al. 2003; Sorrentino et al. 2008; Yamashita et al. 2008). Interestingly, the TRAF6-mediated activation of p38 is independent of the kinase activities of TβRI and TβRII (Sorrentino et al. 2008).

In T cells, the de-ubiquitinase CYLD negatively regulates the activation of TAK1 and p38 (Zhao et al. 2011). Moreover, TGF-β signaling via TAK1 is important for the function of regulatory T cells (Gu et al. 2012). Together, these findings indicate a key role for TAK1 in TGF-β signaling.

1.14.2 Src, PI3-Kinase, and mTOR

In addition to apoptotic pathways, TGF-β activates pro-survival pathways, including the PI3-kinase/Akt pathway TGF-β (Yi et al. 2005). The tyrosine kinase Src has also been shown to be activated by TGF-β and to be important for activation of PI3-kinase (Park et al. 2004; Tanaka et al. 2004). Moreover, TGF-β rapidly induces activation of mTOR complex 1 (mTORC1) and the downstream S6 kinase in a PI3-kinase-dependent manner, leading to increased protein synthesis, cell size, motility, and invasion (Lamouille and Derynck 2007). TGF-β also activates the mTOR complex 2 (mTORC2), which has also been shown to be important for EMT and invasion (Lamouille et al. 2012).

1.14.3 Rho GTPases

TGF-β induces rapid actin reorganization and stress fiber formation by activation of the small GTPases RhoA and RhoB and the downstream effectors ROCK, Lim kinase 2 and cofilin (Vardouli et al. 2005). TGF-β upregulates NET1, a guanine nucleotide exchange factor for RhoA, in a Smad3-dependent manner (Lee et al. 2010; Shen et al. 2001). In keratinocytes, TGF-β was found to induce another member of the NET1 family, NET1A, in a manner which is dependent both on Smads and Erk MAP-kinase (Papadimitriou et al. 2012). Upon prolonged TGF-β stimulation, NET1A is subject to proteasomal degradation and translational silencing by miR-24, contributing to EMT (Papadimitriou et al. 2012). In prostate cancer cells, TGF-β activates RhoA and Cdc42 leading to reorganization of the actin filament system (Edlund et al. 2002), in a Smad7-dependent manner (Edlund et al. 2004). A mechanism for degradation of RhoA in epithelial cells has been demonstrated. Thus, after TGF-β stimulation, in addition to phosphorylating TβRI, TβRII also phosphorylates the polarity protein PAR6, leading to the recruitment of Smurf1 and subsequent degradation of RhoA contributing to the dissolution of tight junctions (Ozdamar et al. 2005).

1.14.4 Nuclear TβRI

TRAF6 was recently shown to ubiquitinate, in addition to TAK1, also TβRI which makes the receptor susceptible for cleavage by the metalloprotease ADAM17; this liberates the intracellular domain of TβRI which is translocated to the nucleus where it interacts with the co-activator p300 and induces several genes involved in cell migration and invasiveness (Mu et al. 2011). Full-length TβRI has also been found to accumulate in the nucleus under certain conditions. Thus, in ErbB2 transformed cells, which have high amounts of the GTPase Ran that is important for nuclear translocation, TβRI was shown to enter the nucleus in a Smad2/3-dependent manner (Chandra et al. 2012). Nuclear TβRI was found to associate with purine-rich RNA sequences synergistically with the RNA-binding factor hnRNPA1 and may thus affect RNA processing.

1.15 Crosstalk with Other Pathways

TGF-β signaling is modulated by crosstalk with several other signaling pathways, which contributes to the characteristic context-dependency of TGF-β signaling.

1.15.1 Wnt

Wnt is a large family of factors that are implicated in stimulation of cell proliferation during embryonal development and tumorigenesis. Key molecules in the Wnt signaling pathway are the transcription factors β-catenin, T cell factor (TCF), and lymphoid enhancer factor (LEF). Smads form complexes with both LEF1 (Vincent et al. 2009) and β-catenin (Kim et al. 2009; Zhou et al. 2012), which enhance the induction of EMT. In addition, Smad7 forms a complex with β-catenin, which was found to be important for TGF-β-induced apoptosis (Edlund et al. 2005).

1.15.2 Notch

The Notch pathway specifies cell fate determination during development. TGF-β induces several Notch receptor ligands, including Jagged1 (Niimi et al. 2007; Zavadil et al. 2004), and Notch signaling induces TGF-β (Aoyagi-Ikeda et al. 2011). The cooperation between TGF-β and Notch signaling enhances EMT. However, there are reports that in certain cell types, e.g. esophageal epithelial cells, Notch signaling counteracts EMT by induction of miR200 which targets ZEB and TGF-β (Ohashi et al. 2011).

1.15.3 Tyrosine Kinase Receptors

A major pathway induced by tyrosine kinase receptors is the Ras pathway. Cooperation between Ras and TGF-β signaling is particularly important during EMT (Gotzmann et al. 2006). In hepatocarcinoma cells, TGF-β induces both platelet-derived growth factor (PDGF) and PDGF receptors, which enhances PI3-kinase and β-catenin signaling and promotes the survival and invasion of the cancer cells (Fischer et al. 2007). Enhanced PI3-kinase signaling also activates Akt, which phosphorylates and activates Twist, promoting EMT (Xue et al. 2012).

1.15.4 Hippo

The Hippo pathway senses cell density and controls cell growth via the transcriptional regulators TAZ and YAP. TAZ/YAP binds Smad complexes and sequesters them in the cytoplasm in high density cell cultures, thereby attenuating TGF-β signaling (Varelas et al. 2008). Moreover, the Crumbs polarity complex interacts with TAZ/YAP and promotes their phosphorylation and cytoplasmic retention; disruption of the Crumbs complex enhances TGF-β signaling and promotes EMT (Varelas et al. 2010).

1.15.5 Parathyroid Hormone

Parathyroid hormone (PTH) regulates calcium homeostasis and bone metabolism by binding to and activating a G protein-coupled receptor. TβRII forms a complex with and phosphorylates the PTH receptor which modulates the internalization of the receptor complex (Qiu et al. 2010). Through this mechanism TGF-β suppresses PTH signaling.

1.16 Switch in TGF-β Signaling During Tumor Progression

TGF-β acts as a tumor suppressor since it inhibits cell proliferation and induces apoptosis. However, chronic exposure of mammary epithelial NMuMG cells to TGF-β leads to suppression of the anti-proliferative and pro-apoptotic effects of TGF-β and induction of EMT and invasiveness (Gal et al. 2008). Moreover, during tumor progression, TGF-β acquires tumor promoting activities, including promotion of cell invasiveness and metastasis. Whereas the mechanisms behind this switch are not fully understood, some interesting observations have recently been made. Thus, the adaptor protein Dab2, which regulates endocytosis of several receptors, is often downregulated in squamous cell carcinomas, and low levels correlate with poor prognosis (Hannigan et al. 2010). Downregulation of Dab2 blocks TGF-β-mediated cell growth arrest and instead promotes TGF-β-induced cell motility, anchorage-independent growth and tumor growth in vivo.

Another mechanism involves the transcription factor distal-less homeobox 2 (Dlx2), which is upregulated by TGF-β. It attenuates TGF-β-induced growth arrest by downregulating TβRII and promotes cell growth and survival by upregulating the epidermal growth factor family member betacellulin (Yilmaz et al. 2011).

Growing tumors are characterized by hypoxia due to poor vascularization. Smad7 is induced by hypoxia in a hypoxia-inducible factor (HIF)- and von Hippel-Lindau protein (pVHL)-dependent manner (Heikkinen et al. 2010a). Interestingly, the inhibitory effect of Smad7 on TGF-β signaling during normoxic condition is converted to a promoting effect of Smad7 on tumor invasion during hypoxia (Heikkinen et al. 2010b). It has also been observed that Smad7 promotes liver metastases of colorectal tumors (Halder et al. 2008). The activity of HIF is controlled by HIF prolyl hydroxylases (PHDs). Knockdown of PHD2 was found to prevent the switch of TGF-β from being a tumor suppressor to being a tumor promoter (Ameln et al. 2011).

1.17 Future Perspectives

Recent work has given ample examples of mechanisms that control TGF-β signaling on essentially all levels. The fact that such an elaborate machinery has evolved probably reflects the importance of TGF-β signaling during embryogenesis and tissue homeostasis, with the concomitant need to carefully titrate its signaling level. The importance of Smads in TGF-β signaling is well established. The activity of both TGF-β receptors and Smads is controlled by a number of posttranslational modifications, although the list of modifying enzymes and modified amino acid residues is already very long, it is likely that additional modifications will be discovered in the future. The availability of sensitive and accurate mass spectrometry techniques will facilitate the search for additional posttranslational modifications.

In most cells, TGF-β activates both Smad2 and Smad3. Although these molecules are structurally very similar, they have very different effects. Some of the differences can be explained by the fact that Smad3, but not Smad2, binds DNA. However, the detailed mechanism of involvement of Smad2 versus Smad3 in TGF-β signaling remains to be elucidated.

In addition to Smads, a number of other signaling pathways are activated in TGF-β stimulated cells, and a number of pathways activated by other growth factors and cytokines are modulated by TGF-β signaling. An important aim for future research will be to determine the mechanisms whereby such pathways are activated by TGF-β, as well as their importance for the various cellular effects of TGF-β.

A remarkable feature of TGF-β signaling is that it is very context-dependent, i.e. certain responses are seen only in certain cell types and under certain conditions. Some insights into the mechanisms for context-dependency have come from the finding that Smads cooperate with several master regulators of transcription and thus contribute to the establishment of different transcriptional programs in different cell types. However, it is likely that there are additional mechanisms involved in the context-dependence of TGF-β signaling, which remain to be discovered. One functionally important aspect of the context-dependency is the switch of TGF-β signaling from being tumor suppressive to being tumor promoting that occurs during tumor progression. Although some mechanisms explaining this switch have been elucidated, additional work is needed to get a more complete picture.

In conclusion, despite the fact that the TGF-β signaling research field is now becoming rather mature, important questions still remain to be answered.

Acknowledgment Ingegärd Schiller is gratefully acknowledged for valuable help in the preparation of this manuscript.

References

Agricola E, Randall RA, Gaarenstroom T et al (2011) Recruitment of TIF1γ to chromatin via its PHD finger-bromodomain activates its ubiquitin ligase and transcriptional repressor activities. Mol Cell 43:85–96

Ahn SM, Cha JY, Kim J et al (2012) Smad3 regulates E-cadherin via miRNA-200 pathway. Oncogene 31:3051–3059

Akiyoshi S, Inoue H, Hanai J et al (1999) c-Ski acts as a transcriptional co-repressor in transforming growth factor-β signaling through interaction with Smads. J Biol Chem 274:35269–35277

Alarcón C, Zaromytidou AI, Xi Q et al (2009) Nuclear CDKs drive Smad transcriptional activation and turnover in BMP and TGF-β pathways. Cell 139:757–769

Ameln AK, Muschter A, Mamlouk S et al (2011) Inhibition of HIF prolyl hydroxylase-2 blocks tumor growth in mice through the antiproliferative activity of TGFβ. Cancer Res 71:3306–3316

Annes JP, Munger JS, Rifkin DB (2003) Making sense of latent TGFβ activation. J Cell Sci 116:217–224

Aoyagi-Ikeda K, Maeno T, Matsui H et al (2011) Notch induces myofibroblast differentiation of alveolar epithelial cells via transforming growth factor-β-Smad3 pathway. Am J Respir Cell Mol Biol 45:136–144

Aragon E, Goerner N, Zaromytidou AI et al (2011) A Smad action turnover switch operated by WW domain readers of a phosphoserine code. Genes Dev 25:1275–1288

Bandyopadhyay B, Han A, Dai J et al (2011) TβRI/Alk5-independent TβRII signaling to ERK1/2 in human skin cells according to distinct levels of TβRII expression. J Cell Sci 124:19–24

Batut J, Howell M, Hill CS (2007) Kinesin-mediated transport of Smad2 is required for signaling in response to TGF-β ligands. Dev Cell 12:261–274

Batut J, Schmierer B, Cao J et al (2008) Two highly related regulatory subunits of PP2A exert opposite effects on TGF-β/Activin/Nodal signalling. Development 135:2927–2937

Bengoechea-Alonso MT, Ericsson J (2010) Tumor suppressor Fbxw7 regulates TGFβ signaling by targeting TGIF1 for degradation. Oncogene 29:5322–5328

Benus GF, Wierenga AT, de Gorter DJ et al (2005) Inhibition of the transforming growth factor β (TGFβ) pathway by interleukin-1β is mediated through TGFβ-activated kinase 1 phosphorylation of SMAD3. Mol Biol Cell 16:3501–3510

Bizet AA, Tran-Khanh N, Saksena A et al (2012) CD109-mediated degradation of TGF-β receptors and inhibition of TGF-β responses involve regulation of SMAD7 and Smurf2 localization and function. J Cell Biochem 113:238–246

Buscemi L, Ramonet D, Klingberg F et al (2011) The single-molecule mechanics of the latent TGF-β1 complex. Curr Biol 21:2046–2054

Chandra M, Zang S, Li H et al (2012) Nuclear translocation of type I transforming growth factor β receptor confers a novel function in RNA processing. Mol Cell Biol 32:2183–2195

Chang CC, Lin DY, Fang HI et al (2005) Daxx mediates the small ubiquitin-like modifier-dependent transcriptional repression of Smad4. J Biol Chem 280:10164–10173

Chaudhury A, Hussey GS, Ray PS et al (2010) TGF-β-mediated phosphorylation of hnRNP E1 induces EMT via transcript-selective translational induction of Dab2 and ILEI. Nat Cell Biol 12:286–293

Clarke DC, Brown ML, Erickson RA et al (2009) Transforming growth factor β depletion is the primary determinant of Smad signaling kinetics. Mol Cell Biol 29:2443–2455

Dai F, Lin X, Chang C, Feng XH (2009) Nuclear export of Smad2 and Smad3 by RanBP3 facilitates termination of TGF-β signaling. Dev Cell 16:345–357

Daly AC, Vizan P, Hill CS (2010) Smad3 protein levels are modulated by Ras activity and during the cell cycle to dictate transforming growth factor-β responses. J Biol Chem 285:6489–6497

Davis BN, Hilyard AC, Lagna G, Hata A (2008) SMAD proteins control DROSHA-mediated microRNA maturation. Nature 454:56–61

Davis BN, Hilyard AC, Nguyen PH et al (2010) Smad proteins bind a conserved RNA sequence to promote microRNA maturation by Drosha. Mol Cell 39:373–384

Davis-Dusenbery BN, Chan MC, Reno KE et al (2011) Down-regulation of Krüppel-like factor-4 (KLF4) by microRNA-143/145 is critical for modulation of vascular smooth muscle cell phenotype by transforming growth factor-β and bone morphogenetic protein 4. J Biol Chem 286:28097–28110

de Caestecker MP, Parks WT, Frank CJ et al (1998) Smad2 transduces common signals from receptor serine-threonine and tyrosine kinases. Genes Dev 12:1587–1592

de Graauw M, van Miltenburg MH, Schmidt MK et al (2010) Annexin A1 regulates TGF-β signaling and promotes metastasis formation of basal-like breast cancer cells. Proc Natl Acad Sci USA 107:6340–6345

Dennler S, Itoh S, Vivien D et al (1998) Direct binding of Smad3 and Smad4 to critical TGF-β-inducible elements in the promoter of human plasminogen activator inhibitor-type 1 gene. EMBO J 17:3091–3100

Derynck R, Miyazono K (eds) (2007) The TGF-β family. Cold Spring Harbor Laboratory Press, New York

Descargues P, Sil AK, Sano Y et al (2008) IKKα is a critical coregulator of a Smad4 independent TGFβ-Smad2/3 signaling pathway that controls keratinocyte differentiation. Proc Natl Acad Sci USA 105:2487–2492

Di Guglielmo GM, Le Roy C, Goodfellow AF, Wrana JL (2003) Distinct endocytic pathways regulate TGF-β receptor signalling and turnover. Nat Cell Biol 5:410–421

Dong C, Li Z, Alvarez R Jr et al (2000) Microtubule binding to Smads may regulate TGFβ activity. Mol Cell 5:27–34
Duan X, Liang YY, Feng XH, Lin X (2006) Protein serine/threonine phosphatase PPM1A dephosphorylates Smad1 in the bone morphogenetic protein signaling pathway. J Biol Chem 281:36526–36532
Dupont S, Zacchigna L, Cordenonsi M et al (2005) Germ-layer specification and control of cell growth by Ectodermin, a Smad4 ubiquitin ligase. Cell 121:87–99
Dupont S, Mamidi A, Cordenonsi M et al (2009) FAM/USP9x, a deubiquitinating enzyme essential for TGFβ signaling, controls Smad4 monoubiquitination. Cell 136:123–135
Ebisawa T, Fukuchi M, Murakami G et al (2001) Smurf1 interacts with transforming growth factor-β type I receptor through Smad7 and induces receptor degradation. J Biol Chem 276:12477–12480
Edlund S, Landström M, Heldin C-H, Aspenström P (2002) Transforming growth factor-β-induced mobilization of actin cytoskeleton requires signaling by small GTPases Cdc42 and RhoA. Mol Biol Cell 13:902–914
Edlund S, Bu S, Schuster N et al (2003) Transforming growth factor-β1 (TGF-β)-induced apoptosis of prostate cancer cells involves Smad7-dependent activation of p38 by TGF-β-activated kinase 1 and mitogen-activated protein kinase kinase 3. Mol Biol Cell 14:529–544
Edlund S, Landström M, Heldin C-H, Aspenström P (2004) Smad7 is required for TGF-β-induced activation of the small GTPase Cdc42. J Cell Sci 117:1835–1847
Edlund S, Lee SY, Grimsby S et al (2005) Interaction between Smad7 and β-catenin: importance for transforming growth factor β-induced apoptosis. Mol Cell Biol 25:1475–1488
Eichhorn PJ, Rodon L, Gonzalez-Junca A et al (2012) USP15 stabilizes TGF-β receptor I and promotes oncogenesis through the activation of TGF-β signaling in glioblastoma. Nat Med 18:429–435
Ferrand N, Atfi A, Prunier C (2010) The oncoprotein c-ski functions as a direct antagonist of the transforming growth factor-β type I receptor. Cancer Res 70:8457–8466
Fischer AN, Fuchs E, Mikula M et al (2007) PDGF essentially links TGF-β signaling to nuclear β-catenin accumulation in hepatocellular carcinoma progression. Oncogene 26:3395–3405
Fukuchi M, Imamura T, Chiba T et al (2001) Ligand-dependent degradation of Smad3 by a ubiquitin ligase complex of ROC1 and associated proteins. Mol Biol Cell 12:1431–1443
Gal A, Sjöblom T, Fedorova L et al (2008) Sustained TGF-β exposure suppresses Smad and non-Smad signalling in mammary epithelial cells, leading to EMT and inhibition of growth arrest and apoptosis. Oncogene 27:1218–1230
Gao S, Alarcon C, Sapkota G et al (2009) Ubiquitin ligase Nedd4L targets activated Smad2/3 to limit TGF-β signaling. Mol Cell 36:457–468
Gomis RR, Alarcon C, He W et al (2006) A FoxO-Smad synexpression group in human keratinocytes. Proc Natl Acad Sci USA 103:12747–12752
Gotzmann J, Fischer AN, Zojer M et al (2006) A crucial function of PDGF in TGF-β-mediated cancer progression of hepatocytes. Oncogene 25:3170–3185
Goumans M-J, Valdimarsdottir G, Itoh S et al (2003) Activin receptor-like kinase (ALK)1 is an antagonistic mediator of lateral TGFβ/ALK5 signaling. Mol Cell 12:817–828
Grönroos E, Hellman U, Heldin C-H, Ericsson J (2002) Control of Smad7 stability by competition between acetylation and ubiquitination. Mol Cell 10:483–493
Gu AD, Wang Y, Lin L et al (2012) Requirements of transcription factor Smad-dependent and independent TGF-β signaling to control discrete T-cell functions. Proc Natl Acad Sci USA 109:905–910
Guo X, Ramirez A, Waddell DS et al (2008) Axin and GSK3-control Smad3 protein stability and modulate TGF-β signaling. Genes Dev 22:106–120
Häger M, Pedersen CC, Larsen MT et al (2011) MicroRNA-130a-mediated down-regulation of Smad4 contributes to reduced sensitivity to TGF-β1 stimulation in granulocytic precursors. Blood 118:6649–6659
Halder SK, Rachakonda G, Deane NG, Datta PK (2008) Smad7 induces hepatic metastasis in colorectal cancer. Br J Cancer 99:957–965

Hannigan A, Smith P, Kalna G et al (2010) Epigenetic downregulation of human disabled homolog 2 switches TGFβ from a tumor suppressor to a tumor promoter. J Clin Invest 120:2842–2857
Hayashi H, Abdollah S, Qiu Y et al (1997) The MAD-related protein Smad7 associates with the TGFβ receptor and functions as an antagonist of TGFβ signaling. Cell 89:1165–1173
Hayes S, Chawla A, Corvera S (2002) TGF β receptor internalization into EEA1-enriched early endosomes: role in signaling to Smad2. J Cell Biol 158:1239–1249
He W, Dorn DC, Erdjument-Bromage H et al (2006) Hematopoiesis controlled by distinct TIF1γ and Smad4 branches of the TGFβ pathway. Cell 125:929–941
Heikkinen PT, Nummela M, Jokilehto T et al (2010a) Hypoxic conversion of SMAD7 function from an inhibitor into a promoter of cell invasion. Cancer Res 70:5984–5993
Heikkinen PT, Nummela M, Leivonen SK et al (2010b) Hypoxia-activated Smad3-specific dephosphorylation by PP2A. J Biol Chem 285:3740–3749
Heldin C-H, Vanlandewijck M, Moustakas A (2012) Regulation of EMT by TGFβ in cancer. FEBS Lett 586:1959–1970
Hesling C, Fattet L, Teyre G et al (2011) Antagonistic regulation of EMT by TIF1γ and Smad4 in mammary epithelial cells. EMBO Rep 12:665–672
Ho J, Cocolakis E, Dumas VM et al (2005) The G protein-coupled receptor kinase-2 is a TGF-β-inducible antagonist of TGFβ signal transduction. EMBO J 24:3247–3258
Hofmann TG, Stollberg N, Schmitz ML, Will H (2003) HIPK2 regulates transforming growth factor-β-induced c-Jun NH(2)-terminal kinase activation and apoptosis in human hepatoma cells. Cancer Res 63:8271–8277
Holm TM, Habashi JP, Doyle JJ et al (2011) Noncanonical TGFβ signaling contributes to aortic aneurysm progression in Marfan syndrome mice. Science 332:358–361
Horiguchi K, Sakamoto K, Koinuma D et al (2012) TGF-β drives epithelial-mesenchymal transition through δEF1-mediated downregulation of ESRP. Oncogene 31:3190–3201
Hu D, Wan Y (2011) Regulation of Kruppel-like factor 4 by the anaphase promoting complex pathway is involved in TGF-β signaling. J Biol Chem 286:6890–6901
Hu H, Milstein M, Bliss JM et al (2008) Integration of transforming growth factor β and RAS signaling silences a RAB5 guanine nucleotide exchange factor and enhances growth factor-directed cell migration. Mol Cell Biol 28:1573–1583
Hua F, Mu R, Liu J et al (2011) TRB3 interacts with SMAD3 promoting tumor cell migration and invasion. J Cell Sci 124:3235–3246
Huang T, David L, Mendoza V et al (2011) TGF-β signalling is mediated by two autonomously functioning TβRI:TβRII pairs. EMBO J 30:1263–1276
Hyytiäinen M, Penttinen C, Keski-Oja J (2004) Latent TGF-β binding proteins: extracellular matrix association and roles in TGF-β activation. Crit Rev Clin Lab Sci 41:233–264
Ikushima H, Komuro A, Isogaya K et al (2008) An Id-like molecule, HHM, is a synexpression group-restricted regulator of TGF-β signalling. EMBO J 27:2955–2965
Imoto S, Ohbayashi N, Ikeda O et al (2008) Sumoylation of Smad3 stimulates its nuclear export during PIASy-mediated suppression of TGF-β signaling. Biochem Biophys Res Commun 370:359–365
Inman GJ, Nicolas FJ, Hill CS (2002) Nucleocytoplasmic shuttling of Smads 2, 3, and 4 permits sensing of TGF-β receptor activity. Mol Cell 10:283–294
Inui M, Manfrin A, Mamidi A et al (2011) USP15 is a deubiquitylating enzyme for receptor-activated SMADs. Nat Cell Biol 13:1368–1375
Itoh S, Landström M, Hermansson A et al (1998) Transforming growth factor β1 induces nuclear export of inhibitory Smad7. J Biol Chem 273:29195–29201
Kamaraju AK, Roberts AB (2005) Role of Rho/ROCK and p38 MAP kinase pathways in transforming growth factor-β-mediated Smad-dependent growth inhibition of human breast carcinoma cells in vivo. J Biol Chem 280:1024–1036
Kamiya Y, Miyazono K, Miyazawa K (2010) Smad7 inhibits transforming growth factor-β family type I receptors through two distinct modes of interaction. J Biol Chem 285:30804–30813
Kaneko S, Chen X, Lu P et al (2011) Smad inhibition by the Ste20 kinase Misshapen. Proc Natl Acad Sci USA 108:11127–11132

Kang JS, Saunier EF, Akhurst RJ, Derynck R (2008) The type I TGF-β receptor is covalently modified and regulated by sumoylation. Nat Cell Biol 10:654–664
Kavsak P, Rasmussen RK, Causing CG et al (2000) Smad7 binds to Smurf2 to form an E3 ubiquitin ligase that targets the TGFβ receptor for degradation. Mol Cell 6:1365–1375
Kim KK, Wei Y, Szekeres C et al (2009) Epithelial cell α3β1 integrin links β-catenin and Smad signaling to promote myofibroblast formation and pulmonary fibrosis. J Clin Invest 119:213–224
Koinuma D, Shinozaki M, Komuro A et al (2003) Arkadia amplifies TGF-β superfamily signalling through degradation of Smad7. EMBO J 22:6458–6470
Koinuma D, Tsutsumi S, Kamimura N et al (2009) Chromatin immunoprecipitation on microarray analysis of Smad2/3 binding sites reveals roles of ETS1 and TFAP2A in transforming growth factor β signaling. Mol Cell Biol 29:172–186
Koinuma D, Shinozaki M, Nagano Y et al (2011) RB1CC1 protein positively regulates transforming growth factor-β signaling through the modulation of Arkadia E3 ubiquitin ligase activity. J Biol Chem 286:32502–32512
Komuro A, Imamura T, Saitoh M et al (2004) Negative regulation of transforming growth factor-β (TGF-β) signaling by WW domain-containing protein 1 (WWP1). Oncogene 23:6914–6923
Kowanetz M, Valcourt U, Bergström R et al (2004) Id2 and Id3 define the potency of cell proliferation and differentiation responses to transforming growth factor β and bone morphogenetic protein. Mol Cell Biol 24:4241–4254
Kowanetz M, Lönn P, Vanlandewijck M et al (2008) TGFβ induces SIK to negatively regulate type I receptor kinase signaling. J Cell Biol 182:655–662
Kretzschmar M, Doody J, Timokhina I, Massagué J (1999) A mechanism of repression of TGF-β/Smad signaling by oncogenic Ras. Genes Dev 13:804–816
Kuratomi G, Komuro A, Goto K et al (2005) NEDD4-2 (neural precursor cell expressed, developmentally down-regulated 4–2) negatively regulates TGF-β (transforming growth factor-β) signalling by inducing ubiquitin-mediated degradation of Smad2 and TGF-β type I receptor. Biochem J 386:461–470
Kurisaki A, Kose S, Yoneda Y et al (2001) Transforming growth factor-β induces nuclear import of Smad3 in an importin-β1 and Ran-dependent manner. Mol Biol Cell 12:1079–1091
Kurisaki A, Kurisaki K, Kowanetz M et al (2006) The mechanism of nuclear export of Smad3 involves exportin 4 and Ran. Mol Cell Biol 26:1318–1332
Kwon S, Christian JL (2011) Sortilin associates with transforming growth factor-β family proteins to enhance lysosome-mediated degradation. J Biol Chem 286:21876–21885
Lallemand F, Seo SR, Ferrand N et al (2005) AIP4 restricts transforming growth factor-β signaling through a ubiquitination-independent mechanism. J Biol Chem 280:27645–27653
Lamouille S, Derynck R (2007) Cell size and invasion in TGF-β-induced epithelial to mesenchymal transition is regulated by activation of the mTOR pathway. J Cell Biol 178:437–451
Lamouille S, Connolly E, Smyth JW et al (2012) TGF-β-induced activation of mTOR complex 2 drives epithelial-mesenchymal transition and cell invasion. J Cell Sci 125:1259–1273
Le Scolan E, Zhu Q, Wang L et al (2008) Transforming growth factor-β suppresses the ability of Ski to inhibit tumor metastasis by inducing its degradation. Cancer Res 68:3277–3285
Lee PS, Chang C, Liu D, Derynck R (2003) Sumoylation of Smad4, the common Smad mediator of transforming growth factor-β family signaling. J Biol Chem 278:27853–27863
Lee B-H, Chen W, Stippec S, Cobb MH (2007a) Biological cross-talk between WNK1 and the transforming growth factor β-Smad signaling pathway. J Biol Chem 282:17985–17996
Lee MK, Pardoux C, Hall MC et al (2007b) TGF-β activates Erk MAP kinase signalling through direct phosphorylation of ShcA. EMBO J 26:3957–3967
Lee J, Moon HJ, Lee JM, Joo CK (2010) Smad3 regulates Rho signaling via NET1 in the transforming growth factor-β-induced epithelial-mesenchymal transition of human retinal pigment epithelial cells. J Biol Chem 285:26618–26627
Levy L, Howell M, Das D et al (2007) Arkadia activates Smad3/Smad4-dependent transcription by triggering signal-induced SnoN degradation. Mol Cell Biol 27:6068–6083

Liang M, Liang YY, Wrighton K et al (2004) Ubiquitination and proteolysis of cancer-derived Smad4 mutants by SCFSkp2. Mol Cell Biol 24:7524–7537

Lin X, Liang M, Feng XH (2000) Smurf2 is a ubiquitin E3 ligase mediating proteasome-dependent degradation of Smad2 in transforming growth factor-β signaling. J Biol Chem 275:36818–36822

Lin X, Liang M, Liang YY et al (2003) SUMO-1/Ubc9 promotes nuclear accumulation and metabolic stability of tumor suppressor Smad4. J Biol Chem 278:31043–31048

Lin X, Duan X, Liang YY et al (2006) PPM1A functions as a Smad phosphatase to terminate TGFβ signaling. Cell 125:915–928

Liu Z, Lin X, Cai Z et al (2011) Global identification of SMAD2 target genes reveals a role for multiple co-regulatory factors in zebrafish early gastrulas. J Biol Chem 286:28520–28532

Liu YN, Abou-Kheir W, Yin JJ et al (2012) Critical and reciprocal regulation of KLF4 and SLUG in transforming growth factor β-initiated prostate cancer epithelial-mesenchymal transition. Mol Cell Biol 32:941–953

Long X, Miano JM (2011) Transforming growth factor-β1 (TGF-β1) utilizes distinct pathways for the transcriptional activation of microRNA 143/145 in human coronary artery smooth muscle cells. J Biol Chem 286:30119–30129

Lönn P, van der Heide L, Dahl M et al (2010) PARP-1 attenuates Smad-mediated transcription. Mol Cell 40:521–532

Lönn P, Vanlandewijck M, Raja E et al (2012) Transcriptional induction of salt-inducible kinase 1 by transforming growth factor β leads to negative regulation of type I receptor signaling in cooperation with the Smurf2 ubiquitin ligase. J Biol Chem 287:12867–12878

López-Casillas F, Payne HM, Andres JL, Massagué J (1994) Betaglycan can act as a dual modulator of TGF-β access to signaling receptors: mapping of ligand binding and GAG attachment sites. J Cell Biol 124:557–568

Louafi F, Martinez-Nunez RT, Sanchez-Elsner T (2010) MicroRNA-155 targets SMAD2 and modulates the response of macrophages to transforming growth factor-β. J Biol Chem 285:41328–41336

Luo K, Stroschein SL, Wang W et al (1999) The Ski oncoprotein interacts with the Smad proteins to repress TGFβ signaling. Genes Dev 13:2196–2206

Massagué J (2004) G1 cell-cycle control and cancer. Nature 432:298–306

Matsuura I, Denissova NG, Wang G et al (2004) Cyclin-dependent kinases regulate the antiproliferative function of Smads. Nature 430:226–231

Matsuura I, Chiang KN, Lai CY et al (2010) Pin1 promotes transforming growth factor-β-induced migration and invasion. J Biol Chem 285:1754–1764

Millet C, Yamashita M, Heller M et al (2009) A negative feedback control of transforming growth factor-β signaling by glycogen synthase kinase 3-mediated Smad3 linker phosphorylation at Ser-204. J Biol Chem 284:19808–19816

Miyake T, Alli NS, McDermott JC (2010) Nuclear function of Smad7 promotes myogenesis. Mol Cell Biol 30:722–735

Mizutani A, Koinuma D, Tsutsumi S et al (2011) Cell type-specific target selection by combinatorial binding of Smad2/3 proteins and hepatocyte nuclear factor 4α in HepG2 cells. J Biol Chem 286:29848–29860

Morén A, Hellman U, Inada Y et al (2003) Differential ubiquitination defines the functional status of the tumor suppressor Smad4. J Biol Chem 278:33571–33582

Morén A, Raja E, Heldin C-H, Moustakas A (2011) Negative regulation of TGFβ signaling by the kinase LKB1 and the scaffolding protein LIP1. J Biol Chem 286:341–353

Mori S, Matsuzaki K, Yoshida K et al (2004) TGF-β and HGF transmit the signals through JNK-dependent Smad2/3 phosphorylation at the linker regions. Oncogene 23:7416–7429

Moustakas A, Heldin C-H (2009) The regulation of TGFβ signal transduction. Development 136:3699–3714

Moustakas A, Heldin C-H (2012) Induction of epithelial-mesenchymal transition by transforming growth factor β. Semin Cancer Biol 22:446–454

Mu Y, Sundar R, Thakur N et al (2011) TRAF6 ubiquitinates TGFβ type I receptor to promote its cleavage and nuclear translocation in cancer. Nat Commun 2:330
Mulder KM (2000) Role of Ras and Mapks in TGFβ signaling. Cytokine Growth Factor Rev 11:23–35
Mullen AC, Orlando DA, Newman JJ et al (2011) Master transcription factors determine cell-type-specific responses to TGF-β signaling. Cell 147:565–576
Nagano Y, Mavrakis KJ, Lee KL et al (2007) Arkadia induces degradation of SnoN and c-Ski to enhance transforming growth factor-β signaling. J Biol Chem 282:20492–20501
Nakano A, Koinuma D, Miyazawa K et al (2009) Pin1 down-regulates transforming growth factor-β (TGF-β) signaling by inducing degradation of Smad proteins. J Biol Chem 284:6109–6115
Nakao A, Afrakhte M, Morén A et al (1997) Identification of Smad7, a TGFβ-inducible antagonist of TGF-β signalling. Nature 389:631–635
Nicolas FJ, De Bosscher K, Schmierer B, Hill CS (2004) Analysis of Smad nucleocytoplasmic shuttling in living cells. J Cell Sci 117:4113–4125
Niimi H, Pardali K, Vanlandewijck M et al (2007) Notch signaling is necessary for epithelial growth arrest by TGF-β. J Cell Biol 176:695–707
Nishimura SL (2009) Integrin-mediated transforming growth factor-β activation, a potential therapeutic target in fibrogenic disorders. Am J Pathol 175:1362–1370
Ohashi S, Natsuizaka M, Naganuma S et al (2011) A NOTCH3-mediated squamous cell differentiation program limits expansion of EMT-competent cells that express the ZEB transcription factors. Cancer Res 71:6836–6847
Ohshima T, Shimotohno K (2003) Transforming growth factor-β-mediated signaling via the p38 MAP kinase pathway activates Smad-dependent transcription through SUMO-1 modification of Smad4. J Biol Chem 278:50833–50842
Ozdamar B, Bose R, Barrios-Rodiles M et al (2005) Regulation of the polarity protein Par6 by TGFβ receptors controls epithelial cell plasticity. Science 307:1603–1609
Papadimitriou E, Vasilaki E, Vorvis C et al (2012) Differential regulation of the two RhoA-specific GEF isoforms Net1/Net1A by TGF-β and miR-24: role in epithelial-to-mesenchymal transition. Oncogene 31:2862–2875
Pardali E, Goumans MJ, ten Dijke P (2010) Signaling by members of the TGF-β family in vascular morphogenesis and disease. Trends Cell Biol 20:556–567
Park SS, Eom YW, Kim EH et al (2004) Involvement of c-Src kinase in the regulation of TGF-β1-induced apoptosis. Oncogene 23:6272–6281
Penheiter SG, Mitchell H, Garamszegi N et al (2002) Internalization-dependent and -independent requirements for transforming growth factor β receptor signaling via the Smad pathway. Mol Cell Biol 22:4750–4759
Perlman R, Schiemann WP, Brooks MW et al (2001) TGF-β-induced apoptosis is mediated by the adapter protein Daxx that facilitates JNK activation. Nat Cell Biol 3:708–714
Pierreux CE, Nicolas FJ, Hill CS (2000) Transforming growth factor β-independent shuttling of Smad4 between the cytoplasm and nucleus. Mol Cell Biol 20:9041–9054
Pulaski L, Landström M, Heldin C-H, Souchelnytskyi S (2001) Phosphorylation of Smad7 at Ser-249 does not interfere with its inhibitory role in transforming growth factor-β-dependent signaling but affects Smad7-dependent transcriptional activation. J Biol Chem 276:14344–14349
Qiu T, Wu X, Zhang F et al (2010) TGF-β type II receptor phosphorylates PTH receptor to integrate bone remodelling signalling. Nat Cell Biol 12:224–234
Roelen BA, Cohen OS, Raychowdhury MK et al (2003) Phosphorylation of threonine 276 in Smad4 is involved in transforming growth factor-β-induced nuclear accumulation. Am J Physiol Cell Physiol 285:C823–C830
Sadej R, Romanska H, Kavanagh D et al (2010) Tetraspanin CD151 regulates transforming growth factor β signaling: implication in tumor metastasis. Cancer Res 70:6059–6070
Sapkota G, Knockaert M, Alarcon C et al (2006) Dephosphorylation of the linker regions of Smad1 and Smad2/3 by small C-terminal domain phosphatases has distinct outcomes for bone

morphogenetic protein and transforming growth factor-β pathways. J Biol Chem 281:40412–40419
Schmierer B, Hill CS (2005) Kinetic analysis of Smad nucleocytoplasmic shuttling reveals a mechanism for transforming growth factor β-dependent nuclear accumulation of Smads. Mol Cell Biol 25:9845–9858
Schmierer B, Hill CS (2007) TGFβ-SMAD signal transduction: molecular specificity and functional flexibility. Nat Rev Mol Cell Biol 8:970–982
Schmierer B, Tournier AL, Bates PA, Hill CS (2008) Mathematical modeling identifies Smad nucleocytoplasmic shuttling as a dynamic signal-interpreting system. Proc Natl Acad Sci USA 105:6608–6613
Seo SR, Lallemand F, Ferrand N et al (2004) The novel E3 ubiquitin ligase Tiul1 associates with TGIF to target Smad2 for degradation. EMBO J 23:3780–3792
Seong H-A, Jung H, Ha H (2010) Murine protein serine/threonine kinase 38 stimulates TGF-β signaling in a kinase-dependent manner via direct phosphorylation of Smad proteins. J Biol Chem 285:30959–30970
Sflomos G, Kostaras E, Panopoulou E et al (2011) ERBIN is a new SARA-interacting protein: competition between SARA and SMAD2 and SMAD3 for binding to ERBIN. J Cell Sci 124:3209–3222
Sharma V, Antonacopoulou AG, Tanaka S et al (2011) Enhancement of TGF-β signaling responses by the E3 ubiquitin ligase Arkadia provides tumor suppression in colorectal cancer. Cancer Res 71:6438–6449
Shen X, Li J, Hu PP et al (2001) The activity of guanine exchange factor NET1 is essential for transforming growth factor-β-mediated stress fiber formation. J Biol Chem 276:15362–15368
Shi Y, Wang YF, Jayaraman L et al (1998) Crystal structure of a Smad MH1 domain bound to DNA: insights on DNA binding in TGF-β signaling. Cell 94:585–594
Shi W, Sun C, He B et al (2004) GADD34-PP1c recruited by Smad7 dephosphorylates TGFβ type I receptor. J Cell Biol 164:291–300
Shi M, Zhu J, Wang R et al (2011) Latent TGF-β structure and activation. Nature 474:343–349
Shirakihara T, Horiguchi K, Miyazawa K et al (2011) TGF-β regulates isoform switching of FGF receptors and epithelial-mesenchymal transition. EMBO J 30:783–795
Simonsson M, Heldin C-H, Ericsson J, Grönroos E (2005) The balance between acetylation and deacetylation controls Smad7 stability. J Biol Chem 280:21797–21803
Simonsson M, Kanduri M, Grönroos E et al (2006) The DNA binding activities of Smad2 and Smad3 are regulated by coactivator-mediated acetylation. J Biol Chem 281:39870–39880
Soond SM, Chantry A (2011) Selective targeting of activating and inhibitory Smads by distinct WWP2 ubiquitin ligase isoforms differentially modulates TGFβ signalling and EMT. Oncogene 30:2451–2462
Sorrentino A, Thakur N, Grimsby S et al (2008) The type I TGF-β receptor engages TRAF6 to activate TAK1 in a receptor kinase-independent manner. Nat Cell Biol 10:1199–1207
Stroschein SL, Wang W, Zhou SL et al (1999) Negative feedback regulation of TGF-β signaling by the SnoN oncoprotein. Science 286:771–774
Sun H, Li D, Chen S et al (2010) Zili inhibits transforming growth factor-β signaling by interacting with Smad4. J Biol Chem 285:4243–4250
Tajima Y, Goto K, Yoshida M et al (2003) Chromosomal region maintenance 1 (CRM1)-dependent nuclear export of Smad ubiquitin regulatory factor 1 (Smurf1) is essential for negative regulation of transforming growth factor-β signaling by Smad7. J Biol Chem 278:10716–10721
Tam BY, Finnson KW, Philip A (2003) Glycosylphosphatidylinositol-anchored proteins regulate transforming growth factor-β signaling in human keratinocytes. J Biol Chem 278:49610–49617
Tanaka Y, Kobayashi H, Suzuki M et al (2004) Transforming growth factor-β1-dependent urokinase up-regulation and promotion of invasion are involved in Src-MAPK-dependent signaling in human ovarian cancer cells. J Biol Chem 279:8567–8576
Tang LY, Yamashita M, Coussens NP et al (2011) Ablation of Smurf2 reveals an inhibition in TGF-β signalling through multiple mono-ubiquitination of Smad3. EMBO J 30:4777–4789

Thillainadesan G, Chitilian JM, Isovic M et al (2012) TGF-β-dependent active demethylation and expression of the p15^{ink4b} tumor suppressor are impaired by the ZNF217/CoREST complex. Mol Cell 46:636–649
Tsukazaki T, Chiang TA, Davison AF et al (1998) SARA, a FYVE domain protein that recruits Smad2 to the TGFβ receptor. Cell 95:779–791
Tu AW, Luo K (2007) Acetylation of Smad2 by the co-activator p300 regulates activin and transforming growth factor β response. J Biol Chem 282:21187–21196
Vardouli L, Moustakas A, Stournaras C (2005) LIM-kinase 2 and cofilin phosphorylation mediate actin cytoskeleton reorganization induced by transforming growth factor-β. J Biol Chem 280:11448–11457
Varelas X, Sakuma R, Samavarchi-Tehrani P et al (2008) TAZ controls Smad nucleocytoplasmic shuttling and regulates human embryonic stem-cell self-renewal. Nat Cell Biol 10:837–848
Varelas X, Samavarchi-Tehrani P, Narimatsu M et al (2010) The Crumbs complex couples cell density sensing to Hippo-dependent control of the TGF-β-SMAD pathway. Dev Cell 19:831–844
Vincent T, Neve EPA, Johnson JR et al (2009) A SNAIL1-SMAD3/4 transcriptional repressor complex promotes TGF-β mediated epithelial-mesenchymal transition. Nat Cell Biol 11:943–950
Wakabayashi Y, Tamiya T, Takada I et al (2011) Histone 3 lysine 9 (H3K9) methyltransferase recruitment to the interleukin-2 (IL-2) promoter is a mechanism of suppression of IL-2 transcription by the transforming growth factor-β-Smad pathway. J Biol Chem 286:35456–35465
Wan M, Cao X, Wu Y et al (2002) Jab1 antagonizes TGF-β signaling by inducing Smad4 degradation. EMBO Rep 3:171–176
Wan M, Tang Y, Tytler EM et al (2004) Smad4 protein stability is regulated by ubiquitin ligase SCF β-TrCP1. J Biol Chem 279:14484–14487
Warzecha CC, Sato TK, Nabet B et al (2009) ESRP1 and ESRP2 are epithelial cell-type-specific regulators of FGFR2 splicing. Mol Cell 33:591–601
Watanabe M, Masuyama N, Fukuda M, Nishida E (2000) Regulation of intracellular dynamics of Smad4 by its leucine-rich nuclear export signal. EMBO Rep 1:176–182
Watanabe Y, Itoh S, Goto T et al (2010) TMEPAI, a transmembrane TGF-β-inducible protein, sequesters Smad proteins from active participation in TGF-β signaling. Mol Cell 37:123–134
Wicks SJ, Lui S, Abdel-Wahab N et al (2000) Inactivation of smad-transforming growth factor β signaling by Ca(2+)-calmodulin-dependent protein kinase II. Mol Cell Biol 20:8103–8111
Wicks SJ, Haros K, Maillard M et al (2005) The deubiquitinating enzyme UCH37 interacts with Smads and regulates TGF-β signalling. Oncogene 24:8080–8084
Wotton D, Lo RS, Lee S, Massagué J (1999a) A Smad transcriptional corepressor. Cell 97:29–39
Wotton D, Lo RS, Swaby LA, Massague J (1999b) Multiple modes of repression by the Smad transcriptional corepressor TGIF. J Biol Chem 274:37105–37110
Wotton D, Knoepfler PS, Laherty CD et al (2001) The Smad transcriptional corepressor TGIF recruits mSin3. Cell Growth Differ 12:457–463
Wrana JL, Attisano L, Wieser R et al (1994) Mechanism of activation of the TGF-β receptor. Nature 370:341–347
Wrighton KH, Willis D, Long J et al (2006) Small C-terminal domain phosphatases dephosphorylate the regulatory linker regions of Smad2 and Smad3 to enhance transforming growth factor-β signaling. J Biol Chem 281:38365–38375
Wrighton KH, Lin X, Feng XH (2008) Critical regulation of TGFβ signaling by Hsp90. Proc Natl Acad Sci USA 105:9244–9249
Wu MY, Hill CS (2009) TGF-β superfamily signaling in embryonic development and homeostasis. Dev Cell 16:329–343
Wu JW, Krawitz AR, Chai J et al (2002) Structural mechanism of Smad4 recognition by the nuclear oncoprotein Ski: insights on Ski-mediated repression of TGF-β signaling. Cell 111:357–367

Xi Q, He W, Zhang XH et al (2008) Genome-wide impact of the BRG1 SWI/SNF chromatin remodeler on the transforming growth factor β transcriptional program. J Biol Chem 283:1146–1155
Xi Q, Wang Z, Zaromytidou AI et al (2011) A poised chromatin platform for TGF-β access to master regulators. Cell 147:1511–1524
Xiao Z, Liu X, Henis YI, Lodish HF (2000a) A distinct nuclear localization signal in the N terminus of Smad 3 determines its ligand-induced nuclear translocation. Proc Natl Acad Sci USA 97:7853–7858
Xiao Z, Liu X, Lodish HF (2000b) Importin β mediates nuclear translocation of Smad 3. J Biol Chem 275:23425–23428
Xiao Z, Latek R, Lodish HF (2003) An extended bipartite nuclear localization signal in Smad4 is required for its nuclear import and transcriptional activity. Oncogene 22:1057–1069
Xin H, Xu X, Li L et al (2005) CHIP controls the sensitivity of transforming growth factor-β signaling by modulating the basal level of Smad3 through ubiquitin-mediated degradation. J Biol Chem 280:20842–20850
Xu L, Chen YG, Massagué J (2000) The nuclear import function of Smad2 is masked by SARA and unmasked by TGFβ-dependent phosphorylation. Nat Cell Biol 2:559–562
Xu L, Kang Y, Col S, Massagué J (2002) Smad2 nucleocytoplasmic shuttling by nucleoporins CAN/Nup214 and Nup153 feeds TGFβ signaling complexes in the cytoplasm and nucleus. Mol Cell 10:271–282
Xu L, Alarcon C, Col S, Massague J (2003) Distinct domain utilization by Smad3 and Smad4 for nucleoporin interaction and nuclear import. J Biol Chem 278:42569–42577
Xu L, Yao X, Chen X et al (2007) Msk is required for nuclear import of TGFβ/BMP-activated Smads. J Cell Biol 178:981–994
Xue G, Restuccia DF, Lan Q et al (2012) Akt/PKB-mediated phosphorylation of Twist1 promotes tumor metastasis via mediating cross-talk between PI3K/Akt and TGFβ signaling axes. Cancer Discov 2:248–259
Yagi K, Goto D, Hamamoto T et al (1999) Alternatively spliced variant of Smad2 lacking exon 3. Comparison with wild-type Smad2 and Smad3. J Biol Chem 274:703–709
Yakymovych I, ten Dijke P, Heldin C-H, Souchelnytskyi S (2001) Regulation of Smad signaling by protein kinase C. FASEB J 15:553–555
Yamashita M, Fatyol K, Jin C et al (2008) TRAF6 mediates Smad-independent activation of JNK and p38 by TGF-β. Mol Cell 31:918–924
Yan X, Zhang J, Pan L et al (2011) TSC-22 promotes transforming growth factor β-mediated cardiac myofibroblast differentiation by antagonizing Smad7 activity. Mol Cell Biol 31: 3700–3709
Yao X, Chen X, Cottonham C, Xu L (2008) Preferential utilization of Imp7/8 in nuclear import of Smads. J Biol Chem 283:22867–22874
Yi JY, Shin I, Arteaga CL (2005) Type I transforming growth factor β receptor binds to and activates phosphatidylinositol 3-kinase. J Biol Chem 280:10870–10876
Yilmaz M, Maass D, Tiwari N et al (2011) Transcription factor Dlx2 protects from TGFβ-induced cell-cycle arrest and apoptosis. EMBO J 30:4489–4499
Yu J, Pan L, Qin X et al (2010) MTMR4 attenuates transforming growth factor β (TGFβ) signaling by dephosphorylating R-Smads in endosomes. J Biol Chem 285:8454–8462
Zavadil J, Cermak L, Soto-Nieves N, Bottinger EP (2004) Integration of TGF-β/Smad and Jagged1/Notch signalling in epithelial-to-mesenchymal transition. EMBO J 23:1155–1165
Zawel L, Dai JL, Buckhaults P et al (1998) Human Smad3 and Smad4 are sequence-specific transcription activators. Mol Cell 1:611–617
Zhang Y, Chang C, Gehling DJ et al (2001) Regulation of Smad degradation and activity by Smurf2, an E3 ubiquitin ligase. Proc Natl Acad Sci USA 98:974–979
Zhang L, Fujita T, Wu G et al (2011) Phosphorylation of the anaphase-promoting complex/Cdc27 is involved in TGF-β signaling. J Biol Chem 286:10041–10050
Zhang L, Huang H, Zhou F et al (2012a) RNF12 controls embryonic stem cell fate and morphogenesis in zebrafish embryos by targeting Smad7 for degradation. Mol Cell 46:650–661

Zhang L, Zhou F, Drabsch Y et al (2012b) USP4 is regulated by AKT phosphorylation and directly deubiquitylates TGF-β type I receptor. Nat Cell Biol 14:717–726
Zhao Y, Thornton AM, Kinney MC et al (2011) The deubiquitinase CYLD targets Smad7 protein to regulate transforming growth factor β (TGF-β) signaling and the development of regulatory T cells. J Biol Chem 286:40520–40530
Zhou Q, Fan J, Ding X et al (2010) TGF-β-induced MiR-491-5p expression promotes Par-3 degradation in rat proximal tubular epithelial cells. J Biol Chem 285:40019–40027
Zhou B, Liu Y, Kahn M et al (2012) Interactions between β-catenin and transforming growth factor-β signaling pathways mediate epithelial-mesenchymal transition and are dependent on the transcriptional co-activator cAMP-response element-binding protein (CREB)-binding protein (CBP). J Biol Chem 287:7026–7038
Zhu H, Kavsak P, Abdollah S et al (1999) A Smad ubiquitin ligase targets the BMP pathway and affects embryonic pattern formation. Nature 400:687–693
Zhu S, Wang W, Clarke DC, Liu X (2007) Activation of Mps1 promotes transforming growth factor-β-independent Smad signaling. J Biol Chem 282:18327–18338

Chapter 2
TGF-β Signaling in Stem Cell Fate Determination

Zhongwei Li, Danny Huylebroeck, and Ye-Guang Chen

Abstract Stem cells, which can self-renew and can differentiate to various cell types, have great potential applications in regenerative medicine. The signaling pathways employed by TGF-β family members (including TGF-β, Activin, Nodal, BMPs, and others) are evolutionarily conserved and regulate embryogenesis and adult tissue homeostasis and repair. Disturbance of production or bioactivity of the ligands and of intracellular signal transduction elicited from TGF-β family members (generally called TGF-β signaling) can lead to various diseases, including cancer, fibrosis, and cardiovascular diseases. Consistent with their critical roles in directing normal development, TGF-β family members have been established as key extrinsic signals that regulate fate commitment in both embryonic and adult stem cells. In this chapter, we review the functions and the underlying molecular mechanisms of TGF-β family members in determining embryonic stem cell fate choices between self-renewal, commitment, and subsequent differentiation to three germ layers and their progenies. We also summarize how TGF-β-related factors control fate commitment of five extensively investigated adult stem cells—intestinal, hair follicle, neural, hematopoietic, and mesenchymal stem cells. The principles gained from these studies illustrate on how stem cell fates can be regulated by extrinsic signals and pave a road for the potential medical applications of stem cells, either as transplanted cells after ex vivo expansion or as in vivo resident stem cells activated in endogenous stem cell compartments.

Z. Li • Y.-G. Chen (✉)
The State Key Laboratory of Biomembrane and Membrane Biotechnology,
Tsinghua-Peking Center for Life Sciences, School of Life Sciences,
Tsinghua University, Beijing 100084, China
e-mail: ygchen@tsinghua.edu.cn

D. Huylebroeck
Department of Development and Regeneration, Faculty of Medicine,
University of Leuven (KU Leuven), Leuven B-3000, Belgium

A. Moustakas and K. Miyazawa (eds.), *TGF-β in Human Disease*,
DOI 10.1007/978-4-431-54409-8_2, © Springer Japan 2013

Keywords Activin • Adult stem cells • BMP • Differentiation • Embryonic stem cells • Nodal • Self-renewal • TGF-β

2.1 Introduction

Stem cells can be derived from both early embryos and various adult tissues. These cells can self-renew and have the ability to differentiate to multiple cell types, thus holding great promise in regenerative medicine. Embryonic stem (ES) cells are derived from the inner cell mass of blastocyst stage embryos (Evans and Kaufman 1981; Martin 1981; Thomson et al. 1998). ES cells can be propagated indefinitely in vitro while maintaining their ability of self-renewal and pluripotency (the ability to differentiate to all three embryonic germ layers and derivatives). Understanding how ES cells sustain self-renewal and differentiate—the ES cell fate determination mechanisms—will pave the road for medical applications of pluripotent stem cells in general. ES cell fates are determined at different levels by extrinsic signals, intrinsic core transcription factors, and epigenetic regulators (Young 2011). Transforming growth factor-β (TGF-β) and related factors play critical roles in embryonic development and in adult tissue homeostasis and repair (Shi and Massague 2003; Massague and Chen 2000). The TGF-β family can be divided into two major groups—the TGF-β/Activin/Nodal and the bone morphogenetic proteins (BMPs)—based amongst others on their signal transduction and regulation mechanisms, which are discussed in depth elsewhere in this book. Briefly, as an extrinsic signal, TGF-β signaling is initiated by binding of the dimeric ligands to the complex of transmembrane receptors, leading to activation of the serine/threonine kinase in the intracellular domain of the type I receptors in the complex. These activated receptors subsequently phosphorylate and activate the cytoplasmic signaling mediators—the R-Smads (Smad1/5/8 for BMP signaling and Smad2/3 for TGF-β/Activin/Nodal signaling), and then the activated R-Smads form a complex with Smad4, the common Smad (co-Smad), and accumulate in the nucleus to directly participate in the regulation of target gene expression (Massague and Chen 2000; Shi and Massague 2003; Feng and Derynck 2005; Heldin et al. 1997; Moustakas et al. 2001). Consistent with its importance in developmental biology, emerging evidence also indicates that TGF-β signaling has profound effects both on the maintenance of ES cell self-renewal state and on the ES cell differentiation into multiple cell lineages. In addition to ES cells, somatic stem cells from different types of tissues have been identified, and they have been documented to play critical roles in tissue homeostasis. TGF-β signaling is also critical in fate determination of these adult stem cells (Watabe and Miyazono 2009; Seuntjens et al. 2009). In this chapter, we summarize and update the current understanding of TGF-β signaling in the fate regulation of ES cells as well as five different kinds of somatic stem cells and discuss their potential medical applications.

2.2 TGF-β and BMP Signaling in ES Cell Fate Determination

2.2.1 *BMP Signaling in ES Cell Fate Determination*

2.2.1.1 BMP Signaling in ES Cell Self-Renewal

BMP supports the self-renewal state of mouse ES cells. The first definitive evidence for the importance of BMP signaling in mouse ES cell self-renewal was provided by Austin Smith's group in 2003. They found that in feeder- and serum-free N2B27 chemically defined medium (CDM), addition of leukemia inhibitory factor (LIF) and BMP4 can support long-term self-renewal of mouse ES cells (Ying et al. 2003a). In the absence of LIF and BMP4, mouse ES cells differentiate to *Sox1*$^{+}$ neural precursor cells by default (Ying et al. 2003a, b). LIF alone can only partially inhibit neural differentiation and is not sufficient to sustain the ES cell self-renewal, whereas BMP4 alone induces ES differentiation to flat, epithelial-like cells (Ying et al. 2003a, b). Interestingly, in the presence of both LIF and BMP4, mouse ES cell self-renewal is sustained for multiple passages as evidenced by cell morphology, marker gene expression, and chimera contributions (Ying et al. 2003a). Further investigation indicated that transcriptional upregulation of inhibitor of differentiation (*Id*) family genes by BMP signaling mediates some of BMP4's functions and overexpression of *Id1* can bypass the requirement of BMP4 and sustains mouse ES cell self-renewal in cooperation with LIF. *Id* family proteins may inhibit neural differentiation via antagonizing precociously expressed neurogenic DNA-binding basic helix-loop-helix (bHLH) transcriptional activators in mouse ES cells (Ying et al. 2003a) (Fig. 2.1).

To investigate how BMP signaling determines mouse ES cell fate from a genome-wide perspective, Fei et al. explored genomic promoter enrichment of Smad1/5 and Smad4 proteins in mouse ES cells by using a chip-based chromatin immunoprecipitation (ChIP) method (Fei et al. 2010a). They found that Smad1/5 and Smad4 have a large portion of overlapping target genes and these genes include mainly developmental regulators. Their promoters are enriched for both histone H3K27 and H3K4 trimethylation marks (the so-called "bivalent-mark"), which is believed to maintain these genes in the poised state—repressed in self-renewal while rapidly expressed upon differentiation. This co-occupancy pattern suggested that BMP/Smad signaling functions via repressing a large cohort of differentiation-related genes (Fei et al. 2010a). To get a better understanding of the transcriptional regulatory network of external signaling and internal core transcription factors, Chen et al. carried out ChIP-seq to map the locations of 15 transcription factors in mouse ES cells (Chen et al. 2008b). These transcription factors include STAT3 as a mediator of LIF signaling, Smad1 as a mediator of BMP signaling and also factors that play critical roles as ES cell core transcription factors or somatic reprogramming factors. They observed the frequent overlapping binding patterns of STAT3, Smad1, OCT4,

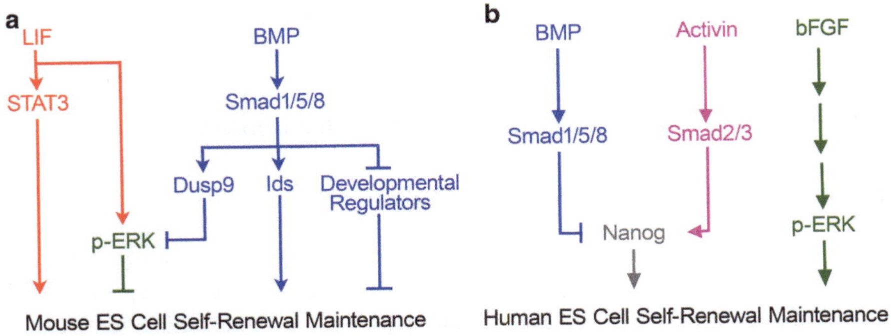

Fig. 2.1 Self-Renewal maintenance of mouse and human ES Cells regulated by TGF-β family members. (**a**) BMP and LIF signaling together maintain mouse ES cell self-renewal. LIF signaling, via activation of downstream STAT3, regulates target gene expression directly and contributes to mouse ES cell self-renewal. The activation of p-ERK1/2 by LIF signaling is suppressed by BMP signaling through transcriptional upregulation of DUSP9. Balanced p-ERK1/2 activity ensures mouse ES cell self-renewal. BMP signaling also enhances mouse ES cell self-renewal by induction of Ids and Nanog and by inhibiting expression of developmental regulators. (**b**) Activin and bFGF signaling support human ES cell self-renewal. Activin upregulates Nanog expression while bFGF mainly exerts its function through ERK1/2 activation. BMP signaling induces robust differentiation, partially by antagonizing Nanog induction by Activin signaling

NANOG, and SOX2, which suggest that external LIF and BMP signaling are connected to the core transcription factor network at the transcriptional level. They also found that the binding of Smad1 and STAT3 in the co-bound sites is dependent on OCT4 binding, suggesting a hierarchy of regulatory interactions between OCT4, Smad1 and STAT3 in mouse ES cells (Chen et al. 2008b).

Recently, a novel mechanism was described which links BMP signaling with ERK activity control for achieving maintenance of self-renewal of cultured mouse ES cells (Li and Chen 2012; Li et al. 2012). ERK is a critical intrinsic regulator to balance self-renewal vs. differentiation of mouse ES cells. High ERK activity triggers mouse ES differentiation (Yoshida-Koide et al. 2004) while low ERK activity sustains self-renewal (Kunath et al. 2007; Stavridis et al. 2007; Ying et al. 2008). Li et al. found that BMP signaling can transcriptionally upregulate dual specificity phosphatase 9 (*Dusp9*), an ERK-preferring phosphatase. The elevated expression of *Dusp9* represses ERK signals and balances the ERK-activating function of LIF signaling, thus contributing to the maintenance of mouse ES cell self-renewal (Li and Chen 2012; Li et al. 2012). Therefore, mouse ES cell self-renewal is sustained by the interplay between extrinsic signals and intrinsic factors. Considering the large amount of downstream target genes of BMP signaling, it is even likely that BMP signaling fully exerts its self-renewal-supportive function through the coordinated regulation of *Id* family genes, *Dusp9*, and the differentiation-related genes.

BMP signaling and LIF signaling may interplay with each other via Nanog (Suzuki et al. 2006). Nanog expression needs both STAT3 and T (Brachyury). STAT3 is activated by LIF while T is induced by Smad1-mediated BMP signaling. Interestingly, Nanog in turn suppresses BMP-induced mesoderm differentiation of mouse ES cells by physically interacting with Smad1.

In complete contrast to the role in mouse ES cells, BMP has been established as a strong differentiation-inducing signal in human ES cells (Fig. 2.1). Several studies have indicated that inhibition of BMP signaling can enhance human ES cell self-renewal (James et al. 2005; Xu et al. 2005; Wang et al. 2005). For feeder-free growth, human ES cells are routinely cultured in medium that contains basic fibroblast growth factor (bFGF, also known as FGF2) and knockout serum replacement (KSR), which is conditioned in the presence of mouse embryonic fibroblasts (MEFs) before feeding hES cells (the so-called "conditioned medium," CM) (Xu et al. 2001). Unconditioned medium (UM) with only bFGF and KSR but without factors secreted from MEFs cannot fully support hES cell self-renewal. It was observed that compared to the ones in CM, human ES cells cultured in UM showed higher p-Smad1/5/8 activity, suggesting activation of BMP signaling may drive human ES cell differentiation (Xu et al. 2005; James et al. 2005). Consistently, addition of high doses of the BMP inhibitor Noggin to UM reduced this p-Smad1/5/8 activation, resulting in long-term self-renewal of pluripotent-state human ES cells (Xu et al. 2005; Wang et al. 2005). It is noteworthy that different from BMP signaling, TGF-β/Activin/Nodal signaling contributes to human ES cell self-renewal maintenance, and this may be achieved in part through antagonizing BMP signaling (James et al. 2005; Wu et al. 2008; Xiao et al. 2006) (Fig. 2.1). Like Noggin, Activin A also inhibited p-Smad1/5/8 activation and hence could provoke sustained human ES cell self-renewal (James et al. 2005). Consistently, inhibition of endogenous TGF-β/Activin/Nodal signaling in CM resulted in trophoblast differentiation, similar to the effect of added BMP4 (Wu et al. 2008). The mutually antagonistic effects of TGF-β/Activin/Nodal signaling and BMP signaling likely converge at the transcriptional regulation of *Nanog* expression. Smad2/3 has been shown to bind to the *Nanog* proximal promoter region in self-renewal-state human ES cells and up-hold *Nanog* mRNA steady-state expression (Xu et al. 2008). BMP4 treatment induces the substitution of Smad2/3 with Smad1/5/8 in this proximal promoter, resulting in a decreased expression of *Nanog* and thus initiating differentiation. This transcriptional regulation is also dependent on the Oct/Sox-binding regulatory elements in the *Nanog* promoter region, indicating the involvement of core transcription factors in this process and again the convergence of external signals and intrinsic regulators on human ES cell fate determination (Xu et al. 2008). The different functions of BMP signal in mouse and human ES cells were proposed to be due to the two different developmental stages that mouse and human ES cells represent—mouse ES cells equivalent to the more primitive inner cell mass of the blastocyst stage while human ES cells resembling the cells in the elongated epiblast stage (Nichols and Smith 2009).

2.2.1.2 BMP Signaling in Extraembryonic Differentiation of ES Cells

In human ES cells cultured in monolayer, activation of BMP signaling drives efficient differentiation to cells of the trophoblast lineage (Xu et al. 2002). Further investigation indicated that differentiation starts from the periphery to the central region of the ES colonies, and this process takes place more efficiently in the absence of FGF2 and under atmospheric oxygen levels (Das et al. 2007). The trophoblast-inducing activity of BMP signaling is also supported by other studies. For instance, in human ES cell clones deficient in gene expression of *PIG-A* (phosphatidyl-inositol-glycan class A), the cells cannot undergo efficient trophoblast differentiation upon treatment with BMP4 or during spontaneous embryoid body (EB) formation (Chen et al. 2008a). Investigation into the underlying mechanisms revealed that PIG-A deficiency impairs the normal function of GPI-anchored BMP co-receptors. Hence, the lack of full BMP signaling activation may very well lead to the resistance to enter trophoblast differentiation (Chen et al. 2008a). These observations indicate that BMP is a robust trophoblast inducer in human ES cells.

2.2.1.3 BMP Signaling in Germ Cell Differentiation of ES Cells

Primordial germ cells (PGCs) are specified from epiblast cells at the onset of the gastrulation stage. It has been well documented that a group of BMP proteins, including BMP4, BMP7 and BMP8b, and their downstream Smads Smad1 and Smad5, and likely their partner transcription factors, play an inductive role in PGC formation during mouse embryonic development (Zhao 2003; Saitou and Yamaji 2010; Saitou 2009). Consistent with its germ cell-inducing function in vivo (Okamura et al. 2005; Hayashi et al. 2002; Tremblay et al. 2001; Chang and Matzuk 2001), BMP signaling also facilitates germ cell differentiation of ES cells in vitro. Toyooka et al. (2003) generated mouse vasa homolog (*Mvh*, a marker gene for germ cells) knock-in ES cells carrying GFP or LacZ and observed a dramatic increase of MVH+ cells when ES cells were co-cultured in BMP4 expressing cells. Importantly, after purification and transplantation, these MVH^+ cells contributed to testicular tubules and even mature sperm in vivo. Similarly, using GFP knock-in mouse ES cells based on another germ cell marker gene, i.e. *Stella*, Wei et al. (2008) also observed the promoting effect of BMP on germ cell differentiation of ES cells, and this effect was enhanced dose-dependently by treatment with BMP4. BMP signaling also promotes germ cell differentiation in human ES cells. Kee et al. (2006) found that addition of BMP4 to differentiating EBs induced germ-cell specific marker gene expression, and BMP7 and BMP8b showed additive effects when used in combination with BMP4, resulting in dramatically enhanced $VASA^+$ germ cells. These data indicate that BMP signaling enhances germ cell differentiation in both mouse and human ES cells.

2.2.1.4 BMP Signaling in Neural Differentiation of ES Cells

Consistent with its inhibitory role in neural lineage commitment in early embryonic development, BMP signaling inhibits mouse ES cell differentiation to neural precursor cells in a wide variety of in vitro differentiation systems that were used. In the monolayer differentiation condition with N2B27 culture medium, mouse ES cells efficiently differentiate into *Sox1*+ neural precursor cells, and BMP signaling dramatically inhibits this process, including via *Id* family proteins to repress the precocious expression of neurogenic bHLH transcription factors (Ying et al. 2003a, b), via *Dusp9* to suppress ERK1/2 activity (Li and Chen 2012; Li et al. 2012) and via other downstream target genes such as *Dpysl2* and *Jmjd3* (Fei et al. 2010a). In the PA6 stromal cell co-culture system to induce neural differentiation, treatment with BMP4 in the early phase also inhibits neural differentiation of mouse ES cells (Mizuseki et al. 2003). Interestingly, when BMP4 is added at appropriate doses in a later phase, it causes the differentiation to neural crest cells and dorsalmost CNS cells, showing a time window-dependent effect (Mizuseki et al. 2003). Consistent with this notion, in the EB-derived neural differentiation system, it was found that early exposure to BMP2 inhibits differentiation to neuroectoderm while later treatment of BMP2 in ES-derived neural precursor cells supports further differentiation to neural crest cells (Gossrau et al. 2007). Zhang et al. (2010) further investigated in detail the time windows of neural inhibition by BMP4 in the serum-free EB (SFEB) neural differentiation system and proposed a step-wise neural differentiation model wherein mouse ES cells first differentiate to an epiblast stem cell (EpiSC)-like state, the so-called ES-derived EpiSCs (ESD-EpiSCs), and then ESD-EpiSCs commit to the neural lineage. In the first step, BMP signaling prevents the formation of ESD-EpiSCs, and in the second step BMP signaling inhibits neural commitment and facilitates non-neural differentiation. Interestingly, the functions of BMP signaling are also partially mediated by *Id* family proteins in both steps and by inhibition of ERK activity specifically in the first step. In this SFEB differentiation system, BMP signaling was further found to repress neural differentiation by inhibiting the expression of *Zfp521*, which seems to be essential and sufficient to drive the intrinsic neural differentiation (Kamiya et al. 2011). However, whether *Zfp521* is also a direct target gene of BMP signaling remains to be investigated. Taken together, current evidence supports that BMP signaling inhibits early neural commitment of mouse ES cells via regulation of a cohort of downstream target genes.

BMP signaling also plays key roles in neural commitment of human ES cells. Similar to what has been observed in mouse ES cells, BMP4 represses neural differentiation induced in the PA6 co-culture system (Zeng et al. 2004) and in serum-free suspension culture conditions (Schulz et al. 2004). On the contrary, inhibition of BMP signaling by Noggin (Pera et al. 2004) or by the small molecule receptor kinase inhibitor dorsomorphin (Zhou et al. 2010) facilitates neural commitment. Interestingly, it was observed that dual inhibition of both BMP signaling and TGF-β/Activin/Nodal signaling by Noggin/SB431542 (Chambers et al. 2009) or by dorsomorphin/SB431542 (Morizane et al. 2011) further enhanced the efficiency of neural differentiation. Indeed, inhibition of TGF-β/Activin/Nodal signaling impairs self-renewal of human ES cells and rapidly drives them into transition from self-renewal to differentiation. During the differentiation process, inhibition of BMP signaling and TGF-β/Activin/Nodal signaling represses trophectoderm differentiation and mesendoderm differentiation, respectively. Thus, by blockage of alternative cell lineages, dual inhibition of BMP and TGF-β/Activin/Nodal signaling drives efficient differentiation of human ES cells to the neural lineage (Chambers et al. 2009). It was also reported that additional inhibition of FGF/ERK signaling further enhanced neural differentiation efficiency (Greber et al. 2011). Thus, in human ES cells, BMP signaling inhibits neural differentiation and further blockage of self-renewal signals from TGF-β/Activin/Nodal signaling and FGF/ERK signaling can drive efficient neural commitment. These findings favor the default model in human ES cell neural differentiation, consistent with the concept established in embryonic development (Weinstein and Hemmati-Brivanlou 1997; Munoz-Sanjuan and Brivanlou 2002).

2.2.1.5 BMP Signaling in ES Cell Differentiation to Mesendoderm and its Derivatives

BMP signaling is well documented to regulate the differentiation of mouse ES cells to mesendoderm (the precursor cells of both mesoderm and definitive endoderm, the counterpart of primitive streak during early mouse development) and its derivatives. In serum-free suspension cultures, treatment with BMP4 can induce posterior primitive streak (PS)-like cells with high expression of the pan-PS marker *Brachyury* (*Bra*) and posterior PS markers *Mesp1* and *Hoxb1* (Nostro et al. 2008). Interestingly, BMP significantly enhances the expression of *Nodal* and *Wnt3*, and the effects of BMP were found to depend on both TGF-β/Activin/Nodal signaling and canonical Wnt signaling, as the addition of SB431542 or DKK1 (Wnt inhibitor) can abolish the expression of these PS marker genes (Nostro et al. 2008). In contrast, Activin A and low concentrations of Wnt3a induce anterior PS-like cells with high expression of *Bra* and *Foxa2*. In this scenario, if BMP4 is also added, the anterior PS marker gene expression was reduced while the posterior marker genes were dramatically induced. These findings highlight a dominant effect of BMP signaling on posterior PS fate commitment (Nostro et al. 2008).

Flk1^{+} hematopoietic mesoderm can be derived from the PS-like cells and BMP signaling was found to be necessary for this fate commitment in cooperation with TGF-β/Activin/Nodal and Wnt signaling (Nostro et al. 2008). In the EB differentiation system, BMP, Wnt, and Notch signaling were found to be necessary for *Flk1*$^{+}$ hematopoietic mesoderm differentiation, and the Ets-type transcription factor *Er71* was found to function downstream of these signaling cascades and regulate *Flk1* expression directly (Lee et al. 2008). BMP signaling also functions in the differentiation process from *Flk1*$^{+}$ mesoderm to blood cells by activating Wnt signaling. Activated Wnt signaling then cooperates with BMP signaling to activate the Cdx-Hox pathway, thus directing the blood cell fate commitment (Lengerke et al. 2008). In the presence of Activin A in serum-free conditions, mouse ES cells can generate endoderm progenitor cells efficiently, and BMP signaling was found to drive hepatic differentiation together with bFGF and Activin (Gouon-Evans et al. 2006). Thus, in mouse ES cells, BMP signaling contributes widely to cell fate commitment processes from mesendoderm to hematopoietic and hepatic derivatives.

Similar promoting functions of BMP in mesendoderm differentiation have been observed in human ES cells. In the spin-EB differentiation system in which EB formation was performed under a centrifugal force, BMP4 first transiently induces PS marker gene expression, and then—with PS gene expression declining—mesodermal and endodermal marker gene expression are upregulated (Davis et al. 2008). In a serum-free monolayer differentiation system, short-term treatment with BMP4 transiently induced posterior but not anterior PS genes (Zhang et al. 2008). This is consistent with the strong posterior PS patterning effect of BMP signaling on mouse ES cells (Nostro et al. 2008). Interestingly, long-term treatment with BMP4 in the same system dramatically enhanced the expression of genes of the trophoblast lineage (Zhang et al. 2008). The posterior patterning effect of BMP is also supported by another observation that in ectopic expression of β-catenin-induced PS differentiation process the blockage of BMP signaling significantly decreased the expression of posterior PS genes whereas anterior PS gene expression increased (Sumi et al. 2008). Using a chemically defined medium (mTeSR), Yu et al. (2011) observed mesendoderm induction of BMP, which depends on active FGF signaling. They also revealed that FGF signaling via the MEK-ERK cascade prolongs the expression of *Nanog* during BMP-induced differentiation. This sustained *Nanog* expression ensures *bona fide* induction of *Bra* by BMP and thus contributes to efficient mesendoderm differentiation (Yu et al. 2011). Like in mouse ES cells, BMP signaling also contributes to the differentiation of mesoderm derivatives in human ES cells, such as hematopoietic (Pick et al. 2007; Park et al. 2010) and cardiac lineages (Yang et al. 2008; Kattman et al. 2011). Activin signaling has been well documented as a strong inducer of definitive endoderm (D'Amour et al. 2005). Interestingly, it was found that BMP can enhance the definitive endoderm differentiation efficiency in cooperation with Activin, and BMP also plays key roles in the differentiation from definitive endoderm to its progenies. Culture of ES-derived definitive endoderm in the presence of BMP2 and FGF4 efficiently drives hepatic commitment (Cai et al. 2007), inhibition of BMP signaling exerted at the appropriate time is essential to generate insulin-expressing pancreatic cells (Nostro et al. 2011), and dual inhibition of both TGF-β and BMP signaling can generate anterior foregut endoderm from definitive endoderm (Green et al. 2011). Taken together, BMP signaling has broad effects on mesendoderm commitment and on the differentiation to its derivatives in both mouse and human ES cells.

2.2.2 TGF-β/Activin/Nodal Signaling in ES Cell Fate Determination

2.2.2.1 TGF-β/Activin/Nodal Signaling in ES Cell Self-Renewal

TGF-β/Activin/Nodal signaling ensures human ES cell self-renewal. Human ES cells are maintained in the undifferentiated state on MEF-derived feeder cells in the presence of bFGF and KSR. In the absence of feeder cells, human ES cells can be cultured using conditioned medium harvested from feeder cells whereas unconditioned medium will lead to human ES cell differentiation (Xu et al. 2001; Besser 2004; James et al. 2005; Xiao et al. 2006). TGF-β/Activin/Nodal signaling is dramatically repressed upon differentiation in this unconditioned medium, as evidenced by decreased levels of phosphorylated Smad2/3 (p-Smad2/3) and its downstream target genes *Nodal*, *Lefty-A* and *Lefty-B* (Xiao et al. 2006; Besser 2004; James et al. 2005). Inhibition of TGF-β/Activin/Nodal signaling in conditioned medium drives human ES cell differentiation (Xiao et al. 2006; Besser 2004; James et al. 2005). On the contrary, addition of Activin A in unconditioned medium maintains similar p-Smad2/3 levels as in conditioned medium and can sustain long-term human ES cell self-renewal (Xiao et al. 2006). In a CDM, Activin A and bFGF can cooperatively support human ES cell self-renewal (Vallier et al. 2005). Mechanistically, TGF-β/Activin/Nodal signaling was found to transcriptionally upregulate the expression of *Nanog* in a Smad-dependent manner, thus contributing to self-renewal maintenance (Xu et al. 2008; Vallier et al. 2009; Singh et al. 2012).

From a genome-wide perspective, Smad2/3 can directly bind to multiple pluripotency gene promoters and regulate their expression transcriptionally (Brown et al. 2011). Smad2/3 also share a large proportion of binding sites that are co-bound by OCT4 and NANOG, suggesting cooperation between extrinsic Activin signaling and intrinsic transcription factors for maintaining human ES cells self-renewal (Brown et al. 2011). TGF-β/Activin/Nodal signaling is hierarchically regulated by the WW-domain containing protein TAZ. In human ES cells, TAZ interacts with the Smad complex, facilitates the translocation of the Smad complex from the cytoplasm to the nucleus, and ensures Smad transcriptional activities (Varelas et al. 2008). Conversely, human ES cell self-renewal cannot be maintained when TAZ function is disturbed, due to the inefficiency of TGF-β/Activin/Nodal signaling (Varelas et al. 2008). Interestingly, the functions of TGF-β/Activin/Nodal signaling are quite different in mouse ES cells, where it is not necessary for self-renewal maintenance (James et al. 2005; Fei et al. 2010b). Inhibition of TGF-β/Activin/Nodal signaling by SB431542 (James et al. 2005) or knockdown of Smad2 (Fei et al. 2010b) has no effects on mouse ES cell self-renewal, but TGF-β/Activin/Nodal signaling seems to be important for normal proliferation of mouse ES cells (Ogawa et al. 2007).

2.2.2.2 TGF-β/Activin/Nodal Signaling in ES Cell Differentiation

In mouse ES cells, TGF-β/Activin/Nodal signaling plays critical roles in mesendoderm and subsequent endoderm commitment. In serum-free conditions, treatment with Activin A induces efficient mesendoderm marker gene expression in both adherent and suspension cultures (Tada et al. 2005; Yasunaga et al. 2005; Gadue et al. 2006; Nostro et al. 2008; Fei et al. 2010b). Interestingly, this effect relies on the cooperation with Wnt signaling. Both Activin A and Wnt3a can induce the expression of mesendoderm marker genes. Inhibition of either Activin or Wnt signaling can abolish the differentiation process induced by Wnt3a and Activin A, showing the interdependence of the two signaling pathways in promoting mesendoderm differentiation (Gadue et al. 2006). Furthermore, Activin A has been shown to induce *Wnt3* expression, while Wnt can upregulate *Nodal* (Gadue et al. 2006).

Using a genome-wide approach, Fei et al. investigated how Smad2 mediates the induction of mesendoderm by Activin and identified Smad2 binding sites in promoter regions across the genome (Fei et al. 2010b). Smad2 binds to promoters of a large number of development-related genes, and one of the Activin/Smad2-induced genes, *Tapbp*, was found to partially mediate the function of Activin/Smad signaling in mesendoderm differentiation (Fei et al. 2010b). Mesendoderm is a bi-potent state that can further generate mesoderm and endoderm derivatives. Continuous Activin signaling has been implicated as necessary for definitive endoderm differentiation from mesendoderm (Yasunaga et al. 2005; Gadue et al. 2006), and the cooperation of Activin, Wnt and BMP signaling drives efficient differentiation of mesendoderm to *Flk1*$^{+}$ mesodermal cells (Nostro et al. 2008), i.e. hematopoietic and cardiac progenitor cells. TGF-β/Activin/Nodal signaling has conserved functions in human ES cells in inducing mesendoderm and subsequent definitive endoderm (D'Amour et al. 2005; McLean et al. 2007; Singh et al. 2012; Sumi et al. 2008) and is also implicated in inhibiting neuroectoderm differentiation (Vallier et al. 2004; Smith et al. 2008).

2.3 TGF-β and BMP Signaling in Adult Stem Cell Fate Determination

TGF-β family members, particularly BMPs, play critical roles in self-renewal and differentiation of adult stem cells. Here, we summarize the current understanding on cell fate determination of intestinal, hair follicle, neural, hematopoietic and mesenchymal stem cells (MSCs) by BMPs, as selected examples (Fig. 2.2).

2.3.1 Intestinal Stem Cell Fate Determination

Intestinal epithelium is vigorously and constantly renewed from a small number of adult stem cells—the intestinal stem cells (ISCs). These ISCs are located at the fourth or fifth cell position from the base of each of the many intestinal crypts ("+4" ISCs) and can generate the entire crypt-villus structure, giving rise to absorptive enterocytes, mucin-producing goblet cells, hormone-secreting enteroendocrine cells and exocrine Paneth cells (van der Flier and Clevers 2009; Scoville et al. 2008). For maintaining intestinal homeostasis, ISC self-renewal and differentiation must be precisely regulated. Multiple signaling pathways in the ISC niche have been identified as critical in this process, including Wnt, Notch, and BMP signaling (van der Flier and Clevers 2009; Scoville et al. 2008). BMP4 is expressed in the intravillus mesenchyme whereas Noggin, the antagonist which can bind and inactivate BMP4, is synthesized in and secreted from cells that are adjacent to the bottom of the crypt and around the ISCs (He et al. 2004). Furthermore, the BMP type I receptor BMPR1A exhibits a graded distribution along the crypt-villus, with ISCs displaying the highest level (He et al. 2004). Conditional gene inactivation of BMPR1A in the intestine (He et al. 2004) or transgenic expression of Noggin (Haramis et al. 2004)

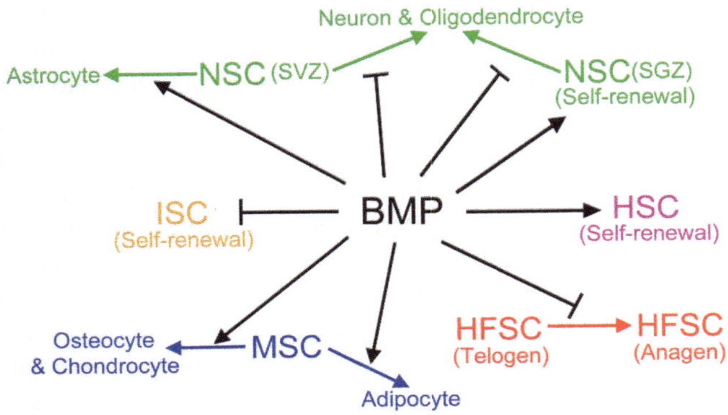

Fig. 2.2 BMP signaling on adult stem cell fate determination. BMP signaling sustains hematopoietic stem cell (HSC) self-renewal while inhibiting intestinal stem cell (ISC) self-renewal. The BMP signal barrier should be overcome for hair follicle stem cells (HFSCs) to transit from the telogen phase to the anagen phase and thus enhancing generation of new hair follicles. BMP signaling also directs mesenchymal stem cell (MSC) differentiation to osteocyte, chondrocyte and adipocyte lineages. Self-renewal of neural stem cells (NSCs) in SGZ is supported by BMP signaling whereas neurogenesis is suppressed. In NSCs from SVZ, BMP signaling induces astrocyte differentiation, at the expense of alternative commitment to neuron and oligodendrocyte cell fates

leads to the formation of ectopic polyps similar to what is seen in human juvenile polyposis syndrome. Further analysis indicated that expansion of stem and progenitor cell populations is responsible for the development of the polyps and that inhibition of BMP signaling by the PI3K-PTEN-AKT pathway enhanced the nuclear β-catenin level and promoted ectopic stem cell proliferation (He et al. 2004). Although BMP signaling plays a negative role on ISC maintenance, it seems to be important for differentiation of ISCs to mature intestinal cell types as the intestinal epithelium-specific knockout of BMPRIA resulted in defects in goblet, enteroendocrine, and Paneth cell maturation (Auclair et al. 2007). Recently, the cycling columnar cells in the crypt base, which are marked by the Wnt target gene Lgr5, have been identified as a novel population of ISCs (Barker et al. 2007). These stem cells can generate crypts-villus structures in vitro, in which $Lgr5^+$ ISCs rapidly self-renew and give rise to all types of intestinal epithelial cells (Sato et al. 2009), including "+4" ISCs (Yan et al. 2012). The presence of Noggin in the culture system is essential for the long-term maintenance of $Lgr5^+$ ISCs, indicating inhibition of BMP signaling is also required for maintenance of $Lgr5^+$ ISCs (Sato et al. 2009).

2.3.2 Hair Follicle Stem Cell Fate Determination

In the adult, hair follicles (HFs) undergo continuous and synchronized cycles to regenerate novel follicles (Blanpain and Fuchs 2009). Such cycle consists of three phases: telogen, the quiescence phase which lasts for months with quiescent hair follicle stem cells (HFSCs) residing in the bulge; anagen, the regeneration phase when HFSCs are activated and generate novel hair follicles; and catagen, the degradation phase which includes apoptotic retraction (Blanpain and Fuchs 2009). In the telogen phase, BMP mRNAs are expressed by dermal cells, including fibroblasts, adipocytes, and dermal papillae (Plikus et al. 2008). BMP ensures the stem cells to stay quiescent (Horsley et al. 2008). This effect of BMP is partially mediated by directly inducing the NFATc1-encoding gene, which subsequently suppresses *Cdk4* and thus leads to maintenance of stem cell quiescence (Horsley et al. 2008). When follicle stem cells are activated, NFATc1 gene expression is downregulated, leading to a high level of *Cdk4* and thus promoting cell proliferation (Horsley et al. 2008). BMP signaling should be suppressed in the transition from the telogen phase to the anagen phase to enable the successful transition (Horsley et al. 2008; Blessing et al. 1993; Botchkarev et al. 1999, 2001). Ectopic expression of *Bmp4* or gene inactivation of *Noggin* leads to a dramatic delay in hair follicle induction (Blessing et al. 1993; Botchkarev et al. 1999), whereas overexpression of *Noggin* induces transition to the anagen phase (Botchkarev et al. 1999, 2001). Inhibition of BMP signaling may contribute to the anagen transition partially by activation of β-catenin signaling via regulation of PI3K-AKT activity (Zhang et al. 2006), resembling the situation in ISCs (He et al. 2004). In an RNAi screen, *Tbx1* was identified to be critical for long-term HFSC regeneration. Interestingly, *Tbx1* exerts its functions mainly by enhancing the sensitivity of the cells to BMP, further highlighting the gatekeeper role of BMP signaling in HFSC maintenance (Chen et al. 2012). Recently, TGF-β2 was found to be a critical paracrine factor secreted from dermal papillae, and via Smad2/3 signaling, to regulate the expression of *Tmeff1*, which subsequently leads to inhibition of BMP signaling and thereby facilitates the transition from telogen to anagen (Oshimori and Fuchs 2012).

2.3.3 Neural Stem Cell Fate Determination

Neural stem cells (NSCs) are defined as progenitor cells that can generate neural tissues—neurons and glial cells (astrocytes and oligodendrocytes) while at the same time retaining some capacity to self-renew (Gage 2000; Temple 2001). In the adult brain, NSCs mainly reside in two specialized areas: the subventricular zone (SVZ) and the hippocampal subgranular zone (SGZ). NSCs in the SGZ divide infrequently and are mainly maintained in a quiescent state. BMP signaling is active in these NSCs in mice as its blockage by in vivo infusion of Noggin or conditional ablation of *Bmpr1a* or *Smad4* drives these cells into the cell cycle and increased neurogenesis (Mira et al. 2010). However, in the long run, conditional deletion of *Bmpr1a* impaired stem cell division, and precursor cells and newborn neurons were lost, indicating loss of stem cell activity (Mira et al. 2010).

In the case of neural stem cells in the SVZ, BMP signaling specifies differentiation to astrocytes while repressing neuron and oligodendrocyte commitment (Lim et al. 2000; Gomes et al. 2003; Samanta and Kessler 2004; Nakashima et al. 2001;

Kohyama et al. 2010). Although intact BMP signaling components are present in SVZ cells, the adjacent ependymal cells produce Noggin and inhibit BMP signaling, thus providing a niche that enables neurogenesis (Lim et al. 2000). Consistently, in vivo ectopic expression of *Bmp4* resulted in a significant increase of astrocytes and decrease of oligodendrocytes (Gomes et al. 2003), and in various in vitro cultures, the robust astrocyte induction by BMP signaling has been established and shown to occur at the expense of neuron and oligodendrocyte fates (Samanta and Kessler 2004; Nakashima et al. 2001; Kohyama et al. 2010). Target genes of BMP signaling mediate these functions. *Id1*, *Id3*, and *Rest* were indicated to interfere with neurogenic bHLH transcription factors to inhibit neurogenesis (Nakashima et al. 2001; Kohyama et al. 2010), while *Id2* and *Id4* were found to complex with and subsequently repress bHLH transcription factors that are required for oligodendrocyte differentiation (Samanta and Kessler 2004). Thus, by coordinated functions of its direct target genes, BMP signaling restricts alternative fates of neuron and oligodendrocyte commitment from SVZ neural stem cells and directs efficient astrocyte differentiation. In contrast, there is also evidence showing that conditional deletion of Smad4 in the SVZ neural stem cells or infusion of Noggin led to the impairment of neurogenesis and increased the number of oligodentrocytes (Colak et al. 2008). These observations add complexity to the physiological roles of BMP signaling in determining the cell fates. Probably the intensities of BMP signaling in this context may explain the discrepancy, which will be interesting for future study. Using the cultures of fetal neuroepithelial cells which contain neural precursors, Nakashima et al. also observed the synergistic effect between BMP and LIF in inducing astrocyte differentiation. Mechanistically, upon cytokine stimulation, p300 bridges the interaction between Smad1 and STAT3, and the Smad1-STAT3-p300 complex thereafter drives the transcription of the astrocyte marker gene—glial fibrillary acidic protein (GFAP) (Nakashima et al. 1999). Apart from BMP signaling, TGF-β signaling is also involved in NSC biology. Conditional inactivation of *Tgfbr2* results in neuroepithelial cell expansion in the dorsal caudal midbrain and in increased sphere formation and self-renewal of neuroepithelial stem cells. Decreased cell-cycle exit and shortened cell-cycle length are responsible for these phenotypes, and FGF and Wnt signaling were implicated to mediate TGF-β signaling functions, indicating the crosstalk among these signal pathways for coordinating NSC functions appropriately (Falk et al. 2008).

2.3.4 Hematopoietic Stem Cell Fate Determination

Hematopoietic stem cells (HSCs) reside in the bone marrow (BM) of the adult body. HSCs are rare in the BM and they stay at the top of a hierarchy of progenitor blood lineages, capable of reconstituting the whole blood system of a recipient, including lymphocytes (T and B cells), myeloid cells (monocytes and neutrophils), megakaryocytes/platelets, and red blood cells (Orkin 2000; Orkin and Zon 2008). Both BMP and TGF-β signaling are involved in the proliferation and fate control of HSCs. BMP has been identified as a critical factor to regulate the commitment from mesoderm to hematopoietic progenitor cells in embryonic development from various species (Huber et al. 1998; Maeno et al. 1996; Winnier et al. 1995; Dzierzak and Speck 2008) and has meanwhile also been employed to obtain blood lineage cells from ES cells and from induced pluripotent cells (iPSCs) (Nostro et al. 2008; Pearson et al. 2008; Irion et al. 2010; Chadwick et al. 2003; Park et al. 2010). In in vitro cultured HSCs, high concentrations of BMP4 extend the time of repopulation ability whereas low concentrations of BMP4 induce proliferation and differentiation (Bhatia et al. 1999). To investigate the functions of BMP4 in vivo in the HSC niche, Goldman et al. employed BMP4 hypomorph mice in which mature BMP4 ligands were dramatically decreased in specific tissues due to the S2G point mutation (*Bmp*$^{S2G/S2G}$) (Goldman et al. 2009). They found that BMP4 is present in various types of cells that constitute the HSC niche, including osteoblasts, endothelial cells, and megakaryocytes. In bone marrow, BMP4 levels were much lower in *Bmp*$^{S2G/S2G}$ than wild-type mice, and the HSC-enriched c-kit^{+}, Sca-1^{+}, Lineage^{-} (KSL) cell number was dramatically lower in *Bmp*$^{S2G/S2G}$ mice. When wild-type HSCs were transplanted to BMP4-deficient recipient mice, the HSC activity was greatly reduced (Goldman et al. 2009). These observations suggest the essential functions of paracrine BMP4 effects in HSC homeostasis. However, Singbrant et al. generated conditional *Smad1*/*Smad5* double knockout mice and found that the BM cells from these adult mice showed normal reconstitution capabilities after transplantation as compared to wild-type mice, arguing against the observations made in the BMP4 polymorph mice (Singbrant et al. 2010). Explanations to this discrepancy include that downstream non-Smad signaling may be involved in mediating BMP4 functions (Zhang 2009) or Smad redundancies may account for the mild phenotype in adult hematopoiesis.

TGF-β signaling has been shown to be a potent inhibitor for HSC proliferation (Sitnicka et al. 1996; Batard et al. 2000), and it is regarded as a key extrinsic signal that keeps HSC quiescent. Cyclin-dependent kinase inhibitors, including *p21* and *p57*, are implicated to mediate TGF-β functions in this process (Ducos et al. 2000; Scandura et al. 2004). Apart from inhibiting HSC proliferation, TGF-β signaling also promotes erythroid differentiation (hemoglobin + cells) in various in vitro studies (Shiozaki et al. 1998; Schwall and Lai 1991; Okafuji et al. 1995; Liu et al. 2000). It has been reported that the different functions of TGF-β can be achieved by distinct Smad complexes as signaling mediators—the classic Smad2/3-Smad4 complex mediates the antiproliferative response while Smad2/3-TIF1γ complex is necessary for erythroid differentiation (He et al. 2006). However, when TGF-β receptor *Tgfbr1* was conditionally knocked out, no abnormal phenotypes were observed for HSCs (Larsson et al. 2005), in contrast to the in vitro inhibitory functions of TGF-β. Moreover, ectopic expression of Smad7, which antagonizes both TGF-β and BMP signaling, even enhanced the self-renewal of HSCs in vivo (Blank et al. 2006). Conversely, in a *Tgfbr2* conditional knockout mouse model, HSC activity was severely impaired (Yamazaki et al. 2011), similar to that observed in conditional knockout of *Smad4* (Karlsson et al. 2007). Thus, whether TGF-β signaling is an essential negative regulator in HSC proliferation is still unclear in a physiological context, and further investigations are needed to fully characterize the functions of TGF-β signaling in HSC functions and homeostasis in vivo.

2.3.5 MSCs Fate Determination

MSCs are multipotent stem cells isolated mainly from adult bone marrow, which can maintain multilineage potential in vitro and differentiate to mesenchymal tissues, including bone, cartilage, fat, muscle, and marrow stroma (Pittenger et al. 1999; Jiang et al. 2002; Barry and Murphy 2004). The proliferation and differentiation of MSCs are regulated by TGF-β family members, often in cooperation with other signaling pathways. Treatment with TGF-β1 (Jian et al. 2006) can promote proliferation of MSCs whereas it will inhibit osteogenic differentiation. Further investigation indicated that TGF-β1 induces nuclear accumulation of β-catenin in MSCs via the physical interaction of the latter with Smad3 (Jian et al. 2006). Consistently, inhibition of TGF-β signaling by SB431542 induces osteoblastic maturation and the anti-osteoblastic function of TGF-β signaling is mediated by inhibitory Smads to antagonize BMP signaling (Maeda et al. 2004). BMP, however, inhibits MSC proliferation and induces osteogenic and chondrogenic differentiation (Partridge et al. 2002; Tsuda et al. 2003; Schmitt et al. 2003; Shea et al. 2003; Noel et al. 2004; Lavery et al. 2008; Liu et al. 2006). Interestingly, BMP signaling also converges to Wnt signaling to exert its anti-proliferative function. In the presence of BMP2, the interaction of Smad1 and Dvl-1 is increased, leading to decreased β-catenin levels in the nucleus and restricted β-catenin activity, thus inhibiting MSC proliferation (Liu et al. 2006). Thus, balanced TGF-β and BMP signaling dictates MSC fates. Bone remodeling requires coordinated bone resorption and subsequent bone formation. Active TGF-β1 is released from osteoclasts upon bone resorption and stimulates the migration of MSCs to the resorptive sites, thus contributing to bone formation (Tang et al. 2009).

Adipose tissues control energy balance, with white adipose tissue storing triglyceride and brown adipose tissue specialized for energy expenditure. Both white and brown adipose tissues can be generated from MSCs, and BMPs play dominant roles in this process. BMP2 and BMP4 specify MSCs to white adipose tissue (Wang et al. 1993; Tang et al. 2004; Jin et al. 2006). By the cooperation of Smad1/4,C/EBPα and Shn-2, BMP2 upregulates the transcription of PPARγ2, the key transcription factor for adipogenesis, to contribute to the white adipose commitment (Jin et al. 2006). BMP7, however, is required for brown fat adipose differentiation in vivo (Loncar 1992; Kalluri and Zeisberg 2003), and multiple lines of evidence also indicate its inductive role in vitro (Tang et al. 2004; Tseng et al. 2008; Schulz and Tseng 2009). A recent study examined the mechanisms and found that BMP7 induces the MSC commitment to brown adipose tissues by activating a series of regulators of the brown adipose fate via the p38 MAPK and PGC-1 pathways (Tseng et al. 2008).

2.4 Perspectives

During embryogenesis BMP signaling regulates multiple lineage specification processes, including neural inhibition, mesoderm commitment, and hematopoietic and cardiac development. Consistent with the in vivo functions in development, BMP retains these activities in cell fate determination of in vitro cultured pluripotent embryonic stem cells (ES cells), making these cells a great and as it turns out a robust system to investigate the detailed molecular mechanisms that are more difficult to study in embryos-, especially in human embryos. It is also important to note that TGF-β and BMP have been found to function differently in mouse vs. human ES cells. In mouse ES cells, BMP signaling promotes self-renewal by restricting neural differentiation whereas in human ES cells BMP signaling has strong differentiation-initiating effects. This functional difference is likely due to the different developmental stages mouse and human ES cells represent. Indeed, human ES cells resemble more the elongated epiblast stage whereas mouse ES cells are derived from prospective epiblast cells from the inner cell mass of the blastocyst stage and represent a more primitive "naive" state. In contrast to the opposite effects on self-renewal, BMP signaling exerts similar functions in differentiation of mouse and human ES cells.

Both TGF-β and BMP signaling act in a context-dependent manner in stem cell fate determination. For example, BMP signaling sustains mouse ES cell self-renewal in the presence of LIF signaling, while it also promotes differentiation to multiple other lineages in the presence of other extrinsic signals but without LIF. How cells read TGF-β or BMP signals and react accordingly under different scenarios is interesting but as yet not well understood. Three types of contextual determinants were proposed to explain this—the difference of signaling intensity in different contexts, the involvement of different transcriptional cofactors in specific contexts and the different states of epigenetic landscapes that pre-determine the access of Smad proteins to specific genes (for more details, see Massague 2012). These mechanisms may individually or cooperatively mediate the context-dependent effects of TGF-β and BMP signaling in different conditions.

TGF-β family members dictate both embryonic stem and adult stem cell fate specification in cooperation with several other signaling pathways, such as FGF and Wnt signaling. For example, in different signaling contexts, BMP signaling can exert different effects and direct differentiation to different lineages: the FGF-MEK-ERK cascade can shift BMP functions from initiating trophoblast lineage to inducing mesendoderm cells. Different signaling inputs may converge on the transcriptional regulation of downstream target genes or directly fine-tune the intensity and duration of critical signaling. However, how TGF-β and BMP signaling, respectively, integrate with other signaling systems to determine stem cell fates is still an important topic of future research. The intensified application of transcriptomics as well as proteomics (with emphasis on protein–protein interaction) and ChIP-sequencing (protein–DNA interaction) will contribute significantly to identifying the underlying molecular mechanisms of the crosstalk between different signaling pathways, and how extrinsic signaling directs stem cell commitment to specified cell types.

Dissecting the underlying mechanisms of how TGF-β and BMP signaling determine stem cell fates is fundamental to the understanding of stem cell biology, and more importantly, the knowledge obtained from these research could lay the basis for regenerative medicine. The ultimate goal of regenerative medicine is to replace malfunctioned cells/tissues/organs with the good ones. Thus, the most urgent task is to find ways to direct stem cell differentiation to specific cell types. TGF-β and BMP signaling, whose roles have been well established in embryogenesis, are also found to be key players in a broad spectrum of stem cell fate decision. Considering the convenience to manipulate these signaling (with various ligands, antagonists, chemical inhibitors) and the advantages of signaling-mediated differentiation (virus-free and higher efficiency) compared to the transcription factors-mediated ones, undoubtedly TGF-β and BMP signaling will contribute to future applications of stem cells in regenerative medicine.

Acknowledgments We are grateful to Bing Zhao for critical comments. The work in YGC's lab was supported by grants from the National Natural Science Foundation of China (30930050, 30921004), and the 973 Program (2006CB943401, 2010CB833706). Work in the DH lab on stem cells was supported by grants from KU Leuven (GOA-11/012), Fund of Scientific Research-Flanders (FWO-V GA.094.11N), PhD grants from the Agency for Innovation by Science and Technology (IWT) and the Interuniversity Attraction Pole Network funding (Belspo IUAP6 and IUAP7 5-year funding cycles). This work was also partially supported by grant BIL-08/10 in the framework of Tsinghua University-KU Leuven Bilateral Scientific Cooperation.

References

Auclair BA, Benoit YD, Rivard N, Mishina Y, Perreault N (2007) Bone morphogenetic protein signaling is essential for terminal differentiation of the intestinal secretory cell lineage. Gastroenterology 133:887–896. doi:S0016-5085(07)01297-8

Barker N, van Es JH, Kuipers J, Kujala P, van den Born M, Cozijnsen M, Haegebarth A, Korving J, Begthel H, Peters PJ, Clevers H (2007) Identification of stem cells in small intestine and colon by marker gene Lgr5. Nature 449:1003–1007. doi:nature06196

Barry FP, Murphy JM (2004) Mesenchymal stem cells: clinical applications and biological characterization. Int J Biochem Cell Biol 36:568–584. doi:10.1016/j.biocel.2003.11.001

Batard P, Monier MN, Fortunel N, Ducos K, Sansilvestri-Morel P, Phan T, Hatzfeld A, Hatzfeld JA (2000) TGF-β1 maintains hematopoietic immaturity by a reversible negative control of cell cycle and induces CD34 antigen up-modulation. J Cell Sci 113:383–390

Besser D (2004) Expression of nodal, lefty-A, and lefty-B in undifferentiated human embryonic stem cells requires activation of Smad2/3. J Biol Chem 279:45076–45084. doi:10.1074/jbc.M404979200

Bhatia M, Bonnet D, Wu D, Murdoch B, Wrana J, Gallacher L, Dick JE (1999) Bone morphogenetic proteins regulate the developmental program of human hematopoietic stem cells. J Exp Med 189:1139–1148

Blank U, Karlsson G, Moody JL, Utsugisawa T, Magnusson M, Singbrant S, Larsson J, Karlsson S (2006) Smad7 promotes self-renewal of hematopoietic stem cells. Blood 108:4246–4254. doi:blood-2006-02-005611

Blanpain C, Fuchs E (2009) Epidermal homeostasis: a balancing act of stem cells in the skin. Nat Rev Mol Cell Biol 10:207–217. doi:nrm2636

Blessing M, Nanney LB, King LE, Jones CM, Hogan BL (1993) Transgenic mice as a model to study the role of TGF-β-related molecules in hair follicles. Genes Dev 7:204–215

Botchkarev VA, Botchkareva NV, Roth W, Nakamura M, Chen LH, Herzog W, Lindner G, McMahon JA, Peters C, Lauster R, McMahon AP, Paus R (1999) Noggin is a mesenchymally derived stimulator of hair-follicle induction. Nat Cell Biol 1:158–164. doi:10.1038/11078

Botchkarev VA, Botchkareva NV, Nakamura M, Huber O, Funa K, Lauster R, Paus R, Gilchrest BA (2001) Noggin is required for induction of the hair follicle growth phase in postnatal skin. FASEB J 15:2205–2214. doi:10.1096/fj.01-0207

Brown S, Teo A, Pauklin S, Hannan N, Cho CH, Lim B, Vardy L, Dunn NR, Trotter M, Pedersen R, Vallier L (2011) Activin/Nodal signaling controls divergent transcriptional networks in human embryonic stem cells and in endoderm progenitors. Stem Cells 29:1176–1185. doi:10.1002/stem.666

Cai J, Zhao Y, Liu Y, Ye F, Song Z, Qin H, Meng S, Chen Y, Zhou R, Song X, Guo Y, Ding M, Deng H (2007) Directed differentiation of human embryonic stem cells into functional hepatic cells. Hepatology 45:1229–1239. doi:10.1002/hep.21582

Chadwick K, Wang L, Li L, Menendez P, Murdoch B, Rouleau A, Bhatia M (2003) Cytokines and BMP-4 promote hematopoietic differentiation of human embryonic stem cells. Blood 102:906–915. doi:10.1182/blood-2003-03-0832

Chambers SM, Fasano CA, Papapetrou EP, Tomishima M, Sadelain M, Studer L (2009) Highly efficient neural conversion of human ES and iPS cells by dual inhibition of SMAD signaling. Nat Biotechnol 27:275–280. doi:nbt.1529

Chang H, Matzuk MM (2001) Smad5 is required for mouse primordial germ cell development. Mech Dev 104:61–67. doi:S0925477301003677

Chen G, Ye Z, Yu X, Zou J, Mali P, Brodsky RA, Cheng L (2008a) Trophoblast differentiation defect in human embryonic stem cells lacking PIG-A and GPI-anchored cell-surface proteins. Cell Stem Cell 2:345–355. doi:S1934-5909(08)00068-4

Chen X, Xu H, Yuan P, Fang F, Huss M, Vega VB, Wong E, Orlov YL, Zhang W, Jiang J, Loh YH, Yeo HC, Yeo ZX, Narang V, Govindarajan KR, Leong B, Shahab A, Ruan Y, Bourque G, Sung WK, Clarke ND, Wei CL, Ng HH (2008b) Integration of external signaling pathways with the core transcriptional network in embryonic stem cells. Cell 133:1106–1117

Chen T, Heller E, Beronja S, Oshimori N, Stokes N, Fuchs E (2012) An RNA interference screen uncovers a new molecule in stem cell self-renewal and long-term regeneration. Nature 485:104–108. doi:nature10940

Colak D, Mori T, Brill MS, Pfeifer A, Falk S, Deng C, Monteiro R, Mummery C, Sommer L, Gotz M (2008) Adult neurogenesis requires Smad4-mediated bone morphogenic protein signaling in stem cells. J Neurosci 28:434–446. doi:28/2/434

D'Amour KA, Agulnick AD, Eliazer S, Kelly OG, Kroon E, Baetge EE (2005) Efficient differentiation of human embryonic stem cells to definitive endoderm. Nat Biotechnol 23:1534–1541. doi:nbt1163

Das P, Ezashi T, Schulz LC, Westfall SD, Livingston KA, Roberts RM (2007) Effects of fgf2 and oxygen in the bmp4-driven differentiation of trophoblast from human embryonic stem cells. Stem Cell Res 1:61–74. doi:10.1016/j.scr.2007.09.004

Davis RP, Ng ES, Costa M, Mossman AK, Sourris K, Elefanty AG, Stanley EG (2008) Targeting a GFP reporter gene to the MIXL1 locus of human embryonic stem cells identifies human primitive streak-like cells and enables isolation of primitive hematopoietic precursors. Blood 111:1876–1884. doi:blood-2007-06-093609

Ducos K, Panterne B, Fortunel N, Hatzfeld A, Monier MN, Hatzfeld J (2000) p21(cip1) mRNA is controlled by endogenous transforming growth factor-β1 in quiescent human hematopoietic stem/progenitor cells. J Cell Physiol 184:80–85. doi:10.1002/(SICI)1097-4652

Dzierzak E, Speck NA (2008) Of lineage and legacy: the development of mammalian hematopoietic stem cells. Nat Immunol 9:129–136. doi:ni1560

Evans MJ, Kaufman MH (1981) Establishment in culture of pluripotential cells from mouse embryos. Nature 292:154–156

Falk S, Wurdak H, Ittner LM, Ille F, Sumara G, Schmid MT, Draganova K, Lang KS, Paratore C, Leveen P, Suter U, Karlsson S, Born W, Ricci R, Gotz M, Sommer L (2008) Brain area-specific effect of TGF-β signaling on Wnt-dependent neural stem cell expansion. Cell Stem Cell 2:472–483. doi:S1934-5909(08)00118-5

Fei T, Xia K, Li Z, Zhou B, Zhu S, Chen H, Zhang J, Chen Z, Xiao H, Han JD, Chen YG (2010a) Genome-wide mapping of SMAD target genes reveals the role of BMP signaling in embryonic stem cell fate determination. Genome Res 20:36–44

Fei T, Zhu S, Xia K, Zhang J, Li Z, Han JD, Chen YG (2010b) Smad2 mediates Activin/Nodal signaling in mesendoderm differentiation of mouse embryonic stem cells. Cell Res 20: 1306–1318

Feng XH, Derynck R (2005) Specificity and versatility in TGF-β signaling through Smads. Annu Rev Cell Dev Biol 21:659–693

Gadue P, Huber TL, Paddison PJ, Keller GM (2006) Wnt and TGF-β signaling are required for the induction of an in vitro model of primitive streak formation using embryonic stem cells. Proc Natl Acad Sci USA 103(45):16806–16811. doi:0603916103

Gage FH (2000) Mammalian neural stem cells. Science 287:1433–1438. doi:8300

Goldman DC, Bailey AS, Pfaffle DL, Al Masri A, Christian JL, Fleming WH (2009) BMP4 regulates the hematopoietic stem cell niche. Blood 114:4393–4401. doi:blood-2009-02-206433

Gomes WA, Mehler MF, Kessler JA (2003) Transgenic overexpression of BMP4 increases astroglial and decreases oligodendroglial lineage commitment. Dev Biol 255:164–177. doi:S0012160602000374

Gossrau G, Thiele J, Konang R, Schmandt T, Brustle O (2007) Bone morphogenetic protein-mediated modulation of lineage diversification during neural differentiation of embryonic stem cells. Stem Cells 25:939–949. doi:2006–0299

Gouon-Evans V, Boussemart L, Gadue P, Nierhoff D, Koehler CI, Kubo A, Shafritz DA, Keller G (2006) BMP-4 is required for hepatic specification of mouse embryonic stem cell-derived definitive endoderm. Nat Biotechnol 24:1402–1411. doi:nbt1258

Greber B, Coulon P, Zhang M, Moritz S, Frank S, Muller-Molina AJ, Arauzo-Bravo MJ, Han DW, Pape HC, Scholer HR (2011) FGF signalling inhibits neural induction in human embryonic stem cells. EMBO J 30:4874–4884. doi:emboj2011407

Green MD, Chen A, Nostro MC, d'Souza SL, Schaniel C, Lemischka IR, Gouon-Evans V, Keller G, Snoeck HW (2011) Generation of anterior foregut endoderm from human embryonic and induced pluripotent stem cells. Nat Biotechnol 29:267–272. doi:nbt.1788

Haramis AP, Begthel H, van den Born M, van Es J, Jonkheer S, Offerhaus GJ, Clevers H (2004) De novo crypt formation and juvenile polyposis on BMP inhibition in mouse intestine. Science 303:1684–1686. doi:10.1126/science.1093587

Hayashi K, Kobayashi T, Umino T, Goitsuka R, Matsui Y, Kitamura D (2002) SMAD1 signaling is critical for initial commitment of germ cell lineage from mouse epiblast. Mech Dev 118: 99–109. doi:S092547730200237X

He XC, Zhang J, Tong WG, Tawfik O, Ross J, Scoville DH, Tian Q, Zeng X, He X, Wiedemann LM, Mishina Y, Li L (2004) BMP signaling inhibits intestinal stem cell self-renewal through suppression of Wnt-β-catenin signaling. Nat Genet 36(10):1117–1121. doi:10.1038/ng1430

He W, Dorn DC, Erdjument-Bromage H, Tempst P, Moore MA, Massague J (2006) Hematopoiesis controlled by distinct TIF1γ and Smad4 branches of the TGFβ pathway. Cell 125:929–941. doi:S0092-8674(06)00518-6

Heldin CH, Miyazono K, ten Dijke P (1997) TGF-β signalling from cell membrane to nucleus through SMAD proteins. Nature 390:465–471

Horsley V, Aliprantis AO, Polak L, Glimcher LH, Fuchs E (2008) NFATc1 balances quiescence and proliferation of skin stem cells. Cell 132:299–310. doi:S0092-8674(07)01614-5

Huber TL, Zhou Y, Mead PE, Zon LI (1998) Cooperative effects of growth factors involved in the induction of hematopoietic mesoderm. Blood 92:4128–4137

Irion S, Clarke RL, Luche H, Kim I, Morrison SJ, Fehling HJ, Keller GM (2010) Temporal specification of blood progenitors from mouse embryonic stem cells and induced pluripotent stem cells. Development 137:2829–2839. doi:dev.042119

James D, Levine AJ, Besser D, Hemmati-Brivanlou A (2005) TGFβ/activin/nodal signaling is necessary for the maintenance of pluripotency in human embryonic stem cells. Development 132:1273–1282. doi:dev.01706

Jian H, Shen X, Liu I, Semenov M, He X, Wang XF (2006) Smad3-dependent nuclear translocation of β-catenin is required for TGF-β1-induced proliferation of bone marrow-derived adult human mesenchymal stem cells. Genes Dev 20:666–674. doi:20/6/666

Jiang Y, Jahagirdar BN, Reinhardt RL, Schwartz RE, Keene CD, Ortiz-Gonzalez XR, Reyes M, Lenvik T, Lund T, Blackstad M, Du J, Aldrich S, Lisberg A, Low WC, Largaespada DA, Verfaillie CM (2002) Pluripotency of mesenchymal stem cells derived from adult marrow. Nature 418:41–49. doi:10.1038/nature00870

Jin W, Takagi T, Kanesashi SN, Kurahashi T, Nomura T, Harada J, Ishii S (2006) Schnurri-2 controls BMP-dependent adipogenesis via interaction with Smad proteins. Dev Cell 10: 461–471. doi:S1534-5807(06)00113-4
Kalluri R, Zeisberg M (2003) Exploring the connection between chronic renal fibrosis and bone morphogenic protein-7. Histol Histopathol 18:217–224
Kamiya D, Banno S, Sasai N, Ohgushi M, Inomata H, Watanabe K, Kawada M, Yakura R, Kiyonari H, Nakao K, Jakt LM, Nishikawa S, Sasai Y (2011) Intrinsic transition of embryonic stem-cell differentiation into neural progenitors. Nature 470:503–509. doi:nature09726
Karlsson G, Blank U, Moody JL, Ehinger M, Singbrant S, Deng CX, Karlsson S (2007) Smad4 is critical for self-renewal of hematopoietic stem cells. J Exp Med 204:467–474. doi:jem.20060465
Kattman SJ, Witty AD, Gagliardi M, Dubois NC, Niapour M, Hotta A, Ellis J, Keller G (2011) Stage-specific optimization of activin/nodal and BMP signaling promotes cardiac differentiation of mouse and human pluripotent stem cell lines. Cell Stem Cell 8:228–240. doi:S1934-5909(10)00703-4
Kee K, Gonsalves JM, Clark AT, Pera RA (2006) Bone morphogenetic proteins induce germ cell differentiation from human embryonic stem cells. Stem Cells Dev 15:831–837. doi:10.1089/scd.2006.15.831
Kohyama J, Sanosaka T, Tokunaga A, Takatsuka E, Tsujimura K, Okano H, Nakashima K (2010) BMP-induced REST regulates the establishment and maintenance of astrocytic identity. J Cell Biol 189:159–170. doi:jcb.200908048
Kunath T, Saba-El-Leil MK, Almousailleakh M, Wray J, Meloche S, Smith A (2007) FGF stimulation of the Erk1/2 signalling cascade triggers transition of pluripotent embryonic stem cells from self-renewal to lineage commitment. Development 134:2895–2902
Larsson J, Blank U, Klintman J, Magnusson M, Karlsson S (2005) Quiescence of hematopoietic stem cells and maintenance of the stem cell pool is not dependent on TGF-β signaling in vivo. Exp Hematol 33:592–596. doi:S0301-472X(05)00091-3
Lavery K, Swain P, Falb D, Alaoui-Ismaili MH (2008) BMP-2/4 and BMP-6/7 differentially utilize cell surface receptors to induce osteoblastic differentiation of human bone marrow-derived mesenchymal stem cells. J Biol Chem 283:20948–20958. doi:M800850200
Lee D, Park C, Lee H, Lugus JJ, Kim SH, Arentson E, Chung YS, Gomez G, Kyba M, Lin S, Janknecht R, Lim DS, Choi K (2008) ER71 acts downstream of BMP, notch, and Wnt signaling in blood and vessel progenitor specification. Cell Stem Cell 2:497–507. doi:S1934-5909(08)00120-3
Lengerke C, Schmitt S, Bowman TV, Jang IH, Maouche-Chretien L, McKinney-Freeman S, Davidson AJ, Hammerschmidt M, Rentzsch F, Green JB, Zon LI, Daley GQ (2008) BMP and Wnt specify hematopoietic fate by activation of the Cdx-Hox pathway. Cell Stem Cell 2:72–82. doi:S1934-5909(07)00239-1
Li Z, Chen YG (2012) Fine-tune of intrinsic ERK activity by extrinsic BMP signaling in mouse embryonic stem cells. Protein Cell 3:401–404. doi:10.1007/s13238-012-2925-5
Li Z, Fei T, Zhang J, Zhu G, Wang L, Lu D, Chi X, Teng Y, Hou N, Yang X, Zhang H, Han JD, Chen YG (2012) BMP4 signaling acts via dual-specificity phosphatase 9 to control ERK activity in mouse embryonic stem cells. Cell Stem Cell 10:171–182. doi:S1934-5909(11)00597-2
Lim DA, Tramontin AD, Trevejo JM, Herrera DG, Garcia-Verdugo JM, Alvarez-Buylla A (2000) Noggin antagonizes BMP signaling to create a niche for adult neurogenesis. Neuron 28:713–726. doi:S0896-6273(00)00148-3
Liu F, Shao LE, Yu J (2000) Truncated activin type II receptor inhibits erythroid differentiation in K562 cells. J Cell Biochem 78:24–33. doi:10.1002/(SICI)1097-4644
Liu Z, Tang Y, Qiu T, Cao X, Clemens TL (2006) A dishevelled-1/Smad1 interaction couples WNT and bone morphogenetic protein signaling pathways in uncommitted bone marrow stromal cells. J Biol Chem 281:17156–17163. doi:M513812200
Loncar D (1992) Brown adipose tissue as a derivative of mesoderm grafted below the kidney capsule. A model for differentiation of isolated rat mesoderm. Int J Dev Biol 36:265–274
Maeda S, Hayashi M, Komiya S, Imamura T, Miyazono K (2004) Endogenous TGF-β signaling suppresses maturation of osteoblastic mesenchymal cells. EMBO J 23:552–563. doi:10.1038/sj.emboj.7600067

Maeno M, Mead PE, Kelley C, Xu RH, Kung HF, Suzuki A, Ueno N, Zon LI (1996) The role of BMP-4 and GATA-2 in the induction and differentiation of hematopoietic mesoderm in Xenopus laevis. Blood 88:1965–1972
Martin GR (1981) Isolation of a pluripotent cell line from early mouse embryos cultured in medium conditioned by teratocarcinoma stem cells. Proc Natl Acad Sci USA 78:7634–7638
Massague J (2012) TGFβ signalling in context. Nat Rev Mol Cell Biol 13:616–630. doi:nrm3434
Massague J, Chen YG (2000) Controlling TGF-β signaling. Genes Dev 14:627–644
McLean AB, D'Amour KA, Jones KL, Krishnamoorthy M, Kulik MJ, Reynolds DM, Sheppard AM, Liu H, Xu Y, Baetge EE, Dalton S (2007) Activin a efficiently specifies definitive endoderm from human embryonic stem cells only when phosphatidylinositol 3-kinase signaling is suppressed. Stem Cells 25:29–38. doi:25/1/29
Mira H, Andreu Z, Suh H, Lie DC, Jessberger S, Consiglio A, San Emeterio J, Hortiguela R, Marques-Torrejon MA, Nakashima K, Colak D, Gotz M, Farinas I, Gage FH (2010) Signaling through BMPR-IA regulates quiescence and long-term activity of neural stem cells in the adult hippocampus. Cell Stem Cell 7:78–89. doi:S1934-5909(10)00171-2
Mizuseki K, Sakamoto T, Watanabe K, Muguruma K, Ikeya M, Nishiyama A, Arakawa A, Suemori H, Nakatsuji N, Kawasaki H, Murakami F, Sasai Y (2003) Generation of neural crest-derived peripheral neurons and floor plate cells from mouse and primate embryonic stem cells. Proc Natl Acad Sci USA 100:5828–5833. doi:10.1073/pnas.1037282100
Morizane A, Doi D, Kikuchi T, Nishimura K, Takahashi J (2011) Small-molecule inhibitors of bone morphogenic protein and activin/nodal signals promote highly efficient neural induction from human pluripotent stem cells. J Neurosci Res 89:117–126. doi:10.1002/jnr.22547
Moustakas A, Souchelnytskyi S, Heldin CH (2001) Smad regulation in TGF-β signal transduction. J Cell Sci 114:4359–4369
Munoz-Sanjuan I, Brivanlou AH (2002) Neural induction, the default model and embryonic stem cells. Nat Rev Neurosci 3:271–280. doi:10.1038/nrn786
Nakashima K, Yanagisawa M, Arakawa H, Kimura N, Hisatsune T, Kawabata M, Miyazono K, Taga T (1999) Synergistic signaling in fetal brain by STAT3-Smad1 complex bridged by p300. Science 284:479–482
Nakashima K, Takizawa T, Ochiai W, Yanagisawa M, Hisatsune T, Nakafuku M, Miyazono K, Kishimoto T, Kageyama R, Taga T (2001) BMP2-mediated alteration in the developmental pathway of fetal mouse brain cells from neurogenesis to astrocytogenesis. Proc Natl Acad Sci USA 98:5868–5873. doi:10.1073/pnas.101109698
Nichols J, Smith A (2009) Naive and primed pluripotent states. Cell Stem Cell 4:487–492. doi:S1934-5909(09)00224-0
Noel D, Gazit D, Bouquet C, Apparailly F, Bony C, Plence P, Millet V, Turgeman G, Perricaudet M, Sany J, Jorgensen C (2004) Short-term BMP-2 expression is sufficient for in vivo osteochondral differentiation of mesenchymal stem cells. Stem Cells 22:74–85. doi:10.1634/stemcells.22-1-74
Nostro MC, Cheng X, Keller GM, Gadue P (2008) Wnt, activin, and BMP signaling regulate distinct stages in the developmental pathway from embryonic stem cells to blood. Cell Stem Cell 2:60–71. doi:S1934-5909(07)00228-7
Nostro MC, Sarangi F, Ogawa S, Holtzinger A, Corneo B, Li X, Micallef SJ, Park IH, Basford C, Wheeler MB, Daley GQ, Elefanty AG, Stanley EG, Keller G (2011) Stage-specific signaling through TGFβ family members and WNT regulates patterning and pancreatic specification of human pluripotent stem cells. Development 138:861–871. doi:dev.055236
Ogawa K, Saito A, Matsui H, Suzuki H, Ohtsuka S, Shimosato D, Morishita Y, Watabe T, Niwa H, Miyazono K (2007) Activin-Nodal signaling is involved in propagation of mouse embryonic stem cells. J Cell Sci 120:55–65. doi:120/1/55
Okafuji K, Kaku K, Seguchi M, Tanaka H, Azuno Y, Kaneko T (1995) Effects of activin A/erythroid differentiation factor on erythroid and megakaryocytic differentiations of mouse erythroleukemia (Friend) cells: evidence for two distinct modes of cell response. Exp Hematol 23:210–216

Okamura D, Hayashi K, Matsui Y (2005) Mouse epiblasts change responsiveness to BMP4 signal required for PGC formation through functions of extraembryonic ectoderm. Mol Reprod Dev 70:20–29. doi:10.1002/mrd.20136

Orkin SH (2000) Diversification of haematopoietic stem cells to specific lineages. Nat Rev Genet 1:57–64. doi:10.1038/35049577

Orkin SH, Zon LI (2008) Hematopoiesis: an evolving paradigm for stem cell biology. Cell 132:631–644. doi:S0092-8674(08)00125-6

Oshimori N, Fuchs E (2012) Paracrine TGF-β signaling counterbalances BMP-mediated repression in hair follicle stem cell activation. Cell Stem Cell 10:63–75. doi:S1934-5909(11)00533-9

Park SW, Jun Koh Y, Jeon J, Cho YH, Jang MJ, Kang Y, Kim MJ, Choi C, Sook Cho Y, Chung HM, Young Koh G, Han YM (2010) Efficient differentiation of human pluripotent stem cells into functional CD34+ progenitor cells by combined modulation of the MEK/ERK and BMP4 signaling pathways. Blood 116:5762–5772. doi:blood-2010-04-280719

Partridge K, Yang X, Clarke NM, Okubo Y, Bessho K, Sebald W, Howdle SM, Shakesheff KM, Oreffo RO (2002) Adenoviral BMP-2 gene transfer in mesenchymal stem cells: in vitro and in vivo bone formation on biodegradable polymer scaffolds. Biochem Biophys Res Commun 292:144–152. doi:S0006291X02966234

Pearson S, Sroczynska P, Lacaud G, Kouskoff V (2008) The stepwise specification of embryonic stem cells to hematopoietic fate is driven by sequential exposure to Bmp4, activin A, bFGF and VEGF. Development 135:1525–1535. doi:dev.011767

Pera MF, Andrade J, Houssami S, Reubinoff B, Trounson A, Stanley EG, Ward-van Oostwaard D, Mummery C (2004) Regulation of human embryonic stem cell differentiation by BMP-2 and its antagonist noggin. J Cell Sci 117:1269–1280. doi:10.1242/jcs.00970

Pick M, Azzola L, Mossman A, Stanley EG, Elefanty AG (2007) Differentiation of human embryonic stem cells in serum-free medium reveals distinct roles for bone morphogenetic protein 4, vascular endothelial growth factor, stem cell factor, and fibroblast growth factor 2 in hematopoiesis. Stem Cells 25:2206–2214. doi:2006–0713

Pittenger MF, Mackay AM, Beck SC, Jaiswal RK, Douglas R, Mosca JD, Moorman MA, Simonetti DW, Craig S, Marshak DR (1999) Multilineage potential of adult human mesenchymal stem cells. Science 284:143–147

Plikus MV, Mayer JA, de la Cruz D, Baker RE, Maini PK, Maxson R, Chuong CM (2008) Cyclic dermal BMP signalling regulates stem cell activation during hair regeneration. Nature 451:340–344. doi:nature06457

Saitou M (2009) Germ cell specification in mice. Curr Opin Genet Dev 19:386–395. doi:S0959-437X(09)00113-0

Saitou M, Yamaji M (2010) Germ cell specification in mice: signaling, transcription regulation, and epigenetic consequences. Reproduction 139:931–942. doi:REP-10-0043

Samanta J, Kessler JA (2004) Interactions between ID and OLIG proteins mediate the inhibitory effects of BMP4 on oligodendroglial differentiation. Development 131:4131–4142. doi:10.1242/dev.01273

Sato T, Vries RG, Snippert HJ, van de Wetering M, Barker N, Stange DE, van Es JH, Abo A, Kujala P, Peters PJ, Clevers H (2009) Single Lgr5 stem cells build crypt-villus structures in vitro without a mesenchymal niche. Nature 459:262–265. doi:nature07935

Scandura JM, Boccuni P, Massague J, Nimer SD (2004) Transforming growth factor β-induced cell cycle arrest of human hematopoietic cells requires p57KIP2 up-regulation. Proc Natl Acad Sci USA 101:15231–15236. doi:0406771101

Schmitt B, Ringe J, Haupl T, Notter M, Manz R, Burmester GR, Sittinger M, Kaps C (2003) BMP2 initiates chondrogenic lineage development of adult human mesenchymal stem cells in high-density culture. Differentiation 71:567–577. doi:S0301-4681(09)60538-7

Schulz TJ, Tseng YH (2009) Emerging role of bone morphogenetic proteins in adipogenesis and energy metabolism. Cytokine Growth Factor Rev 20:523–531. doi:S1359-6101(09)00094-X

Schulz TC, Noggle SA, Palmarini GM, Weiler DA, Lyons IG, Pensa KA, Meedeniya AC, Davidson BP, Lambert NA, Condie BG (2004) Differentiation of human embryonic stem

cells to dopaminergic neurons in serum-free suspension culture. Stem Cells 22:1218–1238. doi:22/7/1218
Schwall RH, Lai C (1991) Erythroid differentiation bioassays for activin. Methods Enzymol 198:340–346
Scoville DH, Sato T, He XC, Li L (2008) Current view: intestinal stem cells and signaling. Gastroenterology 134:849–864. doi:S0016-5085(08)00185-6
Seuntjens E, Umans L, Zwijsen A, Sampaolesi M, Verfaillie CM, Huylebroeck D (2009) Transforming growth factor type β and Smad family signaling in stem cell function. Cytokine Growth Factor Rev 20:449–458. doi:S1359-6101(09)00080-X
Shea CM, Edgar CM, Einhorn TA, Gerstenfeld LC (2003) BMP treatment of C3H10T1/2 mesenchymal stem cells induces both chondrogenesis and osteogenesis. J Cell Biochem 90:1112–1127. doi:10.1002/jcb.10734
Shi Y, Massague J (2003) Mechanisms of TGF-β signaling from cell membrane to the nucleus. Cell 113:685–700. doi:S009286740300432X
Shiozaki M, Kosaka M, Eto Y (1998) Activin A: a commitment factor in erythroid differentiation. Biochem Biophys Res Commun 242:631–635. doi:S0006-291X(97)98020-7
Singbrant S, Karlsson G, Ehinger M, Olsson K, Jaako P, Miharada K, Stadtfeld M, Graf T, Karlsson S (2010) Canonical BMP signaling is dispensable for hematopoietic stem cell function in both adult and fetal liver hematopoiesis, but essential to preserve colon architecture. Blood 115:4689–4698. doi:blood-2009-05-220988
Singh AM, Reynolds D, Cliff T, Ohtsuka S, Mattheyses AL, Sun Y, Menendez L, Kulik M, Dalton S (2012) Signaling network crosstalk in human pluripotent cells: a Smad2/3-regulated switch that controls the balance between self-renewal and differentiation. Cell Stem Cell 10:312–326. doi:S1934-5909(12)00016-1
Sitnicka E, Ruscetti FW, Priestley GV, Wolf NS, Bartelmez SH (1996) Transforming growth factor β 1 directly and reversibly inhibits the initial cell divisions of long-term repopulating hematopoietic stem cells. Blood 88:82–88
Smith JR, Vallier L, Lupo G, Alexander M, Harris WA, Pedersen RA (2008) Inhibition of Activin/Nodal signaling promotes specification of human embryonic stem cells into neuroectoderm. Dev Biol 313:107–117. doi:S0012-1606(07)01414-5
Stavridis MP, Lunn JS, Collins BJ, Storey KG (2007) A discrete period of FGF-induced Erk1/2 signalling is required for vertebrate neural specification. Development 134:2889–2894
Sumi T, Tsuneyoshi N, Nakatsuji N, Suemori H (2008) Defining early lineage specification of human embryonic stem cells by the orchestrated balance of canonical Wnt/β-catenin, Activin/Nodal and BMP signaling. Development 135:2969–2979. doi:dev.021121
Suzuki A, Raya A, Kawakami Y, Morita M, Matsui T, Nakashima K, Gage FH, Rodriguez-Esteban C, Izpisua Belmonte JC (2006) Nanog binds to Smad1 and blocks bone morphogenetic protein-induced differentiation of embryonic stem cells. Proc Natl Acad Sci USA 103:10294–10299. doi:0506945103
Tada S, Era T, Furusawa C, Sakurai H, Nishikawa S, Kinoshita M, Nakao K, Chiba T (2005) Characterization of mesendoderm: a diverging point of the definitive endoderm and mesoderm in embryonic stem cell differentiation culture. Development 132:4363–4374. doi:dev.02005
Tang QQ, Otto TC, Lane MD (2004) Commitment of C3H10T1/2 pluripotent stem cells to the adipocyte lineage. Proc Natl Acad Sci USA 101:9607–9611. doi:10.1073/pnas.0403100101
Tang Y, Wu X, Lei W, Pang L, Wan C, Shi Z, Zhao L, Nagy TR, Peng X, Hu J, Feng X, Van Hul W, Wan M, Cao X (2009) TGF-β1-induced migration of bone mesenchymal stem cells couples bone resorption with formation. Nat Med 15:757–765. doi:nm.1979
Temple S (2001) The development of neural stem cells. Nature 414:112–117. doi:10.1038/35102174
Thomson JA, Itskovitz-Eldor J, Shapiro SS, Waknitz MA, Swiergiel JJ, Marshall VS, Jones JM (1998) Embryonic stem cell lines derived from human blastocysts. Science 282:1145–1147
Toyooka Y, Tsunekawa N, Akasu R, Noce T (2003) Embryonic stem cells can form germ cells in vitro. Proc Natl Acad Sci USA 100:11457–11462. doi:10.1073/pnas.1932826100
Tremblay KD, Dunn NR, Robertson EJ (2001) Mouse embryos lacking Smad1 signals display defects in extra-embryonic tissues and germ cell formation. Development 128:3609–3621

Tseng YH, Kokkotou E, Schulz TJ, Huang TL, Winnay JN, Taniguchi CM, Tran TT, Suzuki R, Espinoza DO, Yamamoto Y, Ahrens MJ, Dudley AT, Norris AW, Kulkarni RN, Kahn CR (2008) New role of bone morphogenetic protein 7 in brown adipogenesis and energy expenditure. Nature 454:1000–1004. doi:nature07221
Tsuda H, Wada T, Ito Y, Uchida H, Dehari H, Nakamura K, Sasaki K, Kobune M, Yamashita T, Hamada H (2003) Efficient BMP2 gene transfer and bone formation of mesenchymal stem cells by a fiber-mutant adenoviral vector. Mol Ther 7:354–365. doi:S152500160200062X
Vallier L, Reynolds D, Pedersen RA (2004) Nodal inhibits differentiation of human embryonic stem cells along the neuroectodermal default pathway. Dev Biol 275:403–421. doi: S0012-1606(04)00587-1
Vallier L, Alexander M, Pedersen RA (2005) Activin/Nodal and FGF pathways cooperate to maintain pluripotency of human embryonic stem cells. J Cell Sci 118:4495–4509. doi:118/19/4495
Vallier L, Mendjan S, Brown S, Chng Z, Teo A, Smithers LE, Trotter MW, Cho CH, Martinez A, Rugg-Gunn P, Brons G, Pedersen RA (2009) Activin/Nodal signalling maintains pluripotency by controlling Nanog expression. Development 136:1339–1349. doi:dev.033951
van der Flier LG, Clevers H (2009) Stem cells, self-renewal, and differentiation in the intestinal epithelium. Annu Rev Physiol 71:241–260. doi:10.1146/annurev.physiol.010908.163145
Varelas X, Sakuma R, Samavarchi-Tehrani P, Peerani R, Rao BM, Dembowy J, Yaffe MB, Zandstra PW, Wrana JL (2008) TAZ controls Smad nucleocytoplasmic shuttling and regulates human embryonic stem-cell self-renewal. Nat Cell Biol 10:837–848. doi:ncb1748
Wang EA, Israel DI, Kelly S, Luxenberg DP (1993) Bone morphogenetic protein-2 causes commitment and differentiation in C3H10T1/2 and 3T3 cells. Growth Factors 9:57–71
Wang G, Zhang H, Zhao Y, Li J, Cai J, Wang P, Meng S, Feng J, Miao C, Ding M, Li D, Deng H (2005) Noggin and bFGF cooperate to maintain the pluripotency of human embryonic stem cells in the absence of feeder layers. Biochem Biophys Res Commun 330:934–942. doi:S0006-291X(05)00531-0
Watabe T, Miyazono K (2009) Roles of TGF-β family signaling in stem cell renewal and differentiation. Cell Res 19:103–115. doi:cr2008323
Wei W, Qing T, Ye X, Liu H, Zhang D, Yang W, Deng H (2008) Primordial germ cell specification from embryonic stem cells. PLoS One 3:e4013. doi:10.1371/journal.pone.0004013
Weinstein DC, Hemmati-Brivanlou A (1997) Neural induction in Xenopus laevis: evidence for the default model. Curr Opin Neurobiol 7:7–12. doi:S0959-4388(97)80114-6
Winnier G, Blessing M, Labosky PA, Hogan BL (1995) Bone morphogenetic protein-4 is required for mesoderm formation and patterning in the mouse. Genes Dev 9:2105–2116
Wu Z, Zhang W, Chen G, Cheng L, Liao J, Jia N, Gao Y, Dai H, Yuan J, Xiao L (2008) Combinatorial signals of activin/nodal and bone morphogenic protein regulate the early lineage segregation of human embryonic stem cells. J Biol Chem 283:24991–25002. doi:M803893200
Xiao L, Yuan X, Sharkis SJ (2006) Activin A maintains self-renewal and regulates fibroblast growth factor, Wnt, and bone morphogenic protein pathways in human embryonic stem cells. Stem Cells 24(6):1476–1486. doi:2005–0299
Xu C, Inokuma MS, Denham J, Golds K, Kundu P, Gold JD, Carpenter MK (2001) Feeder-free growth of undifferentiated human embryonic stem cells. Nat Biotechnol 19(10):971–974. doi:10.1038/nbt1001-971
Xu RH, Chen X, Li DS, Li R, Addicks GC, Glennon C, Zwaka TP, Thomson JA (2002) BMP4 initiates human embryonic stem cell differentiation to trophoblast. Nat Biotechnol 20:1261–1264. doi:10.1038/nbt761
Xu RH, Peck RM, Li DS, Feng X, Ludwig T, Thomson JA (2005) Basic FGF and suppression of BMP signaling sustain undifferentiated proliferation of human ES cells. Nat Methods 2:185–190. doi:nmeth744
Xu RH, Sampsell-Barron TL, Gu F, Root S, Peck RM, Pan G, Yu J, Antosiewicz-Bourget J, Tian S, Stewart R, Thomson JA (2008) NANOG is a direct target of TGFβ/activin-mediated SMAD signaling in human ESCs. Cell Stem Cell 3:196–206. doi:S1934-5909(08)00335-4

Yamazaki S, Ema H, Karlsson G, Yamaguchi T, Miyoshi H, Shioda S, Taketo MM, Karlsson S, Iwama A, Nakauchi H (2011) Nonmyelinating Schwann cells maintain hematopoietic stem cell hibernation in the bone marrow niche. Cell 147:1146–1158. doi:S0092-8674(11)01269-4

Yan KS, Chia LA, Li X, Ootani A, Su J, Lee JY, Su N, Luo Y, Heilshorn SC, Amieva MR, Sangiorgi E, Capecchi MR, Kuo CJ (2012) The intestinal stem cell markers Bmi1 and Lgr5 identify two functionally distinct populations. Proc Natl Acad Sci USA 109:466–471. doi:1118857109

Yang L, Soonpaa MH, Adler ED, Roepke TK, Kattman SJ, Kennedy M, Henckaerts E, Bonham K, Abbott GW, Linden RM, Field LJ, Keller GM (2008) Human cardiovascular progenitor cells develop from a KDR+ embryonic-stem-cell-derived population. Nature 453:524–528. doi:nature06894

Yasunaga M, Tada S, Torikai-Nishikawa S, Nakano Y, Okada M, Jakt LM, Nishikawa S, Chiba T, Era T (2005) Induction and monitoring of definitive and visceral endoderm differentiation of mouse ES cells. Nat Biotechnol 23:1542–1550. doi:nbt1167

Ying QL, Nichols J, Chambers I, Smith A (2003a) BMP induction of Id proteins suppresses differentiation and sustains embryonic stem cell self-renewal in collaboration with STAT3. Cell 115:281–292

Ying QL, Stavridis M, Griffiths D, Li M, Smith A (2003b) Conversion of embryonic stem cells into neuroectodermal precursors in adherent monoculture. Nat Biotechnol 21:183–186

Ying QL, Wray J, Nichols J, Batlle-Morera L, Doble B, Woodgett J, Cohen P, Smith A (2008) The ground state of embryonic stem cell self-renewal. Nature 453:519–523

Yoshida-Koide U, Matsuda T, Saikawa K, Nakanuma Y, Yokota T, Asashima M, Koide H (2004) Involvement of Ras in extraembryonic endoderm differentiation of embryonic stem cells. Biochem Biophys Res Commun 313:475–481

Young RA (2011) Control of the embryonic stem cell state. Cell 144:940–954. doi:S0092-8674(11)00071-7

Yu P, Pan G, Yu J, Thomson JA (2011) FGF2 sustains NANOG and switches the outcome of BMP4-induced human embryonic stem cell differentiation. Cell Stem Cell 8:326–334. doi:S1934-5909(11)00002-6

Zeng X, Cai J, Chen J, Luo Y, You ZB, Fotter E, Wang Y, Harvey B, Miura T, Backman C, Chen GJ, Rao MS, Freed WJ (2004) Dopaminergic differentiation of human embryonic stem cells. Stem Cells 22:925–940. doi:22/6/925

Zhang YE (2009) Non-Smad pathways in TGF-β signaling. Cell Res 19:128–139. doi:cr2008328

Zhang J, He XC, Tong WG, Johnson T, Wiedemann LM, Mishina Y, Feng JQ, Li L (2006) Bone morphogenetic protein signaling inhibits hair follicle anagen induction by restricting epithelial stem/progenitor cell activation and expansion. Stem Cells 24:2826–2839. doi:2005–0544

Zhang P, Li J, Tan Z, Wang C, Liu T, Chen L, Yong J, Jiang W, Sun X, Du L, Ding M, Deng H (2008) Short-term BMP-4 treatment initiates mesoderm induction in human embryonic stem cells. Blood 111:1933–1941. doi:blood-2007-02-074120

Zhang K, Li L, Huang C, Shen C, Tan F, Xia C, Liu P, Rossant J, Jing N (2010) Distinct functions of BMP4 during different stages of mouse ES cell neural commitment. Development 137:2095–2105

Zhao GQ (2003) Consequences of knocking out BMP signaling in the mouse. Genesis 35:43–56. doi:10.1002/gene.10167

Zhou J, Su P, Li D, Tsang S, Duan E, Wang F (2010) High-efficiency induction of neural conversion in human ESCs and human induced pluripotent stem cells with a single chemical inhibitor of transforming growth factor β superfamily receptors. Stem Cells 28:1741–1750. doi:10.1002/stem.504

Chapter 3
Emerging Roles of TGF-β Co-receptors in Human Disease

Alison E. Meyer, Karthikeyan Mythreye, and Gerard C. Blobe

Abstract TGF-β signaling is both regulated and mediated by signaling co-receptors. Several TGF-β co-receptors have been identified including endoglin (CD105), the type III TGF-β receptor (TβRIII, betaglycan), neuropilin-1/2, syndecan-2, CD109, and LRP1. These co-receptors serve diverse functions including the regulation of ligand access to other TGF-β receptors and receptor trafficking. The TGF-β co-receptors can also signal directly. The TGF-β co-receptors are broadly expressed, have essential roles in embryonic development, and are frequently altered during disease progression. TGF-β co-receptors regulate cancer initiation and progression through effects on cell growth, migration, invasion, proliferation, and angiogenesis. In addition to their roles in cancer, these co-receptors are dysregulated during development, in vascular disease and fibrotic disorders. Collectively, the TGF-β co-receptors influence disease biology through complex mechanisms involving the regulation of growth factor-dependent and independent signaling events as well as through interactions with diverse scaffolding protein partners.

Keywords Angiogenesis • Betaglycan • Cancer • Co-receptors • Development • Disease • Endoglin • Fibrosis • Neuropilin • Syndecan • TGF-β • TβRIII

3.1 Introduction

TGF-β signaling pathways have essential roles in multiple cellular processes including proliferation, differentiation, and apoptosis (Massague 1998). While canonical TGF-β signaling is mediated by the type I and type II TGF-β receptors, TGF-β

A.E. Meyer • K. Mythreye • G.C. Blobe (✉)
Division of Medical Oncology, Departments of Medicine, Pharmacology and Cancer Biology, Duke University Medical Center, 450 Research Dr. LSRC B354, Durham, NC 27708, USA
e-mail: gerard.blobe@duke.edu

A. Moustakas and K. Miyazawa (eds.), *TGF-β in Human Disease*,
DOI 10.1007/978-4-431-54409-8_3,

signaling is both regulated and mediated by signaling co-receptors. Multiple TGF-β co-receptors have been identified including endoglin (CD105) (Cheifetz et al. 1992), the type III TGF-β receptor (TβRIII, betaglycan) (Andres et al. 1991; Lopez-Casillas et al. 1991; Wang et al. 1991), neuropilin-1/2 (Glinka and Prud'homme 2008), syndecan-2 (Chen et al. 2004b), CD109 (Finnson et al. 2006), and LRP1 (Cabello-Verrugio and Brandan 2007). These co-receptors serve diverse functions including the regulation of ligand access to other TGF-β receptors and receptor trafficking. The co-receptors can also signal directly. The TGF-β co-receptors are broadly expressed, have essential roles in embryonic development, and are frequently altered during disease progression (Maring et al. 2012; Bernabeu et al. 2009; Staton et al. 2007; Li et al. 1999; Wild et al. 2012; Stenvers et al. 2003; Theocharis et al. 2010; Gatza et al. 2010). Here we summarize the emerging roles of TGF-β co-receptors in human disease with a focus on endoglin, TβRIII, neuropilin-1/2, and syndecan-2.

3.1.1 Structural Features and Ligand Binding Properties of TGF-β Co-receptors

Structurally, endoglin, TβRIII, neuropilin-1/2, and syndecan-2 all have a large extracellular domain, a short, single-pass transmembrane region, and a short cytoplasmic domain (Essner et al. 2006; Gougos et al. 1992; Mythreye and Blobe 2009a; Lopez-Casillas et al. 1991; Wild et al. 2012). The extracellular domains of TβRIII and syndecan-2 are modified with glycosaminoglycan chains. The extracellular domains of all co-receptors interact with multiple ligand classes (see Table 3.1). Furthermore, these co-receptors all undergo ectodomain shedding, with proteolytic cleavage releasing their soluble extracellular domains from the cell surface. The soluble ectodomains have diverse functions including sequestering ligand to inhibit signaling, binding ligand to facilitate signaling in *trans*, and potentially functioning as ligands themselves (Essner et al. 2006; Gatza et al. 2010; Bernabeu et al. 2009; Wild et al. 2012).

3.1.2 TGF-β Co-receptor Function

Endoglin, which is expressed predominantly in proliferating vascular endothelial cells and smooth muscle cells, has a critical role in angiogenesis (Bourdeau et al. 1999; Fonsatti et al. 2003). Endoglin null mice are embryonic lethal due to defects in vascular development (Li et al. 1999), while mutations of the endoglin gene, ENG, cause the autosomal dominant vascular disease, hereditary hemorrhagic telangiectasia (HHT; discussed in detail below) (McAllister et al. 1994). In endothelial cells, endoglin promotes proliferation through the endothelial-specific type I TGF-β receptor ALK1 and the Smad 1/5/8 pathway, while inhibiting proliferation through

Table 3.1 Ligand specificity of TGF-β co-receptors

Co-receptor	Ligand specificity	References
Endoglin	TGF-β1,3 ActivinA BMP 2,7,9 and 10	Bernabeu et al. (2009), Castonguay et al. (2011)
TβRIII (betaglycan)	TGF-β1, 2 and 3 Inhibin BMP 2, 4, 7 GDF-5 FGF2	Gatza et al. (2010)
Neuropilins 1/2	TGF-β1 Sema3A, 3F VEGF HGF FGF PDGF Shh	Wild et al. (2012)
Syndecan-2	TGF-β1 FGF VEGF PDGF	Essner et al. (2006), Chen et al. (2004b)

BMP bone morphogenetic protein, *FGF* fibroblast growth factor, *GDF* growth/differentiation factor, *HGF* hepatocyte growth factor, *PDGF* platelet-derived growth factor, *Shh* sonic hedgehog, *TGF-β* transforming growth factor-β, *VEGF* vascular endothelial growth factor

the canonical type I TGF-β receptor ALK5 and the Smad2/3 pathway (Goumans et al. 2002). Endoglin has also been shown to participate in Erk (Lee and Blobe 2007), H-Ras (Santibanez et al. 2010), and PI3K/Akt signaling (Lee et al. 2012). Despite its predominant role in angiogenesis, Endoglin is also expressed in some non-endothelial cell types (see Cancer section below). However, the signaling pathways that endoglin utilizes in these cell types have not been well characterized.

TβRIII is an important regulator of cell migration, invasion, growth, and angiogenesis (Gatza et al. 2010). TβRIII null mice are embryonic lethal and exhibit liver and heart defects (Stenvers et al. 2003). TβRIII mediates ligand presentation, whereby ligand is transferred from TβRIII to TβRII to potentiate signaling (Lopez-Casillas et al. 1991; Lopez-Casillas et al. 1993; Wang et al. 1991). In addition to facilitating signaling through the canonical Smad pathways, TβRIII regulates NFκB (Criswell and Arteaga 2007; You et al. 2009) , p38 (Santander and Brandan 2006), and Cdc42 signaling (Mythreye and Blobe 2009b).

Neuropilin-1/2 are vertebrate-specific proteins that are essential for the normal embryological development of the nervous and cardiovascular systems (Staton et al. 2007). Neuropilin-1/2 also regulate angiogenesis, with neuropilin-1 predominating in arterial endothelial cells and neuropilin-2 predominating in vein and lymphatic endothelial cells (Staton et al. 2007). Upon ligand binding, neuropilin-1/2 signal through the p38, Akt, focal adhesion kinase (FAK), and MAPK pathways to regulate numerous diverse processes including proliferation, survival, migration, adhesion, and vascular permeability (Wild et al. 2012).

Syndecan-2 has roles in numerous processes including adhesion, cytoskeletal organization, vesicle transport, synaptic transmission, and axonal migration (Essner et al. 2006). This co-receptor regulates binding to the extracellular matrix (ECM) components laminin, collagen, and fibronectin and can also regulate actin dynamics and FAK (Essner et al. 2006).

3.1.3 Co-receptor Interacting Proteins

While their extracellular domains largely mediate ligand binding, the cytoplasmic domains of the TGF-β co-receptors interact with numerous scaffold proteins to regulate receptor trafficking and signaling. Interestingly, endoglin, TβRIII, neuropilin-1/2, and syndecan-2 all interact with the PDZ domain-containing protein GIPC (GAIP-interacting protein C-terminus; also known as synectin or NIP) (Essner et al. 2006; Wang et al. 2006; Blobe et al. 2001; Lee et al. 2008; Cai and Reed 1999). In the case of neuropilin-1/2, this interaction is important for proper arterial branching morphogenesis and can promote the internalization of integrin to regulate cell adhesion to fibronectin (Chittenden et al. 2006; Valdembri et al. 2009). In contrast to integrin, both endoglin and TβRIII are stabilized on the cell surface through their interaction with GIPC (Blobe et al. 2001; Lee et al. 2008). In addition to GIPC, endoglin and TβRIII can associate with β-arrestin2, resulting in their internalization (Chen et al. 2003; Lee and Blobe 2007). Endoglin can also interact with zyxin (Conley et al. 2004) and zyxin-related protein 1 (Sanz-Rodriguez et al. 2004) to regulate cell adhesion, as well as with the cytosolic dynein light chain family member Tctex2β (Meng et al. 2006), which links endoglin with the microtubule-based transport machinery. Interestingly, the cytoplasmic domain of syndecan-2 has been shown to specifically associate with TβRIII, although the consequences of this interaction are unknown (Chen et al. 2004b). Syndecan-2 can also interact with the cell adhesion molecule fibronectin, α5β1 integrin, the PDZ domain proteins GIPC, CASK/LIN-2, synbindin, and syntenin, and the ERM family proteins ezrin, radixin, and moesin, with most of these interactions facilitating actin reorganization and cell-substrate adhesion (Essner et al. 2006).

3.2 TGF-β Co-receptors in Cancer

In normal epithelial cells and in the early stages of tumorigenesis, TGF-β acts as a tumor suppressor by inhibiting growth and inducing apoptosis or differentiation, when appropriate (Elliott and Blobe 2005). However, during cancer progression, cancer cells lose this inhibitory growth response and respond instead with increased migration and invasion (Elliott and Blobe 2005). TGF-β also functions to promote cancer progression by inhibiting the immune system and promoting angiogenesis (Elliott and Blobe 2005). Much like TGF-β ligands, the co-receptors regulate both cancer initiation and progression. Below, we discuss the roles of the endoglin, TβRIII, neuropilin-1/2, and syndecan-2 in tumor initiation, progression and angiogenesis, as well as their influence on the tumor stroma.

3.2.1 Cancer Initiation and Progression

3.2.1.1 Endoglin

Germline mutations in the ENG gene have been reported in patients with early onset Juvenile polyposis (JP), a disease that is associated with the development of gastrointestinal malignancies (Sweet et al. 2005). However, a larger follow-up study failed to confirm this finding (Howe et al. 2007). While endoglin is primarily expressed on cells derived from the hemangioblast (i.e., endothelial and hematopoietic cells), there have been numerous reports of altered endoglin expression in cancer cells (see Table 3.2). While some studies suggest endoglin expression is increased in a variety of cancers, others have observed a decrease in endoglin expression. Collectively, these studies suggest that the role of endoglin in cancer initiation and progression may be context specific.

Endoglin has been reported to have complex roles in TGF-β-mediated processes, including cancer cell adhesion, proliferation, and migration (Conley et al. 2004; Sanz-Rodriguez et al. 2004; Lakshman et al. 2011; Romero et al. 2010). In many cancers, endoglin appears to act as a suppressor of cancer progression. Overexpression of endoglin in prostate cancer cells decreases proliferation, invasion, and metastasis in vivo by influencing TGF-β signaling (Lakshman et al. 2011). Similarly, overexpression of endoglin was reported to reduce the invasiveness and tumorigenicity of esophageal squamous cell carcinoma cells, although the effect of endoglin expression on TGF-β signaling was not examined (Wong et al. 2008). In a breast cancer metastasis model, endoglin overexpression was reported to block TGF-β-enhanced cell motility/invasion, resulting in reduced lung colonization (Henry et al. 2011). In contrast to these studies, endoglin expression has been reported to promote the anchorage-independent growth of Ewing sarcoma cells (Pardali et al. 2011).

How might endoglin influence cancer progression? Endoglin is able to regulate the adhesion and migration of many cell types, in some cases by influencing TGF-β signaling (Sanz-Rodriguez et al. 2004; Conley et al. 2004; Lee et al. 2008; Lee and Blobe 2007; Ray et al. 2010; Guerrero-Esteo et al. 1999). While different cell types utilize different mechanisms to regulate adhesion and migration, insights into the function of endoglin in cancer may be derived from studies on non-tumorigenic cells. The cytosolic domain of endoglin likely plays a crucial role in endoglin's functions, as it mediates interactions with several proteins. Endoglin binds zyxin and zyxin-related protein 1, which are cytoskeleton-interacting proteins that localize to focal adhesions (Conley et al. 2004; Sanz-Rodriguez et al. 2004). Therefore, endoglin may regulate the organization of the actin cytoskeleton and the adhesive properties of cancer cells by interacting with these two proteins. Like TβRIII, the

Table 3.2 Expression of TGF-β co-receptors in human cancers

Cancer type	Increased expression	Decreased expression	References
Bladder	Syndecan-2		Marzioni et al. (2009)
Breast	Neuropilin-1, Neuropilin-2, soluble Endoglin	TβRIII, Endoglin	Ghosh et al. (2008), Yasuoka et al. (2009), Staton et al. (2007), Dong et al. (2007), Hempel et al. (2007), Henry et al. (2011), Calabro et al. (2003), Davidson et al. (2010), Li et al. (2000), Takahashi et al. (2001)
Colon	Neuropilin-1, Neuropilin-2, TβRIII, soluble Endoglin, Syndecan-2	Neuropilin-1	Parikh et al. (2004), Grandclement et al. (2011), Gray et al. (2008), Gatza et al. (2012), Takahashi et al. (2001), Park et al. (2002), Ryu et al. (2009)
Endometrial	Endoglin	TβRIII	Florio et al. (2005)
Esophageal		Endoglin, Syndecan-2	Wong et al. (2008), Huang et al. (2009)
Ewing sarcoma	Endoglin		Pardali et al. (2011)
GI Tract	Neuropilin-1	Neuropilin-2	Hansel et al. (2004), Cohen et al. (2001)
Granulosa tumor		TβRIII	Bilandzic et al. (2009)
Kidney		TβRIII	Copland et al. (2003), Cooper et al. (2010)
Leukemia	Neuropilin-1, TβRIII, soluble Endoglin		Younan et al. (2012), Klein et al. (2001), Jelinek et al. (2003), Calabro et al. (2003)
Liver	Neuropilin-1	TβRIII	Berge et al. (2011), Bae et al. (2009)
Melanoma	Neuropilin-1, Neuropilin-2, Endoglin, Syndecan-2		Lacal et al. (2000), Rushing et al. (2012), Pardali et al. (2011), Lee et al. (2009)
Mesothelioma	Syndecan-2		Gulyas and Hjerpe (2003)
Multiple myeloma		TβRIII	Lambert et al. (2011)
Neuroblastoma		TβRIII	Iolascon et al. (2000)
Non-Hodgkin's lymphoma	TβRIII		Woszczyk et al. (2004)
Non-small cell lung	Neuropilin-1, Neuropilin-2, soluble Endoglin	TβRIII	Kawakami et al. (2002), Finger et al. (2008), Kopczynska et al. (2012)
Osteosarcoma	Neuropilin-2	Syndecan-2	Handa et al. (2000), Orosco et al. (2007)
Ovarian	Neuropilin-1, Neuropilin-2, soluble Endoglin	TβRIII	Drenberg et al. (2009), Baba et al. (2007), Bock et al. (2011), Hempel et al. (2008)

(continued)

Table 3.2 (continued)

Cancer type	Increased expression	Decreased expression	References
Pancreatic	Neuropilin-1, Neuropilin-2, Syndecan-2	TβRIII	Parikh et al. (2004), Cohen et al. (2002), Li et al. (2004), Gordon et al. (2008)
Prostate	Neuropilin-1, Neuropilin-2, Endoglin, soluble Endoglin, Syndecan-2	TβRIII, Endoglin	Latil et al. (2000), Yacoub et al. (2009), Turley et al. (2007), Lakshman et al. (2011), Kassouf et al. (2004), Shariat et al. (2008), Popovic et al. (2010)
Salivary adenoid cystic carcinoma	Neuropilin-2		Cai et al. (2010)
Squamous cell carcinoma		TβRIII, Endoglin	Wong et al. (2008), Meng et al. (2011)
Thyroid	Neuropilin-2		Finley et al. (2004)

cytosolic domain of endoglin can associate with GIPC and β-arrestin 2 (Lee et al. 2008; Lee and Blobe 2007; Chen et al. 2003). The interaction between endoglin and GIPC was recently shown to regulate TGF-β pathway signaling events, endothelial cell migration, and capillary tube formation via the recruitment of PI3K and Akt (Lee et al. 2012). β-arrestin 2, in contrast, mediates the internalization of endoglin, resulting in changes in Erk activation and localization as well as endothelial cell migration in a TGF-β-dependent manner (Lee and Blobe 2007). Whether similar pathways mediate endoglin function in cancer cells remains to be defined. In addition, the role of endoglin in mediating the effects of BMPs and Activin in the context of cancer progression has not been explored.

3.2.1.2 TβRIII

While TβRIII mutations have not been reported in human cancers, a recent study demonstrated that myxoinflammatory fibroblastic sarcomas and hemosiderotic fibrolipomatous tumors contained a recurrent chromosome t(1:10) rearrangement, which involves the TGFBR3 gene (Antonescu et al. 2011), suggesting a potential role for TβRIII in the pathogenesis of these diseases. Additionally, the progressive loss of TβRIII has been documented in multiple cancers (see Table 3.2).

Importantly, loss of TβRIII expression during cancer progression is associated with increasing cancer grade and stage and a poorer patient prognosis, supporting an important functional role for loss of TβRIII expression (Lambert et al. 2011; Dong et al. 2007; Meng et al. 2011; Turley et al. 2007). In many of these cases, the loss of TβRIII leads to disrupted TGF-β signaling. Loss of TβRIII expression in human cancers can occur by multiple mechanisms including loss of heterozygosity (Dong et al. 2007; Turley et al. 2007; Finger et al. 2008), gene silencing by promoter methylation (Turley et al. 2007; Cooper et al. 2010; Hempel et al. 2007), and the negative regulation of TGFBR3 promoter by TGF-β or BMP ligands (Hempel et al. 2008; Gordon et al. 2009). As many cancers are associated with increased TGF-ligand production (Massague 2008), the downregulation of TβRIII expression by ligand is likely a common event. Indeed, the ability of TGF-β ligand to downregulate TβRIII expression has been documented in breast (Dong et al. 2007; Hempel et al. 2007), ovarian (Hempel et al. 2007), and non-small cell lung cancer cell lines (Finger et al. 2008).

Consistent with its frequent loss of expression in a broad spectrum of human cancers, TβRIII has been demonstrated to inhibit both ligand-dependent and independent cancer initiation and cancer progression in preclinical models of these cancers. siRNA-mediated silencing of TβRIII expression in normal murine mammary gland cells (NMuMG) was sufficient to induce tumorigenicity in mice (Criswell and Arteaga 2007). In a reciprocal manner, restoring TβRIII expression decreased tumorigenicity in breast (Sun and Chen 1997; Bandyopadhyay et al. 2002a; Bandyopadhyay et al. 1999a), prostate (Turley et al. 2007), non-small cell lung cancer (Finger et al. 2008), and renal cell carcinoma models (Copland et al. 2003).

Mechanistically, TβRIII has been reported to influence different aspects of tumor biology in a context-dependent manner. Thus, in the context of breast cancer (Sun and Chen 1997), renal cell carcinoma (Copland et al. 2003; Margulis et al. 2008), and multiple myeloma (Lambert et al. 2011), TβRIII inhibited proliferation by mediating TGF-β signaling, while in the context of renal cell carcinoma (Copland et al. 2003; Margulis et al. 2008), TβRIII stimulated apoptosis. Most consistently, in the context of multiple myeloma (Lambert et al. 2011) and cancers of the breast (Dong et al. 2007; Lee et al. 2010), pancreatic (Gordon et al. 2008), ovary (Hempel et al. 2008; Bilandzic et al. 2009) prostate (Turley et al. 2007), and lung (Finger et al. 2008), TβRIII inhibited cancer cell migration and invasion by attenuating TGF-β signaling.

TβRIII inhibits migration and invasion through multiple mechanisms. In some contexts, TβRIII inhibits migration/invasion through the generation of soluble TβRIII, which is thought to bind and sequester free TGF-β ligand in the extracellular space, thereby inhibiting its pro-migratory effects (Bandyopadhyay et al. 2002a; Bandyopadhyay et al. 2002b; Bandyopadhyay et al. 2005; Bandyopadhyay et al. 1999b; Dong et al. 2007; Gordon et al. 2008; Sun and Chen 1997; Finger et al. 2008). TβRIII also inhibits migration and invasion through soluble TβRIII-independent mechanisms, with the interaction between TβRIII and GIPC mediating the anti-migratory and anti-invasive effects of TβRIII in the context of breast cancer (Lee et al. 2010). Through its interaction with β-arrestin2, TβRIII can also negatively regulate the pro-migratory and pro-invasive transcription factor NFκB (Criswell and Arteaga 2007; You et al. 2009). Additionally, TβRIII inhibits the intrinsic ability of epithelial cells to migrate through its interaction with β-arrestin2 (Mythreye and Blobe 2009b), with TβRIII mediating β-arrestin2-dependent constitutive activation of Cdc42 and alterations in the actin cytoskeleton (Mythreye and Blobe 2009b). Interestingly, these effects are also independent of TβRIII's role in ligand presentation (Mythreye and Blobe 2009b). TβRIII also functions to inhibit TGF-β and BMP-mediated EMT and EMT-associated increases in migration and invasion (Gordon et al. 2008; Gordon et al. 2009; Criswell and Arteaga 2007; Mythreye and Blobe 2009b; Hempel et al. 2007). Finally, TβRIII may influence cell migration and invasion by regulating cell adhesion. Indeed, TβRIII has been shown to influence the interaction between multiple myeloma cells and bone marrow stromal cells and to regulate the adhesive properties of ovarian cancer, breast cancer, and granulosa cell tumors (Lambert et al. 2011; Mythreye et al. 2012; Bilandzic et al. 2009). Mechanistically, TβRIII affects focal adhesion dynamics by regulating the localization of α5 integrin (Mythreye et al. 2012).

In contrast to most cancers, TβRIII expression is increased at the protein level in colon cancer, where TβRIII is associated with an inhibition of chemotherapy-induced apoptosis, increased TGF-β signaling, colony formation in soft agar, migration, and ligand-stimulated proliferation in vitro and tumorigenicity in vivo (Gatza et al. 2012). The reasons for the opposing effects of TβRIII on colon versus other cancers are currently unknown. However, as K-Ras is mutated in up to 50 % of colon cancer patients (Benhattar et al. 1993), crosstalk between K-Ras and the TGF-β pathway may be involved. Interestingly, increased TβRIII expression has also been observed in non-Hodgkin's lymphoma and B-cell chronic lymphocytic leukemia (Woszczyk et al. 2004; Klein et al. 2001; Jelinek et al. 2003), although the effect of TβRIII on the biology of these cancers is unknown

3.2.1.3 Neuropilin-1/2

As co-receptors for VEGF and semaphorin ligands, the neuropilins have important roles in regulating angiogenesis. However, the neuropilins are also expressed on cancer cells (see Table 3.2) and function as TGF-β co-receptors in regulating the TGF-β responsiveness of cancer cells (Staton et al. 2007; Mak et al. 2010; Grandclement et al. 2011). Like endoglin, neuropilin-1/2 has context-dependent effects in regulating cancer progression.

Mechanistically, neuropilin-1/2 may promote cancer progression by activating latent TGF-β and enhancing signaling. This has been shown to occur in regulatory T-cells as well as breast cancer cells (Glinka and Prud'homme 2008). Furthermore, VEGF and neuropilin-1 have been shown to directly promote epithelial-mesenchymal transition (EMT) in prostate cancer cells (Mak et al. 2010), and the knockdown of neuropilin-2 in colon cancer cells has also been associated with EMT (Grandclement et al. 2011). As EMT can also be triggered by TGF-β, the neuropilins may also aid in TGF-β-mediated EMT, although this has not yet been directly demonstrated

3.2.1.4 Syndecan-2

While syndecan-2 does bind to TGF-β, it is currently unknown what roles this binding may play in cancer development and progression. However, altered syndecan-2 expression has been observed in several cancers (see Table 3.2), and syndecan-2 has been shown to positively regulate cancer cell migration and invasion in a variety of cancer types (Choi et al. 2010; Lee et al. 2009; Orosco et al. 2007; Ryu et al. 2009).

Syndecan-2's role in colon cancer has been most thoroughly investigated. Increased mRNA and protein expression of syndecan-2 have been observed in human colon adenocarcinoma tissue samples in comparison with neighboring normal epithelium (Ryu et al. 2009). Additionally, high expression of syndecan-2 has been observed in colon cancer cell lines. This increased expression is associated with increased proliferation, migration, and invasion (Ryu et al. 2009; Park et al. 2002; Choi et al. 2010). Syndecan-2 regulates colon and fibrosarcoma cancer cell migration and invasion through Tiam1-dependent Rac activation (Park et al. 2005; Choi et al. 2010). Additionally, MMP7 activity and integrin α2 are upregulated upon syndecan-2 expression in colon cancer cells, which may influence adhesion, migration, and invasion (Ryu et al. 2009; Choi et al. 2010).

3.2.2 Cancer Stem Cells

In some cancers, a small percentage of tumor cells have been isolated that possess the unique capability of sustained self-renewal and differentiation (Grotenhuis et al. 2012). As such features are common to stem cells, these cells are referred to as cancer stem cells (CSCs) or tumor-initiating cells. CSCs are unique in that, unlike the majority of tumor cells, they are typically more chemoresistant and are more effective in generating xenografts in mouse models (Grotenhuis et al. 2012). CSCs are also capable of giving rise to non-CSC populations and thus can recapitulate the heterogeneity of the tumor from which they were isolated. Therefore, it has been suggested that CSCs can drive tumor initiation or recurrence. For a recent review on CSCs, please see Grotenhuis et al. 2012.

While endoglin, TβRIII, and the neuropilins have been shown to have roles primarily in TGF-β-mediated angiogenesis and/or cancer cell migration and invasion, a few reports have identified these co-receptors as CSC markers. In a cohort of matched primary/recurrent ovarian cancer specimens, along with the expression of several known markers of ovarian CSCs, endoglin was found to be significantly increased in persistent tumors compared to matched primary tumors (Steg et al. 2012). Additionally, endoglin-positive cells isolated from a rhabdoid meningioma surgical sample exhibited increased proliferation and had an increased ability to form single-cell tumor spheres in vitro compared to cells lacking endoglin expression (Hu et al. 2012). Interestingly, the endoglin-positive cells were found to have several cell surface markers in common with mesenchymal progenitor cells, which are multipotent and are the precursor cell type for several tissues including bone, muscle, and fat (Hu et al. 2012; Houthuijzen et al. 2012).

With regard to TβRIII, a lack of TβRIII expression has been associated with prostate CSCs. Sharifi et al. showed that prostate cancer cells lacking TβRIII expression had higher levels of the prostate CSC marker CD133 than tumor cells that were positive for TβRIII expression, suggesting that the loss of TβRIII may increase the population of stem-like cells (Sharifi et al. 2007).

Notably, skin squamous cell carcinoma CSCs were found to lack endoglin expression; however, these cells did express high levels of neuropilin-1 (Beck et al. 2011). The expression of neuropilin-1 was shown to be critical for the ability of these CSCs to promote VEGF-stimulated tumor growth in a mouse model (Beck et al. 2011). Hamerlik et al. also detected neuropilin-1 expression on CSCs isolated from glioblastoma multiforme tumors (Hamerlik et al. 2012). Here, neuropilin-1 was found to stabilize VEGFR2 expression, thereby enhancing the self-renewal capacity and viability of the CSCs (Hamerlik et al. 2012). While it is unknown what role TGF-β family ligands play in CSCs, the expression of these TGF-β co-receptors on CSCs provides one more mechanism by which co-receptors can regulate the pathogenesis of human cancers.

3.2.3 Cancer-Associated Angiogenesis

3.2.3.1 Endoglin

Endoglin is intricately involved in the process of angiogenesis, which is required to sustain primary tumor growth and metastasis (Folkman et al. 1989; Neri and Bicknell 2005; Molema and Griffioen 1998). Endoglin expression is higher in nascent blood vessels, where it controls cell proliferation, migration, and capillary tube formation (Li et al. 1999; Fansatti et al. 2001; Lebrin et al. 2004; Goumans et al. 2002; Ma et al. 2000; Cheifetz et al. 1992; Li et al. 2000; Bourdeau et al. 1999; McAllister et al. 1994). While endoglin is highly expressed during angiogenesis in multiple cancer types, its expression is typically absent on vessels in normal tissues (Fansatti et al. 2001). Because of its lack of expression in normal vasculature, endoglin has been extensively used as a marker for measuring microvessel density (MVD), which is inversely correlated with prognosis (Fansatti et al. 2001; Brewer et al. 2000; Saad et al. 2004; Wikstrom et al. 2002; Kumar et al. 1999; Shariat et al. 2008; Schimming and Marme 2001; Tanaka et al. 2001). Endoglin has emerged as a powerful marker of tumor-associated neovascularization, with anti-endoglin antibodies being investigated to visualize tumors and their metastases (Fonsatti et al. 2010).

The dependence of tumors on the pro-angiogenic functions of endoglin has also made it an attractive target for anti-tumor therapies (Dallas et al. 2008). Several in vivo studies have revealed endoglin to be an important therapeutic target. First, endoglin haploinsufficient mice were shown to have decreased lung tumor vascularization and growth compared to their endoglin +/+ littermates (Duwel et al. 2007). This may be due to increased endothelial cell apoptosis, as endoglin was also shown to regulate eNOS levels and to prevent apoptosis in hypoxic endothelial cells (Duwel et al. 2007). Additionally, the suppression of tumor growth and metastasis and an increased survival rate have been observed using anti-endoglin antibodies alone or antibodies conjugated to immunotoxins or radiolabels in mice (Seon et al. 1997; Matsuno et al. 1999; Tabata et al. 1999; Tsujie et al. 2008; Takahashi et al. 2001;

Uneda et al. 2009). A phase I clinical trial using an anti-endoglin antibody (TRC105) in patients with advanced refractory cancers was recently reported, with the treatment is well tolerated, and resulting in disease control in nearly half of the patients (Rosen et al. 2008; Rosen et al. 2012) (NCT00582985).

3.2.3.2 TβRIII

Like endoglin, a few studies have implicated TβRIII in tumor-associated angiogenesis. In a xenograft mouse model of breast cancer, decreased tumor-associated angiogenesis was observed in mice injected with TβRIII-expressing cells vs. control cells (Dong et al. 2007). The administration of recombinant soluble TβRIII also decreased tumor vascularization in xenograft models of breast and prostate cancer (Bandyopadhyay et al. 2005; Bandyopadhyay et al. 2002a), and soluble TβRIII has been shown to inhibit the ability of human endothelial cells to form tubes in an in vitro Matrigel assay (Bandyopadhyay et al. 2002b). In contrast to these reports, SERPINE1, which is a known inhibitor of angiogenesis, was found to be downregulated in a gene array performed on prostate cancer cells lacking TβRIII expression (Sharifi et al. 2007). Taken together, these studies suggest an anti-angiogenic function for TβRIII.

3.2.3.3 The Neuropilins

Most of the work regarding neuropilin-1/2 and tumor angiogenesis has focused on their role as VEGF co-receptors, and little attention has been given to their potential role in angiogenesis as TGF-β co-receptors. The neuropilins are, however, clearly involved in tumor-associated angiogenesis. In breast cancer specimens, increased neuropilin-1 expression was observed in both smooth muscle and endothelial cells within the areas of invasion (Stephenson et al. 2002). Additionally, non-small cell lung cancer patients whose tumors expressed both neuropilins had increased vessel counts compared to those patients that lacked co-expression (Kawakami et al. 2002). The neuropilin ligands Sema3A and Sema3F (class 3 semaphorins) may also regulate tumor angiogenesis as they have been shown to inhibit the pro-angiogenic effects of VEGF in endothelial cells (Staton et al. 2007; Wild et al. 2012). Both of these ligands were found to inhibit endothelial cell migration, proliferation, and tube formation (Narazaki and Tosato 2006; Serini et al. 2003). As the interaction between neuropilin-1 and VEGF is important for tumor angiogenesis, several current strategies for anti-tumor therapy revolve around the inhibition of neuropilin-1, VEGF, both, or their interaction (Geretti and Klagsbrun 2007; Pan et al. 2007; Jia et al. 2010; Jia et al. 2006; Soker et al. 1997; von Wronski et al. 2006). The role of the neuropilins in angiogenesis as it relates to their role as TGF-β co-receptors requires additional investigation.

3.2.3.4 Syndecan-2

Very little is known regarding syndecan-2's role in tumor-associated angiogenesis. However, high syndecan-2 expression has been reported in the microvasculature of mouse gliomas (Fears et al. 2006), while downregulation of syndecan-2 resulted in decreased endothelial cell motility and a reduced formation of capillary tube-like structures (Fears et al. 2006). The ability of syndecan-2 to promote angiogenesis was found to be due to increased levels of soluble syndecan-2, which were increased during glioma progression (Fears et al. 2006).

3.2.4 TGF-β Co-receptors and Stromal Components

3.2.4.1 Endoglin

Endoglin is weakly expressed in some stromal cell types (Rokhlin et al. 1995; Gougos et al. 1992). In a mouse model of prostate cancer progression, endoglin (+/−)-derived tumors lacked the typically observed infiltration of cancer-associated fibroblasts (Romero et al. 2011). Therefore, endoglin may act to support the viability of CAFs in the tumor microenvironment by promoting neovascularization and growth. In a co-culture model of the prostate cancer cell line C4-2B and the bone marrow stromal cell line HS-27a, endoglin was significantly downregulated in the stromal cells only upon co-culture (O'Connor et al. 2007). This loss of endoglin resulted in attenuated TGF-β signaling and decreased stromal cell proliferation (O'Connor et al. 2007). Thus the prostate cancer cells, upon interaction with bone marrow stromal cells, may alter stromal cell TGF-β signaling to promote the formation of bone metastases (O'Connor et al. 2007).

3.2.4.2 TβRIII

Similar to cancer cells, the expression of TβRIII mRNA is significantly reduced in CAFs in oral squamous cell carcinoma (Meng et al. 2011). As increased amounts of TGF-β ligand are present within the oral cancer microenvironment, ligand-induced downregulation of TβRIII expression was thought to be a potential mechanism for downregulation (Meng et al. 2011). TβRIII also has a potential role in exosomes, which are secreted vesicles that participate in intercellular communication (Webber et al. 2010). Exosomes can deliver protein, mRNA, or microRNAs to recipient cells. Interestingly, TGF-β was recently shown to be expressed by prostate and breast cancer-derived exosomes (Webber et al. 2010). This exosomal TGF-β was capable of promoting fibroblast–myofibroblast differentiation (Webber et al. 2010). The association of TGF-β with the exosomal membrane was linked with TβRIII expression, suggesting that TβRIII acts as a TGF-β anchor on cancer-associated exosomes (Webber et al. 2010).

3.2.4.3 The Neuropilins

Neuropilin-1 has been shown to promote the activation of fibroblasts to myofibroblasts, which are important regulators of cancer progression (De Wever and Mareel 2003). Decreasing neuropilin-1 expression in fibroblasts enhanced Smad1/5 signaling but reduced Smad2/3 signaling (Cao et al. 2010b). Mechanistically, an interaction between neuropilin-1 and endogenous TβRII was associated with decreased fibroblast proliferation and a reversion to a quiescent state (Cao et al. 2010b). In the immune system, neuropilin-1 is expressed by regulatory T cells and has been shown to activate latent TGF-β in this cell type (Glinka and Prud'homme 2008). Indeed, T cells expressing neuropilin-1 could efficiently bind latent TGF-β1. However, this was not observed in cells lacking neuropilin-1 (Glinka and Prud'homme 2008). Similar results were observed in a breast cancer cell line, suggesting that this role of neuropilin-1 is common and may be important in cancer progression (Glinka and Prud'homme 2008).

3.3 TGF-β Co-receptors in Vascular Disease

As TGF-β plays a key role in physiological angiogenesis, it is not surprising that animal models and human studies have implicated TGF-β superfamily signaling in the initiation and progression of many vascular disorders. Due to the role of the co-receptors in regulating the TGF-β pathway, direct mutations or disruptions in their function have been tied to these disorders as well.

3.3.1 Endoglin

A critical role for endoglin in angiogenesis has come from studies of mice lacking this co-receptor. Endoglin-deficient mice die at midgestation (day 10.5) of angiogenic and cardiovascular defects (Sorensen et al. 2003). During development, endoglin-deficient mice lose structural, molecular, and functional distinctions between arteries and veins (Li et al. 1999). Additionally, these mice fail to form mature blood vessels in the yolk sac (Li et al. 1999). Such studies indicate that altered TGF-β signaling is the cause of the phenotypes observed in endoglin null mice and also establish a functional link between endoglin and the TGF-β pathway. Vasculogenesis in the endoglin null mice is normal, but angiogenesis and remodeling of the primary vascular plexus are impaired. The mice exhibit poor vascular smooth muscle cell (VSMC) development that results in dilation and rupture of the vascular channels. In the heterozygous mice, disease severity increases with age and can include the rupture of major vessels.

Mutations in endoglin can also lead to the vascular disease hereditary hemorrhagic telangiectasia (HHT), or Rendu–Osler–Weber syndrome. HHT is a rare autosomal dominant disorder that is characterized by arteriovenous malformations or localized abnormal arteriovenous connections that affect both the microvasculature and large vessels. HHT patients typically manifest spontaneous epistaxis (nose bleeds), cutaneous telangiectases (small dilated blood vessels), and arteriovenous malformations in internal organs. Additional features include highly variable expression and variable, age-dependent penetrance. More than 150 mutations have been reported in endoglin including deletions, insertions, missense mutations, and splice site changes, all of which result in reduced levels of endoglin on the surface of endothelial cells (McAllister et al. 1994). Patients with HHT also have significantly lower amounts of endoglin in peripheral monocytes compared to normal individuals (Abdalla and Letarte 2006). Interestingly, most of the known mutations in endoglin are within exons 1–12, which encode the extracellular domain, and very few mutations have been observed in exons 13 and 14, which encode the transmembrane and cytoplasmic domains. Many of these alterations result in deficient angiogenesis, i.e. the sprouting of new vessels from preexisting ones. One proposed explanation for the development of HHT is that mutations in endoglin result in misfolded proteins that are either incapable of heretodimerizing or are unable to reach the cell surface, both of which would account for the halpoinsufficiency that is observed in HHT1.

Preeclampsia, a systemic, pregnancy-associated hypertensive disorder, has also been linked with endoglin. Preeclampsia appears to originate in the placenta and is characterized by endothelial cell dysfunction. As such, several authors have reported changes in the expression of placental anti-angiogenic factors in preeclamptic patients [reviewed in (Maynard and Karumanchi 2011)]. Soluble endoglin has been found to be elevated in preeclampsia and has been proposed as a viable marker for this disorder as well. However, the molecular basis for preeclampsia and the exact role of soluble endoglin in this disorder remain unknown.

3.3.2 The Neuropilins

Neuropilin-1 mediates normal developmental angiogenesis, which has been observed in mouse and zebrafish models. Neuropilin-1 has also been shown to mediate pathological angiogenesis in tumors and retinal disease, particularly in the context of VEGF signaling, wherein neuropilin 1, which is highly expressed in the retinal endothelial cells, is upregulated by both VEGF and hypoxia to regulate a positive feedback mechanism in retinal neovascularization. As such, most of neuropilin-1/2 functions in angiogenesis and vascular biology are related to its roles as a receptor for VEGF and as a co-receptor for VEGFR2 (Soker et al. 1997). Currently, much less is known about the ability of neuropilin-1 to regulate TGF-β's function in angiogenesis.

3.3.3 Syndecan-2

Syndecan-2 has been shown to regulate endothelial cell functions in vitro, including migration and cytoskeletal organization (Noguer et al. 2009). These functions may in turn contribute to its roles in capillary tube formation. Studies in zebrafish have demonstrated specific roles for syndecan-2 in regulating sprouting during angiogenesis (Chen et al. 2004a). Notably, the role of syndecan-2 in regulating TGF-β signaling in the context of endothelial biology has not been established and requires further exploration.

3.4 TGF-β Co-receptors and Developmental Defects

3.4.1 Endoglin

Genes that are required for proper vascular development also often affect blood flow. In addition, genes that are expressed in the endocardial lining of the heart can affect cardiac development, as in the case for the TGF-β co-receptors including endoglin, TβRIII, neuropilin-1, and syndecan-2. In addition to endoglin's well-studied role in vascular development, endoglin knockout mice have revealed a role for endoglin in cardiovascular development. In endoglin null mice, the atrioventricular canal endocardium fails to undergo mesenchymal transformation and does not form cushions (Li et al. 1999). These heart valve formation defects may be due to a perturbation in TGF-β signaling. Interestingly, endoglin has also been detected in hematopoietic stem cells in adult bone marrow and in the AGM (aorta–gonal mesonephros) at E11.5 (Chen et al. 2002) as well as in fetal liver cells (Pimanda et al. 2008). In addition to cardiovascular defects, endoglin null mouse embryos also display anemia of the yolk sac, which was, until recently, assumed to be an indirect result of insufficient blood flow. However, recent evidence demonstrates that endoglin plays a direct role in hematopoiesis. Endoglin is expressed in the first hematopoietic progenitor cells during embryonic development where it has been shown to regulate TGF-β/BMP signaling during the initiation of hematopoiesis (Borges et al. 2012).

3.4.2 TβRIII

Similar to endoglin, the deletion of TβRIII in mice results in embryonic lethality due to failed coronary vessel development and palate fusion (Compton et al. 2007). TβRIII null mice exhibit significantly impaired coronary vasculogenesis, defects in the epicardium, and dysmorphic and distended vessels. Collectively, these phenotypes demonstrate a requirement for TβRIII during coronary vessel development, which is essential for embryonic viability (Compton et al. 2007). More recently, an analysis of E13.5 TβRIII null embryos revealed a lower rate of epicardial cell proliferation and decreased epicardial-derived cell invasion. This effect is mediated by the interaction of TβRIII's cytoplasmic domain with GIPC (Sanchez et al. 2011), with loss of responsiveness to TGF-β and FGF2, two important regulators of epicardial cell behavior (Sanchez et al. 2011). In addition to its role in proper heart development, TβRIII has also been shown to participate in cord formation, fetal Leydig cell development, and the establishment of fetal endocrine testis function. Collectively, the observations implicate TGF-β superfamily members as regulators of early fetal testis structure and function (Sarraj et al. 2010). TβRIII mRNA has a distinct expression pattern during gonadogenesis, where it is expressed at higher levels in developing testes compared to developing ovaries. TβRIII is predominantly expressed by Leydig cells within the fetal testis interstitium; however, at birth, expression is shifted to the seminiferous cords, suggesting possible roles for TβRIII in the gonadogenesis (Sarraj et al. 2007). Interestingly, TβRIII can bind inhibin and can regulate ovarian and granulosa carcinogenesis through this interaction.

3.4.3 Neuropilin-1

Neuropilin-1 plays key roles in vascular biology. Neuropilin-1 mutant mouse embryos have impaired arterial differentiation, which appears to be independent of blood flow patterns (Jones et al. 2008). Neuropilin-1's role in arterial specification during vascular development is primarily due to its ability to regulate VEGF/VEGFR2 function. The function of neuropilin-1 in development can be additionally regulated at the transcriptional level. As such, the Notch signaling pathway, which directs cell fate decisions during embryonic development, has been shown to regulate the neuropilin-1 promoter to alter cell responsiveness to several ligands including VEGF (Sorensen et al. 2009) and, potentially, TGF-β.

In addition to its roles in vascular development (see angiogenesis), neuropilin-1 plays pivotal roles in the development of the neuronal system as a receptor for members of the class-3 semaphorin family of axonal guidance factors [reviewed in (Geretti and Klagsbrun 2007)]. Much less is known about its role in regulating TGF-β's role in developmental processes.

3.4.4 Syndecan-2

The role of syndecan-2 in development has been best studied in Xenopus embryos. Here, syndecan-2 has been shown to be a regulator of left–right development in the embryonic mesoderm (Kramer and Yost 2002). This function of syndecan-2 appears to be mediated via the functional interactions of its heparan sulfate glycosaminoglycan chains with the TGF-β superfamily cell–cell signaling molecule Vg1, which is known to regulate early left–right development. Syndecan-2 is expressed in the ectoderm and likely presents Vg1 to the migrating mesoderm in a unilateral manner, demonstrating that syndecan-2 can function in cell signaling by directly affecting neighboring cells or in a cell-autonomous manner. Additional mechanistic studies indicate that PKCγ can phosphorylate syndecan-2 and that this phosphorylation event is required in the right-side ectodermal cells (Kramer et al. 2002). While studies in zebrafish have revealed a similar role for syndecan-2 in left–right development, knockdown experiments have also demonstrated a requirement for syndecan-2 in embryonic angiogenesis. During both zebrafish and mouse development, syndecan-2 is expressed in the mesenchymal cell layer surrounding the axial blood vessels, suggesting a potential role for this molecule in coordinating and organizing vascular development (Chen et al. 2004a). Syndecan 2 and VEGF are thought to act synergistically to facilitate angiogenesis.

3.5 TGF-β Co-receptors in Fibrosis

Fibrosis, which is characterized by the fibroblast-mediated deposition of the ECM proteins collagen and fibronectin, can play roles in diverse pathological conditions including the autoimmune disease systemic sclerosis (SSc), wound healing (particularly after myocardial infarction), Crohn's disease, pulmonary hypertension, and diabetic nephropathy. Hepatic fibrosis can also set the stage for the development of cirrhosis and, in some cases, hepatocellular carcinoma (HCC). TGF-β is a key regulator of the fibrotic processes [reviewed in (Hawinkels and Ten Dijke 2011)]. Not all of the TGF-β co-receptors have established roles in each of the pathological conditions associated with fibrosis; however, some common themes exist regarding the ability of the co-receptors to modulate TGF-β signaling during the fibrotic process.

3.5.1 Endoglin

Endoglin has been shown to be a negative regulator of TGF-β1 signaling in the intestinal fibroblast, where it modulates Smad3 phosphorylation, Smad binding element (SBE) promoter activity, connective tissue growth factor (CTGF) production, and collagen contraction (Burke et al. 2009). Endoglin is also upregulated in patients suffering from fibrosis or scleroderma, where increased soluble endoglin levels are present as well (Wipff et al. 2008; Coral-Alvarado et al. 2010; Dharmapatni et al. 2001). Endoglin can both stimulate and inhibit TGF-β-induced fibrosis, depending on the context. For instance, in human and rat cell line models using myoblasts and myofibroblasts, endoglin was shown to reduce cell responsiveness to TGF-β, thereby reducing collagen production via an ERK-dependent mechanism (Rodriguez-Barbero et al. 2006). In addition, fibroblasts derived from patients with fibrosis or SSc have higher basal levels of CTGF and collagen relative to healthy individuals, and the overexpression of endoglin in control fibroblasts (from healthy donors) was shown to block TGF-β-induced CTGF expression. Collectively, these studies suggest an inhibitory role for endoglin in fibrosis (Holmes et al. 2011). In constrast, endoglin has also been shown to positively regulate fibrosis, especially in the context of cardiac fibroblasts. Here, endoglin expression is upregulated by TGF-β and angiotensin II, and blocking this upregulation resulted in reduced collagen II expression (Shyu et al. 2010), reviewed in (Maring et al. 2012). Fibroblasts from patients suffering from Crohn's disease have also been shown to express high levels of endoglin. Future studies on how endoglin expression is regulated should shed more light on the dichotomous roles of this co-receptor in fibrosis.

3.5.2 TβRIII

Given the structural and functional similarities between endoglin and TβRIII, similar roles for these two co-receptors in fibrosis could be predicted. Consistent with this idea, soluble TβRIII has been shown to inhibit fibrous airway obliteration (Liu et al. 2002) and can prevent myocardial fibrosis in spontaneously hypertensive rats (Hermida et al. 2009) via the sequestration of TGF-β. Like endoglin, TβRIII levels are elevated in patients with SSc, where TGF-β activity is enhanced (Holmes et al. 2011). Some evidence indicates that TβRIII expression is lost during fibroblast differentiation, suggesting that TβRIII may be required for suppressing lung fibrosis; however, the precise mechanism behind TβRIII's effects on fibroblastic differentiation remains to be elucidated (Ahn et al. 2010).

3.5.3 Neuropilin-1/2

Neuropilin-1 has been implicated as a regulator of the fibrotic response, as it has been suggested to amplify TGF-β and PDGF signaling in hepatic stellate cells (HSC) and in HCC. In two rat models of liver fibrosis, neuropilin-1 was found to be upregulated in activated HSCs, which exhibit the highly motile myofibroblast phenotype (Cao et al. 2010a). Neuropilin-1 overexpression has also been shown to increase cell motility and TGF-β-dependent collagen production (Cao et al. 2010a). While these studies reveal a role for neuropilin-1 in liver fibrosis as a modulator of multiple growth factor targets, additional in vivo fibrosis studies must be performed to clarify the exact effects of neuropilin-1 on TGF-β, PDGF, and VEGF signaling during this process.

3.5.4 Syndecan-2

Syndecan-2 levels are elevated in renal interstitial cells under diabetic conditions. In addition, syndecan-2 has been shown to modulate TGF-β's ability to increase matrix deposition in several cell lines, possibly via effecting TβRIII expression levels (Chen et al. 2004b). Whether syndecan-2's role in fibrosis is mediated via TβRIII or via a direct modulation of TGF-β signaling remains to be determined.

3.6 Conclusions and Perspectives

The TGF-β signaling pathway has essential roles in many physiological processes including development, growth, differentiation, cell cycle regulation, cytokine and ECM production. Deregulation of these events results in a broad spectrum of human disease. The TGF-β co-receptors have diverse roles in regulating and mediating the TGF-β response. They may perform their actions through the regulation of signaling either in a cell autonomous or a non-cell autonomous manner through regulated ectodomain shedding, by localizing signaling in a spatio-temporal manner, or by orchestrating signaling with other growth factor receptors. Through a combination of these functions, the TGF-β co-receptors play essential roles in regulating physiological and pathophysiological processes in a tissue- and context-dependent manner. While the current body of literature supports complex roles for the co-receptors in mediating TGF-β's effects, many questions remain in terms of how these receptors function under physiological circumstances and during disease progression. Additional complexity exists due to the ability of the TGF-β co-receptors to bind multiple ligands through their extracellular domains as well as other proteins through their cytoplasmic domains. Defining the precise nature and relationship of these interactions and their functional consequences remains a challenge, particularly in determining which interactions mediate the biological functions of individual co-receptors. Further refining the structural domains mediating these interactions and assessing the role of loss of function mutants in vitro and in vivo should provide further insight into TGF-β co-receptor function. Despite these challenges, additional studies regarding the roles and mechanisms of action of the TGF-β co-receptors in signaling, biology, and human disease should advance our ability to target the TGF-β signaling pathway and these co-receptors in human disease.

References

Abdalla SA, Letarte M (2006) Hereditary haemorrhagic telangiectasia: current views on genetics and mechanisms of disease. J Med Genet 43(2):97–110. doi:10.1136/jmg.2005.030833

Ahn JY, Park S, Yun YS, Song JY (2010) Inhibition of type III TGF-β receptor aggravates lung fibrotic process. Biomed Pharmacother 64(7):472–476. doi:10.1016/j.biopha.2010.01.006

Andres JL, Ronnstrand L, Cheifetz S, Massague J (1991) Purification of the transforming growth factor-β (TGF-β) binding proteoglycan betaglycan. J Biol Chem 266(34):23282–23287

Antonescu CR, Zhang L, Nielsen GP, Rosenberg AE, Dal Cin P, Fletcher CD (2011) Consistent t(1;10) with rearrangements of TGFBR3 and MGEA5 in both myxoinflammatory fibroblastic sarcoma and hemosiderotic fibrolipomatous tumor. Genes Chromosomes Cancer 50(10):757–764. doi:10.1002/gcc.20897

Baba T, Kariya M, Higuchi T, Mandai M, Matsumura N, Kondoh E, Miyanishi M, Fukuhara K, Takakura K, Fujii S (2007) Neuropilin-1 promotes unlimited growth of ovarian cancer by evading contact inhibition. Gynecol Oncol 105(3):703–711

Bae HJ, Eun JW, Noh JH, Kim JK, Jung KH, Xie HJ, Park WS, Lee JY, Nam SW (2009) Down-regulation of transforming growth factor β receptor type III in hepatocellular carcinoma is not directly associated with genetic alterations or loss of heterozygosity. Oncol Rep 22(3): 475–480

Bandyopadhyay A, Lopez-Casillas F, Malik SN (2002a) Antitumor activity of a recombinant soluble betaglycan in human breast cancer xenograft. Cancer Res 63:4690–4695
Bandyopadhyay A, Wang L, Lopez-Casillas F, Mendoza V, Yeh IT, Sun L (2005) Systemic administration of a soluble betaglycan suppresses tumor growth, angiogenesis, and matrix metalloproteinase-9 expression in a human xenograft model of prostate cancer. Prostate 63(1):81–90
Bandyopadhyay A, Zhu Y, Cibull LB (1999a) A soluble transforming growth factor β type III receptor suppresses tumorigenicity and metastasis of human breast MDA-MB-231 cells. Cancer Res 59:5041–5046
Bandyopadhyay A, Zhu Y, Cibull LB, Chen C, Sun L-Z (1999b) A soluble transforming growth factor β type III receptor suppresses tumorigenicity and metastasis of human breast cancer. Cancer Res 59:5041–5046
Bandyopadhyay A, Zhu Y, Malik SN, Kreisberg J, Brattain MG, Sprague E, Luo J, Lopez-Casillas F, Sun L-Z (2002b) Extracellular domain of TGFβ type III receptor inhibits angiogenesis and tumor growth in human cancer cells. Oncogene 21:3541–3551
Beck B, Driessens G, Goossens S, Youssef KK, Kuchnio A, Caauwe A, Sotiropoulou PA, Loges S, Lapouge G, Candi A, Mascre G, Drogat B, Dekoninck S, Haigh JJ, Carmeliet P, Blanpain C (2011) A vascular niche and a VEGF-Nrp1 loop regulate the initiation and stemness of skin tumours. Nature 478(7369):399–403. doi:10.1038/nature10525
Benhattar J, Losi L, Chaubert P, Givel JC, Costa J (1993) Prognostic significance of K-ras mutations in colorectal carcinoma. Gastroenterology 104(4):1044–1048
Berge M, Allanic D, Bonnin P, de Montrion C, Richard J, Suc M, Boivin JF, Contreres JO, Lockhart BP, Pocard M, Levy BI, Tucker GC, Tobelem G, Merkulova-Rainon T (2011) Neuropilin-1 is upregulated in hepatocellular carcinoma and contributes to tumour growth and vascular remodeling. J Hepatol 55(4):866–875
Bernabeu C, Lopez-Novoa JM, Quintanilla M (2009) The emerging role of TGF-β superfamily coreceptors in cancer. Biochim Biophys Acta 1792(10):954–973. doi:10.1016/j.bbadis.2009.07.003
Bilandzic M, Chu S, Farnworth PG, Harrison C, Nicholls P, Wang Y, Escalona RM, Fuller PJ, Findlay JK, Stenvers KL (2009) Loss of betaglycan contributes to the malignant properties of human granulosa tumor cells. Mol Endocrinol 23(4):539–548. doi:me.2008-0300 [pii] 10.1210/me.2008-0300
Blobe GC, Liu X, Fang SJ, How T, Lodish HF (2001) A novel mechanism for regulating transforming growth factor β (TGF-β) signaling. Functional modulation of type III TGF-β receptor expression through interaction with the PDZ domain protein, GIPC. J Biol Chem 276(43):39608–39617
Bock AJ, Tuft Stavnes H, Kaern J, Berner A, Staff AC, Davidson B (2011) Endoglin (CD105) expression in ovarian serous carcinoma effusions is related to chemotherapy status. Tumour Biol 32(3):589–596
Borges L, Iacovino M, Mayerhofer T, Koyano-Nakagawa N, Baik J, Garry D, Kyba M, Letarte M, Perlingeiro RCR (2012) A critical role for endoglin in the emergence of blood during embryonic development. Blood 119:5417–5428
Bourdeau A, Dumont DJ, Letarte M (1999) A murine model of hereditary hemorrhagic telangiectasia. J Clin Invest 104:1343–1351
Brewer CA, Setterdahl JJ, Li MJ, Johnston JM, Mann JL, McAsey ME (2000) Endoglin expression as a measure of microvessel density in cervical cancer. Obstet Gynecol 96(2):224–228
Burke JP, Watson RW, Murphy M, Docherty NG, Coffey JC, O'Connell PR (2009) Simvastatin impairs smad-3 phosphorylation and modulates transforming growth factor β1-mediated activation of intestinal fibroblasts. Br J Surg 96(5):541–551. doi:10.1002/bjs.6577
Cabello-Verrugio C, Brandan E (2007) A novel modulatory mechanism of transforming growth factor-β signaling through decorin and LRP-1. J Biol Chem 282(26):18842–18850. doi:10.1074/jbc.M700243200
Cai Y, Wang R, Zhao YF, Jia J, Sun ZJ, Chen XM (2010) Expression of Neuropilin-2 in salivary adenoid cystic carcinoma: its implication in tumor progression and angiogenesis. Pathol Res Pract 206(12):793–799

Cai H, Reed RR (1999) Cloning and characterization of neuropilin-1-interacting protein: a PSD-95/Dlg/ZO-1 domain-containing protein that interacts with the cytoplasmic domain of neuropilin-1. J Neurosci 19(15):6519–6527

Calabro L, Fonsatti E, Bellomo G, Alonci A, Colizzi F, Sigalotti L, Altomonte M, Musolino C, Maio M (2003) Differential levels of soluble endoglin (CD105) in myeloid malignancies. J Cell Physiol 194(2):171–175

Cao S, Yaqoob U, Das A, Shergill U, Jagavelu K, Huebert RC, Routray C, Abdelmoneim S, Vasdev M, Leof E, Charlton M, Watts RJ, Mukhopadhyay D, Shah VH (2010a) Neuropilin-1 promotes cirrhosis of the rodent and human liver by enhancing PDGF/TGF-β signaling in hepatic stellate cells. J Clin Invest 120(7):2379–2394. doi:10.1172/JCI41203

Cao Y, Szabolcs A, Dutta SK, Yaqoob U, Jagavelu K, Wang L, Leof EB, Urrutia RA, Shah VH, Mukhopadhyay D (2010b) Neuropilin-1 mediates divergent R-Smad signaling and the myofibroblast phenotype. J Biol Chem 285(41):31840–31848. doi:10.1074/jbc.M110.151696

Castonguay R, Werner ED, Matthews RG, Presman E, Mulivor AW, Solban N, Sako D, Pearsall RS, Underwood KW, Seehra J, Kumar R, Grinberg AV (2011) Soluble endoglin specifically binds bone morphogenetic proteins 9 and 10 via its orphan domain, inhibits blood vessel formation, and suppresses tumor growth. J Biol Chem 286(34):30034–30046. doi:10.1074/jbc.M111.260133

Cheifetz S, Bellon T, Cales C, Vera S, Bernabeu C, Massague J, Letarte M (1992) Endoglin is a component of the transforming growth factor-β receptor system in human endothelial cells. J Biol Chem 267(27):19027–19030

Chen CZ, Li M, de Graaf D, Monti S, Gottgens B, Sanchez MJ, Lander ES, Golub TR, Green AR, Lodish HF (2002) Identification of endoglin as a functional marker that defines long-term repopulating hematopoietic stem cells. Proc Natl Acad Sci USA 99(24):15468–15473. doi:10.1073/pnas.202614899

Chen E, Hermanson S, Ekker SC (2004a) Syndecan-2 is essential for angiogenic sprouting during zebrafish development. Blood 103(5):1710–1719. doi:10.1182/blood-2003-06-1783

Chen L, Klass C, Woods A (2004b) Syndecan-2 regulates transforming growth factor-β signaling. J Biol Chem 279(16):15715–15718. doi:10.1074/jbc.C300430200 C300430200 [pii]

Chen W, Kirkbride KC, How T, Nelson CD, Mo J, Frederick JP, Wang XF, Lefkowitz RJ, Blobe GC (2003) Beta-arrestin 2 mediates endocytosis of type III TGF-β receptor and down-regulation of its signaling. Science 301(5638):1394–1397

Chittenden TW, Claes F, Lanahan AA, Autiero M, Palac RT, Tkachenko EV, Elfenbein A, Ruiz de Almodovar C, Dedkov E, Tomanek R, Li W, Westmore M, Singh JP, Horowitz A, Mulligan-Kehoe MJ, Moodie KL, Zhuang ZW, Carmeliet P, Simons M (2006) Selective regulation of arterial branching morphogenesis by synectin. Dev Cell 10(6):783–795

Choi Y, Kim H, Chung H, Hwang JS, Shin JA, Han IO, Oh ES (2010) Syndecan-2 regulates cell migration in colon cancer cells through Tiam1-mediated Rac activation. Biochem Biophys Res Commun 391(1):921–925. doi:10.1016/j.bbrc.2009.11.165

Cohen T, Gluzman-Poltorak Z, Brodzky A, Meytal V, Sabo E, Misselevich I, Hassoun M, Boss JH, Resnick M, Shneyvas D, Eldar S, Neufeld G (2001) Neuroendocrine cells along the digestive tract express neuropilin-2. Biochem Biophys Res Commun 284(2):395–403

Cohen T, Herzog Y, Brodzky A, Greenson JK, Eldar S, Gluzman-Poltorak Z, Neufeld G, Resnick MB (2002) Neuropilin-2 is a novel marker expressed in pancreatic islet cells and endocrine pancreatic tumours. J Pathol 198(1):77–82

Compton LA, Potash DA, Brown CB, Barnett JV (2007) Coronary vessel development is dependent on the type III transforming growth factor β receptor. Circ Res 101(8):784–791. doi:CIRCRESAHA.107.152082 [pii] 10.1161/CIRCRESAHA.107.152082

Conley BA, Koleva R, Smith JD, Kacer D, Zhang D, Bernabeu C, Vary CP (2004) Endoglin controls cell migration and composition of focal adhesions: function of the cytosolic domain. J Biol Chem 279(26):27440–27449. doi:10.1074/jbc.M312561200

Cooper SJ, Zou H, Legrand SN, Marlow LA, von Roemeling CA, Radisky DC, Wu KJ, Hempel N, Margulis V, Tun HW, Blobe GC, Wood CG, Copland JA (2010) Loss of type III transforming growth factor-β receptor expression is due to methylation silencing of the transcription factor

GATA3 in renal cell carcinoma. Oncogene 29(20):2905–2915. doi:onc201064 [pii] 10.1038/onc.2010.64

Copland JA, Luxon BA, Ajani L, Maity T, Campagnaro E, Guo H, LeGrand SN, Tamboli P, Wood CG (2003) Genomic profiling identifies alterations in TGFβ signaling through loss of TGFβ receptor expression in human renal cell carcinogenesis and progression. Oncogene 22(39): 8053–8062. doi:10.1038/sj.onc.1206835 1206835 [pii]

Coral-Alvarado PX, Garces MF, Caminos JE, Iglesias-Gamarra A, Restrepo JF, Quintana G (2010) Serum endoglin levels in patients suffering from systemic sclerosis and elevated systolic pulmonary arterial pressure. Int J Rheumatol. doi:10.1155/2010/969383

Criswell TL, Arteaga CL (2007) Modulation of NFκB activity and E-cadherin by the type III transforming growth factor β receptor regulates cell growth and motility. J Biol Chem 282(44): 32491–32500. doi:10.1074/jbc.M704434200

Dallas NA, Samuel S, Xia L, Fan F, Gray MJ, Lim SJ, Ellis LM (2008) Endoglin (CD105): a marker of tumor vasculature and potential target for therapy. Clin Cancer Res 14(7):1931–1937. doi:10.1158/1078-0432.CCR-07-4478

Davidson B, Stavnes HT, Forsund M, Berner A, Staff AC (2010) CD105 (endoglin) expression in breast carcinoma effusions is a marker of poor survival. Breast 19(6):493–498

De Wever O, Mareel M (2003) Role of tissue stroma in cancer cell invasion. J Pathol 200(4): 429–447

Dharmapatni AA, Smith MD, Ahern MJ, Simpson A, Li C, Kumar S, Roberts-Thomson PJ (2001) The TGF β receptor endoglin in systemic sclerosis. Asian Pac J Allergy Immunol 19(4):275–282

Dong M, How T, Kirkbride KC, Gordon KJ, Lee JD, Hempel N, Kelly P, Moeller BJ, Marks JR, Blobe GC (2007) The type III TGF-β receptor suppresses breast cancer progression. J Clin Invest 117(1):206–217

Drenberg CD, Livingston S, Chen R, Kruk PA, Nicosia SV (2009) Expression of Semaphorin 3F and its receptors in epithelial ovarian cancer, fallopian tubes, and secondary mullerian tissues. Obstet Gynecol Int (epub). doi:10.1155/2009/730739

Duwel A, Eleno N, Jerkick M, Arevalo M, Blolanos JP, Bernabeu C, Lopez-Novoa JM (2007) Reduced tumor growth and angiogenesis in endoglin-haploinsufficient mice. Tumour Biol 28(1):1–8

Elliott RL, Blobe GC (2005) Role of transforming growth factor β in human cancer. J Clin Oncol 23(9):2078–2093

Essner JJ, Chen E, Ekker SC (2006) Syndecan-2. Int J Biochem Cell Biol 38(2):152–156. doi:10.1016/j.biocel.2005.08.012

Fansatti E, Vecchio LD, Altomonte M, Sigalotti L, Nicotra MR, Coral S, Natali PG, Maio M (2001) Endoglin: an accessory component of the TGF-β-binding receptor-complex with diagnostic, prognostic, and bioimmunotherapeutic potential in human malignancies. J Cell Physiol 188:1–7

Fears CY, Gladson CL, Woods A (2006) Syndecan-2 is expressed in the microvasculature of gliomas and regulates angiogenic processes in microvascular endothelial cells. J Biol Chem 281(21):14533–14536. doi:C600075200 [pii] 10.1074/jbc.C600075200

Finley DJ, Arora N, Zhu B, Gallagher L, Fahey TJ 3rd (2004) Molecular profiling distinguishes papillary carcinoma from benign thyroid nodules. J Clin Endcrinol Metab 89(7):3214–3223

Finger EC, Turley RS, Dong M, How T, Fields TA, Blobe GC (2008) TβRIII suppresses non-small cell lung cancer invasiveness and tumorigenicity. Carcinogenesis 29(3):528–535. doi:bgm289 [pii] 10.1093/carcin/bgm289

Finnson KW, Tam BY, Liu K, Marcoux A, Lepage P, Roy S, Bizet AA, Philip A (2006) Identification of CD109 as part of the TGF-β receptor system in human keratinocytes. FASEB J 20(9):1525–1527. doi:10.1096/fj.05-5229fje

Florio P, Ciarmela P, Reis FM, Toti P, Galleri L, Santopietro R, Tiso E, Tosi P, Petraglia F (2005) Inhibin alpha-subunit and the inhibin coreceptor betaglycan are downregulated in endometrial carcinoma. Eur J Endocrinol 152(2):277–284

Folkman J, Watson K, Ingber D, Hanahan D (1989) Induction of angiogenesis during the transition from hyperplasia to neoplasia. Nature 339(6219):58–61

Fonsatti E, Altomonte M, Nicotra MR, Natali PG, Maio M (2003) Endoglin (CD105): a powerful therapeutic target on tumor-associated angiogenetic blood vessels. Oncogene 22(42):6557–6563. doi:10.1038/sj.onc.1206813

Fonsatti E, Nicolay HJ, Altomonte M, Covre A, Maio M (2010) Targeting cancer vasculature via endoglin/CD105: a novel antibody-based diagnostic and therapeutic strategy in solid tumours. Cardiovasc Res 86(1):12–19. doi:10.1093/cvr/cvp332

Gatza CE, Holtzhausen A, Kirkbride KC, Morton A, Gatza ML, Datto MB, Blobe GC (2012) Type III TGF-β receptor enhances colon cancer cell migration and anchorage-independent growth. Neoplasia 13(8):758–770

Gatza CE, Oh SY, Blobe GC (2010) Roles for the type III TGF-β receptor in human cancer. Cell Signal 22(8):1163–1174. doi:S0898-6568(10)00033-1 [pii] 10.1016/j.cellsig.2010.01.016

Geretti E, Klagsbrun M (2007) Neuropilins: novel targets for anti-angiogenesis therapies. Cell Adh Migr 1(2):56–61

Glinka Y, Prud'homme GJ (2008) Neuropilin-1 is a receptor for transforming growth factor β-1, activates its latent form, and promotes regulatory T cell activity. J Leukoc Biol 84(1):302–310. doi:10.1189/jlb.0208090

Gordon KJ, Dong M, Chislock EM, Fields TA, Blobe GC (2008) Loss of type III transforming growth factor β receptor expression increases motility and invasiveness associated with epithelial to mesenchymal transition during pancreatic cancer progression. Carcinogenesis 29(2):252–262. doi:bgm249 [pii] 10.1093/carcin/bgm249

Gordon KJ, Kirkbride KC, How T, Blobe GC (2009) Bone morphogenetic proteins induce pancreatic cancer cell invasiveness through a Smad1-dependent mechanism that involves matrix metalloproteinase-2. Carcinogenesis 30(2):238–248. doi:10.1093/carcin/bgn274

Ghosh S, Sullivan CA, Zerkowski MP, Molinaro AM, Rimm DL, Camp RL, Chung GG (2008) High levels of vascular endothelial growth factor and its receptors (VEGFR-1, VEGFR-2, neuropilin-1) are associated with worse outcome in breast cancer. Hum Pathol 39(12): 1835–1843

Gougos A, St Jacques S, Greaves A, O'Connell PJ, d'Apice AJ, Burhring HJ, Bernabeu C, van Mourik JA, Letarte M (1992) Identification of distinct epitopes of endoglin, an RGD-containing glycoprotein of endothelial cells, leukemic cells, and syncytiotrophoblasts. Int Immunol 4(1):83–92

Goumans M-J, Valdimarsdottir G, Itoh S, Rosendahl A, Sideras P, ten Dijke P (2002) Balancing the activation state of the endothelium via two distinct TGF-β type I receptors. EMBO J 21(7):1743–1752

Grandclement C, Pallandre JR, Degano SV, Viel E, Bouard A, Balland J, Remy-Martin J-P, Simon B, Rouleau A, Boireau W, Klagsbrun M, Ferrand C, Borg C (2011) Neuropilin-2 expression promotes TGF-β1-mediated epithelial to mesenchymal transition in colorectal cancer cells. PLoS One 6(7):e20444. doi:10.1371/journal.pone.0020444.t001

Gray MJ, Van Buren G, Dallas NA, Xia L, Wang X, Yang AD, Somcio RJ, Lin YG, Lim S, Fan F, Mangala LS, Arumugam T, Logsdon CD, Lopez-Berestein G, Sood AK, Ellis LM (2008) Therapeutic targeting of neuropilin-2 on colorectal carcinoma cells implanted in the murine liver. J Natl Cancer Inst 100(2):109–120

Grotenhuis BA, Wijnhoven BP, van Lanschot JJ (2012) Cancer stem cells and their potential implications for the treatment of solid tumors. J Surg Oncol. doi:10.1002/jso.23069

Guerrero-Esteo M, Lastres P, Letamendía A, Pérez-Alvarez MJ, Langa C, López LA, Fabra A, García-Pardo A, Vera S, Letarte M, Bernabéu C (1999) Endoglin overexpression modulates cellular morphology, migration, and adhesion of mouse fibroblasts. Eur J Cell Biol 78(9):614–623. doi:10.1016/s0171-9335(99)80046-6

Gulyas M, Hjerpe A (2003) Proteoglycans and WT1 as markers for distinguishing adenocarcinoma, epithelioid mesothelioma, and benign mesothelium. J Pathol 199(4):179–187

Hamerlik P, Lathia JD, Rasmussen R, Wu Q, Bartkova J, Lee M, Moudry P, Bartek J Jr, Fischer W, Lukas J, Rich JN, Bartek J (2012) Autocrine VEGF-VEGFR2-Neuropilin-1 signaling promotes glioma stem-like cell viability and tumor growth. J Exp Med 209(3):507–520. doi:10.1084/jem.20111424

Handa A, Tokunaga T, Tsuchida T, Lee YH, Kijima H, Yamazaki H, Ueyama Y, Fukuda H, Nakamura M (2000) Neuropilin-2 expression affects the increased vascularization and is a prognostic factor in osteosarcoma. Int J Oncol 17(2):291–296

Hansel DE, Wilentz RE, Yeo CJ, Schulick RD, Montgomery E, Maitra A (2004) Expression of neuropilin-1 in high-grade dysplasia, invasive cancer, and metastases of the human gastrointestinal tract. Am J Surg Pathol 28:347–356

Hawinkels LJ, Ten Dijke P (2011) Exploring anti-TGF-β therapies in cancer and fibrosis. Growth Factors 29(4):140–152. doi:10.3109/08977194.2011.595411

Hempel N, How T, Cooper SJ, Green TR, Dong M, Copland JA, Wood CG, Blobe GC (2008) Expression of the type III TGF-β receptor is negatively regulated by TGF-β. Carcinogenesis 29(5):905–912. doi:10.1093/carcin/bgn049

Hempel N, How T, Dong M, Murphy SK, Fields TA, Blobe GC (2007) Loss of betaglycan expression in ovarian cancer: role in motility and invasion. Cancer Res 67(11):5231–5238

Henry LA, Johnson DA, Sarrio D, Lee S, Quinlan PR, Crook T, Thompson AM, Reis-Filho JS, Isacke CM (2011) Endoglin expression in breast tumor cells suppresses invasion and metastasis and correlates with improved clinical outcome. Oncogene 30(9):1046–1058. doi:10.1038/onc.2010.488

Hermida N, Lopez B, Gonzalez A, Dotor J, Lasarte JJ, Sarobe P, Borras-Cuesta F, Diez J (2009) A synthetic peptide from transforming growth factor-β1 type III receptor prevents myocardial fibrosis in spontaneously hypertensive rats. Cardiovasc Res 81(3):601–609. doi:10.1093/cvr/cvn315

Holmes AM, Ponticos M, Shi-Wen X, Denton CP, Abraham DJ (2011) Elevated CCN2 expression in scleroderma: a putative role for the TGFβ accessory receptors TGFβRIII and endoglin. J Cell Commun Signal 5(3):173–177. doi:10.1007/s12079-011-0140-4

Houthuijzen JM, Daenen LG, Roodhart JM, Voest EE (2012) The role of mesenchymal stem cells in anti-cancer drug resistance and tumour progression. Br J Cancer 106(12):1901–1906. doi:10.1038/bjc.2012.201

Howe JR, Haidle JL, Lal G, Bair J, Song C, Pechman B, Chinnathambi S, Lynch HT (2007) ENG mutations in MADH4/BMPR1A mutation negative patients with juvenile polyposis. Clin Genet 71(1):91–92. doi:10.1111/j.1399-0004.2007.00734.x

Hu D, Wang X, Mao Y, Zhou L (2012) Identification of CD105 (endoglin)-positive stem-like cells in rhabdoid meningioma. J Neurooncol 106(3):505–517. doi:10.1007/s11060-011-0705-3

Huang X, Xiao D-W, Xu L-Y, Zhong H-J, Liao L-D, Xie Z-F, Li E-M (2009) Prognostic significance of altered expression of SDC2 and CYR61 in esophageal squamous cell carcinoma. Oncol Rep 21(4):1123–1129

Iolascon A, Giordani L, Borriello A, Carbone R, Izzo A, Tonini GP, Gambini C, Della Ragione F (2000) Reduced expression of transforming growth factor-β receptor type III in high stage neuroblastomas. Br J Cancer 82(6):1171–1176

Jelinek DF, R.C. T, Stolovitzky GA (2003) Identification of a global gene expression signature of B-chronic lymphocytic leukemia. Mol Cancer Res 1:346–361

Jia H, Bagherzadeh A, Hartzoulakis B, Jarvis A, Lohr M, Shaikh S, Aqil R, Cheng L, Tickner M, Esposito D, Harris R, Driscoll PC, Selwood DL, Zachary IC (2006) Characterization of a bicyclic peptide neuropilin-1 (NP-1) antagonist (EG3287) reveals importance of vascular endothelial growth factor exon 8 for NP-1 binding and role of NP-1 in KDR signaling. J Biol Chem 281(19):13493–13502. doi:10.1074/jbc.M512121200

Jia H, Cheng L, Tickner M, Bagherzadeh A, Selwood D, Zachary I (2010) Neuropilin-1 antagonism in human carcinoma cells inhibits migration and enhances chemosensitivity. Br J Cancer 102(3):541–552. doi:10.1038/sj.bjc.6605539

Jones EA, Yuan L, Breant C, Watts RJ, Eichmann A (2008) Separating genetic and hemodynamic defects in neuropilin 1 knockout embryos. Development 135(14):2479–2488. doi:10.1242/dev.014902

Kassouf W, Ismail HR, Aprikian AG, Chevalier S (2004) Whole-mount prostate sections reveal differential endoglin expression in stromal, epithelial, and endothelial cells with the development of prostate cancer. Prostate Cancer Prostatic Dis 7(2):105–110

Kawakami T, Tokunaga T, Hatanaka H, Kijima H, Yamazaki H, Abe Y, Osamura Y, Inoue H, Ueyama Y, Nakamura M (2002) Neuropilin 1 and neuropilin 2 co-expression is significantly correlated with increased vascularity and poor prognosis in nonsmall cell lung carcinoma. Cancer 95(10):2196–2201. doi:10.1002/cncr.10936

Klein U, Tu Y, Stolovitzky GA, Mattioli M, Cattoretti G, Husson H, Freedman A, Inghirami G, Cro L, Baldini L, Neri A, Califano A, Dalla-Favera R (2001) Gene expression profiling of B cell chronic lymphocytic leukemia reveals a homogeneous phenotype related to memory B cells. J Exp Med 194(11):1625–1638

Kopczynska E, Dancewicz M, Kowalewski J, Makarewicz R, Kardymowicz H, Kaczmarczyk A, Tyrakowski T (2012) Influence of surgical resection on plasma endoglin (CD105) level in non-small cell lung cancer patients. Exp Oncol 34(1):53–56

Kramer KL, Barnette JE, Yost HJ (2002) PKCγ regulates syndecan-2 inside-out signaling during xenopus left-right development. Cell 111(7):981–990

Kramer KL, Yost HJ (2002) Ectodermal syndecan-2 mediates left-right axis formation in migrating mesoderm as a cell-nonautonomous Vg1 cofactor. Dev Cell 2(1):115–124

Kumar S, Ghellal A, Li C, Byrne G, Haboubi N, Wang JM, Bundred N (1999) Breast carcinoma: vascular density determined using CD105 antibody correlates with tumor prognosis. Cancer Res 59:856–861

Lacal PM, Failla CM, Pagani E, Odorisio T, Schietroma C, Falcinelli S, Zambruno G, D'Atri S (2000) Human melanoma cells secrete and respond to placenta growth factor and vascular endothelial growth factor. J Invest Dermatol 115:1000–1007

Lakshman M, Huang X, Ananthanarayanan V, Jovanovic B, Liu Y, Craft CS, Romero D, Vary CP, Bergan RC (2011) Endoglin suppresses human prostate cancer metastasis. Clin Exp Metastasis 28(1):39–53. doi:10.1007/s10585-010-9356-6

Lambert KE, Huang H, Mythreye K, Blobe GC (2011) The type III transforming growth factor-β receptor inhibits proliferation, migration, and adhesion in human myeloma cells. Mol Biol Cell 22(9):1463–1472. doi:10.1091/mbc.E10-11-0877

Latil A, Bieche I, Pesche S, Valeri A, Fournier G, Cussenot O, Lidereau R (2000) VEGF overexpression in clinically localized prostate tumors and neuropilin-1 overexpression in metastatic forms. Int J Cancer 89:167–171

Lebrin F, Goumans M-J, Jonker L, Carvalho RLC, Valdimarsdottir G, Thorikay M, Mummery C, Arthur HM, ten Dijke P (2004) Endoglin promotes endothelial cell proliferation and TGF-β/ALK1 signal transduction. EMBO J 23:4018–4028. doi:10.1038/

Lee JD, Hempel N, Lee NY, Blobe GC (2010) The type III TGF-β receptor suppresses breast cancer progression through GIPC-mediated inhibition of TGF-β signaling. Carcinogenesis 31(2):175–183. doi:bgp271 [pii] 10.1093/carcin/bgp271

Lee JH, Park H, Chung H, Choi S, Kim Y, Yoo H, Kim TY, Hann HJ, Seong I, Kim J, Kang KG, Han IO, Oh ES (2009) Syndecan-2 regulates the migratory potential of melanoma cells. J Biol Chem 284(40):27167–27175. doi:10.1074/jbc.M109.034678

Lee NY, Blobe GC (2007) The interaction of endoglin with β-arrestin2 regulates transforming growth factor-β-mediated ERK activation and migration in endothelial cells. J Biol Chem 282(29):21507–21517. doi:M700176200 [pii] 10.1074/jbc.M700176200

Lee NY, Golzio C, Gatza CE, Sharma A, Katsanis N, Blobe GC (2012) Endoglin regulates PI3-kinase/Akt trafficking and signaling to alter endothelial capillary stability during angiogenesis. Mol Biol Cell 23(13):2412–2423

Lee NY, Ray B, How T, Blobe GC (2008) Endoglin promotes transforming growth factor β-mediated Smad 1/5/8 signaling and inhibits endothelial cell migration through its association with GIPC. J Biol Chem 283(47):32527–32533. doi:M803059200 [pii] 10.1074/jbc.M803059200

Li C, Hampson IN, Hampson L, Kumar P, Bernabeu C, Kumar S (2000) CD105 antagonizes the inhibitory signaling of transforming growth factor β1 on human vascular endothelial cells. FASEB J 14:55–64

Li DY, Sorensen LK, Brooke BS, Urness LD, Davis EC, Taylor DG, Boak BB, Wendel DP (1999) Defective angiogenesis in mice lacking endoglin. Science 284(5419):1534–1537

Li Q, Shirabe K, Kuwada JY (2004) Chemokine signaling regulates sensory cell migration in zebrafish. Dev Biol 269(1):123–136

Liu M, Suga M, Maclean AA, St George JA, Souza DW, Keshavjee S (2002) Soluble transforming growth factor-β type III receptor gene transfection inhibits fibrous airway obliteration in a rat model of Bronchiolitis obliterans. Am J Respir Crit Care Med 165(3):419–423

Lopez-Casillas F, Cheifetz S, Doody J, Andres JL, Lane WS, Massague J (1991) Structure and expression of the membrane proteoglycan betaglycan, a component of the TGF-β receptor system. Cell 67(4):785–795

Lopez-Casillas F, Wrana JL, Massague J (1993) Betaglycan presents ligand to the TGF β signaling receptor. Cell 73(7):1435–1444. doi:0092-8674(93)90368-Z [pii]

Ma X, Labinaz M, Goldstein J, Miller H, Keon WJ, Letarte M, O'Brien E (2000) Endoglin is overexpressed after arterial injury and is required for transforming growth factor-β-induced inhibition of smooth muscle cell migration. Arterioscler Thromb Vasc Biol 20(12):2546–2552. doi:10.1161/01.atv.20.12.2546

Mak P, Leav I, Pursell B, Bae D, Yang X, Taglienti CA, Gouvin LM, Sharma VM, Mercurio AM (2010) ERβ impedes prostate cancer EMT by destabilizing HIF-1α and inhibiting VEGF-mediated snail nuclear localization: implications for Gleason grading. Cancer Cell 17(4): 319–332. doi:10.1016/j.ccr.2010.02.030

Margulis V, Maity T, Zhang XY, Cooper SJ, Copland JA, Wood CG (2008) Type III transforming growth factor-β (TGF-β) receptor mediates apoptosis in renal cell carcinoma independent of the canonical TGF-β signaling pathway. Clin Cancer Res 14(18):5722–5730. doi:14/18/5722 [pii] 10.1158/1078-0432.CCR-08-0546

Marzioni D, Lorenzi T, Mazzucchelli R, Capparuccia L, Morroni M, Fiorini R, Bracalenti C, Catalano A, David G, Castellucci M, Muzzonigro G, Montironi R (2009) Expression of basic fibroblast growth factor, its receptors and syndecans in bladder cancer. Int J Immunopathol Pharmacol 22(3):627–638

Maring JA, Trojanowska M, ten Dijke P (2012) Role of endoglin in fibrosis and scleroderma. Int Rev Cell Mol Biol 297:295–308

Massague J (1998) TGF-β signal transduction. Annu Rev Biochem 67:753–791

Massague J (2008) TGFβ in cancer. Cell 134:215–230. doi:10.1016/j.cell.2008.07.001

Matsuno F, Haruta Y, Kondo M, Tsai IY, Barcos M, Seon BK (1999) Induction of lasting complete regression of preformed distinct solid tumors by targeting the tumor vasculature using two new anti-endoglin monoclonal antibodies. Clin Cancer Res 5:371–382

Maynard SE, Karumanchi SA (2011) Angiogenic factors and preeclampsia. Semin Nephrol 31(1):33–46. doi:10.1016/j.semnephrol.2010.10.004

McAllister KA, Grogg KM, Johnson DW, Gallione CJ, Baldwin MA, Jackson CE, Helmbodl EA, Markel DS, McKinnon WC, Murrell J, McCormick MK, Pericak-Vance MA, Heutink P, Oostra BA, Haitjema T, Westerman CJJ, Porteous ME, Guttmacher AE, Letarte M, Marchuk DA (1994) Endoglin, a TGF-β binding protein of endothelial cells, is the gene for hereditary haemorrhagic telangiectasia type 1. Nat Genet 8:345–351

Meng Q, Lux A, Holloschi A, Li J, Hughes JM, Foerg T, McCarthy JE, Heagerty AM, Kioschis P, Hafner M, Garland JM (2006) Identification of Tctex2β, a novel dynein light chain family member that interacts with different transforming growth factor-β receptors. J Biol Chem 281(48):37069–37080. doi:10.1074/jbc.M608614200

Meng W, Xia Q, Wu L, Chen S, He X, Zhang L, Gao Q, Zhou H (2011) Downregulation of TGF-β receptor types II and III in oral squamous cell carcinoma and oral carcinoma-associated fibroblasts. BMC Cancer 11:88. doi:10.1186/1471-2407-11-88

Molema G, Griffioen AW (1998) Rocking the foundations of solid tumor growth by attacking the tumor's blood supply. Immunol Today 19(9):392–394

Mythreye K, Blobe GC (2009a) Proteoglycan signaling co-receptors: roles in cell adhesion, migration and invasion. Cell Signal 21:1548–1558. doi:S0898-6568(09)00154-5 [pii] 10.1016/j.cellsig.2009.05.001

Mythreye K, Blobe GC (2009b) The type III TGF-β receptor regulates epithelial and cancer cell migration through β-arrestin2-mediated activation of Cdc42. Proc Natl Acad Sci USA 106(20):8221–8226. doi:10.1073/pnas.0812879106

Mythreye K, Knelson EH, Gatza CE, Gatza ML, Blobe GC (2012) TβRIII/β-arrestin2 regulates integrin α5β1 trafficking, function, and localization in epithelial cells. Oncogene. doi:10.1038/onc.2012.157

Narazaki M, Tosato G (2006) Ligand-induced internalization selects use of common receptor neuropilin-1 by VEGF165 and semaphorin3A. Blood 107:3892–3901. doi:10.1182/blood-2005-10-4113 10.1182/blood-2005-104113

Neri D, Bicknell R (2005) Tumour vascular targeting. Nat Rev Cancer 5(6):436–446. doi:10.1038/nrc1627

Noguer O, Villena J, Lorita J, Vilaro S, Reina M (2009) Syndecan-2 downregulation impairs angiogenesis in human microvascular endothelial cells. Exp Cell Res 315(5):795–808. doi:10.1016/j.yexcr.2008.11.016

O'Connor JC, Farach-Carson MC, Schneider CJ, Carson DD (2007) Coculture with prostate cancer cells alters endoglin expression and attenuates transforming growth factor-β signaling in reactive bone marrow stromal cells. Mol Cancer Res 5(6):585–603. doi:10.1158/1541-7786.MCR-06-0408

Orosco A, Fromigue O, Bazille C, Entz-Werle N, Levillain P, Marie PJ, Modrowski D (2007) Syndecan-2 affects the basal and chemotherapy-induced apoptosis in osteosarcoma. Cancer Res 67(8):3708–3715. doi:10.1158/0008-5472.CAN-06-4164

Pan Q, Chanthery Y, Liang WC, Stawicki S, Mak J, Rathore N, Tong RK, Kowalski J, Yee SF, Pacheco G, Ross S, Cheng Z, Le Couter J, Plowman G, Peale F, Koch AW, Wu Y, Bagri A, Tessier-Lavigne M, Watts RJ (2007) Blocking neuropilin-1 function has an additive effect with anti-VEGF to inhibit tumor growth. Cancer Cell 11(1):53–67. doi:S1535-6108(06)00367-9 [pii] 10.1016/j.ccr.2006.10.018

Pardali E, van der Schaft DW, Wiercinska E, Gorter A, Hogendoorn PC, Griffioen AW, ten Dijke P (2011) Critical role of endoglin in tumor cell plasticity of Ewing sarcoma and melanoma. Oncogene 30(3):334–345. doi:10.1038/onc.2010.418

Parikh AA, Fan F, Liu WB, Ahmad SA, Stoeltzing O, Reinmuth N, Bielenberg D, Bucana CD, Klagsbrun M, Ellis LM (2004) Neuropilin-1 in human colon cancer: expression, regulation, and role in induction of angiogenesis. Am J Pathol 164(6):2139–2151

Park H, Han I, Kwon HJ, Oh ES (2005) Focal adhesion kinase regulates syndecan-2-mediated tumorigenic activity of HT1080 fibrosarcoma cells. Cancer Res 65(21):9899–9905. doi:10.1158/0008-5472.CAN-05-1386

Park H, Kim Y, Lim Y, Han I, Oh ES (2002) Syndecan-2 mediates adhesion and proliferation of colon carcinoma cells. J Biol Chem 277(33):29730–29736. doi:10.1074/jbc.M202435200

Pimanda JE, Chan WY, Wilson NK, Smith AM, Kinston S, Knezevic K, Janes ME, Landry JR, Kolb-Kokocinski A, Frampton J, Tannahill D, Ottersbach K, Follows GA, Lacaud G, Kouskoff V, Gottgens B (2008) Endoglin expression in blood and endothelium is differentially regulated by modular assembly of the Ets/Gata hemangioblast code. Blood 112(12):4512–4522. doi:10.1182/blood-2008-05-157560

Popovic A, Demirovic A, Spajic B, Stimac G, Kruslin B, Tomas D (2010) Expression and prognostic role of syndecan-2 in prostate cancer. Prostate Cancer Prostate Dis 13(1):78–82

Ray BN, Lee NY, How T, Blobe GC (2010) ALK5 phosphorylation of the endoglin cytoplasmic domain regulates Smad1/5/8 signaling and endothelial cell migration. Carcinogenesis 31(3):435–441. doi:bgp327 [pii] 10.1093/carcin/bgp327

Rodriguez-Barbero A, Obreo J, Alvarez-Munoz P, Pandiella A, Bernabeu C, Lopez-Novoa JM (2006) Endoglin modulation of TGF-β1-induced collagen synthesis is dependent on ERK1/2 MAPK activation. Cell Physiol Biochem 18(1–3):135–142. doi:10.1159/000095181

Rokhlin OW, Cohen MB, Kubagawa H, Letarte M, Cooper MD (1995) Differential expression of endoglin on fetal and adult hematopoietic cells in human bone marrow. J Immunol 154(9):4456–4465

Romero D, O'Neill C, Terzic A, Contois L, Young K, Conley BA, Bergan RC, Brooks PC, Vary CP (2011) Endoglin regulates cancer-stromal cell interactions in prostate tumors. Cancer Res 71(10):3482–3493. doi:10.1158/0008-5472.CAN-10-2665

Romero D, Terzic A, Conley BA, Craft CS, Jovanovic B, Bergan RC, Vary CP (2010) Endoglin phosphorylation by ALK2 contributes to the regulation of prostate cancer cell migration. Carcinogenesis 31(3):359–366. doi:10.1093/carcin/bgp217

Rosen LS, Gordon MS, Hurwitz HI, Mendelson DS, Kleinzweig D, Adams BJ, C.P. T (2008) Early evidence of toelrability and clinical activity from a phase 1 study of TRC105 (anti-CD105 antibody) in patients with advanced refractory cancer. Eur J Cancer Supplements 6(126)

Rosen LS, Hurwitz HI, Wong MK, Goldman J, Mendelson DS, Fig. WD, Spencer S, Adams BJ, Alvarez D, Seon BK, Theuer CP, Leigh B, Gordon MS (2012) A phase 1 first-in-human study of TRC105 (anti-endoglin antibody) in patients with advanced cancer. Clin Cancer Res 18(17):4820–4829. doi:10.1158/1078-0432.CCR-12-0098

Rushing EC, Stine MJ, Hahn SJ, Shea S, Eller MS, Naif A, Khanna S, Westra WH, Jungbluth AA, Busam KJ, Mahalingam M, Alani RM (2012) Neuropilin-2: a novel biomarker for malignant melanoma? Hum Pathol 43(3):381–389

Ryu HY, Lee J, Yang S, Park H, Choi S, Jung KC, Lee ST, Seong JK, Han IO, Oh ES (2009) Syndecan-2 functions as a docking receptor for pro-matrix metalloproteinase-7 in human colon cancer cells. J Biol Chem 284(51):35692–35701. doi:10.1074/jbc.M109.054254

Saad RS, Liu YL, Nathan G, Celebrezze J, Medich D, Silverman JF (2004) Endoglin (CD105) and vascular endothelial growth factor as prognostic markers. Mod Pathol 17(2):197–203

Sanchez NS, Hill CR, Love JD, Soslow JH, Craig E, Austin AF, Brown CB, Czirok A, Camenisch TD, Barnett JV (2011) The cytoplasmic domain of TGFβR3 through its interaction with the scaffolding protein, GIPC, directs epicardial cell behavior. Dev Biol 358(2):331–343. doi:10.1016/j.ydbio.2011.08.008

Santander C, Brandan E (2006) Betaglycan induces TGF-β signaling in a ligand-independent manner, through activation of the p38 pathway. Cell Signal 18(9):1482–1491. doi:S0898-6568(05)00327-X [pii] 10.1016/j.cellsig.2005.11.011

Santibanez JF, Perez-Gomez E, Fernandez LA, Garrido-Martin EM, Carnero A, Malumbres M, Vary CP, Quintanilla M, Bernabeu C (2010) The TGF-β co-receptor endoglin modulates the expression and transforming potential of H-Ras. Carcinogenesis 31(12):2145–2154. doi:10.1093/carcin/bgq199

Sanz-Rodriguez F, Guerrero-Esteo M, Botella LM, Banville D, Vary CP, Bernabeu C (2004) Endoglin regulates cytoskeletal organization through binding to ZRP-1, a member of the Lim family of proteins. J Biol Chem 279(31):32858–32868. doi:10.1074/jbc.M400843200

Sarraj MA, Chua HK, Umbers A, Loveland KL, Findlay JK, Stenvers KL (2007) Differential expression of TGFBR3 (betaglycan) in mouse ovary and testis during gonadogenesis. Growth Factors 25(5):334–345. doi:789171518 [pii] 10.1080/08977190701833619

Sarraj MA, Escalona RM, Umbers A, Chua HK, Small C, Griswold M, Loveland K, Findlay JK, Stenvers KL (2010) Fetal testis dysgenesis and compromised Leydig cell function in Tgfbr3 (β glycan) knockout mice. Biol Reprod 82(1):153–162. doi:biolreprod.109.078766 [pii]10.1095/biolreprod.109.078766

Schimming R, Marme D (2001) Endoglin (CD105) expression in squamous cell carcinoma of the oral cavity. Head Neck 24(2):151–156

Seon BK, Matsuno F, Haruta Y, Kondo M, Barcos M (1997) Long-lasting complete inhibition of human solid tumors in SCID mice by targeting endothelial cells of tumor vasculature with antihuman endoglin immunotoxin. Clin Cancer Res 3:1031–1044

Serini G, Valdembri D, Zanivan S, Morterra G, Burkhardt C, Caccavari F, Zammataro L, Primo L, Tamagnone L, Logan M, Tessier-Lavigne M, Taniguchi M, Puschel AW, Bussolino F (2003) Class 3 semaphorins control vascular morphogenesis by inhibiting integrin function. Nature 424(6947):391–397. doi:10.1038/nature01784 nature01784 [pii]

Shariat SF, Karam JA, Walz J, Roehrborn CG, Montorsi F, Margulis V, Saad F, Slawin KM, Karakiewicz PI (2008) Improved prediction of disease relapse after radical prostatectomy through a panel of preoperative blood-based biomarkers. Clin Cancer Res 14(12):3785–3791. doi:10.1158/1078-0432.CCR-07-4969

Sharifi N, Hurt EM, Kawasaki BT, Farrar WL (2007) TGFBR3 loss and consequences in prostate cancer. Prostate 67(3):301–311. doi:10.1002/pros.20526

Shyu HY, Fong CS, Fu YP, Shieh JC, Yin JH, Chang CY, Wang HW, Cheng CW (2010) Genotype polymorphisms of GGCX, NQO1, and VKORC1 genes associated with risk susceptibility in patients with large-artery atherosclerotic stroke. Clin Chim Acta 411(11–12):840–845. doi:10.1016/j.cca.2010.02.071

Soker S, Gollamudi-Payne S, Fidder H, Charmahelli H, Klagsbrun M (1997) Inhibition of vascular endothelial growth factor (VEGF)-induced endothelial cell proliferation by a peptide corresponding to the exon 7-encoded domain of VEGF165. J Biol Chem 272(50):31582–31588

Sorensen I, Adams RH, Gossler A (2009) DLL1-mediated Notch activation regulates endothelial identity in mouse fetal arteries. Blood 113(22):5680–5688. doi:10.1182/blood-2008-08-174508

Sorensen LK, Brooke BS, Li DY, Urness LD (2003) Loss of distinct arterial and venous boundaries in mice lacking endoglin, a vascular-specific TGFβ coreceptor. Dev Biol 261(1): 235–250

Staton CA, Kumar I, Reed MW, Brown NJ (2007) Neuropilins in physiological and pathological angiogenesis. J Pathol 212(3):237–248. doi:10.1002/path.2182

Steg AD, Bevis KS, Katre AA, Ziebarth A, Dobbin ZC, Alvarez RD, Zhang K, Conner M, Landen CN (2012) Stem cell pathways contribute to clinical chemoresistance in ovarian cancer. Clin Cancer Res 18(3):869–881. doi:10.1158/1078-0432.CCR-11-2188

Stenvers KL, Tursky ML, Harder KW, Kountouri N, Amatayakul-Chantler S, Grail D, Small C, Weinberg RA, Sizeland AM, Zhu HJ (2003) Heart and liver defects and reduced transforming growth factor β2 sensitivity in transforming growth factor β type III receptor-deficient embryos. Mol Cell Biol 23(12):4371–4385

Stephenson JM, Banerjee S, Saxena NK, Cherian R, Banerjee SK (2002) Neuropilin-1 is differentially expressed in myoepithelial cells and vascular smooth muscle cells in preneoplastic and neoplastic human breast: a possible marker for the progression of breast cancer. Int J Cancer 101(5):409–414. doi:10.1002/ijc.10611

Sun L, Chen C (1997) Expression of transforming growth factor β type III receptor suppresses tumorigenicity of human breast cancer MDA-MB-231 cells. J Biol Chem 272(40): 25367–25372

Sweet K, Willis J, Zhou X, Gallione C, Sawada T, Alhopuro P, Khoo S, Patocs A, Martin C, Bridgeman S, Heinz J, Pilarski R, Lehtonen R, Prior T, Rebourg T, Teh B, Marchuk D, Aaltonen L, Eng C (2005) Molecular classification of patients with unexplained hamartomatous and hyperplastic polyposis. JAMA 294(19):2465–2473

Tabata M, Kondo M, Haruta Y, Seon BK (1999) Antiangiogenic radioimmunotherapy of human solid tumors in SCID mice using 125I-labeled anti-endoglin monoclonal antibodies. Int J Cancer 82:737–742

Takahashi N, Haba A, Matsuno F, Seon BK (2001) Antiangiogenic therapy of established tumors in human skin/severe combined immunodeficiency mouse chimerias by anti-endoglin (CD105) monoclonal antibodies, and synergy between anti-endoglin antibody and cyclophosphamide. Cancer Res 61:7846–7854

Tanaka F, Otake Y, Yanagihara K, Kawano Y, Miyahara R, Li M, Yamada T, Hanaoka N, Inui K, Wada H (2001) Evaluation of angiogenesis in non-small cell lung cancer: comparison between anti-CD35 antibody and anti-CD105 antibody. Clin Cancer Res 7:3410–3415

Theocharis AD, Skandalis SS, Tzanakakis GN, Karamanos NK (2010) Proteoglycans in health and disease: novel roles for proteoglycans in malignancy and their pharmacological targeting. FEBS J 277(19):3904–3923. doi:10.1111/j.1742-4658.2010.07800.x

Tsujie M, Tsujie T, Toi H, Uneda S, Shiozaki K, Tsai H, Seon BK (2008) Anti-tumor activity of an anti-endoglin monoclonal antibody is enhanced in immunocompetent mice. Int J Cancer 122(10):2266–2273. doi:10.1002/ijc.23314

Turley RS, Finger EC, Hempel N, How T, Fields TA, Blobe GC (2007) The type III transforming growth factor-β receptor as a novel tumor suppressor gene in prostate cancer. Cancer Res 67(3):1090–1098. doi:10.1158/0008-5472.CAN-06-3117

Uneda S, Toi H, Tsujie T, Tsujie M, Harada N, Tsai H, Seon BK (2009) Anti-endoglin monoclonal antibodies are effective for suppressing metastasis and the primary tumors by targeting tumor vasculature. Int J Cancer 125(6):1446–1453. doi:10.1002/ijc.24482

Valdembri D, Caswell PT, Anderson KI, Schwarz JP, Konig I, Astanina E, Caccavari F, Norman JC, Humphries MJ, Bussolino F, Serini G (2009) Neuropilin-1/GIPC1 signaling regulates α5β1 integrin traffic and function in endothelial cells. PLoS Biol 7(1):e25. doi:08-PLBI-RA-0162 [pii] 10.1371/journal.pbio.1000025

von Wronski MA, Raju N, Pillai R, Bogdan NJ, Marinelli ER, Nanjappan P, Ramalingam K, Arunachalam T, Eaton S, Linder KE, Yan F, Pochon S, Tweedle MF, Nunn AD (2006) Tuftsin binds neuropilin-1 through a sequence similar to that encoded by exon 8 of vascular endothelial growth factor. J Biol Chem 281(9):5702–5710. doi:10.1074/jbc.M511941200

Wang L, Mukhopadhyay D, Xu X (2006) C terminus of RGS-GAIP-interacting protein conveys neuropilin-1-mediated signaling during angiogenesis. FASEB J 20(9):1513–1515. doi:fj.05-5504fje [pii] 10.1096/fj.05-5504fje

Wang XF, Lin HY, Ng-Eaton E, Downward J, Lodish HF, Weinberg RA (1991) Expression cloning and characterization of the TGF-β type III receptor. Cell 67(4):797–805. doi:0092-8674(91)90074-9 [pii]

Webber J, Steadman R, Mason MD, Tabi Z, Clayton A (2010) Cancer exosomes trigger fibroblast to myofibroblast differentiation. Cancer Res 70(23):9621–9630. doi:10.1158/0008-5472.CAN-10-1722

Wikstrom P, Lissbrant IF, Stattin P, Egevad L, Bergh A (2002) Endoglin (CD105) is expressed on immature blood vessels and is a marker for survival in prostate cancer. Prostate 51(4):268–275. doi:10.1002/pros.10083

Wild JR, Staton CA, Chapple K, Corfe BM (2012) Neuropilins: expression and roles in the epithelium. Int J Exp Pathol 93(2):81–103. doi:10.1111/j.1365-2613.2012.00810.x

Wipff J, Avouac J, Borderie D, Zerkak D, Lemarechal H, Kahan A, Boileau C, Allanore Y (2008) Disturbed angiogenesis in systemic sclerosis: high levels of soluble endoglin. Rheumatology (Oxford) 47(7):972–975. doi:10.1093/rheumatology/ken100

Wong VC, Chan PL, Bernabeu C, Law S, Wang LD, Li JL, Tsao SW, Srivastava G, Lung ML (2008) Identification of an invasion and tumor-suppressing gene, Endoglin (ENG), silenced by both epigenetic inactivation and allelic loss in esophageal squamous cell carcinoma. Int J Cancer 123(12):2816–2823. doi:10.1002/ijc.23882

Woszczyk D, Gola J, Jurzak M, Mazurek U, Mykala-Ciesla J, Wilczok T (2004) Expression of TGF β1 genes and their receptor types I, II, and III in low- and high-grade malignancy non-Hodgkin's lyphomas. Med Sci Monit 10(1):CR33–CR37

Yacoub M, Coulon A, Celhay O, Irani J, Cussenot O, Fromont G (2009) Differential expression of the semaphorin 3A pathway in prostatic cancer. Histopathology 55(4):392–398

Yasuoka H, Kodama R, Tsujimoto M, Yoshidome K, Akamatsu H, Nakahara M, Inagaki M, Sanke T, Nakamura Y (2009) Neuropilin-2 expression in breast cancer: correlation with lymph node metastasis, poor prognosis, and regulation of CXCR4 expression. BMC Cancer 9:220

You HJ, How T, Blobe GC (2009) The type III transforming growth factor-β receptor negatively regulates nuclear factor κ B signaling through its interaction with β-arrestin2. Carcinogenesis 30(8):1281–1287. doi:bgp071 [pii] 10.1093/carcin/bgp071

Younan S, Elhoseiny S, Hammam A, Gawdat R, El-Wakil M, Fawzy M (2012) Role of neuropilin-1 and its expression in Egyptian acute myeloid and acute lymphoid leukemia patients. Leuk Res 36(2):169–173

Part II
Cancer, Fibrosis, and Inflammation

Chapter 4
TGF-β in Cancer Stem Cells

Hiroaki Ikushima and Kohei Miyazono

Abstract Cancer stem cells are tumor cells characterized by their ability to induce tumorigenesis and to self-renew. Recent research advance in cancer research field has demonstrated heterogeneity in cancer cell population and established the cancer stem cell model. Simultaneously, this model has shed new light on the mechanism of cancer recurrence, tumor metastasis, and several other cancer-related phenomena. The most important issue of cancer stem cells in the clinical situation is their resistibility against conventional therapies. The mechanisms in which cancer stem cells maintain their stem-like abilities have thus been energetically studied in the last decade. TGF-β is a well-recognized factor responsible for maintenance of normal tissue stem cells. In addition to that, recent studies have unveiled the roles of TGF-β signaling in cancer stem cells. In this chapter, we first consider the basic concept of the cancer stem cell model and describe their special characters in cancer cell population. We then discuss the roles of TGF-β signaling in cancer stem cells, focusing on some types of cancers. We also address perspectives on TGF-β signaling as a potential target against cancer stem cells.

Keywords Cancer stem cells • Heterogeneity • Niche • SP cells • Tumorigenicity

Abbreviations

ABC	ATP-binding cassette
ALDH	Aldehyde dehydrogenase
AML	Acute myeloid leukemia
CML	Chronic myelogenous leukemia

H. Ikushima • K. Miyazono (✉)
Department of Molecular Pathology, Graduate School of Medicine,
University of Tokyo, 7-3-1 Hongo, Bunkyo-ku, Tokyo 113-0033, Japan
e-mail: miyazono@m.u-tokyo.ac.jp

A. Moustakas and K. Miyazawa (eds.), *TGF-β in Human Disease*,
DOI 10.1007/978-4-431-54409-8_4,

EMT	Epithelial–mesenchymal transition
FOXO	Forkhead O transcription factor
LIF	Leukemia inhibitory factor
MP	Main population
ROS	Reactive oxygen species
SP	Side population

4.1 Cancer Stem Cell Model

The basic principle of the cancer stem cell model is that not all cells in tumors are equal. In the cancer stem cell model, like the growth of normal proliferative tissues such as skin, intestinal epithelium, and bone marrow, the growth of tumors is accelerated by limited numbers of cells that have capacity to self-renew and have the exclusive ability to drive tumor progression (Reya et al. 2001; Jordan et al. 2006; Lobo et al. 2007; Visvader and Lindeman 2008). The bulk of a tumor consists of rapidly proliferating cells and differentiated non-dividing cells. A defined subset of cancer cells, or cancer stem cells, can give rise to more differentiated cancer cells (Fig. 4.1).

Two important observations led to the hypothesis that cancer stem cells may be responsible for tumor initiation and progression. One is that not all cells within a tumor are identical. Although most tumors arise from a single transformed cell, there are many different types of cells in a tumor. This observation is known as tumor heterogeneity (Park et al. 1971). The other observation is that a certain number of cancer cells were required to grow a tumor in a recipient mouse (Hamburger and Salmon 1977). In contrast, in the model in which cancer cells are usually clonal in origin and every cancer cell has the potential to form a new tumor (clonal evolution model), even a few cancer cells would be able to form new tumors.

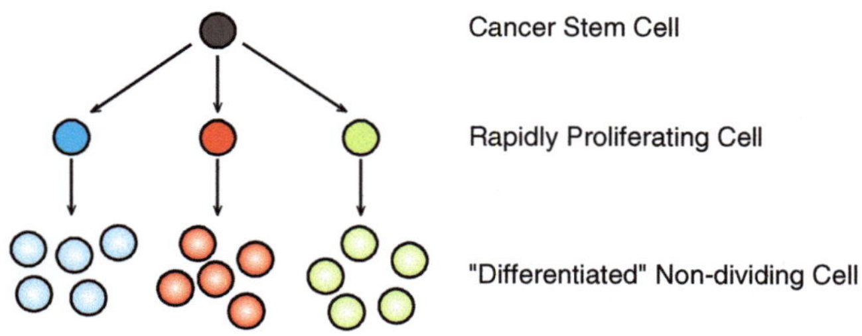

Fig. 4.1 In the cancer stem cell model, only a rare subpopulation of cells within the tumor, the cancer stem cells, has the capacity to generate and maintain a tumor. They can give rise to more differentiated cancer cells, which in turn form a hierarchy

4.2 History of Cancer Stem Cell Model

The concept of a cancer stem cell is not new and the connection between tumors and stem cells has a long history (Clevers 2011; Nguyen et al. 2012). In 1937, Furth and Kahn reported that a single cell from mouse leukemia could initiate a new tumor in a recipient mouse and that the tumor in a recipient mouse shares pathological characters similar to those of the original tumor (Furth and Kahn 1937). In 1941, Jackson and Brues showed that malignant teratocarcinomas contain many types of differentiated cells (Jackson and Brues 1941), suggesting their origin from tumorigenic cells with multilineage potential. Furthermore, in 1963, Bruce and Van Der Gaag reported the first in vivo colony assay for a mouse lymphoma-initiating cell (Bruce and Van Der Gaag 1963). In 1960s, studies by Pierce and his colleagues demonstrated that the morphologically undifferentiated cells in teratocarcinomas showed high tumorigenic activities and could differentiate into multiple differentiated non-tumorigenic cell types (Kleinsmith and Pierce 1964). These studies implicated these undifferentiated cells as candidate teratocarcinoma stem cells. Pierce's group also showed that, in mouse squamous cell carcinoma, well-differentiated cancer cells were derived from undifferentiated cells, using radiolabeling techniques (Pierce and Wallace 1971). From these and other experiments, Pierce defined the cancer stem cell concept as "a concept of neoplasms, based upon developmental and oncological principles, states that carcinomas are caricatures of tissue renewal, in that they are composed of a mixture of malignant stem cells, which have a marked capacity for proliferation and a limited capacity for differentiation under normal homeostatic conditions, and of the differentiated, possibly benign, progeny of these malignant cells" (Pierce and Speers 1988).

In 1970s, epoch-making discoveries were reported in the cancer research field: mutations in proto-oncogenes and tumor suppressor genes were found to cause most human cancers. After the existence of proto-oncogenes and tumor suppressor genes was established, the focus of cancer research shifted to the clonal evolution model as symbolized by the Knudson's two-hit theory and the Vogelstein's multistep carcinogenesis model. In the clonal evolution model, "tumor progression results from acquired genetic variability within the original clone allowing sequential selection of more aggressive sublines" (Nowell 1976; Merlo et al. 2006).

Until 1990s, it had been virtually impossible to directly prove the existence of cancer stem cells in human malignant tumors. However, development of immunocompromised mice, progression of cell sorting techniques, accumulation of knowledge on normal stem cells, and a variety of other scientific progression led Dick and his colleagues to revive the study of functional heterogeneity in leukemias (Lapidot et al. 1994; Bonnet and Dick 1997). They reported that only a subset of acute myeloid leukemia (AML) cells were able to transplant AML into immunodeficient recipient mice. These tumorigenic cells were defined as $CD34^{+}CD38^{-}$, suggesting similarities between leukemia stem cells and normal hematopoietic stem cells. In addition, the xenograft assay allowed measurement of the frequency of the leukemia stem cells; the order of one per million tumor cells. The remaining AML cells that grew from the transplanted cells were in various stages of differentiation. These data suggest that heterogeneity in leukemias reflects the similar organizational hierarchy to that in normal blood cells.

Clarke and his colleagues applied these concepts and approaches to breast cancer cells and reported that only a minor fraction of breast cancer cells can form a tumor (Al-Hajj et al. 2003). This report was the first reported isolation and characterization of cancer stem cells from solid tumors. They isolated breast cancer stem cells on the basis of surface marker expression ($CD44^{+}CD24^{-}$) and tumor regeneration potential in the mammary fat pads of NOD/SCID mice. The tumorigenic subpopulation could be serially passaged and, even after secondary and tertiary transplants, the new tumors showed histopathological characteristics similar to the original tumor. There were no significant differences in cell morphology between the tumorigenic and non-tumorigenic breast cancer cells. This study was followed by similar studies on other solid tumors such as brain tumors (Singh et al. 2003; Singh et al. 2004; Galli et al. 2004), melanomas (Fang et al. 2005; Schatton et al. 2008), prostate cancers (Patrawala et al. 2006; Patrawala et al. 2007; Li et al. 2008), ovarian cancers (Zhang et al. 2008; Curley et al. 2009; Stewart et al. 2011), and colon cancers (O'Brien et al. 2007; Ricci-Vitiani et al. 2007; Dalerba et al. 2007a). In each type of cancer stem cells, small numbers of cells defined by specific markers have ability to form tumors in immunodeficient mice and the transplanted tumors recapitulate the heterogeneity of the original tumors. However, the quantification of cancer stem cells in a population of tumor cells is never absolute, but instead depends on the tumor-initiating assay used to test for their presence (Quintana et al. 2008). Furthermore, recent studies have suggested that reversible regulatory mechanisms allow cancer cells to dynamically cycle between a tumor-initiating "cancer stem cell" state and a non-tumor-initiating cell state (Sharma et al. 2010; Roesch et al. 2010; Quintana et al. 2010; Scheel and Weinberg 2011).

Thus, the clonal evolution model and the cancer stem cell model constitute two major frameworks for interpreting clinical and experimental phenomena of malignant tumors. Of course, these two models are not mutually exclusive, as evidenced by substantial genetic heterogeneity in populations of putative cancer stem cells (Anderson et al. 2011; Notta et al. 2011).

4.3 Characteristics of Cancer Stem Cells

4.3.1 Tumorigenic Activity and Self-Renewal Capacity

The most characteristic feature of cancer stem cells is their high tumorigenic activity. This character is usually proven by the ability to form tumors from a small number of cells in immunodeficient mice, and the newly formed tumor usually shows histopathological characteristics similar to the original tumor. The frequency of cancer stem cells in a tumor can be estimated by limiting dilution analysis. Serial transplantation assay is also used to assess tumorigenic activity and self-renewal capacity of cancer stem cells.

4.3.2 Multilineage Potential

The well-known characters of normal tissue stem cells include multipotency as well as self-renewal capacity. Likewise, cancer stem cells undergo a process similar to differentiation under certain circumstances. In vitro incubation of breast cancer stem cells gives rise to cells positive for cytokeratin 8 (CK8) or mucin 1 (MUC1), both of which are breast epithelial cell markers. Glioma stem cells acquire similar cell morphology and marker expression pattern to astrocytes or oligodendrocytes when cultured in respective differentiation media. These results indicate that cancer stem cells have multilineage potential like normal tissue stem cells, although it is still controversial whether these processes can be defined as "differentiation." Thus, some researchers use the term "cancer-initiating cells" instead of "cancer stem cells" to literally and purely describe their ability to initiate tumor formation.

4.3.3 Resistance to Conventional Cancer Therapy

Another intriguing character of cancer stem cells is their resistance to conventional chemotherapy and radiation therapy (Dean et al. 2005; Baumann et al. 2008). In addition to their ability to self-renew, they are in the dormant phase and are quiescent and divide infrequently. Although our current therapeutic strategies against cancer succeed at eliminating rapidly proliferating bulk cells, they often miss slow-dividing cancer stem cells, which are the source of recurrence and metastasis.

Furthermore, cancer stem cells express high levels of specific ATP-binding cassette (ABC) drug transporters (Dean et al. 2005). The two ABC transporters that have been studied most extensively in cancer stem cells are ABCG2 (also known as breast cancer resistance protein 1, BCRP1) and ABCC2. By using the energy of ATP hydrolysis, these transporters actively efflux drugs from cancer stem cells, serving to protect them from cytotoxic agents (Gottesman et al. 2002).

Cancer stem cells contain lower levels of reactive oxygen species (ROS) than their matched bulk cancer cells (Diehn et al. 2009). Low ROS levels help to protect genomes in cancer stem cells from endogenous and exogenous ROS-mediated damage, including conventional radiation therapy. The mechanism for low ROS levels is at least partially due to the increased production of free radical scavengers. In addition, it has been reported that cancer stem cells show preferential checkpoint response and undergo DNA repair to counteract radiation damage (Bao et al. 2006a).

4.3.4 Invasion and Metastasis

The cancer stem cell model has shed new light on the mechanism of invasion and metastasis, and explain why, despite extensive intratumor heterogeneity, comparison of paired samples of primary tumors and distant metastases usually reveals

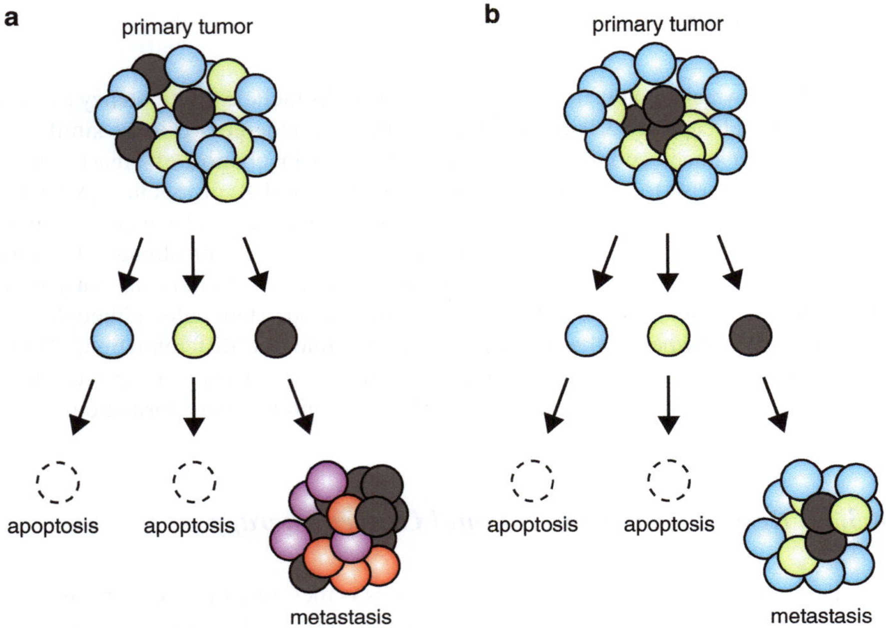

Fig. 4.2 (**a**) Based on the conventional stochastic cancer model, tumors are composed of heterogeneous mixtures of independent subclones with divergent genetic mutations. Different subclones are endowed with different functional properties and only selected clones (*dark gray*) can acquire ability to migrate and form metastases. The metastasis originates from a homogeneous monoclonal expansion of an individual subclone, which in turn can accumulate further mutations and diverge even further from the primary tumor. Therefore, the conventional stochastic cancer model predicts that primary tumors and corresponding metastases are substantially different. (**b**) In the cancer stem cell model, intratumor heterogeneity is mainly caused by cell differentiation and only cancer stem cells (*dark gray*) can migrate and form metastases. The cancer stem cells in metastatic sites undergo differentiation programs that closely resemble those observed in the corresponding primary tumors

striking similarities over a wide range of parameters, including tissue morphology (Brabletz et al. 2001), repertoire of genetic mutations (Losi et al. 1992; Khan et al. 2000; Zauber et al. 2003), and overall transcriptional profile as defined by gene expression arrays (Weigelt et al. 2003; Weigelt et al. 2005; D'Arrigo et al. 2005). These observations are not explained by the clonal evolution model, in which metastases are considered to originate from monoclonal expansions of specific individual tumor subclones endowed with specific genotypic and phenotypic features and therefore are postulated to be substantially different from primary tumors (Fig. 4.2a). However, under the cancer stem cell model, we can predict that, if two lesions share identical genetic abnormalities, they will also undergo similar differentiation programs and display similar patterns of intratumor heterogeneity (Fig. 4.2b).

In addition, C-X-C chemokine receptor type 4 (CXCR4) was reported to be highly expressed in prostate cancer stem cells (Dubrovska et al. 2012), suggesting the role of cancer stem cells as "pacemakers" of tumor metastasis.

4.4 Cancer Stem Cell Markers

Cancer stem cells have been prospectively isolated on the basis of their expression of particular markers such as CD44 and CD133 (Alison et al. 2010), although "non-cancer stem cells" defined as subpopulations negative for certain reported "cancer stem cell markers" also have tumorigenic activities in some contexts (Shmelkov et al. 2008; Quintana et al. 2010). Interestingly, cancer stem cells often share their markers with their normal counterparts, suggesting characteristic similarities between cancer stem cells and normal tissue stem cells.

Stem cells are known to efficiently extrude dyes such as Hoechst 33342. Cells with the capacity to efflux the dye were referred to as side population (SP) cells (Challen and Little 2006). SP cells have been identified in a variety of tissues, and they express high levels of stem-like genes and possess multi-potent differentiation potential (Goodell et al. 1996; Goodell et al. 1997; Pearce et al. 2004). The mechanism regulating the efflux of Hoechst dye can be explained, at least in part, by the expression of ABC transporters, including ABCG2 (Zhou et al. 2001).

SP cells have been also identified in a large variety of cancer cells (Kondo et al. 2004; Chiba et al. 2006; Haraguchi et al. 2006; Ho et al. 2007; Wang et al. 2007). When compared to non-SP population, SP cells isolated from cancer cells are highly enriched for the capacity to initiate tumor formation in immunodeficient mice. They also have the capacity to initiate tumors upon serial transplantation. Furthermore, SP cells in cancer cell population have increased expression of genes which are believed to be involved in the regulation of stem cell function, such as ABCG2 transporter. These data suggest that SP cells in tumors act as cancer stem cells.

4.5 Cancer Stem Cell Niche

Normal stem cells reside in a special microenvironment called "niche." They interact with the niche through adhesion molecules and exchange signals to maintain the specific features of stem cells (Wilson and Trumpp 2006; Martino and Pluchino 2006; Kiel and Morrison 2008; Wang and Wagers 2011). It has been suggested that there is a functional microenvironment to support cancer stem cells as well (Iwasaki and Suda 2009). This should also be considered a niche and is thus called "cancer stem cell niche."

Dick and colleagues demonstrated that anti-CD44 antibody-treated mice transplanted with AML cells exhibited a significantly lower rate of disease onset (Jin et al. 2006). In addition, Van Etten and colleagues showed that there was impaired induction of chronic myelogenous leukemia (CML)-like myeloproliferative disease among recipient mice transplanted with BCR-ABL-transduced CML progenitors from CD44-mutant donors (Krause et al. 2006). These results indicate that CD44 is required for the homing and/or engraftment of leukemia stem cells to the niche, and that the binding of CD44-expressing leukemic stem cells to the niche is crucial for the maintenance of their tumorigenic activities. Interestingly, this role of CD44 in leukemia stem cells resembles that in normal hematopoietic stem cells, suggesting that cancer stem cells and normal stem cells share the maintenance system within their niches.

In brain tumor stem cell research, Gilbertson and colleagues revealed that brain tumor cells co-expressing Nestin and CD133 exist near the capillary vessels in the brain tumor (Calabrese et al. 2007). In addition, the CD133-positive subpopulation in human medulloblastoma developed brain tumors in a recipient nude mouse only when xenografted with endothelial cells. These results suggest that brain tumor stem cells rely on endothelial cells, which form a vascular niche to maintain the stem-like characters of brain tumor stem cells, such as self-renewal capacity and tumorigenic activity. Conversely, brain tumor stem cells themselves can elicit angiogenic effects by secreting factors such as vascular endothelial growth factor (VEGF) and stromal-derived factor 1 (SDF-1) (Bao et al. 2006b; Folkins et al. 2009). Furthermore, under certain conditions, brain tumor stem cells can even directly transdifferentiate into the endothelial lineage (Ricci-Vitiani et al. 2010; Wang et al. 2010), indicating a close connection between cancer stem cells and vascular niches.

4.6 Roles of TGF-β Signaling in Cancer Stem Cell Model

Cancer stem cells have been suggested to make use of a microenvironment similar to that found in normal stem cell niches for the maintenance of their stem cell-like properties. TGF-β signaling has been identified as a niche signal in the control of hematopoietic stem cells and hair follicle stem cells (Yamazaki et al. 2009; Oshimori and Fuchs 2012), and so a broader role for TGF-β signaling in the maintenance of cancer stem cells has been proposed. Recent studies have revealed crucial roles of TGF-β signaling in interaction between cancer stem cell and niche, as well as cancer stem cell-autonomous signaling pathways (Ikushima and Miyazono 2010a).

4.6.1 TGF-β Signaling in Breast Cancer Stem Cells

Breast cancer stem cells are, as described above, the first identified solid tumor stem cells (Dalerba et al. 2007b). Since the prospective identification using $CD44^{+}CD24^{-}$ subpopulation (Al-Hajj et al. 2003), the involvement of TGF-β signaling in breast cancer stem cells has been energetically studied. Recent reports have uncovered bilateral characters of TGF-β signaling in breast cancer stem cells: deprivation of tumorigenic activities in some contexts and maintenance of stem cell-like characters in other contexts.

Wakefield and her colleagues showed that the suppression of the TGF-β pathway increased the size of the putative breast cancer stem cell compartment and promoted tumorigenesis by a mechanism that was independent of direct effects on proliferation (Tang et al. 2007). They demonstrated that the introduction of a dominant negative TGFBR2 enhanced the proliferation of immortalized and transformed human breast epithelial cell line, Ca1h, although the expression level of the cyclin-dependent kinase inhibitor p21 was unchanged. They also showed that TGF-β stimulation resulted in the loss of stem cell-like properties and the ability to form mammospheres. The ability of TGF-β to deprive breast cancer stem cells of tumorigenic activity was dependent on the down-regulation of Id1, which is highly expressed during embryogenesis and has been implicated in the regulation of self-renewal and differentiation. Id family proteins (Id1 through 4) act as inhibitors of differentiation and stimulators of cell growth by antagonizing the function of basic helix-loop-helix (bHLH) transcription factors (Miyazawa et al. 2002). Id1 has been reported to inhibit lineage commitment and differentiation in many cell types (Norton 2000) and to be a direct TGF-β target (Kang et al. 2003). These findings suggest that breast cancer stem cells benefit from similar mechanisms that regulate the function of normal stem cells.

Weinberg and his colleagues demonstrated that TGF-β signaling has an important role in the maintenance of stem cell-like properties and tumorigenic activity through the induction of epithelial–mesenchymal transition (EMT). A $CD44^{+}CD24^{-}$ subpopulation that was isolated from breast cancer specimens exhibited mesenchymal properties, and transformed mammary epithelial cells, in which EMT was induced by TGF-β stimulation, acquired stem cell-like properties, including mammosphere-forming ability (Mani et al. 2008; Scheel et al. 2011). Transformed mammary epithelial cells with TGF-β-induced EMT also showed higher tumorigenic activity in vivo and fewer cells were required to initiate tumor formation than cells without TGF-β treatment. These results connect TGF-β-induced EMT and the gain of epithelial stem cell properties of carcinoma.

4.6.2 TGF-β Signaling in Brain Tumor Stem Cells

Glioma stem cells are characterized by the expression of neural stem cell antigens (Singh et al. 2004; Kondo et al. 2004; Hirschmann-Jax et al. 2004) and share several characteristics with normal neural stem cells (Vescovi et al. 2006). The overexpression of TGF-β commonly seen in malignant glioma has been variously implicated in glioma cell proliferation, decreased apoptosis, migration, and tumor-specific immunosuppression (Golestaneh and Mishra 2005). TGF-β signaling also has important roles in the regulation of the stem cell properties of neural stem cells (Watabe and Miyazono 2009). These facts have shed some light on the role of TGF-β signaling in the maintenance of glioma stem cells.

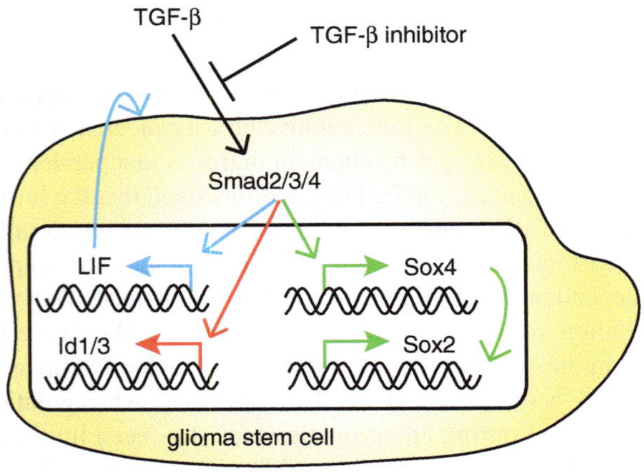

Fig. 4.3 TGF-β-Smad signaling pathway maintains stem cell-like properties and tumorigenic activity of glioma stem cells through many independent pathways, including the induction of the Sox4-Sox2 cascade, the activation of the LIF pathway, and the up-regulation of Id1 and Id3 expressions

TGF-β signaling has been reported to play pivotal roles in the maintenance of tumorigenic activity and stem cell-like properties of glioma stem cells (Peñuelas et al. 2009; Ikushima et al. 2009; Anido et al. 2010; Ikushima et al. 2011). TGF-β inhibitors markedly deprived glioma stem cells of glioma sphere-forming activity and self-renewal capacity in vitro and tumorigenic activity in vivo. Inhibition of TGF-β signaling also suppressed marker expressions that are associated with stem cell-like properties. These results indicate that microenvironmental niche-derived or glioma stem cell-autonomous TGF-β signaling maintains tumor-initiating abilities of glioma stem cells. TGF-β mediates this activity through the direct induction of the leukemia inhibitory factor (LIF) expression (Peñuelas et al. 2009). LIF activates the JAK-STAT pathway in glioma stem cells, leading to their increased tumorigenicity. Independently of this mechanism, TGF-β induces the expression of Sox2, a self-renewal gene that helps to maintain stem cell-like properties in embryonic stem cells and neural stem cells (Kamachi et al. 2000; Graham et al. 2003; Ferri et al. 2004). TGF-β induces the expression of Sox4, and this subsequently induces the expression of Sox2 (Ikushima et al. 2009; Ikushima et al. 2011). The maintenance of tumorigenic activity of glioma stem cells by TGF-β is also mediated by induction of Id1 and Id3 expressions (Anido et al. 2010). TGF-β signaling thus maintains the stem cell-like properties and tumorigenic activities of glioma stem cells through multiple pathways (Fig. 4.3), although interactions among these pathways have been still unclear.

4.6.3 TGF-β Signaling in Leukemia Stem Cells

TGF-β signaling is known to be a candidate niche signal in the control of hematopoietic stem cell hibernation (Yamazaki et al. 2009). Also, it plays important roles in the maintenance of leukemia stem cells.

CML is caused by a t(9;22)(q34;q11) translocation that generates a constitutively active tyrosine kinase, BCR-ABL (Ren 2005). BCR-ABL activates AKT signaling to suppress the forkhead O transcription factors (FOXO) in CML cells (Ghaffari et al. 2003; Essafi et al. 2005). In CML, a rare Lineage$^-$(Lin$^-$)Sca-1$^+$c-Kit$^+$ population was identified as leukemia stem cells (Hu et al. 2006; Neering et al. 2007; Zhao et al. 2009). Although the use of the tyrosine kinase inhibitor imatinib is a breakthrough for CML therapy, it does not deplete the leukemia stem cells, which drive the recurrence of CML (Graham et al. 2002; Michor et al. 2005; Roeder et al. 2006).

TGF-β signaling has crucial roles in the maintenance of leukemia stem cells in CML. TGF-β regulates AKT activation and FOXO3a localization in leukemia stem cells. Furthermore, this TGF-β-FOXO pathway maintains the stem cell-like properties and tumorigenic activities of leukemia stem cells (Naka et al. 2010). It was also demonstrated that a combination of TGF-β inhibition, FOXO3a deficiency, and imatinib treatment led to the efficient depletion of CML cells in vivo. These results indicate the central roles of TGF-β signaling in leukemia stem cells of CML. However, when leukemia stem cells were cultured with TGF-β inhibitors in a stroma-free system, colony formation was not inhibited (Naka et al. 2010), suggesting that the maintenance of leukemia stem cells depends not only on TGF-β produced by leukemia stem cells themselves but also on TGF-β in the surrounding microenvironment.

HOXA9 is involved in human AML caused by the translocation t(7;11)(p15;p15), through which HOXA9 gene is fused with the gene encoding NUP98, a nucleoporin protein (Nakamura et al. 1996; Borrow et al. 1996). Expression of the NUP98-HOXA9 fusion protein enforces expression of HOXA9 target genes and immortalizes hematopoietic myeloid progenitors, resulting in development of AML. In wild-type mouse hematopoietic stem cells transduced with NUP98-HOXA9, Smad4 binds to HOXA9 and inhibits nuclear transportation of HOXA9. In contrast, there is no cytoplasmic accumulation of HOXA9 in Smad4$^{-/-}$ hematopoietic stem cells, and as a consequence increased levels of HOXA9 is observed in the nucleus, leading to increased immortalization in vitro (Quéré et al. 2011). In addition, loss of Smad4 accelerates the development of NUP98-HOXA9-induced AML in vivo, and NUP98-HOXA9-transformed Smad4$^{-/-}$ leukemic cell population contains a higher amount of the leukemia stem cells than wild-type leukemia cells (Quéré et al. 2011). These results indicate that the cytoplasmic binding of Smad4 to HOXA9 is a mechanism to protect NUP98-HOXA9-induced transformation and acquisition of leukemogenic activity.

4.6.4 TGF-β Signaling in Gastric Cancer Stem Cells

Some molecules, including CD44 and aldehyde dehydrogenase 1 (ALDH1), have been reported as markers for gastric cancer stem cells (Takaishi et al. 2009; Katsuno et al. 2012). In addition, flow cytometric analyses using Hoechst 33342 have disclosed that SP cells in gastric carcinomas have higher tumorigenic activities, suggesting their roles as gastric cancer stem cells (Nishii et al. 2009; Ehata et al. 2011).

Mutations of TGFBR2, Smad4, and Smad2 have been reported to be responsible for progression of gastrointestinal tumors, indicating the tumor-suppressive activity of TGF-β signaling in gastrointestinal tumors (Wakefield and Roberts 2002; Bierie and Moses 2006). Besides suppression of cell proliferation and induction of apoptosis, TGF-β decreases the number of SP cells and so attenuates the tumor-forming ability of gastric cancer cells. TGF-β transcriptionally represses ABCG2 expression through direct binding of Smad2/3 to its promoter (Ehata et al. 2011). TGF-β also reduces the expression of ALDH1 and the size of the $ALDH1^+$ cell population in diffuse-type gastric cancer cells. In addition, ALDH1 expression inversely correlates with phosphorylation of Smad3 protein in human diffuse-type gastric cancer tissues (Katsuno et al. 2012). These results suggest that TGF-β reduces the cancer stem cell subpopulation through suppression of ABCG2 and ALDH1 expressions and inhibits tumor-initiating capacity of gastric cancer stem cells.

It has not yet been fully elucidated why TGF-β signaling suppresses tumorigenicity of gastric cancer stem cells while it maintains that of brain tumor stem cells and leukemia stem cells. Possible answers for this question may include the differential roles of TGF-β in the corresponding normal tissue stem cells (See Sect. 8).

4.6.5 TGF-β Signaling in Pancreatic Cancer Stem Cells

When incubated with TGF-β, pancreatic cancer stem cells enriched by sorting for SP cells change their shape into mesenchymal-like spindle-shaped appearance (Kabashima et al. 2009). This change is associated with significant reduction of E-cadherin expression and induction of Snail expression. Furthermore, SP cells show marked invasion activity in response to TGF-β treatment. Interestingly, such invasive activity is not induced by TGF-β stimulation in main population (MP) cells. These results suggest that TGF-β responsiveness is greater in SP cells than in MP cells, resulting in enhanced induction of EMT and invasiveness (Kabashima et al. 2009), although this study did not provide us direct evidences for the effect of TGF-β on tumorigenicity of pancreatic cancer stem cells.

4.7 TGF-β Signaling as Therapeutic Target in Cancer Stem Cell Model

According to the concept of cancer stem cells, therapeutic strategies that do not eradicate the cancer stem cell compartment are likely to achieve little success. They fail to prevent disease relapse and metastatic dissemination even when they might kill the majority of tumor cells and induce temporary regression of gross tumor lesions (Fig. 4.4). As mentioned above, cancer stem cells are considered to be inherently resistant to the toxic effect of conventional chemotherapeutic regimens and

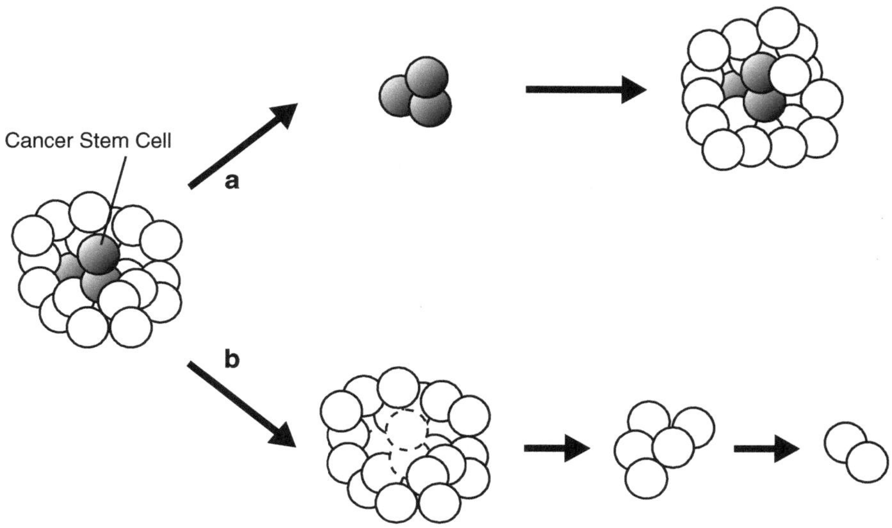

Fig. 4.4 Anti-tumor treatments designed for broad cytotoxic effects (**a**) may kill the majority of cancer cells within a tumor and induce dramatic regression of tumor masses. However, if a certain number of cancer stem cells are spared, tumor tissues can be regenerated, resulting in cancer recurrence. In contrast, anti-tumor treatments specifically designed to target cancer stem cells (**b**) might achieve long-term disease eradication by exhausting growth potential of cancer tissues, even though they are theoretically unable to cause rapid shrinkage of tumor lesions

radiation therapies. Therefore, investigational therapies should be developed, focusing on their ability to target the cancer stem cell subpopulation and its specific signaling pathways. However, there is one important problem in targeting signaling pathways specific for cancer stem cells: these pathways may also be important for normal stem cells. It has not yet fully determined what is the same or what is different between cancer stem cells and normal stem cells. Therefore, we should bear in mind that cancer stem cell-targeting agents could have adverse effects on maintenance of normal tissue stem cells.

The TGF-β pathway has been targeted using multiple strategies, including small-molecule inhibitors of the TGFBR1 kinase domain, TGF-β-specific neutralizing antibodies, and antisense compounds (Yingling et al. 2004). Some of them have been in clinical trials for human cancers (Schlingensiepen et al. 2006; Hau et al. 2007).

4.8 Concluding Remarks

As we have discussed, TGF-β is involved in the maintenance of the tumorigenic activity of cancer stem cells in several types of tumors in a tissue-specific manner. Targeting the pathways that maintain cancer stem cells might ultimately prove to be an effective therapeutic strategy against malignant tumors. However, such pathways could have divergent roles in cancer stem cell populations from different patients.

Several in vitro and in vivo studies have uncovered cellular context-dependent diversity in TGF-β-induced cell responses (Ikushima and Miyazono 2010b). Because of such diversity, TGF-β can be both pro-tumorigenic and tumor-suppressive in a cellular context-dependent fashion. In addition, TGF-β promotes differentiation in some kinds of tissue stem cells and maintains stemness properties in others. Likewise, as we have discussed, TGF-β signaling shows positive effects on some kinds of cancer stem cells and negative effects on others. Such diversity among cancer stem cells could reflect both the differences between the oncogenic mutations expressed by the cells and their progeny and the differences in their origin. These differences will need to be taken into account when developing treatments based on TGF-β signaling for any individual patient. Also, we need to elucidate the exact mechanism by which TGF-β shows such complex effects against cancer stem cells in cellular context-dependent manner.

References

Al-Hajj M, Wicha MS, Benito-Hernandez A, Morrison SJ, Clarke MF (2003) Prospective identification of tumorigenic breast cancer cells. Proc Natl Acad Sci USA 100:3983–3988. doi:10.1073/pnas.0530291100

Alison MR, Islam S, Wright NA (2010) Stem cells in cancer: instigators and propagators? J Cell Sci 123:2357–2368. doi:10.1242/jcs.054296

Anderson K, Lutz C, van Delft FW, Bateman CM, Guo Y, Colman SM, Kempski H, Moorman AV, Titley I, Swansbury J, Kearney L, Enver T, Greaves M (2011) Genetic variegation of clonal architecture and propagating cells in leukemia. Nature 469:356–361. doi:10.1038/nature09650

Anido J, Sáez-Borderías A, Gonzàlez-Juncà A, Rodón L, Folch G, Carmona MA, Prieto-Sánchez RM, Barba I, Martínez-Sáez E, Prudkin L, Cuartas I, Raventós C, Martínez-Ricarte F, Poca MA, García-Dorado D, Lahn MM, Yingling JM, Rodón J, Sahuquillo J, Baselga J, Seoane J (2010) TGF-β receptor inhibitors target the CD44(high)/Id1(high) glioma-initiating cell population in human glioblastoma. Cancer Cell 18:655–668. doi:10.1016/j.ccr.2010.10.023

Bao S, Wu Q, McLendon RE, Hao Y, Shi Q, Hjelmeland AB, Dewhirst MW, Bigner DD, Rich JN (2006a) Glioma stem cells promote radioresistance by preferential activation of the DNA damage response. Nature 444:756–760. doi:10.1038/nature05236

Bao S, Wu Q, Sathornsumetee S, Hao Y, Li Z, Hjelmeland AB, Shi Q, McLendon RE, Bigner DD, Rich JN (2006b) Stem cell-like glioma cells promote tumor angiogenesis through vascular endothelial growth factor. Cancer Res 66:7843–7848. doi:10.1158/0008-5472.CAN-06-1010

Baumann M, Krause M, Hill R (2008) Exploring the role of cancer stem cells in radioresistance. Nat Rev Cancer 8:545–554. doi:10.1038/nrc2419

Bierie B, Moses HL (2006) Tumour microenvironment: TGFβ: the molecular Jekyll and Hyde of cancer. Nat Rev Cancer 6:506–520. doi:10.1038/nrc1926

Bonnet D, Dick JE (1997) Human acute myeloid leukemia is organized as a hierarchy that originates from a primitive hematopoietic cell. Nat Med 3:730–737

Borrow J, Shearman AM, Stanton VP Jr, Becher R, Collins T, Williams AJ, Dubé I, Katz F, Kwong YL, Morris C, Ohyashiki K, Toyama K, Rowley J, Housman DE (1996) The t(7;11)(p15;p15) translocation in acute myeloid leukaemia fuses the genes for nucleoporin NUP98 and class I homeoprotein HOXA9. Nat Genet 12:159–167. doi:10.1038/ng0296-159

Brabletz T, Jung A, Reu S, Porzner M, Hlubek F, Kunz-Schughart LA, Knuechel R, Kirchner T (2001) Variable β-catenin expression in colorectal cancers indicates tumor progression driven by the tumor environment. Proc Natl Acad Sci USA 98:10356–10361. doi:10.1073/pnas.171610498

Bruce WR, Van Der Gaag H (1963) A quantitative assay for the number of murine lymphoma cells capable of proliferation in vivo. Nature 199:79–80

Calabrese C, Poppleton H, Kocak M, Hogg TL, Fuller C, Hamner B, Oh EY, Gaber MW, Finklestein D, Allen M, Frank A, Bayazitov IT, Zakharenko SS, Gajjar A, Davidoff A, Gilbertson RJ (2007) A perivascular niche for brain tumor stem cells. Cancer Cell 11:69–82. doi:10.1016/j.ccr.2006.11.020

Challen GA, Little MH (2006) A side order of stem cells: the SP phenotype. Stem Cells 24:3–12. doi:10.1634/stemcells.2005-0116

Chiba T, Kita K, Zheng YW, Yokosuka O, Saisho H, Iwama A, Nakauchi H, Taniguchi H (2006) Side population purified from hepatocellular carcinoma cells harbors cancer stem cell-like properties. Hepatology 44:240–251. doi:10.1002/hep.21227

Clevers H (2011) The cancer stem cell: premises, promises and challenges. Nat Med 17:313–319. doi:10.1038/nm.2304

Curley MD, Therrien VA, Cummings CL, Sergent PA, Koulouris CR, Friel AM, Roberts DJ, Seiden MV, Scadden DT, Rueda BR, Foster R (2009) CD133 expression defines a tumor initiating cell population in primary human ovarian cancer. Stem Cells 27:2875–2883. doi:10.1002/stem.236

Dalerba P, Dylla SJ, Park IK, Liu R, Wang X, Cho RW, Hoey T, Gurney A, Huang EH, Simeone DM, Shelton AA, Parmiani G, Castelli C, Clarke MF (2007a) Phenotypic characterization of human colorectal cancer stem cells. Proc Natl Acad Sci USA 104:10158–10163. doi:10.1073/pnas.0703478104

Dalerba P, Cho RW, Clarke MF (2007b) Cancer stem cells: models and concepts. Annu Rev Med 58:267–284. doi:10.1146/annurev.med.58.062105.204854

D'Arrigo A, Belluco C, Ambrosi A, Digito M, Esposito G, Bertola A, Fabris M, Nofrate V, Mammano E, Leon A, Nitti D, Lise M (2005) Metastatic transcriptional pattern revealed by gene expression profiling in primary colorectal carcinoma. Int J Cancer 115:256–262. doi:10.1002/ijc.20883

Dean M, Fojo T, Bates S (2005) Tumour stem cells and drug resistance. Nat Rev Cancer 5:275–284. doi:10.1038/nrc1590

Diehn M, Cho RW, Lobo NA, Kalisky T, Dorie MJ, Kulp AN, Qian D, Lam JS, Ailles LE, Wong M, Joshua B, Kaplan MJ, Wapnir I, Dirbas FM, Somlo G, Garberoglio C, Paz B, Shen J, Lau SK, Quake SR, Brown JM, Weissman IL, Clarke MF (2009) Association of reactive oxygen species levels and radioresistance in cancer stem cells. Nature 458:780–783. doi:10.1038/nature07733

Dubrovska A, Elliott J, Salamone RJ, Telegeev GD, Stakhovsky AE, Schepotin IB, Yan F, Wang Y, Bouchez LC, Kularatne SA, Watson J, Trussell C, Reddy VA, Cho CY, Schultz PG (2012) CXCR4 expression in prostate cancer progenitor cells. PLoS One 7:e31226. doi:10.1371/journal.pone.0031226

Ehata S, Johansson E, Katayama R, Koike S, Watanabe A, Hoshino Y, Katsuno Y, Komuro A, Koinuma D, Kano MR, Yashiro M, Hirakawa K, Aburatani H, Fujita N, Miyazono K (2011) Transforming growth factor-β decreases the cancer-initiating cell population within diffuse-type gastric carcinoma cells. Oncogene 30:1693–1705. doi:10.1038/onc.2010.546

Essafi A, Fernández de Mattos S, Hassen YA, Soeiro I, Mufti GJ, Thomas NS, Medema RH, Lam EW (2005) Direct transcriptional regulation of Bim by FoxO3a mediates STI571-induced apoptosis in Bcr-Abl-expressing cells. Oncogene 24:2317–2329. doi:10.1038/sj.onc.1208421

Fang D, Nguyen TK, Leishear K, Finko R, Kulp AN, Hotz S, Van Belle PA, Xu X, Elder DE, Herlyn M (2005) A tumorigenic subpopulation with stem cell properties in melanomas. Cancer Res 65:9328–9337. doi:10.1158/0008-5472.CAN-05-1343

Ferri AL, Cavallaro M, Braida D, Di Cristofano A, Canta A, Vezzani A, Ottolenghi S, Pandolfi PP, Sala M, DeBiasi S, Nicolis SK (2004) Sox2 deficiency causes neurodegeneration and impaired neurogenesis in the adult mouse brain. Development 131:3805–3819. doi:10.1242/dev.01204

Folkins C, Shaked Y, Man S, Tang T, Lee CR, Zhu Z, Hoffman RM, Kerbel RS (2009) Glioma tumor stem-like cells promote tumor angiogenesis and vasculogenesis via vascular endothelial growth factor and stromal-derived factor 1. Cancer Res 69:7243–7251. doi:10.1158/0008-5472.CAN-09-0167

Furth J, Kahn M (1937) The transmission of leukemia of mice with a single cell. Am J Cancer 31:276–282

Galli R, Binda E, Orfanelli U, Cipelletti B, Gritti A, De Vitis S, Fiocco R, Foroni C, Dimeco F, Vescovi A (2004) Isolation and characterization of tumorigenic, stem-like neural precursors from human glioblastoma. Cancer Res 64:7011–7021. doi:10.1158/0008-5472.CAN-04-1364

Ghaffari S, Jagani Z, Kitidis C, Lodish HF, Khosravi-Far R (2003) Cytokines and BCR-ABL mediate suppression of TRAIL-induced apoptosis through inhibition of forkhead FOXO3a transcription factor. Proc Natl Acad Sci USA 100:6523–6528. doi:10.1073/pnas.0731871100

Golestaneh N, Mishra B (2005) TGF-β, neuronal stem cells and glioblastoma. Oncogene 24:5722–5730. doi:10.1038/sj.onc.1208925

Goodell MA, Brose K, Paradis G, Conner AS, Mulligan RC (1996) Isolation and functional properties of murine hematopoietic stem cells that are replicating in vivo. J Exp Med 183:1797–1806

Goodell MA, Rosenzweig M, Kim H, Marks DF, DeMaria M, Paradis G, Grupp SA, Sieff CA, Mulligan RC, Johnson RP (1997) Dye efflux studies suggest that hematopoietic stem cells expressing low or undetectable levels of CD34 antigen exist in multiple species. Nat Med 3:1337–1345

Gottesman MM, Fojo T, Bates SE (2002) Multidrug resistance in cancer: role of ATP-dependent transporters. Nat Rev Cancer 2:48–58. doi:10.1038/nrc706

Graham SM, Jørgensen HG, Allan E, Pearson C, Alcorn MJ, Richmond L, Holyoake TL (2002) Primitive, quiescent, Philadelphia-positive stem cells from patients with chronic myeloid leukemia are insensitive to STI571 in vitro. Blood 99:319–325

Graham V, Khudyakov J, Ellis P, Pevny L (2003) SOX2 functions to maintain neural progenitor identity. Neuron 39:749–765

Hamburger AW, Salmon SE (1977) Primary bioassay of human tumor stem cells. Science 197:461–463

Haraguchi N, Utsunomiya T, Inoue H, Tanaka F, Mimori K, Barnard GF, Mori M (2006) Characterization of a side population of cancer cells from human gastrointestinal system. Stem Cells 24:506–513. doi:10.1634/stemcells.2005-0282

Hau P, Jachimczak P, Schlingensiepen R, Schulmeyer F, Jauch T, Steinbrecher A, Brawanski A, Proescholdt M, Schlaier J, Buchroithner J, Pichler J, Wurm G, Mehdorn M, Strege R, Schuierer G, Villarrubia V, Fellner F, Jansen O, Straube T, Nohria V, Goldbrunner M, Kunst M, Schmaus S, Stauder G, Bogdahn U, Schlingensiepen KH (2007) Inhibition of TGF-β2 with AP 12009 in recurrent malignant gliomas: from preclinical to phase I/II studies. Oligonucleotides 17:201–212. doi:10.1089/oli.2006.0053

Hirschmann-Jax C, Foster AE, Wulf GG, Nuchtern JG, Jax TW, Gobel U, Goodell MA, Brenner MK (2004) A distinct "side population" of cells with high drug efflux capacity in human tumor cells. Proc Natl Acad Sci USA 101:14228–14233. doi:10.1073/pnas.0400067101

Ho MM, Ng AV, Lam S, Hung JY (2007) Side population in human lung cancer cell lines and tumors is enriched with stem-like cancer cells. Cancer Res 67:4827–4833. doi:10.1158/0008-5472.CAN-06-3557

Hu Y, Swerdlow S, Duffy TM, Weinmann R, Lee FY, Li S (2006) Targeting multiple kinase pathways in leukemic progenitors and stem cells is essential for improved treatment of Ph+ leukemia in mice. Proc Natl Acad Sci USA 103:16870–16875. doi:10.1073/pnas.0606509103

Ikushima H, Miyazono K (2010a) TGFβ signalling: a complex web in cancer progression. Nat Rev Cancer 10:415–424. doi:10.1038/nrc2853

Ikushima H, Miyazono K (2010b) Cellular context-dependent "colors" of transforming growth factor-β signaling. Cancer Sci 101:306–312. doi:10.1111/j.1349-7006.2009.01441.x
Ikushima H, Todo T, Ino Y, Takahashi M, Miyazawa K, Miyazono K (2009) Autocrine TGF-β signaling maintains tumorigenicity of glioma-initiating cells through Sry-related HMG-box factors. Cell Stem Cell 5:504–514. doi:10.1016/j.stem.2009.08.018
Ikushima H, Todo T, Ino Y, Takahashi M, Saito N, Miyazawa K, Miyazono K (2011) Glioma-initiating cells retain their tumorigenicity through integration of the Sox axis and Oct4 protein. J Biol Chem 286:41434–41441. doi:10.1074/jbc.M111.300863
Iwasaki H, Suda T (2009) Cancer stem cells and their niche. Cancer Sci 100:1166–1172. doi:10.1111/j.1349-7006.2009.01177.x
Jackson EB, Brues AM (1941) Studies on a transplantable embryoma of the mouse. Cancer Res 1:494–498
Jin L, Hope KJ, Zhai Q, Smadja-Joffe F, Dick JE (2006) Targeting of CD44 eradicates human acute myeloid leukemic stem cells. Nat Med 12:1167–1174. doi:10.1038/nm1483
Jordan CT, Guzman ML, Noble M (2006) Cancer stem cells. N Engl J Med 355:1253–1261. doi:10.1056/NEJMra061808
Kabashima A, Higuchi H, Takaishi H, Matsuzaki Y, Suzuki S, Izumiya M, Iizuka H, Sakai G, Hozawa S, Azuma T, Hibi T (2009) Side population of pancreatic cancer cells predominates in TGF-β-mediated epithelial to mesenchymal transition and invasion. Int J Cancer 124:2771–2779. doi:10.1002/ijc.24349
Kamachi Y, Uchikawa M, Kondoh H (2000) Pairing SOX off: with partners in the regulation of embryonic development. Trends Genet 16:182–187
Kang Y, Chen CR, Massagué J (2003) A self-enabling TGFβ response coupled to stress signaling: Smad engages stress response factor ATF3 for Id1 repression in epithelial cells. Mol Cell 11:915–926
Katsuno Y, Ehata S, Yashiro M, Yanagihara K, Hirakawa K, Miyazono K (2012) Coordinated expression of REG4 and aldehyde dehydrogenase 1 regulating tumorigenic capacity of diffuse-type gastric carcinoma-initiating cells is inhibited by TGF-β. J Pathol. 228:391–404. doi:10.1002/path.4020
Khan ZA, Jonas SK, Le-Marer N, Patel H, Wharton RQ, Tarragona A, Ivison A, Allen-Mersh TG (2000) p53 mutations in primary and metastatic tumors and circulating tumor cells from colorectal carcinoma patients. Clin Cancer Res 6:3499–3504
Kiel MJ, Morrison SJ (2008) Uncertainty in the niches that maintain haematopoietic stem cells. Nat Rev Immunol 8:290–301. doi:10.1038/nri2279
Kleinsmith LJ, Pierce GB Jr (1964) Multipotentiality of single embryonal carcinoma cells. Cancer Res 24:1544–1551
Kondo T, Setoguchi T, Taga T (2004) Persistence of a small subpopulation of cancer stem-like cells in the C6 glioma cell line. Proc Natl Acad Sci USA 101:781–786. doi:10.1073/pnas.0307618100
Krause DS, Lazarides K, von Andrian UH, Van Etten RA (2006) Requirement for CD44 in homing and engraftment of BCR-ABL-expressing leukemic stem cells. Nat Med 12:1175–1180. doi:10.1038/nm1489
Lapidot T, Sirard C, Vormoor J, Murdoch B, Hoang T, Caceres-Cortes J, Minden M, Paterson B, Caligiuri MA, Dick JE (1994) A cell initiating human acute myeloid leukaemia after transplantation into SCID mice. Nature 367:645–648. doi:10.1038/367645a0
Li H, Chen X, Calhoun-Davis T, Claypool K, Tang DG (2008) PC3 human prostate carcinoma cell holoclones contain self-renewing tumor-initiating cells. Cancer Res 68:1820–1825. doi:10.1158/0008-5472.CAN-07-5878
Lobo NA, Shimono Y, Qian D, Clarke MF (2007) The biology of cancer stem cells. Annu Rev Cell Dev Biol 23:675–699. doi:10.1146/annurev.cellbio.22.010305.104154
Losi L, Benhattar J, Costa J (1992) Stability of K-ras mutations throughout the natural history of human colorectal cancer. Eur J Cancer 28A:1115–1120
Mani SA, Guo W, Liao MJ, Eaton EN, Ayyanan A, Zhou AY, Brooks M, Reinhard F, Zhang CC, Shipitsin M, Campbell LL, Polyak K, Brisken C, Yang J, Weinberg RA (2008)

The epithelial-mesenchymal transition generates cells with properties of stem cells. Cell 133:704–715. doi:10.1016/j.cell.2008.03.027
Martino G, Pluchino S (2006) The therapeutic potential of neural stem cells. Nat Rev Neurosci 7:395–406. doi:10.1038/nrn1908
Merlo LM, Pepper JW, Reid BJ, Maley CC (2006) Cancer as an evolutionary and ecological process. Nat Rev Cancer 6:924–935. doi:10.1038/nrc2013
Michor F, Hughes TP, Iwasa Y, Branford S, Shah NP, Sawyers CL, Nowak MA (2005) Dynamics of chronic myeloid leukaemia. Nature 435:1267–1270. doi:10.1038/nature03669
Miyazawa K, Shinozaki M, Hara T, Furuya T, Miyazono K (2002) Two major Smad pathways in TGF-β superfamily signalling. Genes Cells 7:1191–1204
Naka K, Hoshii T, Muraguchi T, Tadokoro Y, Ooshio T, Kondo Y, Nakao S, Motoyama N, Hirao A (2010) TGF-β-FOXO signalling maintains leukaemia-initiating cells in chronic myeloid leukaemia. Nature 463:676–680. doi:10.1038/nature08734
Nakamura T, Largaespada DA, Lee MP, Johnson LA, Ohyashiki K, Toyama K, Chen SJ, Willman CL, Chen IM, Feinberg AP, Jenkins NA, Copeland NG, Shaughnessy JD Jr (1996) Fusion of the nucleoporin gene NUP98 to HOXA9 by the chromosome translocation t(7;11)(p15;p15) in human myeloid leukaemia. Nat Genet 12:154–158. doi:10.1038/ng0296-154
Neering SJ, Bushnell T, Sozer S, Ashton J, Rossi RM, Wang PY, Bell DR, Heinrich D, Bottaro A, Jordan CT (2007) Leukemia stem cells in a genetically defined murine model of blast-crisis CML. Blood 110:2578–2585. doi:10.1182/blood-2007-02-073031
Nguyen LV, Vanner R, Dirks P, Eaves CJ (2012) Cancer stem cells: an evolving concept. Nat Rev Cancer 12:133–143. doi:10.1038/nrc3184
Nishii T, Yashiro M, Shinto O, Sawada T, Ohira M, Hirakawa K (2009) Cancer stem cell-like SP cells have a high adhesion ability to the peritoneum in gastric carcinoma. Cancer Sci 100:1397–1402. doi:10.1111/j.1349-7006.2009.01211.x
Norton JD (2000) ID helix-loop-helix proteins in cell growth, differentiation and tumorigenesis. J Cell Sci 113:3897–3905
Notta F, Mullighan CG, Wang JC, Poeppl A, Doulatov S, Phillips LA, Ma J, Minden MD, Downing JR, Dick JE (2011) Evolution of human BCR-ABL1 lymphoblastic leukaemia-initiating cells. Nature 469:362–367. doi:10.1038/nature09733
Nowell PC (1976) The clonal evolution of tumor cell populations. Science 194:23–28
O'Brien CA, Pollett A, Gallinger S, Dick JE (2007) A human colon cancer cell capable of initiating tumour growth in immunodeficient mice. Nature 445:106–110. doi:10.1038/nature05372
Oshimori N, Fuchs E (2012) Paracrine TGF-β signaling counterbalances BMP-mediated repression in hair follicle stem cell activation. Cell Stem Cell 10:63–75. doi:10.1016/j.stem.2011.11.005
Park CH, Bergsagel DE, McCulloch EA (1971) Mouse myeloma tumor stem cells: a primary cell culture assay. J Natl Cancer Inst 46:411–422
Patrawala L, Calhoun T, Schneider-Broussard R, Li H, Bhatia B, Tang S, Reilly JG, Chandra D, Zhou J, Claypool K, Coghlan L, Tang DG (2006) Highly purified CD44+ prostate cancer cells from xenograft human tumors are enriched in tumorigenic and metastatic progenitor cells. Oncogene 25:1696–1708. doi:10.1038/sj.onc.1209327
Patrawala L, Calhoun-Davis T, Schneider-Broussard R, Tang DG (2007) Hierarchical organization of prostate cancer cells in xenograft tumors: the CD44+α2β1+ cell population is enriched in tumor-initiating cells. Cancer Res 67:6796–6805. doi:10.1158/0008-5472.CAN-07-0490
Pearce DJ, Ridler CM, Simpson C, Bonnet D (2004) Multiparameter analysis of murine bone marrow side population cells. Blood 103:2541–2546. doi:10.1182/blood-2003-09-3281
Peñuelas S, Anido J, Prieto-Sánchez RM, Folch G, Barba I, Cuartas I, García-Dorado D, Poca MA, Sahuquillo J, Baselga J, Seoane J (2009) TGF-β increases glioma-initiating cell self-renewal through the induction of LIF in human glioblastoma. Cancer Cell 15:315–327. doi:10.1016/j.ccr.2009.02.011
Pierce GB, Speers WC (1988) Tumors as caricatures of the process of tissue renewal: prospects for therapy by directing differentiation. Cancer Res 48:1996–2004
Pierce GB, Wallace C (1971) Differentiation of malignant to benign cells. Cancer Res 31: 127–134

Quéré R, Karlsson G, Hertwig F, Rissler M, Lindqvist B, Fioretos T, Vandenberghe P, Slovak ML, Cammenga J, Karlsson S (2011) Smad4 binds Hoxa9 in the cytoplasm and protects primitive hematopoietic cells against nuclear activation by Hoxa9 and leukemia transformation. Blood 117:5918–5930. doi:10.1182/blood-2010-08-301879

Quintana E, Shackleton M, Sabel MS, Fullen DR, Johnson TM, Morrison SJ (2008) Efficient tumour formation by single human melanoma cells. Nature 456:593–598. doi:10.1038/nature07567

Quintana E, Shackleton M, Foster HR, Fullen DR, Sabel MS, Johnson TM, Morrison SJ (2010) Phenotypic heterogeneity among tumorigenic melanoma cells from patients that is reversible and not hierarchically organized. Cancer Cell 18:510–523. doi:10.1016/j.ccr.2010.10.012

Ren R (2005) Mechanisms of BCR-ABL in the pathogenesis of chronic myelogenous leukaemia. Nat Rev Cancer 5:172–183. doi:10.1038/nrc1567

Reya T, Morrison SJ, Clarke MF, Weissman IL (2001) Stem cells, cancer, and cancer stem cells. Nature 414:105–111. doi:10.1038/35102167

Ricci-Vitiani L, Lombardi DG, Pilozzi E, Biffoni M, Todaro M, Peschle C, De Maria R (2007) Identification and expansion of human colon-cancer-initiating cells. Nature 445:111–115. doi:10.1038/nature05384

Ricci-Vitiani L, Pallini R, Biffoni M, Todaro M, Invernici G, Cenci T, Maira G, Parati EA, Stassi G, Larocca LM, De Maria R (2010) Tumour vascularization via endothelial differentiation of glioblastoma stem-like cells. Nature 468:824–828. doi:10.1038/nature09557

Roeder I, Horn M, Glauche I, Hochhaus A, Mueller MC, Loeffler M (2006) Dynamic modeling of imatinib-treated chronic myeloid leukemia: functional insights and clinical implications. Nat Med 12:1181–1184. doi:10.1038/nm1487

Roesch A, Fukunaga-Kalabis M, Schmidt EC, Zabierowski SE, Brafford PA, Vultur A, Basu D, Gimotty P, Vogt T, Herlyn M (2010) A temporarily distinct subpopulation of slow-cycling melanoma cells is required for continuous tumor growth. Cell 141:583–594. doi:10.1016/j.cell.2010.04.020

Schatton T, Murphy GF, Frank NY, Yamaura K, Waaga-Gasser AM, Gasser M, Zhan Q, Jordan S, Duncan LM, Weishaupt C, Fuhlbrigge RC, Kupper TS, Sayegh MH, Frank MH (2008) Identification of cells initiating human melanomas. Nature 451:345–349. doi:10.1038/nature06489

Scheel C, Weinberg RA (2011) Phenotypic plasticity and epithelial-mesenchymal transitions in cancer and normal stem cells? Int J Cancer 129:2310–2314. doi:10.1002/ijc.26311

Scheel C, Eaton EN, Li SH, Chaffer CL, Reinhardt F, Kah KJ, Bell G, Guo W, Rubin J, Richardson AL, Weinberg RA (2011) Paracrine and autocrine signals induce and maintain mesenchymal and stem cell states in the breast. Cell 145:926–940. doi:10.1016/j.cell.2011.04.029

Schlingensiepen KH, Schlingensiepen R, Steinbrecher A, Hau P, Bogdahn U, Fischer-Blass B, Jachimczak P (2006) Targeted tumor therapy with the TGF-β2 antisense compound AP 12009. Cytokine Growth Factor Rev 17:129–139. doi:10.1016/j.cytogfr.2005.09.002

Sharma SV, Lee DY, Li B, Quinlan MP, Takahashi F, Maheswaran S, McDermott U, Azizian N, Zou L, Fischbach MA, Wong KK, Brandstetter K, Wittner B, Ramaswamy S, Classon M, Settleman J (2010) A chromatin-mediated reversible drug-tolerant state in cancer cell subpopulations. Cell 141:69–80. doi:10.1016/j.cell.2010.02.027

Shmelkov SV, Butler JM, Hooper AT, Hormigo A, Kushner J, Milde T, St Clair R, Baljevic M, White I, Jin DK, Chadburn A, Murphy AJ, Valenzuela DM, Gale NW, Thurston G, Yancopoulos GD, D'Angelica M, Kemeny N, Lyden D, Rafii S (2008) CD133 expression is not restricted to stem cells, and both CD133+ and CD133- metastatic colon cancer cells initiate tumors. J Clin Invest 118:2111–2120. doi:10.1172/JCI34401

Singh SK, Clarke ID, Terasaki M, Bonn VE, Hawkins C, Squire J, Dirks PB (2003) Identification of a cancer stem cell in human brain tumors. Cancer Res 63:5821–5828

Singh SK, Hawkins C, Clarke ID, Squire JA, Bayani J, Hide T, Henkelman RM, Cusimano MD, Dirks PB (2004) Identification of human brain tumour initiating cells. Nature 432:396–401. doi:10.1038/nature03128

Stewart JM, Shaw PA, Gedye C, Bernardini MQ, Neel BG, Ailles LE (2011) Phenotypic heterogeneity and instability of human ovarian tumor-initiating cells. Proc Natl Acad Sci USA 108:6468–6473. doi:10.1073/pnas.1005529108

Takaishi S, Okumura T, Tu S, Wang SS, Shibata W, Vigneshwaran R, Gordon SA, Shimada Y, Wang TC (2009) Identification of gastric cancer stem cells using the cell surface marker CD44. Stem Cells 27:1006–1020. doi:10.1002/stem.30

Tang B, Yoo N, Vu M, Mamura M, Nam JS, Ooshima A, Du Z, Desprez PY, Anver MR, Michalowska AM, Shih J, Parks WT, Wakefield LM (2007) Transforming growth factor-β can suppress tumorigenesis through effects on the putative cancer stem or early progenitor cell and committed progeny in a breast cancer xenograft model. Cancer Res 67:8643–8652. doi:10.1158/0008-5472.CAN-07-0982

Vescovi AL, Galli R, Reynolds BA (2006) Brain tumour stem cells. Nat Rev Cancer 6:425–436. doi:10.1038/nrc1889

Visvader JE, Lindeman GJ (2008) Cancer stem cells in solid tumours: accumulating evidence and unresolved questions. Nat Rev Cancer 8:755–768. doi:10.1038/nrc2499

Wakefield LM, Roberts AB (2002) TGF-β signaling: positive and negative effects on tumorigenesis. Curr Opin Genet Dev 12:22–29

Wang LD, Wagers AJ (2011) Dynamic niches in the origination and differentiation of haematopoietic stem cells. Nat Rev Mol Cell Biol 12:643–655. doi:10.1038/nrm3184

Wang J, Guo LP, Chen LZ, Zeng YX, Lu SH (2007) Identification of cancer stem cell-like side population cells in human nasopharyngeal carcinoma cell line. Cancer Res 67:3716–3724. doi:10.1158/0008-5472.CAN-06-4343

Wang R, Chadalavada K, Wilshire J, Kowalik U, Hovinga KE, Geber A, Fligelman B, Leversha M, Brennan C, Tabar V (2010) Glioblastoma stem-like cells give rise to tumour endothelium. Nature 468:829–833. doi:10.1038/nature09624

Watabe T, Miyazono K (2009) Roles of TGF-β family signaling in stem cell renewal and differentiation. Cell Res 19:103–115. doi:10.1038/cr.2008.323

Weigelt B, Glas AM, Wessels LF, Witteveen AT, Peterse JL, van't Veer LJ (2003) Gene expression profiles of primary breast tumors maintained in distant metastases. Proc Natl Acad Sci USA 100:15901–15905. doi:10.1073/pnas.2634067100

Weigelt B, Peterse JL, van't Veer LJ (2005) Breast cancer metastasis: markers and models. Nat Rev Cancer 5:591–602. doi:10.1038/nrc1670

Wilson A, Trumpp A (2006) Bone-marrow haematopoietic-stem-cell niches. Nat Rev Immunol 6:93–106. doi:10.1038/nri1779

Yamazaki S, Iwama A, Takayanagi S, Eto K, Ema H, Nakauchi H (2009) TGF-β as a candidate bone marrow niche signal to induce hematopoietic stem cell hibernation. Blood 113: 1250–1256. doi:10.1182/blood-2008-04-146480

Yingling JM, Blanchard KL, Sawyer JS (2004) Development of TGF-β signalling inhibitors for cancer therapy. Nat Rev Drug Discov 3:1011–1022. doi:10.1038/nrd1580

Zauber P, Sabbath-Solitare M, Marotta SP, Bishop DT (2003) Molecular changes in the Ki-ras and APC genes in primary colorectal carcinoma and synchronous metastases compared with the findings in accompanying adenomas. Mol Pathol 56:137–140

Zhang S, Balch C, Chan MW, Lai HC, Matei D, Schilder JM, Yan PS, Huang TH, Nephew KP (2008) Identification and characterization of ovarian cancer-initiating cells from primary human tumors. Cancer Res 68:4311–4320. doi:10.1158/0008-5472.CAN-08-0364

Zhao C, Chen A, Jamieson CH, Fereshteh M, Abrahamsson A, Blum J, Kwon HY, Kim J, Chute JP, Rizzieri D, Munchhof M, VanArsdale T, Beachy PA, Reya T (2009) Hedgehog signalling is essential for maintenance of cancer stem cells in myeloid leukaemia. Nature 458:776–779. doi:10.1038/nature07737

Zhou S, Schuetz JD, Bunting KD, Colapietro AM, Sampath J, Morris JJ, Lagutina I, Grosveld GC, Osawa M, Nakauchi H, Sorrentino BP (2001) The ABC transporter Bcrp1/ABCG2 is expressed in a wide variety of stem cells and is a molecular determinant of the side-population phenotype. Nat Med 7:1028–1034. doi:10.1038/nm0901-1028

Chapter 5
TGF-β as Tumor Suppressor: In Vitro Mechanistic Aspects of Growth Inhibition

Laurent Bartholin, David F. Vincent, and Ulrich Valcourt

Abstract Over time, TGF-β has emerged as both a tumor promoter and a tumor suppressor, depending on the cell type. Although it was first characterized as having the capacity to transform mesenchymal cells, many studies have shown it to exert an anti-proliferative effect on epithelial, endothelial, and immune cells. Most of the functional studies performed during the two decades following its discovery focused on its tumor-suppressive role and addressed a detailed understanding of the signaling events and molecular mechanisms by which TGF-β orchestrated cell growth arrest and apoptosis. This chapter entitled "TGF-β as tumor suppressor: in vitro mechanistic aspects of growth inhibition" depicts how TGF-β protects cells from transformation at the transcriptional, translational, and posttranslational levels. After a detailed view of the cytostatic program elicited by TGF-β to control the cell cycle, the focus shifts to TGF-β signaling pathways and biochemical events by which this cytokine governs the apoptotic program.

Keywords Apoptosis • Cell cycle regulation • Cytostasis • Posttranslational events • Smad signaling • Smad-independent pathways • Transcriptional regulation

Abbreviations

ATF/CREB	Activation transcription factor/cAMP responsive element-binding
bHLH	Basic helix-loop-helix
CDKs	Cyclin-dependent kinases

These authors have equally contributed to this work

L. Bartholin • D.F. Vincent • U. Valcourt (✉)
TGF-β and Pancreatic Cancer-Lab, Centre de Recherche en Cancérologie de LYON (CRCL), Centre Léon Bérard, UMR INSERM 1052-CNRS 5286-Université Lyon 1052S, 28, rue Laënnec, 69373 LYON Cedex 08, France
e-mail: ulrich.valcourt@lyon.unicancer.fr

A. Moustakas and K. Miyazawa (eds.), *TGF-β in Human Disease*,
DOI 10.1007/978-4-431-54409-8_5, © Springer Japan 2013

ChIP	Chromatin immunoprecipitation
CKIs	Cyclin-dependent kinase inhibitors
CTGF	Connective tissue growth factor
eEF1A1	Eukaryotic elongation factor 1A1
eIF4E	Eukaryotic translation initiation factor-4E
EMT	Epithelial-to-mesenchymal transition
ERK	Extracellular signal-regulated kinase
FHBE	Forkhead-binding element
HDAC1	Histone deacetylase 1
HIPK2	Homeodomain-interacting protein kinase 2
IAP	Inhibitor of apoptosis
ID	Inhibitor of DNA binding/inhibitor of differentiation
INK4	Inhibitor of CDK4
JNK	c-Jun N-terminal kinase
MAP	Mitogen-activated protein
MAPK(K)	Mitogen-activated protein kinase (kinase)
MOMP	Mitochondrial outer membrane permeabilization
NGF	Nerve growth factor
PDGF	Platelet-derived growth factor
RB	Retinoblastoma protein
SBE	Smad binding element
TAK1	TGF-β-activated kinase 1
TGF-β	Transforming growth factor β
TIE	TGF-β inhibitory element
TIEG1	TGF-β-induced early gene 1
TNF	Tumor necrosis factor
TSS	Transcriptional start site
TβRI	TGF-β type I receptor
TβRII	TGF-β type II receptor
4E-BP1	eIF4E-binding protein 1

5.1 Introduction

Transforming growth factor-β (TGF-β) was first discovered as a tumor-promoting factor by supporting anchorage-independent growth of untransformed rat fibroblasts (Roberts et al. 1980; Roberts et al. 1981) and was further demonstrated to stimulate proliferation of different cell types in vitro such as mesenchymal-derived cells, like osteoblasts (Battegay et al. 1990), or glial cells, such as Schwann cells (Guenard et al. 1995; Ridley et al. 1989). However, the proliferative response induced by TGF-β in these cells appeared later on to be mostly mediated indirectly through autocrine secretion of other soluble factors, such as platelet-derived growth factor (PDGF) (Battegay et al. 1990; Leof et al. 1986; Soma and Grotendorst 1989) or connective tissue growth factor (CTGF) (Igarashi et al. 1993), and TGF-β has then been proved to act mainly as a growth-inhibitory factor rather than a growth-promoting cytokine.

The TGF-β-mediated inhibition of cell growth was initially observed in vitro in a study focusing on the purification of growth inhibitors from the culture medium of BSC-1 epithelial kidney cells from African green monkey (Holley et al. 1980). Indeed, BSC-1-cell-conditioned medium was found to inhibit growth of Hs578T human mammary tumor cells (Holley et al. 1983), and TGF-β was identified as the mediator of this anti-proliferative effect (Tucker et al. 1984). It has been clearly established in the same period that TGF-β caused growth inhibitory responses in different cell types including epithelial (Masui et al. 1986; Tucker et al. 1984), endothelial (Baird and Durkin 1986; Frater-Schroder et al. 1986; Heimark et al. 1986), hematopoietic (Kehrl et al. 1986a; Kehrl et al. 1986b), neural cells as well as certain mesenchymal cells (reviewed in (Massague 1990)) and primary embryonic fibroblasts (Anzano et al. 1986). One of the first compelling evidence showing that TGF-β was able to inhibit cell growth in vivo was produced in 1987, when Silberstein et al. reported that implantation of ethylene vinyl acetate copolymer pellets allowing the slow release of TGF-β into small zones of the mammary gland could induce a potent inhibition of ductal growth (Silberstein and Daniel 1987).

In addition to its growth-inhibitory properties, the tumor-suppressive action of TGF-β was also shown to be mediated by its ability to induce apoptosis in different cell types both in vivo (Nguyen and Pollard 2000) and in vitro. Interestingly, TGF-β-mediated apoptosis concerns not only normal (Oberhammer et al. 1992) but also tumor cells (Lafon et al. 1996; Yanagihara and Tsumuraya 1992) and occurs in vitro concomitantly with the TGF-β-induced growth-inhibitory response in a cell culture (Kim et al. 1998; Rotello et al. 1991).

After the initial observation and characterization of the tumor-suppressive functions of TGF-β in the 1980s, the two last decades have been more dedicated to the detailed understanding of the signaling events and molecular mechanisms by which TGF-β orchestrated cell growth arrest and apoptosis. Hence, this chapter entitled "TGF-β as tumor suppressor: in vitro mechanistic aspects of growth inhibition" depicts how TGF-β protects cells from transformation at the transcriptional, translational, and posttranslational levels. After a detailed view of the cytostatic program elicited by TGF-β to control the cell cycle, the focus shifts to TGF-β signaling pathways and biochemical events governing the apoptotic program.

For a comprehensive review of mouse models with impaired TGF-β signaling and what they have taught us about the tumor-suppressive role of TGF-β, the reader is invited to read Chapter 6.

5.2 Control of the Cell Cycle: The Cytostatic Program

The growth-inhibitory response elicited by TGF-β has been extensively explored in epithelial cells. It involves many key cell-cycle regulators directly or indirectly regulated at the transcriptional level via the Smad signaling pathway or by

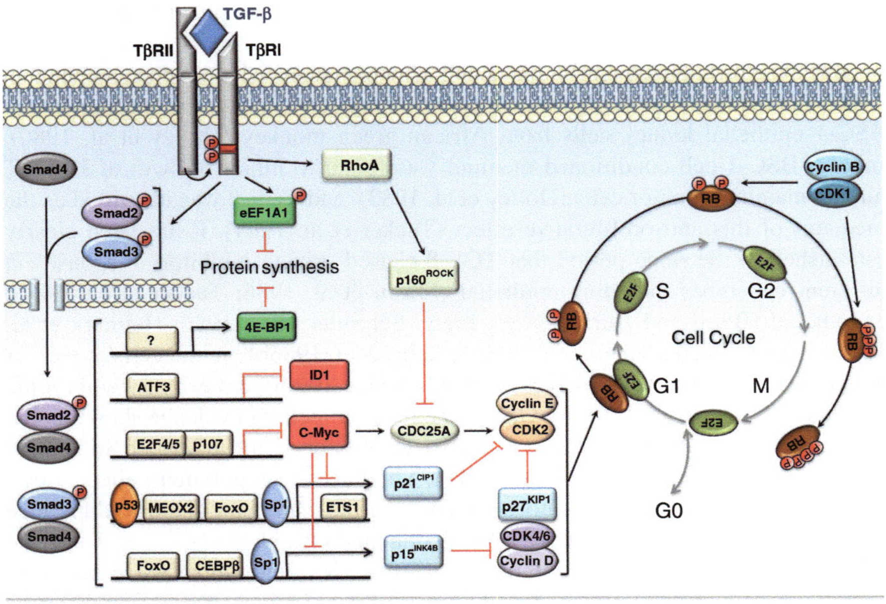

Fig. 5.1 The cytostatic program orchestrated by TGF-β. *Note*: The figure was illustrated using Servier Medical Art

posttranslational modifications mediated by the TGF-β receptor complex. The different targets of the TGF-β/Smad pathway thus cooperate to slow down the cell cycle and to eventually arrest cell growth, as is detailed in the following sections.

5.2.1 Control of the G1/S-Phase Transition by TGF-β

This growth inhibition is achieved by maintaining the retinoblastoma (RB) protein in a hypophosphorylated state (Laiho et al. 1990). Accordingly, ectopic expression of DNA virus oncoproteins sequestering the hypophosphorylated form of RB, such as the Adenovirus 5 protein ElA or the SV40 large T antigen, abolishes TGF-β-mediated growth inhibition (Pietenpol et al. 1990b). The requirement of a functional RB protein for TGF-β-mediated growth suppression has been confirmed in RB-deficient cells, in which TGF-β-mediated growth inhibition is abolished (Herrera et al. 1996). TGF-β inhibits cell growth by arresting the cell cycle at the early G1 phase (Holley et al. 1980; Laiho et al. 1990). Much effort has been devoted to understanding how TGF-β maintains the RB protein in a hypophosphorylated state and to identifying, cloning, and characterizing regulators of the cell cycle machinery (Fig. 5.1).

RB phosphorylation is catalyzed by cyclin-dependent kinases (CDKs), a family of serine/threonine protein kinases that are activated at specific points in the cell cycle. The enzymatic activity of CDKs requires the presence of cell-cycle-dependent regulators, the cyclins. Cell cycle progression from the G1 phase to the S phase is achieved through the successive action of the cyclin D-CDK4 and/or cyclin D-CDK6 and cyclin E/CDK2 complexes. Changes in cyclin levels, selective dephosphorylation of specific residues in CDKs, and interactions with stoichiometric inhibitors (the cyclin-dependent kinase inhibitors, CKIs) all contribute to tightly regulating the activity of these complexes (Sherr and Roberts 1999; Vermeulen et al. 2003). CKIs are classified into two families according to their mode of action: the CIP/KIP family and the INK4 (inhibitor of CDK4) family. The CIP/KIP family includes $p21^{CIP1}$, $p27^{KIP1}$, and $p57^{KIP2}$, whereas the INK4 family comprises $p16^{INK4A}$, $p15^{INK4B}$, $p18^{INK4C}$, and $p19^{INK4D}$. The INK4 proteins prevent the interaction of cyclin D with CDK4/6. The CIP/KIP proteins can form ternary complexes with G1-phase cyclin-CDK complexes, but their role depends on the target: on the one hand, CIP/KIP proteins interact with cyclin E-CDK2 and inhibit its enzymatic activity by physically occluding the catalytic cleft of CDK2; on the other hand, they interact with and stabilize the cyclin D-CDK4/6 complex without necessarily inhibiting cyclin D-associated kinase activity (Sherr and Roberts 1999).

In epithelial cells, TGF-β has been shown to decrease the enzymatic activities of both cyclin D-CDK4/6 (Hannon and Beach 1994) and cyclin E-CDK2 (Koff et al. 1993). In Mv1Lu lung epithelial cells, it has also been shown to induce growth arrest by inhibiting CDK4 synthesis (Ewen et al. 1993). Extensive in vitro studies have been conducted, mostly on epithelial cells, in order to decipher the molecular mechanisms involved in the regulation of CDK activities and/or synthesis. Findings have led to identifying a limited number of factors involved in the growth-suppressive program governed by TGF-β, called the cytostatic program, which is shared by many lineages of epithelial origin in skin, lung, or mammary gland. The cytostatic program governed by TGF-β is orchestrated mainly at the transcriptional level, through the Smad transcription factors, which convey cues from the membrane to the nucleus. TGF-β-mediated transcriptional effects resulting in growth inhibition include (Fig. 5.1):

1. Transcriptional repression of the growth-promoting c-MYC transcription factor and of the differentiation-inhibiting ID nuclear factors (ID1 to −3).
2. Transcriptional activation of the genes encoding the cyclin-dependent kinase inhibitors $p15^{INK4B}$, $p21^{CIP1}$, and $p57^{KIP2}$.

TGF-β also exerts its cytostatic action via posttranslational events, such as protein modifications (phosphorylations) and changes in protein complex stoichiometry (stimulation of the inhibitory action of the $p27^{KIP1}$ CKI). The following paragraph presents the studies having led to major breakthroughs in this field.

5.2.2 *Smad-Mediated Transcriptional Repression of the* c-MYC *and* ID *Genes*

5.2.2.1 Regulation of the *c-MYC* Gene by the Smad Pathway

c-MYC was the first direct transcriptional target gene of TGF-β identified as being involved in the cell growth arrest induced by this cytokine in epithelial cells (Pietenpol et al. 1990a). In proliferating cells, *c-MYC* is transcriptionally active and its proto-oncoprotein product promotes the expression of genes essential to progression from the G1 to the S phase, such as those encoding cyclin D2 (Bouchard et al. 1999) or CDK4 (Hermeking et al. 2000). An important breakthrough in efforts to understand the cytostatic effect of TGF-β on epithelial cells was the finding that TGF-β can rapidly down-regulate *c-MYC* expression (Pietenpol et al. 1990a). Inversely, forced expression of *c-MYC* in epithelial cells impairs the cytostatic effect mediated by TGF-β (Alexandrow et al. 1995).

Repression of *c-MYC* transcription after TGF-β treatment is direct and involves the Smad signaling pathway. The activated Smad3/4 complex is recruited to a TGF-β inhibitory element (TIE) located between the two transcription start sites of the human *c-MYC* promoter, repressing its transcriptional activity (Chen et al. 2001). This event also involves the input of a repressive complex containing the E2F4 or E2F5 transcription factors and an additional co-repressor, the p107 pocket protein (Chen et al. 2002) (Fig. 5.1). Upon TGF-β stimulation and the subsequent phosphorylation of Smad3, the pre-formed Smad3-E2F4/5-p107 complex translocates into the nucleus, interacts with Smad4, and the Smad3/4-E2F4/5-p107 complex binds to the TIE of the *c-MYC* promoter (Chen et al. 2002; Frederick et al. 2004). The C/EBPβ transcription factor, which is a member of the basic leucine-zipper protein family, is also involved in TGF-β-mediated repression of *c-MYC* promoter activity, being recruited to the TIE sequence along with the Smad3/4-E2F4/5-p107 complex (Gomis et al. 2006b). TGF-β thus allows recruitment of a repressive transcriptional complex to the *c-MYC* promoter, causing cell cycle arrest through a rapid drop in the intracellular level of c-MYC protein.

5.2.2.2 Regulation of *ID* Genes by the TGF-β-Smad Pathway

Another set of target genes of the TGF-β/Smad signaling pathway has been implicated in the cytostatic response of epithelial cells: the ID (inhibitor of DNA binding) family of transcriptional regulators. ID proteins are inhibitors of cell differentiation belonging to the family of basic helix-loop-helix (bHLH) transcription factors (Perk et al. 2005). Classically, bHLH proteins bind to DNA as either homodimers or heterodimers and regulate gene transcription, either positively or negatively, depending on the dimer recruited to the DNA. As ID proteins lack a DNA-binding domain but have a dimerization domain, they act as dominant-negative proteins for other bHLH factors, preventing their binding to DNA and thus inhibiting their function (Perk et al. 2005).

The proteins ID1 to −3 are important positive regulators of cell growth (Fig. 5.1):

1. IDs indirectly promote inactivation of the RB protein by down-regulating CKI expression. They inhibit the transcriptional activity of bHLH proteins that positively regulate the CKI-encoding genes $p15^{INK4B}$, $p16^{INK4A}$, and $p21^{CIP1}$ by preventing their interaction with E-box DNA sequences contained in the promoters of CKI genes (Yokota and Mori 2002).
2. Alternatively, ID2 can block the cell-cycle-inhibiting action of retinoblastoma-family pocket proteins. This function correlates with the ability of ID2 to associate physically with active, hypophosphorylated forms of the pocket proteins (Lasorella et al. 1996; Lasorella et al. 2000). Like the E2F factors, ID2 can interact with the pocket domains of hypophosphorylated RB, p107, and p130. The formation of ID2-pocket protein complexes results in release of E2F from its repressive state, allowing progression of the cell cycle from G1 to S.

TGF-β causes a rapid decrease of ID1, −2, and −3 in various cell systems, such as mammary epithelial cells (NMuMG and MCF10A), normal epithelial lung cells (HPL-1), carcinoma-derived lung cells (A549), keratinocytes (HaCaT) (Kang et al. 2003; Kowanetz et al. 2004; Pardali et al. 2005; Siegel et al. 2003), and mouse embryonic fibroblasts (Lasorella et al. 2000). The mechanism by which TGF-β mediates this transcriptional repression is different for *ID1* and *ID2*.

The human *ID1* promoter contains three Smad binding elements (SBE) motifs downstream of an ATF/CREB binding site, suggesting transcriptional cooperation of Smad proteins with a member of the ATF/CREB (*activation transcription factor/cAMP responsive element-binding*) protein family (Kang et al. 2003). ATF-3, being rather a repressor than an activator of gene transcription, appears to be involved in TGF-β-mediated *ID1* gene repression. Interestingly, the Smad pathway directly regulates expression of the *ATF-3* gene, whose product interacts physically with the activated Smad3/4 complex and binds to the *ID1* promoter (Kang et al. 2003) (Fig. 5.1).

TGF-β-mediated regulation of *ID2* gene expression involves a different mechanism (Siegel et al. 2003). The *ID2* promoter contains E-box DNA sequences constituting a consensus binding site for bHLH transcription factors such as c-MYC. In proliferating cells, c-MYC has been shown to induce *ID2* expression through its recruitment to these E-box motifs (Lasorella et al. 2000), along with its partner Max (Siegel et al. 2003). Transcriptional repression of c-MYC and the subsequent drop in the c-MYC protein level might explain TGF-β-mediated inhibition of *ID2* gene transcription. Yet the time course of *ID2* down-regulation does not necessarily correlate with *c-MYC* repression in certain cell types (e.g., mammary epithelial cells), suggesting that other factors are required to negatively control *ID2* transcription. MAD proteins (MAD1-4) act as antagonists of c-MYC by replacing it in c-MYC-MAX transcriptional complexes. This leads to competition between complexes for the same regulatory sequences. TGF-β induces expression of MAD2/4 proteins in epithelial cells, causing increased association of MAD-MAX heterodimers with the *ID2* promoter (Siegel et al. 2003). Together, the TGF-β-mediated down-regulation of c-MYC and increased recruitment of MAD-MAX complexes to the *ID2* promoter contribute to the sustained transcriptional repression of this gene.

To our knowledge, the mechanism by which TGF-β causes transcriptional repression of *ID3* has not been elucidated to date. Unexpectedly, ID3 synthesis is reported to be induced upon TGF-β stimulation in B lymphocyte progenitor cells to mediate growth inhibition (Kee et al. 2001). This suggests that regulation of ID-factor expression by TGF-β might be cell- or context-specific.

5.2.3 *Transcriptional Activation of* $p21^{CIP1}$*- and* $p15^{INK4B}$*-Encoding Genes by Smad Complexes*

Repression of *c-MYC* and *ID* genes is generally not sufficient to ensure TGF-β-mediated growth inhibition and requires the activation of cell-cycle inhibitors. In epithelial cells, TGF-β induces a rapid increase in $p15^{INK4B}$ (Hannon and Beach 1994; Sandhu et al. 1997) and $p21^{CIP1}$ (Datto et al. 1995) protein levels, whereas in hematopoietic cells, it causes growth arrest through $p57^{KIP2}$ induction (Scandura et al. 2004). TGF-β directly stimulates transcriptional activation of the genes encoding the $p15^{INK4B}$, $p21^{CIP1}$, and $p57^{KIP2}$ CKIs through the Smad signaling pathway. The regulation of $p15^{INK4B}$ and $p21^{CIP1}$ promoter activity by the TGF-β/Smad pathway has been extensively analyzed, whereas the mechanism of TGF-β-mediated transcriptional regulation of $p57^{KIP2}$ is not yet fully understood.

Interestingly, the $p15^{INK4B}$ and $p21^{CIP1}$ promoters are similarly regulated. Under growth-promoting conditions, they are repressed by a complex containing c-MYC. This repressive complex must be released to allow recruitment of transactivators interacting with Smad proteins in TGF-β-stimulated cells. These transcriptional regulatory mechanisms are described below.

Forced expression of c-MYC blocks the transcriptional activation of $p15^{INK4B}$ (Warner et al. 1999) as well as the induction of $p21^{CIP1}$ (Claassen and Hann 2000) in epithelial cells, suggesting that c-MYC may hinder transcriptional activation of the genes encoding these CKIs. This appears to be the case, since in proliferating cells, c-MYC is expressed to a high level and tethered to the $p21^{CIP1}$ and $p15^{INK4B}$ promoters via the zinc finger protein Miz-1, thus interfering with transcriptional activation by TGF-β (Seoane et al. 2001; Seoane et al. 2002; Staller et al. 2001). In the $p15^{INK4B}$ promoter, the c-MYC-Miz-1 complex binds directly to the transcriptional initiator element to block transcriptional activation (Staller et al. 2001). It is noteworthy that c-MYC-mediated repression of $p21^{CIP1}$ occurs in other contexts besides TGF-β signaling. Binding of the c-MYC-Miz-1 complex also impedes transcriptional activation of $p21^{CIP1}$ by other transactivators, such as p53 (in the context of the response to DNA damage) (Seoane et al. 2002).

In this context, c-MYC down-regulation mediated by TGF-β signaling (Fig. 5.1) is essential to rendering the $p21^{CIP1}$ and $p15^{INK4B}$ promoters competent for transcriptional activation. Yet relief from inhibition by c-MYC is not sufficient for activation of these promoters. Transcriptional activation of both promoters by TGF-β requires the input of other transactivators. A precise analysis of the $p21^{CIP1}$ and $p15^{INK4B}$

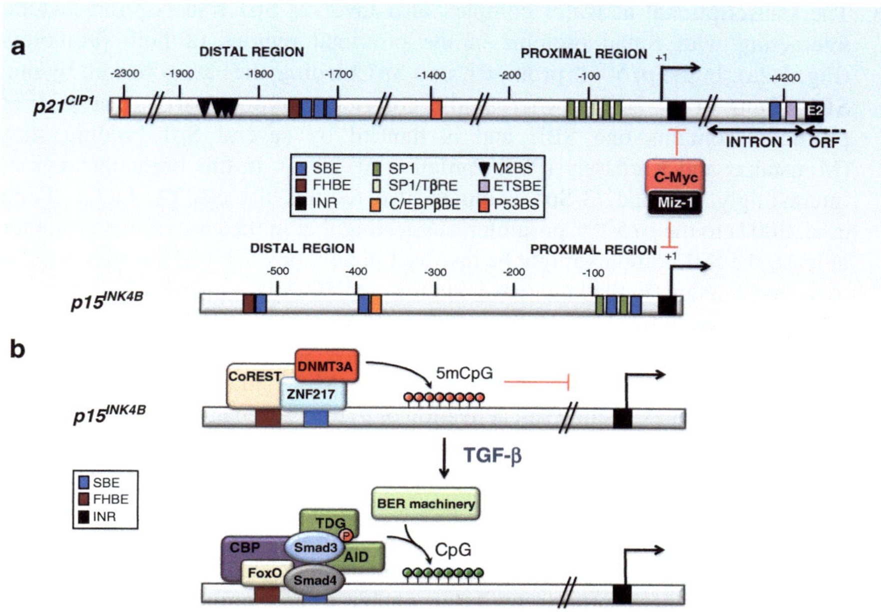

Fig. 5.2 (**a**) Regulatory elements of the *p15*^{INK4B} and *p21*^{CIP1} promoters involved in the TGF-β/Smad-mediated transcriptional control. (**b**) Active demethylation of the *p15*^{INK4B} promoter in response to the TGF-β/Smad signaling pathway. SBE, Smad-binding element; FHBE, Forkhead-binding element; Sp1, Sp1-binding site; TβRE, TGF-β-responsive element; INR, transcriptional initiator element; ETS, ETS-binding element; P53BS, p53-binding site; M2BS, MEOX2-binding site (putative); C/EBPβBE, C/EBPβ-binding element. *Note*: The figure was illustrated using Servier Medical Art

promoters of different species has contributed to identifying the transcriptional regulators that orchestrate transcriptional activation in response to TGF-β.

The mechanisms involved would appear to be the same for *p21*^{CIP1} and *p15*^{INK4B} and to be shared by many epithelial cells:

1. The *p21*^{CIP1} (Moustakas and Kardassis 1998) and *p15*^{INK4B} (Feng et al. 2000) genes contain not only a GC-rich rich region in the proximal promoter region but also SBEs in more distal enhancer elements (Gomis et al. 2006a; Seoane et al. 2004), where activated Smad3/Smad4 complexes can bind upon TGF-β stimulation (Fig. 5.2a).
2. The SBEs in the distal enhancer elements are located close to a consensus site for binding of forkhead-family proteins. The *p21*^{CIP1} promoter contains one forkhead-binding element (FHBE) adjacent to three closely spaced SBEs, whereas the *p15*^{INK4B} promoter contains an FHBE followed by one SBE (Fig. 5.2a). Upon TGF-β stimulation, activated Smad3/4 forms a complex with FoxO proteins (FoxO1, FoxO3, and FoxO4), recruiting them to the promoters and thus activating transcription (Gomis et al. 2006a; Seoane et al. 2004).

3. The transcriptional activator complex also involves Sp1 transcription factors, interacting with Smad proteins in the proximal regions of both promoters (Fig. 5.2a). In the *p15*^{INK4B} promoter, two Sp1 binding sites are followed by one SBE (Feng et al. 2000), whereas the GC-rich proximal part of the *p21*^{CIP1} promoter contains one SBE and is flanked by several Sp1 binding sites (Moustakas and Kardassis 1998; Pardali et al. 2000). In this particular region, interestingly, a Smad2/3-Sp1 complex also recruits the c-MYC factor (Feng et al. 2002) to the *p15*^{INK4B} promoter, suggesting that in the case of this promoter at least, the Sp1 complex might be involved in both positive and negative regulatory mechanisms (in the latter case, via the c-MYC-Miz-1 complex) (Feng et al. 2002). Accordingly, the transcriptional repression mediated by c-MYC is lost when the Sp1-binding sites are mutated. Conversely, when c-MYC-mediated repression is relieved upon TGF-β signaling, Miz-1 may cooperate with the Smad-Sp1 complex to elicit full activation of *p15*^{INK4B} (Seoane et al. 2001). Taken together, these findings suggest that Sp1-binding elements serve as docking sites for both repressor and activator complexes.
4. The *p15*^{INK4B} promoter (but not the *p21*^{CIP1} promoter) contains a third TGF-β-responsive region, in its distal part, where an SBE is immediately followed by a C/EBPβ binding site (Fig. 5.2a). Upon stimulation, a Smad3/4-C/EBPβ complex is recruited to this promoter region to enable transcriptional activation. As a regulator of both c-MYC repression and p15^{INK4} activation, C/EBPβ appears as a crucial mediator of the cytostatic program governed by TGF-β (Gomis et al. 2006b).
5. p53, p63, and p73—major transactivators of *p21*^{CIP1} expression—have also been implicated in TGF-β-mediated transcriptional activation of the *p21*^{CIP1} promoter. Indeed, down-regulation of p53 or p63 in epithelial cells abolishes *p21*^{CIP} gene induction by TGF-β (or activin). The p53 transactivator interacts with activated Smads at the distal region of the *p21*^{CIP} promoter to promote robust induction of *p21*^{CIP} expression (Cordenonsi et al. 2003) (Figs. 5.1 and 5.2).
6. The induction of *p21*^{CIP1} expression in response to TGF-β also requires the involvement of MEOX2 (mesenchyme homeobox2). *MEOX2* is a target gene of the TGF-β signaling pathway, whose product has been shown to interact physically with Smad3 and Smad4 and to bind to the distal region of the *p21*^{CIP1} promoter (Figs. 5.1 and 5.2) (Valcourt et al. 2007).
7. Chromatin immunoprecipitation (ChIP) experiment followed by microarray (ChIP on chip) analysis identified two strong Smad-binding regions in the first intron of the *p21*^{CIP1} gene downstream from the translation start site (Figs. 5.1 and 5.2). The binding of Smad2/3 proteins in those intronic regions is strengthened by the presence of the ETS1 factor, the v-ets erythroblastosis virus E26 oncogene homolog (Koinuma et al. 2009).
8. Both *p15*^{INK4B} and *p21*^{CIP1} promoters contain a CpG island within 1 kb of the transcriptional start site (TSS). In proliferating cells, core CpG islands are highly methylated contributing to the silencing of the two promoters. Upon stimulation, TGF-β rapidly triggers an active demethylation of both promoters (Thillainadesan et al. 2012). In the case of the *p15*^{INK4B} promoter, the CpG island is located

adjacently to SBE/FHBE sequences (Fig. 5.2b). The promoter methylation and subsequent gene silencing is mediated by the constitutive binding of a repressive complex consisting of the oncogene ZNF217, coREST (corepressor of REST), and the DNA methyltransferase DNMT3A. TGF-β induces the displacement of the CoREST-ZNF217-DNMT3A complex, and the subsequent recruitment of the activated Smad3/4 proteins, together with the CBP co-activator and enzymes regulating DNA methylation. Indeed, the Smad3/4 complex recruits the cytidine deaminase AID and the DNA glycolases TDG and MBD4, as well as components of the base excision repair (BER) machinery (Thillainadesan et al. 2012) (Fig. 5.2b). It is important to highlight that this observation is the first showing that epigenetics is a dynamic process that could be regulated by intracellular signaling molecules.

The cytostatic activity of TGF-β also results from additional cooperation between Smad and other signaling pathways. For instance, TGF-β induces Notch signaling to mediate $p21^{CIP1}$ expression (Niimi et al. 2007). Indeed, the TGF-β/Smad pathway induces the expression of the Notch transmembrane ligand Jagged1, which in turn triggers the Notch intracellular signaling pathway in adjacent cells. This process ultimately stimulates the expression of the $p21^{CIP1}$ gene, through the activity of the CLS transcription factor, an effector and a partner of the Notch intracellular signaling domain.

In hematopoietic cells, TGF-β induces robust expression of $p57^{KIP2}$ rather than $p21^{CIP1}$ and $p15^{INK4B}$ (Scandura et al. 2004). TGF-β-mediated up-regulation of $p57^{KIP2}$ expression is direct and requires the action of the Smad3/4 complex on a proximal region of the $p57^{KIP2}$ promoter. Surprisingly, this promoter contains no consensus SBE but includes putative binding sites for other transcription factors, such as YY1, RUNX1, and Sp1, known to interact physically with Smad proteins. Whether these factors cooperate with Smads in regulating $p57^{KIP2}$ expression in hematopoietic cells remains to be elucidated.

Thus, TGF-β/Smad signaling involves (1) several transcription factors, acting both at the proximal promoter and at distal enhancer elements, as well as (2) the DNA demethylation machinery, to mediate rapid and sustained expression of both $p21^{CIP1}$ and $p15^{INK4B}$. The functional consequence of $p15^{INK4B}$, $p21^{CIP1}$, and $p57^{KIP2}$ induction is the abolition of CKD4/6 and CDK2 kinase activities, causing an arrest of G1-to-S cell cycle progression by impairing RB phosphorylation.

5.2.4 *Stimulation by TGF-β of the p27 ^{KIP1} Inhibitory Function*

In growth-stimulated epithelial cells, $p27^{KIP1}$ interacts with active cyclin D-CDK4/6 complexes, and its expression level is not altered by TGF-β in the course of the cell cycle (Reynisdottir et al. 1995). TGF-β regulates $p27^{KIP1}$ activity by increasing its bioavailability for binding to the cyclin E-CDK2 complex. By increasing the pool of $p15^{INK4B}$ protein, TGF-β promotes the formation of $p15^{INK4B}$-CDK4/6 complexes at the expense of ternary p27^{KIP1}-cyclin D-CDK4/6 complexes, and thus causes p27^{KIP1} release. This leads to the recruitment of p27^{KIP1} to available cyclin E-CDK2 complexes, which become inhibited (Reynisdottir et al. 1995). Together with the concomitant induction of p21^{CIP1}, the displacement of p27^{KIP1} protein from active cyclin D-CDK4/6 to cyclin E-CDK2 complexes ensures robust inhibition of CDK2 catalytic activity, prevents phosphorylation of RB, and causes cell cycle arrest at the G1-to-S transition (Fig. 5.1).

5.2.5 TGF-β-Mediated Down-Regulation of CDC25A Prevents CDK2 Activation

TGF-β also regulates the activity of the G1-phase CDKs through posttranslational modifications. The activity of CDKs is negatively regulated by their phosphorylation on specific threonine (Thr14) or tyrosine (Tyr15) residues (Donzelli and Draetta 2003). In the G1 phase, the CDK dual-specificity phosphatase CDC25A dephosphorylates CDK2, allowing subsequent activation of cyclin E-CDK2 complexes (Donzelli and Draetta 2003). This activation step is crucial to subsequent cell cycle progression to the S phase. TGF-β has been shown to decrease the expression of CDC25A in epithelial cells, both in vitro, in cultured mammary epithelial cells (Iavarone and Massague 1997), and in vivo, in mice characterized by specific over-expression of TGF-β1 in hepatocytes (Bouzahzah et al. 2000). More precisely, TGF-β causes direct transcriptional repression of the *CDC25A* gene through recruitment of an E2F4-p130 repressive complex, together with the histone deacetylase 1 (HDAC1), to a region located immediately upstream of the transcription start site of the *CDC25A* promoter (Bouzahzah et al. 2000; Iavarone and Massague 1999). Furthermore, the proto-oncogene c-MYC has been shown to cooperate with its partner Max to directly activate transcription of *CDC25A* (Galaktionov et al. 1996). In this context, TGF-β might also regulate indirectly the transcriptional repression of *CDC25A*, by facilitating Smad-mediated transcriptional repression of the c-*MYC* gene. Alternatively, another TGF-β-dependent mechanism might negatively regulate CDC25A activity. TGF-β activates the RhoA-p160^{ROCK} axis, causing an inhibitory phosphorylation of CDC25A (Bhowmick et al. 2003). In conclusion, CDC25A is targeted by TGF-β via both transcriptional and posttranslational events so as to inhibit CDK2 activation (Fig. 5.1).

5.2.6 TGF-β-Mediated Inhibition of the Translational Machinery is Involved in Cell-Growth Arrest

Protein synthesis is a tightly regulated process involving more than a hundred of different proteins. Regulators of the translational machinery have emerged as new players in cell-autonomous tumor-suppressive function of TGF-β. Two studies have linked actors of the translation process to the cytostatic program governed by TGF-β. As detailed below, the first mechanism involves transcriptional activation of the translation-inhibiting protein 4E-BP1 (eIF4E-binding protein 1), mediated by the Smad signaling pathway (Azar et al. 2009); the second relies on catalytic inactivation of the translation initiation factor eEF1A1 (eukaryotic elongation factor 1A1) by the TGF-β type I receptor (TβRI) (Lin et al. 2010) (Fig. 5.1).

The translation-inhibiting protein 4E-BP1 blocks the function of the cap-binding eukaryotic translation initiation factor-4E (eIF4E), which docks the ribosome at the 5′-end of mRNA in order to initiate protein synthesis. 4E-BP1 is required for TGF-β-mediated growth arrest in pancreatic carcinoma cells, mouse embryonic fibroblasts, and keratinocytes (Azar et al. 2009). The gene encoding 4E-BP1 appears as a direct target of TGF-β signaling, its transcription being activated by TGF-β. Smad4 has been shown to bind to a conserved SBE located in the first exon of this gene. Artificial down-regulation of *4E-BP1* expression abolishes TGF-β-induced cell cycle arrest in pancreatic epithelial cells and embryonic fibroblasts (Azar et al. 2009). Thus, by increasing the level of 4E-BP1, TGF-β contributes to inhibiting the translational machinery and to causing cell growth arrest.

From a phage library screen followed by phosphorylation assays performed with a soluble form of recombinant TGF-β type I receptor, the eukaryotic elongation factor 1A1 (eEF1A1) has emerged as a novel direct substrate of TβRI kinase activity (Lin et al. 2010). TβRI phosphorylates eEF1A1 at serine 300, an amino acid crucial to efficient binding of amino-acyl-tRNA to the elongation factor. In vitro, TGF-β-mediated phosphorylation of serine 300 causes efficient inhibition of protein synthesis and subsequent growth arrest in the breast cancer cell line MCF-7 (Lin et al. 2010). This is the first evidence of a direct link between TGF-β receptors and the regulation of protein synthesis in the control of cell proliferation.

TGF-β has recently been shown to regulate epithelial-to-mesenchymal transition (EMT) by stimulating the translation elongation of selected transcripts (Hussey et al. 2011). More precisely, the translation of several genes essential for EMT is inhibited by the heterogeneous nuclear ribonucleoprotein E1 (hnRNP E1), which blocks eEF1A1 release during translation elongation. Phosphorylation of hnRNP E1 in response to TGF-β signaling disrupts the hnRNP E1-eEF1A1 interaction, triggering EMT. Whether a similar mechanism is employed by TGF-β in order to regulate cytostasis has not been investigated yet.

5.3 The Apoptotic Program

5.3.1 Overview of Apoptotic Pathways

Apoptosis or programmed cell death (Conradt 2009; Joza et al. 2002) is a process leading to ordered disruption of the cell without leakage of cell components. It is absolutely required both during embryogenesis (elimination of interdigital structures, regulation of neuron cell numbers) and in adult life (destruction of immature lymphocytes, regulation of regenerative growth of the liver, involution of the mammary gland after lactation). Apoptosis is characterized by activation of several pathways leading to activation of caspase-family cysteine proteases (Brenner and Mak 2009; Kroemer et al. 2007). Caspases are synthesized as pro-enzymes requiring activation in order to trigger apoptosis.

Two caspase apoptotic pathways have been described:

1. Extrinsic apoptosis is a form of cell death induced by extracellular signals resulting in binding of ligands to specific transmembrane receptors of the tumor necrosis factor/nerve growth factor (TNF/NGF) family, called death receptors. Upon ligand binding, receptor oligomerization allows recruitment of adaptor proteins and subsequent formation of a death initiation signaling complex (DISC) causing activation of the caspase cascade (initiated by caspase-8) (Lavrik and Krammer 2012).
2. The intrinsic pathway is activated in response to internal stresses, such as DNA damage and accumulation of free oxidant species. The various stress-induced signals converge on the mitochondria and determine mitochondrial outer membrane permeabilization (MOMP), which in turn results in dissipation of the mitochondrial membrane potential and hence the release of proteins, such as cytochrome C and the apoptosis-inducing factor (AIF), contributing to caspase-9 activation. The local regulation and occurrence of MOMP involve proteins of the BCL-2 family, which comprises pro-apoptotic factors (BAX, BAD, BIM) whose activity is finely tuned by its anti-apoptotic members (BCL-2, BCL-XL) (Green and Kroemer 2004).

5.3.2 TGF-β-Mediated Apoptosis

TGF-β-induced apoptosis plays important roles in many in vivo biological processes, such as wound healing (Amendt et al. 2002) and involution of the mammary epithelium after lactation (Nguyen and Pollard 2000) (see Chapter 6). It also plays a crucial role in tumor suppression, by eliminating cells with a malignant potential (Guasch et al. 2007; Yang et al. 2006). In vitro, a pro-apoptotic activity of TGF-β has been observed in many different cell types, such as B and T lymphocytes, prostate, mammary, and lung epithelial cells, hepatocytes, and neurons (reviewed in (Schuster and Krieglstein 2002)). Yet TGF-β has also been shown to promote survival of highly differentiated cells, such as cells of the nervous system (Sanchez-Capelo 2005).

TGF-β-induced apoptosis occurs mainly via the mitochondrial pathway (Schuster and Krieglstein 2002) (Fig. 5.3). In cell types where TGF-β (or activin) induces an apoptotic event, programmed cell death is usually associated with (1) down-regulation of anti-apoptotic factors such as BCL-2 or BCL-XL and/or (2) up-regulation of apoptosis-inducing factor such as BAX or BIK (Saltzman et al. 1998; Schuster and Krieglstein 2002). In contrast to the TGF-β cytostatic program, which

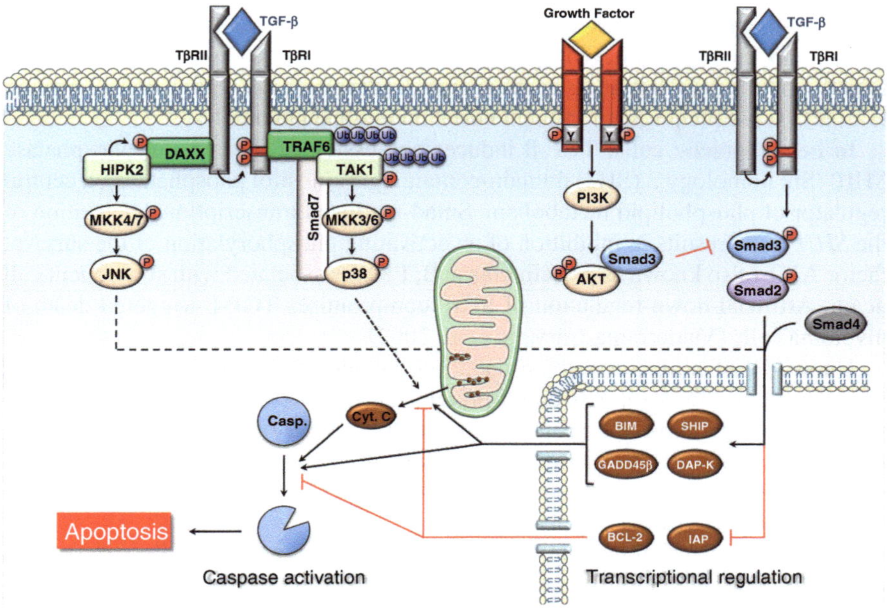

Fig. 5.3 The apoptotic program in response to TGF-β. *Dashed arrows* represent indirect biochemical events. *Note*: The figure was illustrated using Servier Medical Art

is shared by many cell types and involves a limited set of genes and regulators, the TGF-β apoptotic program is mostly cell-type specific and involves a plethora of regulators. It is also less well known and understood.

TGF-β regulates apoptosis via two distinct means that are not mutually exclusive (Fig. 5.3):

1. Smad-dependent transcription control of pro-apoptotic genes whose products are directly involved in the death pathway.
2. Smad-independent pathways initiated by enzymatic or non-catalytic activities of the TGF-β receptor complex.

The following sections present selected studies illustrating the variety of mechanisms that TGF-β can regulate to trigger cell death.

5.3.2.1 The Apoptotic Gene Program Regulated by the Smad Pathway

One of the first evidence of Smad pathway involvement in TGF-β-triggered apoptosis was the identification, in 1997, of the TGF-β-induced early gene 1 (TIEG1) as an immediate target gene of the transcription-regulating action of TGF-β. TIEG1 is an Sp1/Krüppel-like zinc finger transcription factor shown to induce apoptosis in pancreatic epithelial cells (Tachibana et al. 1997), mink lung epithelial cells (Chalaux et al. 1999), and hepatocytes (Ribeiro et al. 1999). The mechanism by which TIEG induces apoptosis is believed to rely on generation of reactive oxygen species, which would be the initiating death signals leading to a decreased BCL-2 protein level and subsequent loss of mitochondrial membrane potential.

In hematopoietic cells, TGF-β induces expression of the inositol phosphatase SHIP (Src homology 2 (SH2) domain-containing 5′-inositol phosphatase), a central regulator of phospholipid metabolism. Smad-mediated transcriptional activation of the *SHIP* gene results in inhibition of an activating phosphorylation of the survival factor AKT (also known as protein kinase B, PKB), associated with subsequent cell death. Artificial down-regulation of SHIP compromises TGF-β-triggered death of myeloma cells (Valderrama-Carvajal et al. 2002).

Upon TGF-β stimulation, the Smad2/3-4 complex activates transcription of the gene encoding the death-associated protein kinase (DAP-K), an enzyme directly involved in regulating apoptosis. By facilitating cytochrome C release from the mitochondria and dissipation of the mitochondrial membrane potential, TGF-β-induced DAP-K expression causes apoptosis of Hep3B hepatocellular carcinoma cells (Jang et al. 2002).

TGF-β stimulates transcription of GADD45β, a member of the stress- and cytokine-induced factor family, via the action of activated Smad3/4 complex on the proximal region of the *GADD45β* promoter (Yoo et al. 2003) and on an enhancer element located in the third intron (Major and Jones 2004). GADD45β was first identified in a complex containing the p38-activating kinase MEKK4/MTK1 (a mitogen-activated protein (MAP) kinase kinase kinase). By phosphorylating the MAPKK MKK3/6, the GADD45β-MEKK4 complex promotes activation of the MAPK p38, thus regulating p38-mediated biological functions such as apoptosis (Takekawa et al. 2002). In hepatocytes, the involvement of the p38 signaling pathway in TGF-β/GADD45β-triggered apoptosis is supported by the fact that apoptosis is inhibited by a small-molecular-weight inhibitor selectively blocking p38 kinase activity (Yoo et al. 2003). Expression of GADD45β in hepatocytes is sufficient to trigger apoptosis in the absence of TGF-β stimulation, whereas down-regulation of GADD45β expression inhibits TGF-β-induced p38 activation and apoptosis (Yoo et al. 2003).

The Smad pathway has been shown to control directly or indirectly the transcriptional expression of BCL-2-family members, key regulators of the intrinsic apoptotic pathway. A first example concerns down-regulation of the prototypic anti-apoptotic protein BCL-2 by TGF-β in vivo. Ectopic expression of Smad3 in transgenic mice protects the liver from chemically induced carcinogenesis (Yang et al. 2006). Indeed, Smad3 exerts its protective role by inducing apoptosis in hepatocytes in order to prevent their malignant conversion. Smad3 triggers apoptosis by down-regulating the transcription of the gene encoding the anti-apoptotic protein BCL-2. This repression is exerted upon binding of a Smad3-containing transcriptional complex to a GC-rich TGF-β-responsive element in the *Bcl-2* promoter (Yang et al. 2006). Interestingly, Smad3-mediated apoptosis occurs concomitantly with p38 phosphorylation in both liver tumors of transgenic mice and primary hepatocytes cultured in the presence of TGF-β (Yang et al. 2006). As p38 MAPK activity appears also to be involved in TGF-β-mediated apoptosis (see above), these findings indicate that Smad-dependent and Smad-independent pathways may cooperate in the same cell to strengthen death-initiating cues.

While down-regulating anti-apoptotic factors, TGF-β also positively regulates pro-apoptotic factors of the BCL-2 family. For example, it increases the level of the pro-apoptotic BIM protein in B lymphocytes (Ramesh et al. 2008; Wildey et al. 2003) and hepatocytes (Wildey and Howe 2009) in a Smad3-dependent manner so as to trigger apoptosis via the mitochondrial pathway. TGF-β can control the BIM protein level via two distinct mechanisms:

1. The canonical TGF-β/Smad3 pathway indirectly regulates transcriptional activation of the *BIM* gene, inducing expression of transcription factor RUNX1, which cooperates with FoxO3 and binds to the *BIM* promoter to trigger transcriptional activation in hepatocytes (Wildey et al. 2003). Despite the presence of a putative SBE in the *BIM* promoter, cooperation of activated Smad3/4 with the RUNX1-FoxO3 complex has not yet been established.
2. TGF-β indirectly promotes BIM protein stabilization. Extracellular signal-regulated kinase (ERK)-regulated phosphorylation of BIM targets it for degradation by the proteasome machinery. This posttranslational event is inhibited by the MAPK phosphatase MKP2, which dephosphorylates and thus inactivates the ERK kinase. The *MKP2* gene responds directly to TGF-β. By stimulating transcriptional activation of *MKP2*, the Smad3/4 complex causes stabilization of the BIM protein (Ramesh et al. 2008).

Survivin, a member of the mammalian inhibitor of apoptosis (IAP) family of endogenous caspase inhibitors, is also transcriptionally down-regulated by TGF-β (Yang et al. 2008). Interestingly, TGF-β represses transcription *of* the *Survivin* gene by a peculiar mechanism. The proximal region of the *Survivin* promoter contains two canonical SBEs and two cell cycle repressor elements: a cell-cycle-dependent element (CDE) and a cell-cycle gene homology region (CHR). Both the CDE and the CHR are indispensable to TGF-β-mediated transcriptional repression in reporter assays. TGF-β causes recruitment of an RB-E2F4 repressive complex to the CDE/CHR elements of the *Survivin* promoter. Although Smad2/3 signaling is absolutely required for TGF-β-mediated repression of *Survivin* and although Smad3 binds to the minimal proximal region of the *Survivin* promoter, these SBEs do not seem to be the key elements mediating repression, since Smad4 does not bind to this particular promoter region and since mutations within the two SBEs do not abolish TGF-β-mediated transcriptional repression in reporter assays. Smad proteins might be required in this transcriptional repression indirectly, for their ability to maintain the RB protein in a hypophosphorylated state in a complex with E2F4 factors (in the framework of the TGF-β-governed cytostatic program). Alternatively, a Smad complex might interact physically with unidentified SBEs located elsewhere in the promoter or in other regulatory elements (Yang et al. 2008).

AKT overexpression is sufficient to block TGF-β-mediated apoptosis, independently of AKT kinase activity (Conery et al. 2004; Remy et al. 2004). This inhibition results from direct interaction of AKT with Smad3, preventing phosphorylation of

Smad3 by TβRI (titration). Hence, the Smad3:AKT ratio is a crucial determinant of the sensitivity to apoptosis in response to TGF-β, stronger apoptotic responses being observed in cells with higher Smad3:AKT ratios (Conery et al. 2004). More interestingly, the growth of all cell lines tested is potently inhibited by TGF-β, while their response to apoptosis depends on the Smad3-to-AKT ratio (Conery et al. 2004). The AKT-Smad3 complex might be part of a regulatory mechanism enabling a given cell to switch from the cytostatic program to the apoptotic program in response to TGF-β (Fig. 5.3).

5.3.2.2 Smad-Independent Pathways Controlling TGF-β-Mediated Apoptosis

TGF-β can also induce apoptosis through Smad-independent signaling pathways. In this case, to promote cell death, TGF-β signals through the p38 and JNK (c-Jun N-terminal kinase) pathways (Fig. 5.3), both classically associated with responses to cell stress and apoptosis. TGF-β has been known for long time to facilitate activation of JNK (Atfi et al. 1997; Parkinson et al. 2001) and p38 (Hyman et al. 2002; Takekawa et al. 2002; Yoo et al. 2003) in various cell types, but for many years the biochemical links between TGF-β receptors and the actors of the MAPK cascade remained unelucidated.

From a two-hybrid screen, the adaptor protein DAXX has emerged as a target protein interacting physically with TGF-β type II receptor (TβRII) used as bait (Perlman et al. 2001). DAXX is a FAS-receptor-associated protein mediating FAS-induced programmed cell death by activating the JNK pathway. In vitro, DAXX interacts directly with the cytoplasmic domain of TβRII, but not with TβRI. In vivo, it has been found to associate with the TβRI/II receptor complex upon TGF-β stimulation. This association requires the kinase activity of the TβRII receptor, since a kinase-dead mutant of TβRII fails to interact with DAXX (Perlman et al. 2001). DAXX is crucial to TGF-β-induced apoptosis in both B-cell lymphoma cells and mouse hepatocytes, since overexpression of a dominant-negative form of DAXX or artificial *DAXX* silencing inhibits TGF-β-mediated apoptosis in these cells (Perlman et al. 2001). Upon TGF-β stimulation, interestingly, DAXX acts as an adaptor protein facilitating the phosphorylation and subsequent activation of JNK and triggering cell death. DAXX can interact physically with HIPK2 (homeodomain-interacting protein kinase 2), a serine/threonine kinase involved in transcriptional regulation and also apoptosis (Hofmann et al. 2003). The HIPK2-DAXX complex facilitates activation of the JNK protein by the upstream mitogen-activated protein kinase kinases MKK4 and MKK7. Overexpression of both DAXX and HIPK2 leads to potentiation of TGF-β-induced cell death. This effect is mediated by the kinase function of HIPK2, since expression of a kinase-deficient HIPK2 protein fails to increase TGF-β-induced cell death in these cells. Interestingly, HIPK2 also phosphorylates the DAXX protein (Hofmann et al. 2003), but whether this phosphorylation event is required for the adaptor function of DAXX in the specific context of TGF-β-induced apoptosis has not been explored. These findings constitute the first evidence of a biochemical link between TGF-β receptors and the JNK pathway, leading to the apoptotic machinery (Fig. 5.3).

Concerning activation of the p38 MAPK by TGF-β, previous studies have indicated that TAK1 (TGF-β-activated kinase 1) might be an indirect link between p38 and the TGF-β receptor complex (Shibuya et al. 1996; Yamaguchi et al. 1995). TAK1 is a member of the MAP kinase kinase kinase family, identified as a TβRI-interacting protein (Shibuya et al. 1996; Yamaguchi et al. 1995) almost concomitantly with the discovery of the Smad proteins (Heldin et al. 1997). TAK1 has been proposed to be involved in TGF-β-induced cell death through its ability to activate p38 in a Smad-independent manner (Sano et al. 1999). Until recently, the biochemical link between the TβRI-TAK1 complex and p38 activation remained unidentified (Sorrentino et al. 2008; Yamashita et al. 2008). The E3-ligase TRAF6 can bind constitutively to TβRI. Upon ligand binding and TGF-β receptor TβRI-TβRII hetero-oligomerization, TRAF6 becomes auto-ubiquitinated by polyubiquitin chains linked via their Lys63. TRAF6 then catalyzes Lys63-mediated polyubiquitination of TAK1, and thus activates it. It is important to stress that unlike Lys48-linked polyubiquitination, Lys63-linked polyubiquitination is not a signal targeting proteins for degradation by the proteasome complex. Although its role in the TAK1 activation process is not fully understood, TAK1 polyubiquitination seems to facilitate TAK1 juxtaposition-induced auto-phosphorylation. Alternatively, it might favor a conformational change allowing exposure of the TAK1 catalytic domain. In turn, TAK1 phosphorylates and activates the MAP kinase kinases: MKK3 and MKK6, which ultimately phosphorylate MAPK p38. In this context, Smad7 might act as a docking protein facilitating the phosphorylation cascade, having previously been shown (*1*) to interact with the TAK1, MKK3, and p38 proteins and (*2*) to be required for TGF-β-induced apoptosis in prostate cancer cells (Edlund et al. 2003) (Fig. 5.3).

Interestingly, activation of TAK1 and p38 does not require TβRI kinase activity (Sorrentino et al. 2008; Yamashita et al. 2008), but the presence of TRAF6 and its E3-ubiquitin ligase activity are indispensable to TGF-β-mediated apoptosis in prostate cancer cells (Sorrentino et al. 2008) and hepatocytes (Yamashita et al. 2008). Artificial down-regulation of *TRAF6* in prostate cancer cells or hepatocytes abolishes TGF-β-triggered apoptosis, and similar observations have been made on TRAF6-deficient mouse embryonic fibroblasts (Sorrentino et al. 2008; Yamashita et al. 2008). Moreover, overexpression of a deletion mutant of TRAF6 lacking E3-ubiquitin ligase activity inhibits TGF-β-mediated TAK1 ubiquitination and subsequent activation and also p38 phosphorylation (Sorrentino et al. 2008).

Another adaptor protein has recently been implicated in this noncanonical TRAF6-TAK1 pathway initiated by TGF-β: TTRAP (TRAF and TNF receptor associated protein). TTRAP forms a ternary complex with TAK1 and TRAF6, which can be recruited to TβRI in the presence of the ligand. The presence of TTRAP potentiates the E3-ubiquitin ligase activity of TRAF6, thereby promoting TAK1 ubiquitination. Artificial down-regulation of TTRAP impairs TGF-β-induced activation of p38 (Varady et al. 2011).

5.4 Conclusion

In conclusion, to the extent that the tumor-suppressive mechanisms orchestrated by TGF-β have been deciphered, it is obvious that the cytostatic and apoptotic responses of a cell to TGF-β involve totally different regulatory events and processes. The cytostatic program, on the one hand, involves regulation of a limited number of genes by Smad proteins, with few additional effectors intervening to block the cell cycle before the G1 restriction point. The apoptotic program, on the other hand, involves many modifying enzymes, notably kinases, and also requires Smad-mediated transcriptional regulation in order to favor pro-apoptotic instead of pro-survival pathways. The mechanism by which a cell switches from one program to the other depends on an intricate dialogue with its microenvironment and is still largely unknown.

References

Alexandrow MG, Kawabata M, Aakre M, Moses HL (1995) Overexpression of the c-Myc oncoprotein blocks the growth-inhibitory response but is required for the mitogenic effects of transforming growth factor-β1. Proc Natl Acad Sci USA 92:3239–3243

Amendt C, Mann A, Schirmacher P, Blessing M (2002) Resistance of keratinocytes to TGF-β-mediated growth restriction and apoptosis induction accelerates re-epithelialization in skin wounds. J Cell Sci 115:2189–2198

Anzano MA, Roberts AB, Sporn MB (1986) Anchorage-independent growth of primary rat embryo cells is induced by platelet-derived growth factor and inhibited by type-β transforming growth factor. J Cell Physiol 126:312–318

Atfi A, Buisine M, Mazars A, Gespach C (1997) Induction of apoptosis by DPC4, a transcriptional factor regulated by transforming growth factor-β through stress-activated protein kinase/c-Jun N-terminal kinase (SAPK/JNK) signaling pathway. J Biol Chem 272:24731–24734

Azar R, Alard A, Susini C, Bousquet C, Pyronnet S (2009) 4E-BP1 is a target of Smad4 essential for TGF-β-mediated inhibition of cell proliferation. EMBO J 28:3514–3522

Baird A, Durkin T (1986) Inhibition of endothelial cell proliferation by type β-transforming growth factor: interactions with acidic and basic fibroblast growth factors. Biochem Biophys Res Commun 138:476–482

Battegay EJ, Raines EW, Seifert RA, Bowen-Pope DF, Ross R (1990) TGF-β induces bimodal proliferation of connective tissue cells via complex control of an autocrine PDGF loop. Cell 63:515–524

Bhowmick NA, Ghiassi M, Aakre M, Brown K, Singh V, Moses HL (2003) TGF-β-induced RhoA and p160ROCK activation is involved in the inhibition of Cdc25A with resultant cell-cycle arrest. Proc Natl Acad Sci USA 100:15548–15553

Bouchard C, Thieke K, Maier A, Saffrich R, Hanley-Hyde J, Ansorge W, Reed S, Sicinski P, Bartek J, Eilers M (1999) Direct induction of cyclin D2 by Myc contributes to cell cycle progression and sequestration of p27. EMBO J 18:5321–5333

Bouzahzah B, Fu M, Iavarone A, Factor VM, Thorgeirsson SS, Pestell RG (2000) Transforming growth factor-β1 recruits histone deacetylase 1 to a p130 repressor complex in transgenic mice in vivo. Cancer Res 60:4531–4537

Brenner D, Mak TW (2009) Mitochondrial cell death effectors. Curr Opin Cell Biol 21:871–877

Chalaux E, Lopez-Rovira T, Rosa JL, Pons G, Boxer LM, Bartrons R, Ventura F (1999) A zinc-finger transcription factor induced by TGF-β promotes apoptotic cell death in epithelial Mv1Lu cells. FEBS Lett 457:478–482

Chen CR, Kang Y, Massague J (2001) Defective repression of c-myc in breast cancer cells: a loss at the core of the transforming growth factor-β growth arrest program. Proc Natl Acad Sci USA 98:992–999

Chen CR, Kang Y, Siegel PM, Massague J (2002) E2F4/5 and p107 as Smad cofactors linking the TGF-β receptor to c-myc repression. Cell 110:19–32

Claassen GF, Hann SR (2000) A role for transcriptional repression of p21CIP1 by c-Myc in overcoming transforming growth factor-β-induced cell-cycle arrest. Proc Natl Acad Sci USA 97:9498–9503

Conery AR, Cao Y, Thompson EA, Townsend CM Jr, Ko TC, Luo K (2004) Akt interacts directly with Smad3 to regulate the sensitivity to TGF-β induced apoptosis. Nat Cell Biol 6:366–372

Conradt B (2009) Genetic control of programmed cell death during animal development. Annu Rev Genet 43:493–523

Cordenonsi M, Dupont S, Maretto S, Insinga A, Imbriano C, Piccolo S (2003) Links between tumor suppressors: p53 is required for TGF-β gene responses by cooperating with Smads. Cell 113:301–314

Datto MB, Li Y, Panus JF, Howe DJ, Xiong Y, Wang XF (1995) Transforming growth factor-β induces the cyclin-dependent kinase inhibitor p21 through a p53-independent mechanism. Proc Natl Acad Sci USA 92:5545–5549

Donzelli M, Draetta GF (2003) Regulating mammalian checkpoints through Cdc25 inactivation. EMBO Rep 4:671–677

Edlund S, Bu S, Schuster N, Aspenstrom P, Heuchel R, Heldin NE, ten Dijke P, Heldin CH, Landstrom M (2003) Transforming growth factor-β1 (TGF-β)-induced apoptosis of prostate cancer cells involves Smad7-dependent activation of p38 by TGF-β-activated kinase 1 and mitogen-activated protein kinase kinase 3. Mol Biol Cell 14:529–544

Ewen ME, Sluss HK, Whitehouse LL, Livingston DM (1993) TGF-β inhibition of Cdk4 synthesis is linked to cell cycle arrest. Cell 74:1009–1020

Feng XH, Lin X, Derynck R (2000) Smad2, Smad3 and Smad4 cooperate with Sp1 to induce p15(Ink4B) transcription in response to TGF-β. EMBO J 19:5178–5193

Feng XH, Liang YY, Liang M, Zhai W, Lin X (2002) Direct interaction of c-Myc with Smad2 and Smad3 to inhibit TGF-β-mediated induction of the CDK inhibitor p15(Ink4B). Mol Cell 9:133–143

Frater-Schroder M, Muller G, Birchmeier W, Bohlen P (1986) Transforming growth factor-β inhibits endothelial cell proliferation. Biochem Biophys Res Commun 137:295–302

Frederick JP, Liberati NT, Waddell DS, Shi Y, Wang XF (2004) Transforming growth factor-β-mediated transcriptional repression of c-myc is dependent on direct binding of Smad3 to a novel repressive Smad binding element. Mol Cell Biol 24:2546–2559

Galaktionov K, Chen X, Beach D (1996) Cdc25 cell-cycle phosphatase as a target of c-myc. Nature 382:511–517

Gomis RR, Alarcon C, He W, Wang Q, Seoane J, Lash A, Massague J (2006a) A FoxO-Smad synexpression group in human keratinocytes. Proc Natl Acad Sci USA 103:12747–12752

Gomis RR, Alarcon C, Nadal C, Van Poznak C, Massague J (2006b) C/EBPbeta at the core of the TGF-β cytostatic response and its evasion in metastatic breast cancer cells. Cancer Cell 10:203–214

Green DR, Kroemer G (2004) The pathophysiology of mitochondrial cell death. Science 305:626–629

Guasch G, Schober M, Pasolli HA, Conn EB, Polak L, Fuchs E (2007) Loss of TGF-β signaling destabilizes homeostasis and promotes squamous cell carcinomas in stratified epithelia. Cancer Cell 12:313–327

Guenard V, Rosenbaum T, Gwynn LA, Doetschman T, Ratner N, Wood PM (1995) Effect of transforming growth factor-β1 and -β2 on Schwann cell proliferation on neurites. Glia 13:309–318

Hannon GJ, Beach D (1994) p15INK4B is a potential effector of TGF-β-induced cell cycle arrest. Nature 371:257–261
Heimark RL, Twardzik DR, Schwartz SM (1986) Inhibition of endothelial regeneration by type-β transforming growth factor from platelets. Science 233:1078–1080
Heldin CH, Miyazono K, ten Dijke P (1997) TGF-β signalling from cell membrane to nucleus through SMAD proteins. Nature 390:465–471
Hermeking H, Rago C, Schuhmacher M, Li Q, Barrett JF, Obaya AJ, O'Connell BC, Mateyak MK, Tam W, Kohlhuber F, Dang CV, Sedivy JM, Eick D, Vogelstein B, Kinzler KW (2000) Identification of CDK4 as a target of c-MYC. Proc Natl Acad Sci USA 97:2229–2234
Herrera RE, Makela TP, Weinberg RA (1996) TGF-β-induced growth inhibition in primary fibroblasts requires the retinoblastoma protein. Mol Biol Cell 7:1335–1342
Hofmann TG, Stollberg N, Schmitz ML, Will H (2003) HIPK2 regulates transforming growth factor-β-induced c-Jun NH(2)-terminal kinase activation and apoptosis in human hepatoma cells. Cancer Res 63:8271–8277
Holley RW, Bohlen P, Fava R, Baldwin JH, Kleeman G, Armour R (1980) Purification of kidney epithelial cell growth inhibitors. Proc Natl Acad Sci USA 77:5989–5992
Holley RW, Armour R, Baldwin JH, Greenfield S (1983) Activity of a kidney epithelial cell growth inhibitor on lung and mammary cells. Cell Biol Int Rep 7:141–147
Hussey GS, Chaudhury A, Dawson AE, Lindner DJ, Knudsen CR, Wilce MC, Merrick WC, Howe PH (2011) Identification of an mRNP complex regulating tumorigenesis at the translational elongation step. Mol Cell 41:419–431
Hyman KM, Seghezzi G, Pintucci G, Stellari G, Kim JH, Grossi EA, Galloway AC, Mignatti P (2002) Transforming growth factor-β1 induces apoptosis in vascular endothelial cells by activation of mitogen-activated protein kinase. Surgery 132:173–179
Iavarone A, Massague J (1997) Repression of the CDK activator Cdc25A and cell-cycle arrest by cytokine TGF-β in cells lacking the CDK inhibitor p15. Nature 387:417–422
Iavarone A, Massague J (1999) E2F and histone deacetylase mediate transforming growth factor-β repression of cdc25A during keratinocyte cell cycle arrest. Mol Cell Biol 19:916–922
Igarashi A, Okochi H, Bradham DM, Grotendorst GR (1993) Regulation of connective tissue growth factor gene expression in human skin fibroblasts and during wound repair. Mol Biol Cell 4:637–645
Jang CW, Chen CH, Chen CC, Chen JY, Su YH, Chen RH (2002) TGF-β induces apoptosis through Smad-mediated expression of DAP- kinase. Nat Cell Biol 4:51–58
Joza N, Kroemer G, Penninger JM (2002) Genetic analysis of the mammalian cell death machinery. Trends Genet 18:142–149
Kang Y, Chen CR, Massague J (2003) A self-enabling TGF-β response coupled to stress signaling: Smad engages stress response factor ATF3 for Id1 repression in epithelial cells. Mol Cell 11:915–926
Kee BL, Rivera RR, Murre C (2001) Id3 inhibits B lymphocyte progenitor growth and survival in response to TGF-β. Nat Immunol 2:242–247
Kehrl JH, Roberts AB, Wakefield LM, Jakowlew S, Sporn MB, Fauci AS (1986a) Transforming growth factor-β is an important immunomodulatory protein for human B lymphocytes. J Immunol 137:3855–3860
Kehrl JH, Wakefield LM, Roberts AB, Jakowlew S, Alvarez-Mon M, Derynck R, Sporn MB, Fauci AS (1986b) Production of transforming growth factor-β by human T lymphocytes and its potential role in the regulation of T cell growth. J Exp Med 163:1037–1050
Kim JW, Kim HS, Kim IK, Kim MR, Cho EY, Kim HK, Lee JM, Namkoong SE (1998) Transforming growth factor-β1 induces apoptosis through down- regulation of c-myc gene and overexpression of p27Kip1 protein in cervical carcinoma. Gynecol Oncol 69:230–236
Koff A, Ohtsuki M, Polyak K, Roberts JM, Massague J (1993) Negative regulation of G1 in mammalian cells: inhibition of cyclin E-dependent kinase by TGF-β. Science 260:536–539
Koinuma D, Tsutsumi S, Kamimura N, Taniguchi H, Miyazawa K, Sunamura M, Imamura T, Miyazono K, Aburatani H (2009) Chromatin immunoprecipitation on microarray analysis of

Smad2/3 binding sites reveals roles of ETS1 and TFAP2A in transforming growth factor-β signaling. Mol Cell Biol 29:172–186

Kowanetz M, Valcourt U, Bergstrom R, Heldin CH, Moustakas A (2004) Id2 and Id3 define the potency of cell proliferation and differentiation responses to transforming growth factor-β and bone morphogenetic protein. Mol Cell Biol 24:4241–4254

Kroemer G, Galluzzi L, Brenner C (2007) Mitochondrial membrane permeabilization in cell death. Physiol Rev 87:99–163

Lafon C, Mathieu C, Guerrin M, Pierre O, Vidal S, Valette A (1996) Transforming growth factor-β1-induced apoptosis in human ovarian carcinoma cells: protection by the antioxidant N-acetylcysteine and bcl-2. Cell Growth Differ 7:1095–1104

Laiho M, DeCaprio JA, Ludlow JW, Livingston DM, Massague J (1990) Growth inhibition by TGF-β linked to suppression of retinoblastoma protein phosphorylation. Cell 62:175–185

Lasorella A, Iavarone A, Israel MA (1996) Id2 specifically alters regulation of the cell cycle by tumor suppressor proteins. Mol Cell Biol 16:2570–2578

Lasorella A, Noseda M, Beyna M, Yokota Y, Iavarone A (2000) Id2 is a retinoblastoma protein target and mediates signalling by Myc oncoproteins. Nature 407:592–598

Lavrik IN, Krammer PH (2012) Regulation of CD95/Fas signaling at the DISC. Cell Death Differ 19:36–41

Leof EB, Proper JA, Goustin AS, Shipley GD, DiCorleto PE, Moses HL (1986) Induction of c-sis mRNA and activity similar to platelet-derived growth factor by transforming growth factor-β: a proposed model for indirect mitogenesis involving autocrine activity. Proc Natl Acad Sci USA 83:2453–2457

Lin KW, Yakymovych I, Jia M, Yakymovych M, Souchelnytskyi S (2010) Phosphorylation of eEF1A1 at Ser300 by TβR-I results in inhibition of mRNA translation. Curr Biol 20:1615–1625

Major MB, Jones DA (2004) Identification of a gadd45β 3′ enhancer that mediates SMAD3- and SMAD4-dependent transcriptional induction by transforming growth factor-β. J Biol Chem 279:5278–5287

Massague J (1990) The transforming growth factor-β family. Annu Rev Cell Biol 6:597–641

Masui T, Wakefield LM, Lechner JF, LaVeck MA, Sporn MB, Harris CC (1986) Type β-transforming growth factor is the primary differentiation- inducing serum factor for normal human bronchial epithelial cells. Proc Natl Acad Sci USA 83:2438–2442

Moustakas A, Kardassis D (1998) Regulation of the human p21/WAF1/Cip1 promoter in hepatic cells by functional interactions between Sp1 and Smad family members. Proc Natl Acad Sci USA 95:6733–6738

Nguyen AV, Pollard JW (2000) Transforming growth factor-β3 induces cell death during the first stage of mammary gland involution. Development 127:3107–3118

Niimi H, Pardali K, Vanlandewijck M, Heldin CH, Moustakas A (2007) Notch signaling is necessary for epithelial growth arrest by TGF-β. J Cell Biol 176:695–707

Oberhammer FA, Pavelka M, Sharma S, Tiefenbacher R, Purchio AF, Bursch W, Schulte-Hermann R (1992) Induction of apoptosis in cultured hepatocytes and in regressing liver by transforming growth factor-β1. Proc Natl Acad Sci USA 89:5408–5412

Pardali K, Kurisaki A, Moren A, ten Dijke P, Kardassis D, Moustakas A (2000) Role of Smad proteins and transcription factor Sp1 in p21(Waf1/Cip1) regulation by transforming growth factor-β. J Biol Chem 275:29244–29256

Pardali K, Kowanetz M, Heldin CH, Moustakas A (2005) Smad pathway-specific transcriptional regulation of the cell cycle inhibitor p21(WAF1/Cip1). J Cell Physiol 204:260–272

Parkinson DB, Dong Z, Bunting H, Whitfield J, Meier C, Marie H, Mirsky R, Jessen KR (2001) Transforming growth factor-β (TGF-β) mediates Schwann cell death in vitro and in vivo: examination of c-Jun activation, interactions with survival signals, and the relationship of TGFβ-mediated death to Schwann cell differentiation. J Neurosci 21:8572–8585

Perk J, Iavarone A, Benezra R (2005) Id family of helix-loop-helix proteins in cancer. Nat Rev Cancer 5:603–614

Perlman R, Schiemann WP, Brooks MW, Lodish HF, Weinberg RA (2001) TGF-β-induced apoptosis is mediated by the adapter protein Daxx that facilitates JNK activation. Nat Cell Biol 3:708–714

Pietenpol JA, Holt JT, Stein RW, Moses HL (1990a) Transforming growth factor-β1 suppression of c-myc gene transcription: role in inhibition of keratinocyte proliferation. Proc Natl Acad Sci USA 87:3758–3762

Pietenpol JA, Stein RW, Moran E, Yaciuk P, Schlegel R, Lyons RM, Pittelkow MR, Munger K, Howley PM, Moses HL (1990b) TGF-β1 inhibition of c-myc transcription and growth in keratinocytes is abrogated by viral transforming proteins with pRB binding domains. Cell 61:777–785

Ramesh S, Qi XJ, Wildey GM, Robinson J, Molkentin J, Letterio J, Howe PH (2008) TGF-β-mediated BIM expression and apoptosis are regulated through SMAD3-dependent expression of the MAPK phosphatase MKP2. EMBO Rep 9:990–997

Remy I, Montmarquette A, Michnick SW (2004) PKB/Akt modulates TGF-β signalling through a direct interaction with Smad3. Nat Cell Biol 6:358–365

Reynisdottir I, Polyak K, Iavarone A, Massague J (1995) Kip/Cip and Ink4 Cdk inhibitors cooperate to induce cell cycle arrest in response to TGF-β. Genes Dev 9:1831–1845

Ribeiro A, Bronk SF, Roberts PJ, Urrutia R, Gores GJ (1999) The transforming growth factor-β(1)-inducible transcription factor TIEG1, mediates apoptosis through oxidative stress. Hepatology 30:1490–1497

Ridley AJ, Davis JB, Stroobant P, Land H (1989) Transforming growth factors-β1 and -β2 are mitogens for rat Schwann cells. J Cell Biol 109:3419–3424

Roberts AB, Lamb LC, Newton DL, Sporn MB, De Larco JE, Todaro GJ (1980) Transforming growth factors: isolation of polypeptides from virally and chemically transformed cells by acid/ethanol extraction. Proc Natl Acad Sci USA 77:3494–3498

Roberts AB, Anzano MA, Lamb LC, Smith JM, Sporn MB (1981) New class of transforming growth factors potentiated by epidermal growth factor: isolation from non-neoplastic tissues. Proc Natl Acad Sci USA 78:5339–5343

Rotello RJ, Lieberman RC, Purchio AF, Gerschenson LE (1991) Coordinated regulation of apoptosis and cell proliferation by transforming growth factor-β1 in cultured uterine epithelial cells. Proc Natl Acad Sci USA 88:3412–3415

Saltzman A, Munro R, Searfoss G, Franks C, Jaye M, Ivashchenko Y (1998) Transforming growth factor-β-mediated apoptosis in the Ramos B-lymphoma cell line is accompanied by caspase activation and Bcl-XL downregulation. Exp Cell Res 242:244–254

Sanchez-Capelo A (2005) Dual role for TGF-β1 in apoptosis. Cytokine Growth Factor Rev 16:15–34

Sandhu C, Garbe J, Bhattacharya N, Daksis J, Pan CH, Yaswen P, Koh J, Slingerland JM, Stampfer MR (1997) Transforming growth factor-β stabilizes p15INK4B protein, increases p15INK4B-cdk4 complexes, and inhibits cyclin D1-cdk4 association in human mammary epithelial cells. Mol Cell Biol 17:2458–2467

Sano Y, Harada J, Tashiro S, Gotoh-Mandeville R, Maekawa T, Ishii S (1999) ATF-2 is a common nuclear target of Smad and TAK1 pathways in transforming growth factor-β signaling. J Biol Chem 274:8949–8957

Scandura JM, Boccuni P, Massague J, Nimer SD (2004) Transforming growth factor-β-induced cell cycle arrest of human hematopoietic cells requires p57KIP2 up-regulation. Proc Natl Acad Sci USA 101:15231–15236

Schuster N, Krieglstein K (2002) Mechanisms of TGF-β-mediated apoptosis. Cell Tissue Res 307:1–14

Seoane J, Pouponnot C, Staller P, Schader M, Eilers M, Massague J (2001) TGF-β influences Myc, Miz-1 and Smad to control the CDK inhibitor p15INK4b. Nat Cell Biol 3:400–408

Seoane J, Le HV, Massague J (2002) Myc suppression of the p21(Cip1) Cdk inhibitor influences the outcome of the p53 response to DNA damage. Nature 419:729–734

Seoane J, Le HV, Shen L, Anderson SA, Massague J (2004) Integration of Smad and forkhead pathways in the control of neuroepithelial and glioblastoma cell proliferation. Cell 117:211–223
Sherr CJ, Roberts JM (1999) CDK inhibitors: positive and negative regulators of G1-phase progression. Genes Dev 13:1501–1512
Shibuya H, Yamaguchi K, Shirakabe K, Tonegawa A, Gotoh Y, Ueno N, Irie K, Nishida E, Matsumoto K (1996) TAB1: an activator of the TAK1 MAPKKK in TGF-β signal transduction. Science 272:1179–1182
Siegel PM, Shu W, Massague J (2003) Mad upregulation and Id2 repression accompany transforming growth factor (TGF)-β-mediated epithelial cell growth suppression. J Biol Chem 278:35444–35450
Silberstein GB, Daniel CW (1987) Reversible inhibition of mammary gland growth by transforming growth factor-β. Science 237:291–293
Soma Y, Grotendorst GR (1989) TGF-β stimulates primary human skin fibroblast DNA synthesis via an autocrine production of PDGF-related peptides. J Cell Physiol 140:246–253
Sorrentino A, Thakur N, Grimsby S, Marcusson A, von Bulow V, Schuster N, Zhang S, Heldin CH, Landstrom M (2008) The type I TGF-β receptor engages TRAF6 to activate TAK1 in a receptor kinase-independent manner. Nat Cell Biol 10:1199–1207
Staller P, Peukert K, Kiermaier A, Seoane J, Lukas J, Karsunky H, Moroy T, Bartek J, Massague J, Hanel F, Eilers M (2001) Repression of p15INK4b expression by Myc through association with Miz-1. Nat Cell Biol 3:392–399
Tachibana I, Imoto M, Adjei PN, Gores GJ, Subramaniam M, Spelsberg TC, Urrutia R (1997) Overexpression of the TGF-β-regulated zinc finger encoding gene, TIEG, induces apoptosis in pancreatic epithelial cells. J Clin Invest 99:2365–2374
Takekawa M, Tatebayashi K, Itoh F, Adachi M, Imai K, Saito H (2002) Smad-dependent GADD45β expression mediates delayed activation of p38 MAP kinase by TGF-β. EMBO J 21:6473–6482
Thillainadesan G, Chitilian JM, Isovic M, Ablack JN, Mymryk JS, Tini M, Torchia J (2012) TGF-β-dependent active demethylation and expression of the p15ink4b tumor suppressor are impaired by the ZNF217/CoREST complex. Mol Cell 46:636–649
Tucker RF, Shipley GD, Moses HL, Holley RW (1984) Growth inhibitor from BSC-1 cells closely related to platelet type-β transforming growth factor. Science 226:705–707
Valcourt U, Thuault S, Pardali K, Heldin CH, Moustakas A (2007) Functional role of Meox2 during the epithelial cytostatic response to TGF-β. Mol Oncol 1:55–71
Valderrama-Carvajal H, Cocolakis E, Lacerte A, Lee EH, Krystal G, Ali S, Lebrun JJ (2002) Activin/TGF-β induce apoptosis through Smad-dependent expression of the lipid phosphatase SHIP. Nat Cell Biol 4:963–969
Varady G, Sarkadi B, Fatyol K (2011) TTRAP is a novel component of the non-canonical TRAF6-TAK1 TGF-β signaling pathway. PLoS One 6:e25548
Vermeulen K, Van Bockstaele DR, Berneman ZN (2003) The cell cycle: a review of regulation, deregulation and therapeutic targets in cancer. Cell Prolif 36:131–149
Warner BJ, Blain SW, Seoane J, Massague J (1999) Myc downregulation by transforming growth factor-β required for activation of the p15(Ink4b) G(1) arrest pathway. Mol Cell Biol 19:5913–5922
Wildey GM, Howe PH (2009) Runx1 is a co-activator with FOXO3 to mediate transforming growth factor-β (TGF-β)-induced Bim transcription in hepatic cells. J Biol Chem 284:20227–20239
Wildey GM, Patil S, Howe PH (2003) Smad3 potentiates transforming growth factor-β (TGFβ)-induced apoptosis and expression of the BH3-only protein Bim in WEHI 231 B lymphocytes. J Biol Chem 278:18069–18077
Yamaguchi M, Endo H, Tasaka K, Miyake A (1995) Mouse growth hormone-releasing factor secretion is activated by inhibin and inhibited by activin in placenta. Biol Reprod 53:368–372
Yamashita M, Fatyol K, Jin C, Wang X, Liu Z, Zhang YE (2008) TRAF6 mediates Smad-independent activation of JNK and p38 by TGF-β. Mol Cell 31:918–924

Yanagihara K, Tsumuraya M (1992) Transforming growth factor-β1 induces apoptotic cell death in cultured human gastric carcinoma cells. Cancer Res 52:4042–4045
Yang YA, Zhang GM, Feigenbaum L, Zhang YE (2006) Smad3 reduces susceptibility to hepatocarcinoma by sensitizing hepatocytes to apoptosis through downregulation of Bcl-2. Cancer Cell 9:445–457
Yang J, Song K, Krebs TL, Jackson MW, Danielpour D (2008) Rb/E2F4 and Smad2/3 link survivin to TGF-β-induced apoptosis and tumor progression. Oncogene 27:5326–5338
Yokota Y, Mori S (2002) Role of Id family proteins in growth control. J Cell Physiol 190:21–28
Yoo J, Ghiassi M, Jirmanova L, Balliet AG, Hoffman B, Fornace AJ Jr, Liebermann DA, Bottinger EP, Roberts AB (2003) Transforming growth factor-β-induced apoptosis is mediated by Smad-dependent expression of GADD45b through p38 activation. J Biol Chem 278:43001–43007

Chapter 6
TGF-β as Tumor Suppressor: Lessons from Mouse Models

Ulrich Valcourt, David F. Vincent, and Laurent Bartholin

Abstract Transforming growth factor-β (TGF-β) signals through serine/threonine-kinase receptors to activate several intracellular signaling pathways, including the "canonical" Smad signaling pathway. TGF-β has tumor-suppressive functions through its capacity to induce growth arrest and apoptosis. Escape from the anti-proliferative effects of TGF-β is a hallmark of almost all tumors of epithelial origin. This chapter first depicts the genetic alterations ablating the TGF-β tumor-suppressive functions in human tumors. Then, the chapter presents the genetically engineered mouse models that have been developed by introducing genetic alterations found in humans. These mouse models demonstrated the tumor-suppressive role of TGF-β in vivo shedding light on its intricate relationship with the tumor microenvironment. In all, the presented mouse models of cancer with an impaired TGF-β signaling pathway provide an integrated view of the complex tumor-suppressive role of TGF-β and represent valuable tools for preclinical studies.

Keywords Cre/lox • Digestive tract • Epithelial-to-mesenchymal transition (EMT) • Homotypic cell cannibalism (HoCC) • Mammary gland • Mutations • Pancreas • Smad • Transgenic mice • Xenograft

Abbreviations

ALK5	Activin receptor-like kinase 5
AOM	Azoxymethane
APC	Adenomatous polyposis coli

U. Valcourt • D.F. Vincent • L. Bartholin (✉)
TGF-β and Pancreatic Cancer Lab, Centre de Recherche en Cancérologie de LYON (CRCL), UMR INSERM 1052, CNRS 5286, Lyon I University UMR S 1052, Centre Léon Bérard, 28, rue Laënnec, 69373 LYON Cedex 08, France
e-mail: laurent.bartholin@lyon.unicancer.fr

A. Moustakas and K. Miyazawa (eds.), *TGF-β in Human Disease*,
DOI 10.1007/978-4-431-54409-8_6,

BMP	Bone morphogenetic protein
BMPR1A	Bone morphogenetic protein receptor, type IA
CTGF	Connective tissue growth factor
DEN	Diethylnitrosamine
DMBA	7,12-Dimethylbenz(α)anthracene
DPC4	Deleted in pancreatic carcinoma 4
ELF	Embryonic liver fodrin
EMT	Epithelial-to-mesenchymal transition
FAP	Familial adenomatous polyposis
HoCC	Homotypic cell cannibalism
IPMN	Intraductal papillary mucinous neoplasms
KSSTT	Pdx1-Cre; $KRAS^{G12D}$; Smad4-$KO^{L/L}$; TIF1γ-$KO^{L/L}$
LOH	Loss of heterozygosity
LRP	Low-density lipoprotein receptor-related protein
LSL	Lox-stop-lox
MAPK	Mitogen-activated protein kinase
MCN	Mucinous cystic neoplasms
Min	Multiple intestinal neoplasia
MMP	Matrix metalloproteinase
MMTV	Mouse mammary tumor virus
MSI	Microsatellite instability
PanIN	Pancreatic intraepithelial neoplasms
PDAC	Pancreatic ductal adenocarcinoma
PDGF	Platelet-derived growth factor
PyVmT	Polyoma virus middle T antigen
SCC	Squamous cell carcinoma
TGF-β	Transforming growth factor β
TGF-α	Transforming growth factor α
TIF1	Transcriptional intermediary factor 1
TPA	12-O-tetradecanoylphorbol-13-acetate
TβRI	TGF-β type I receptor
TβRII	TGF-β type II receptor
WAP	Whey acidic protein

6.1 Introduction: TGF-B Signaling and Genetic Alterations in Human Tumors

Transforming growth factor-β (TGF-β) is a secreted polypeptide belonging to a large family of cytokines and growth factors including TGF-βs, bone morphogenetic proteins (BMPs), and activins. The three isoforms of TGF-β (TGF-β1, TGF-β2, and TGF-β3) are secreted to different levels by most cell types, and they act as positive and negative regulators of differentiation and proliferative programs

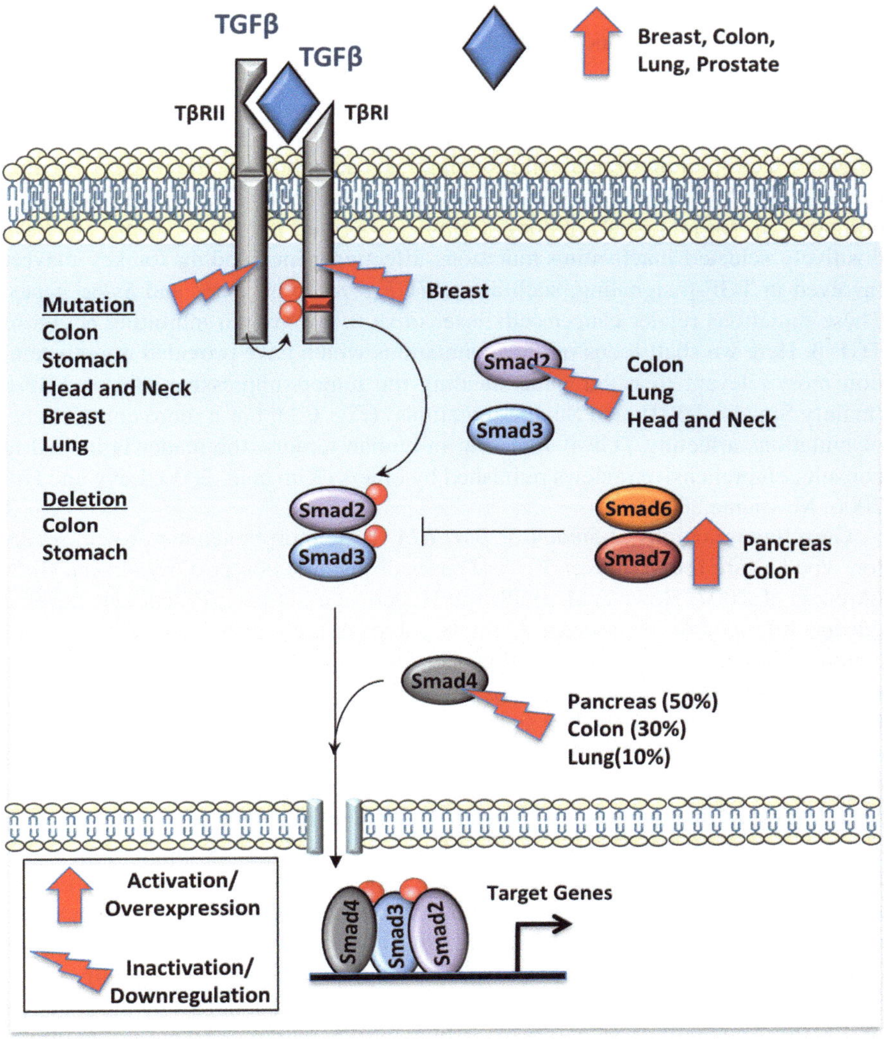

Fig. 6.1 Genetic alterations in human tumors affecting TGF-β signaling

(Letterio and Roberts 1998). TGF-βs play a crucial role in physiological processes involved in embryonic development, immune responses, cell growth control, and wound healing. All three TGF-β isoforms signal through a common heterotetrameric serine/threonine-kinase receptor complex (Massague 1992) comprising TβRII and TβRI (also termed activin receptor-like kinase 5, ALK5) (Fig. 6.1). After binding to its receptors, TGF-β induces phosphorylation of TβRI, which phosphorylates receptor-associated Smad2 and Smad3 (R-Smads). Phosphorylated Smad2 and

Smad3 interact with Smad4. The Smad2, -3, -4 complex accumulates within the nucleus, binds to DNA, and regulates transcription of target genes. In addition to this "canonical" Smad signaling pathway, TGF-β activates Smad-independent pathways (MAPK, RHOA, PI3K/AKT) (Mu et al. 2012).

For a comprehensive review of the signaling events and molecular mechanisms by which TGF-β orchestrated cell growth arrest and apoptosis, the reader is invited to read Chap. 5. Escape from the anti-proliferative effects of TGF-β signaling is an obligate step for almost all tumors of epithelial origin. For this, cancer cells have positively selected inactivating mutations affecting genes coding for key players involved in TGF-β signaling, such as the *TGF-β* receptor genes and *Smad* genes. These mutations render cancer cells insensitive to the growth-inhibiting action of TGF-β. Here we shall focus on those mutations which have provided the information most relevant to better understanding the tumor-suppressive role of TGF-β (mainly Smad4, TβRII, and Smad2 mutations) (Fig. 6.1). For a more complete list of mutations affecting TGF-β signaling in human tumors, the reader is invited to consult comprehensive reviews published by others (Kim et al. 2000; Levy and Hill 2006; Massague 2008).

Germline mutations in *Smad4* or *BMPRIA* (bone morphogenetic protein receptor, type IA) are found in over 50 % of cases of juvenile polyposis syndrome (JPS) (Aretz et al. 2007; Howe et al. 1998; van Hattem et al. 2008). JPS patients develop during childhood or adolescence multiple polyps in the gastrointestinal tract, associated with a high risk of colorectal cancer (50 % by the age of 60 years). Beyond these germline mutations concerning a minority of cancer patients, somatic mutations compromising the TGF-β signaling pathway are recurrently found in many sporadic tumors (Fig. 6.1). For instance, the locus 18q21.1 containing *Smad4* is deleted in over 50 % of human pancreatic cancers (this locus is also known as *DPC4, Deleted in Pancreatic Carcinoma 4*) (Hahn et al. 1996). Mutation of *Smad4* appears, indeed, as a preferential means of tumor escape from the tumor-suppressive functions of TGF-β signaling, since Smad3 is practically never mutated in human neoplasms and Smad2 mutations are rather rare. Why this is so remains unclear, but it has been hypothesized that Smad4, being involved in the signaling of all the TGF-β superfamily members, might additionally inactivate BMP signaling, which is also involved in tumor suppression. This hypothesis is supported by the presence of BMPRIA mutations in a subset of human JPS patients (Howe et al. 2001). Alternatively, Smad2 and Smad3 might be relatively spared by genetic alterations as compared to Smad4 because of an "R-Smad dependency" of many tumors. It is tempting to speculate that retention of Smad2 and Smad3 might confer a selective advantage to the tumor by driving invasive programs (epithelial-to-mesenchymal transition - EMT, immunosuppression, extracellular remodeling). For instance, Smad2 and Smad3 still accumulate in the nucleus in Smad4-deficient human pancreatic cancer cells (Subramanian et al. 2004) suggesting that R-associated Smad proteins have nuclear functions independently of Smad4. This hypothesis is comforted by the Smad4-independent association of Smad2 and Smad3 with inhibitor of nuclear factor κ-B kinase subunit alpha (IKKα) (Descargues et al. 2008) or transcriptional intermediary factor 1 γ (TIF1γ) (He et al. 2006).

TβRII is rarely altered in pancreatic cancers but is deleted in the majority of colorectal cancers associated with microsatellite instability (MSI) (Markowitz et al. 1995). Why is it that in MSI colon cancer mutations occur more frequently in TβRII than in any other gene involved in TGF-β signaling? MSI colon cancers are characterized by defective DNA mismatch repair (MMR), so the tumors accumulate mutations in microsatellite sequences. The TβRII gene contains "microsatellite-like" sequences rendering it particularly vulnerable to mutations in cases of MSI colorectal cancer. Interestingly, TβRII mutations resulting from MSI are positively selected, enabling the tumor to escape the growth-inhibiting effect of TGF-β (Biswas et al. 2008). This is a genuine example of the capacity of human neoplasms to escape the tumor-suppressive effect of TGF-β thanks to preexisting intrinsic tumor properties (microsatellite instability, in the case of MSI colorectal tumors).

Genome-scale exon sequencing has confirmed that TGF-β signaling mutations in the genes encoding the TβRII and Smad proteins are a hallmark of certain types of cancer, as illustrated by the sequencing of 24 pancreatic tumors (Jones et al. 2008) and 276 colorectal cancers (Cancer Genome Atlas Network 2012). In contrast, TGF-β mutations failed to emerge as a tumor signature from a study on 92 medulloblastomas (Pugh et al. 2012) and 100 breast tumors (Stephens et al. 2012). This raises the question: how do cancer cells having no recurrent genomic alteration that compromises TGF-β signaling escape TGF-β-mediated growth inhibition? One should consider, firstly, that some tumors derive from cells that are not sensitive to TGF-β growth inhibition in the first place. This is the case of sarcomas, which derive from mesenchymal cells intrinsically insensitive to the cytostatic action of TGF-β and which therefore do not require impairment of the TGF-β signaling pathway in order to develop. Secondly, tumors displaying a low rate of mutation in genes involved in TGF-β signaling may have escaped the tumor-suppressive effect of TGF-β by other means. For instance, repression of TGF-β receptor gene transcription by DNA methylation has been reported in a human gastric cancer (Kang et al. 1999) and human breast tumors (Hinshelwood et al. 2007). Also reported in human tumors is up-regulation of negative regulators of TGF-β signaling. For instance, SKI/SNON expression is increased in breast cancer (Zhang et al. 2003), colorectal cancer (Buess et al. 2004), melanoma (Medrano 2003; Poser et al. 2005; Reed et al. 2001), and esophageal squamous cell carcinoma (Fukuchi et al. 2004).

6.2 Mouse Models of TGF-β-Induced Tumor Suppression

Many mouse models have been developed to decipher the role of TGF-β in cancer. They rely on introducing into the murine genome the human mutations described in paragraph 1 with the ultimate goal to compromise in vivo the tumor-suppressive functions described in paragraph 2.

Homozygous TGF-$\beta1^{null}$ mice develop autoimmune disorders, culminating in their death by 3 weeks of age (Kulkarni et al. 1993; Shull et al. 1992). Half of the TGF-$\beta1^{null}$ embryos die in utero because of hematopoiesis and vasculogenesis

Table 6.1 Mouse models with whole-body homozygous inactivation of central regulators of the TGF-β signaling

Knockout gene	References	Phenotype
TGF-β1^{null}	Shull et al. (1992)	Multifocal inflammatory disease, death before 1 month
	Kulkarni et al. (1993)	Multifocal inflammatory disease, death before 1 month
	Dickson et al. (1995)	Multifocal inflammatory disease, death before 1 month, 50 % of death at E10.5 because of defective hematopoiesis and vasculogenesis
TGF-β2^{null}	Sanford et al. (1997)	Wide range of developmental defects causing perinatal death (cardiac, lung, craniofacial, limb, spinal column, eye, inner ear, and urogenital defects)
TGF-β3^{null}	Proetzel et al. (1995)	Failure of the palatal shelves to fuse leading to cleft palate
TβRI/ALK5^{null}	Larsson et al. (2001)	Death at E10.5 because of defective hematopoiesis and vasculogenesis
TβRII^{null}	Oshima et al. (1996)	Death at E10.5 because of defective hematopoiesis and vasculogenesis
TβRIII^{null}/Betaglycan	Stenvers et al. (2003)	Embryonic lethality caused by defects in heart and liver
Smad2^{null}	Nomura and Li (1998) Waldrip et al. (1998) Weinstein et al. (1998)	Embryonic death because gastrulation defect
Smad3^{null}	Zhu et al. (1998)	Colorectal cancers
Smad4^{null}	Yang et al. (1998)	Death at E8.5, gastrulation defect
	Sirard et al. (1998)	Death at E8.5, gastrulation defect
	Takaku et al. (1998)	Death at E8.5, gastrulation defect

defects in the yolk sac (Dickson et al. 1995). TGF-β2^{null} (Sanford et al. 1997) and TGF-β3^{null} (Proetzel et al. 1995) mice are viable, but display multiple developmental defects. Homozygous knockouts of TβRI/ALK5 (Larsson et al. 2001) or TβRII (Oshima et al. 1996) are embryonically lethal. Knockout TβRIII (Betaglycan) mice develop embryonic lethal proliferative defects in heart and apoptosis in liver (Stenvers et al. 2003). Smad3^{null} mice spontaneously develop colorectal cancers (Zhu et al. 1998). Inactivation either of the other two Smad proteins involved in TGF-β signaling, Smad2 (Nomura and Li 1998; Waldrip et al. 1998; Weinstein et al. 1998) and Smad4 (Sirard et al. 1998; Takaku et al. 1998; Yang et al. 1998), leads to embryonic death (Table 6.1).

Altogether, these mouse models provide a perfect illustration of the pleiotropic effects of TGF-β and of its crucial role in development and homeostasis, as all these whole-organism mutations lead to embryonic or perinatal lethality (with the exception of Smad3 inactivation, responsible for colorectal tumors). As the severity of the induced phenotypes limits the utility of these models in cancer studies, several strategies have been devised to get around this limitation. In this paragraph, we will describe in detail informative mouse models that have been obtained by targeting the mammary gland, digestive tract, exocrine pancreas. Mouse models of skin cancer, which constitute appropriate systems for exploring the effects of external carcinogens such as chemicals and ultraviolet irradiation, will be described in Chap. 9. Skin models along with models targeting other organs [e.g., the liver (Tang et al. 1998), prostate (Kundu et al. 2000; Pu et al. 2009), or head and neck (Bian et al. 2009; Bornstein et al. 2009)] are listed in Table 6.2 but are not extensively described in the text.

Table 6.2 Mouse models demonstrating the tumor-suppressive role of TGF-β

Organ	Genetic alteration	Organ specificity	Oncogenic cooperation	References
Breast	TGF-β1$^{S223/225}$	Yes	Alone	Pierce et al. (1993)
	TGF-β1$^{S223/225}$	Yes	Alone	Jhappan et al. (1993)
	TGF-β1$^{S223/225}$	Yes	Alone	Kordon et al. (1995)
	TGF-β1$^{S223/225}$	Yes	TGF-α	Pierce et al. (1995)
	TGF-β1$^{S223/225}$	Yes	Alone	Buggiano et al. (2001)
	TGF-β1$^{S223/225}$	Yes	Alone	Boulanger and Smith (2001)
	TGF-β1$^{S223/225}$	Yes	Neu	Muraoka et al. (2003)
	TGF-β1$^{S223/225}$	Yes	Alone	Boulanger et al. (2005)
	TGF-β3-KO	Whole body	Alone	Nguyen and Pollard (2000)
	TβRI^{CA}	Yes	PyVmT	Muraoka-Cook et al. (2004)
	TβRI^{CA}	Yes	Neu	Muraoka-Cook et al. (2006)
	TβRII^{DN}/TβRI^{CA}	Yes	Neu	Siegel et al. (2003)
	TβRII^{DN}	Yes	DMBA	Bottinger et al. (1997a)
	TβRII^{DN}	Yes	Alone	Gorska et al. (1998)
	TβRII^{DN}	Yes	None and TGF-α	Gorska et al. (2003)
	TβRII-KO^{het}	Yes	PyVmT	Fang et al. (2011)
	TβRII-KO^{null}	Yes	PyVmT	Forrester et al. (2005)
	TβRII-KO^{null}	Yes	Alone	Cheng et al. (2005)
	TβRII-KO^{null}	Yes	Alone	Yang et al. (2008)
	TβRII-KO^{null}	Yes	PyVmT	Bierie et al. (2008)
	TβRII-KO^{null}	Yes	PyVmT	Bierie et al. (2009)
	None (treatment with TβRII blocking antibody)	N/A	PyVmT	Muraoka et al. (2002)
	Smad4-KO^{null}	Yes	Alone	Li et al. (2003)
	Smad3-KO^{null}	Whole body	Alone	Yang et al. (2002)
Intestine	TGF-β1^{null}	Whole body	Rag2^{null}	Engle et al. (2002)
	TGF-β1^{null}	Whole body	Rag2^{null}	Engle et al. (1999)
	TβRI-KO^{het}	Whole body	Apc^{Min}	Zeng et al. (2009)

(continued)

Table 6.2 (continued)

Organ	Genetic alteration	Organ specificity	Oncogenic cooperation	References
	TβRII-KO^{null}	Yes	Azoxymethane (AOM)^{null}	Biswas et al. (2004)
	TβRII-KO^{null}	Yes	Apc^{1638N}	Munoz et al. (2006)
	TβRII-KO^{null}	Yes	KRAS^{G12D} and Apc^{1638N}	Trobridge et al. (2009)
	Smad2-KO^{het}	Whole body	ApcΔ^{716}	Takaku et al. (2002)
	Smad2-KO^{het}	Whole body	Apc^{580D}	Hamamoto et al. (2002)
	Smad3-KO^{null}	Whole body	Alone	Zhu et al. (1998)
	Smad3-KO^{null}	Whole body	Helicobacter spp.	Maggio-Price et al. (2006)
	Smad3-KO^{null}	Whole body	ApcMin	Sodir et al. (2006)
	Smad4-KO^{het}	Whole body	ApcΔ^{716}	Takaku et al. (1998)
	Smad4-KO^{het}	Whole body	Alone	Takaku et al. (1999)
	Smad4-KO^{het}	Whole body	Alone	Xu et al. (2000)
	Smad4-KO^{het}	Whole body	Alone	Hohenstein et al. (2003)
	Smad4-KO^{hetE6D}	Whole body	Apc^{1638N}	Alberici et al. (2006)
	Smad4-KO^{het}	Whole body	Elf^{het}	Redman et al. (2005)
	Smad4-KO^{het}	Whole body	Elf^{het}	Tang et al. (2005)
	Smad4-KO^{het}	Whole body	Elf^{het}	Katuri et al. (2006)
	TβRII^{fspKO}	Mesenchymal compartment	Alone	Bhowmick et al. (2004)
	Smad4-KO^{het}	Whole body	ApcΔ^{716}	Kitamura et al. (2007)
	Smad4-KO^{het}	Yes	Apc^{1638N}	Freeman et al. (2012)
	Smad4-KO^{het}	Yes	Alone	Kim et al. (2006)
Skin	TGF-β1$^{S223/225}$	Yes	Alone	Sellheyer et al. (1993)
	TGF-β1$^{S223/225}$	Yes	TPA	Cui et al. (1995)
	TGF-β1$^{S223/225}$	Yes	TPA	Fowlis et al. (1996)
	TGF-β1$^{S223/225}$	Yes	TPA	Cui et al. (1996)
	TGF-β1$^{S223/225}$	Yes	DMBA	Weeks et al. (2001)
	TGF-β1$^{S223/225}$	Yes	Alone	Liu et al. (2001)
	TGF-β1$^{S223/225}$	Yes	Alone	Wang et al. (1999)
	TβRII^{DN}	Yes	Alone	Wang et al. (1997)
	TβRII^{DN}	Yes	Alone	Amendt et al. (1998)
	TβRII^{DN}	Yes	TPA	Go et al. (1999)
	TβRII^{DN}	Yes	TPA,DMBA	Go et al. (2000)
	TβRII^{DN}	Yes	Alone	Amendt et al. (2002)
	TβRII^{DN}+ TβRII-KO^{null}	Yes	DMBA/TPA	Han et al. (2005)
	TβRII-KO^{null}	Yes	Ha-Ras^{V12}	Guasch et al. (2007)
	Smad3-KO^{null}	Whole body	Alone	Ashcroft et al. (1999)
	Smad4-KO^{het}	Whole body	Elf^{het}	Redman et al. (2005)
	Smad4-KO^{null}	Yes	Alone	Li et al. (2003)

(continued)

Table 6.2 (continued)

Organ	Genetic alteration	Organ specificity	Oncogenic cooperation	References
	Smad4-KO^{null}	Yes	$PTEN^{-/-}$	Yang et al. (2005)
	Smad4-KO^{null}	Yes	Alone	Qiao et al. (2006)
Pancreas	$T\beta RII^{DN}$	Whole body	Alone	Bottinger et al. (1997b)
	$T\beta RII$-KO^{null}	Yes	K-Ras^{G12D}	Ijichi et al. (2006)
	$T\beta RII$-KO^{null}	Yes	K-Ras^{G12D}	Ijichi et al. (2011)
	Smad4-KO^{null}	Yes	K-Ras^{G12D}	Bardeesy et al. (2006)
	Smad4-KO^{null}	Yes	K-Ras^{G12D}	Kojima et al. (2007)
	Smad4-KO^{null}	Yes	K-Ras^{G12D}	Izeradjene et al. (2007)
	$Smad7^{Tg}$	Yes	Alone	Kuang et al. (2006)
	TIF1γ-KO^{null}	Yes	K-Ras^{G12D}	Vincent et al. (2009)
	TIF1γ-KO^{null}	Yes	K-Ras^{G12D}	Vincent et al. (2012)
	NupR1-KO^{null}	Whole body	K-Ras^{G12D}	Cano et al. (2012)
Head and Neck	$T\beta RI$-KO^{null}	Yes	DMBA	Bian et al. (2009)
	Smad4-KO^{null}	Yes	Alone	Bornstein et al. (2009)
Prostate	$T\beta RII^{DN}$	Yes	SV40-Large T	Pu et al. (2009)
	$T\beta RII^{fspKO}$	Mesenchymal compartment	none	Bhowmick et al. (2004)
	$T\beta RII^{DN}$	Yes	none	Kundu et al. (2000)
Liver	TGF-$\beta 1^{het}$	Whole body	Diethylnitro-samine (DEN)	Tang et al. (1998)
	$Smad3^{Tg}$	Yes	Diethylnitro-samine (DEN)	Yang et al. (2006)

6.2.1 *The Mammary Gland*

In mice, mammary gland development (Sternlicht 2006) begins shortly after mid-gestation (E10.5). It is characterized by the presence at birth of a rudimentary system of branching ducts opening into the nipple. From birth to puberty, there is but little development of the branching duct network. At puberty (~5 weeks of age), a massive wave of branching occurs, so that the ducts fill the entire mammary fat pad of the young adult. During the estrus cycles, the mammary gland undergoes successive cycles of proliferation and regression. Pregnancy is associated with massive cell proliferation in the mammary gland, along with lobular-alveolar differentiation to produce milk. Lactation in the mother's mammary gland starts when the pups are born. After weaning, lactation stops and the secreting lobular-alveolar network of the mammary gland involutes, an event resulting from massive apoptosis.

In the TGF-$\beta 1^{S223/225}$ mutant, cysteine residues at positions 223 and 225 are replaced with serine residues so that TGF-β1 is secreted in an active form (this mutant is unable to bind the latency-associated protein, LAP). Transgenic mice expressing TGF-$\beta 1^{S223/225}$ have been generated, with expression of the TGF-$\beta 1^{S223/225}$ transgene under the control of either the promoter of the mouse mammary tumor virus (MMTV) gene (associated with early differentiation of the mammary gland) or the promoter of the whey acidic protein (WAP) gene (associated with late pregnancy, lactation, and early involution). In MMTV-TGF-β1$^{S223/225}$ mice (Pierce et al. 1993), duct development is severely compromised, this leading to mammary gland hypoplasia. WAP-TGF-β1$^{S223/225}$ mice develop normal ducts but are not capable to support lactation as they present a severe deficiency in their ability to form secretory lobules during pregnancy (Jhappan et al. 1993) as a consequence of the early senescence of the regenerative capacity of the mammary ductal epithelium (Kordon et al. 1995). TGF-β3 expression from the β-lactoglobulin promoter in the lactating epithelium of mice is likewise reported to induce involution of the mouse mammary gland (Nguyen and Pollard 2000). Models have also been developed to inhibit TGF-β signaling in the mammary gland. As expected, these models give rise to phenotypes opposite to those just described. Thus, virgin MMTV-TβRII^{DN} (dominant negative mutant form of TβRII) (Bottinger et al. 1997b; Gorska et al. 1998) and MMTV-Cre/TβRII-KO$^{L/L}$ (a conditional TβRII-knockout allele) (Forrester et al. 2005) mice display hyperplasia and differentiation of the mammary glands. The MMTV-TβRII^{DN} model shows delayed involution of the mammary gland after forced weaning of pups during lactation (Gorska et al. 2003).

Smad3^{null} mice, which develop colorectal tumors (Zhu et al. 1998), also show under-developed mammary glands and involution defects (Yang et al. 2002). The mammary glands of WAP-Cre/TβRII-KO$^{L/L}$ mice likewise fail to undergo complete involution of the lobular-alveolar secreting structures (Bierie et al. 2009). Defective mammary gland involution is due to compromised apoptosis, a crucial biological process controlled by TGF-β. Unexpectedly, MMTV-Cre/Smad4$^{L/L}$ and WAP-Cre/Smad4$^{L/L}$ mice display no involution defects or obvious developmental defects (Li et al. 2003), but they do show spontaneous mammary tumor formation (Li et al. 2003). The mammary glands of MMTV-Cre/Smad4-KO$^{L/L}$ mice show increased proliferation, alveolar hyperplasia, and trans-differentiation into squamous epithelial cells, and they finally develop squamous cell carcinoma (SCC). To date, this is the only model of spontaneous mammary tumor formation in genetically engineered mice with compromised TGF-β signaling. The development of mammary tumors when Smad4 is knocked down may be due to more efficient inactivation of TGF-β signaling than in TβRII^{DN} mice (Bottinger et al. 1997b; Gorska et al. 1998). Alternatively, it might be due to inactivation of a non-TGF-β pathway such as the BMP pathway, which should occur upon Smad4 knockdown but not in the TβRII-KO model (Forrester et al. 2005).

Altogether, these observations demonstrate that TGF-β exerts anti-proliferative effects constraining mammary duct development in virgin mice and enabling involution of the mammary gland at the end of lactation. They also functionally validate the pivotal role of Smad4 and TβRII in limiting cell proliferation and tumor formation, in line with the high prevalence of genetic alterations affecting both genes in human neoplasms.

Because disruption of TGF-β signaling is not sufficient to induce mammary tumors in most of the models described above (only Smad4 inactivation is sufficient to induce mammary tumors), mutations affecting TGF-β signaling have been combined with various oncogenic factors. This has yielded robust models of predisposition to mammary tumors. These approaches include:

1. Exploring the effects of chemical carcinogens in mice with attenuated TGF-β signaling. It appears that MMTV-Cre/TβRII^{DN} mice are more sensitive to 7,12-dimethylbenz(α)anthracene (DMBA)-induced mammary tumorigenesis than control littermates (Bottinger et al. 1997b). This sensitization indirectly confirms the tumor-suppressive function of TGF-β.
2. Studying the role of TGF-β in relation to mammary tumorigenesis in multiparous females. MMTV-Cre/TβRII^{DN} mice having experienced two pregnancies with partial involution of the secreting lobular-alveolar structures have an increased risk of mammary tumor formation (Gorska et al. 2003). This observation suggests that TGF-β is a potent inhibitor of the transformation of hyperplastic tissues, known to be more likely to evolve towards neoplastic lesions. This observation identifies TGF-β-induced apoptosis, responsible for mammary gland involution after lactation, as a crucial tumor-suppressive mechanism.
3. Assessing susceptibility to the development of mammary tumors after MMTV infection. Milk-transmitted MMTV infection has been extensively used to induce mammary tumors [for reviews, see (Callahan and Smith 2000; Ross 2010)]. This line of research has yielded contradictory results: one group found WAP-TGF-β1$^{S223/225}$ mice to show no change in susceptibility to MMTV infection or subsequent tumor development, despite a severe decrease in lobular alveolar development (Buggiano et al. 2001); another group found WAP-TGF-β1$^{S223/225}$ mice to display a decreased risk of mammary cancer after MMTV infection (Boulanger and Smith 2001), thanks to early senescence in the mammary stem cell population (Boulanger et al. 2005). Although the reason for this discrepancy is unclear, it is important to note that TGF-β may prevent mammary tumors by inducing the senescence of stem cells, in addition to its pro-apoptotic effect on differentiated cells.
4. Exploiting oncogene-induced mouse models of mammary cancer. Three oncogenes are classically used in transgenic mice: transforming growth factor alpha (TGF-α), Her-2/neu (erbB-2), and polyoma virus middle T antigen (PyVmT). MMTV-TGF-β1$^{S223/225}$/MMTV-TGF-α mice develop significantly fewer mammary tumors than MMTV-TGF-α mice (Pierce et al. 1995). Conversely, MMTV-Cre/TβRII-KO$^{L/L}$/MMTV-PyVmT mice develop tumors more quickly than MMTV-PyVmT mice (Forrester et al. 2005), and MMTV-TβRII^{DN}/MMTV-TGF-α develop mammary tumors at a higher frequency than MMTV-TGF-α mice (Gorska et al. 2003). WAP-Cre/TβRII-KO$^{L/L}$/MMTV-PyVmT mice also present an accelerated mammary tumorigenesis (Bierie et al. 2008). Finally, MMTV-TβRI^{CA}/MMTV-Neu mice develop primary tumors later (Siegel et al. 2003) and MMTV-TβRIIDN/MMTV-Neu mice develop them earlier than MMTV-Neu mice (Siegel et al. 2003). Together, these results indicate that TGF-β efficiently opposes TGF-α-, PyVmT-, and Her2/Neu- oncogene-driven transformation of the epithelial mammary tissue, illustrating once again its tumor-suppressor role.

5. Disrupting TGF-β signaling in the stromal compartment instead of the epithelial compartment. This approach relies on a mouse model in which TGF-β signaling is specifically abolished in the mesenchymal compartment and involves associating a Cre-recombinase controlled by the FSP1 (fibroblast specific protein-1; S100A4) promoter with a conditional TβRII-knockout allele in FSP1-Cre/TβRII$^{L/L}$ compound transgenic mice. These mice develop pre-neoplastic lesions resulting in prostate and stomach carcinomas and display severely compromised development of the mammary gland (Bhowmick et al. 2004). Furthermore, it appears that TβRII-deficient fibroblasts can facilitate tumor growth when grafted with PyVmT or 4T1 mammary carcinoma cells (Cheng et al. 2005). This finding has been further validated in MMTV-PyVmT/FSP1-Cre/TβRII$^{L/+}$ mice developing autochthonous mammary tumors (Fang et al. 2011). Finally, xenografts onto the chicken embryo chorioallantoic membrane revealed that the absence of TβRII in fibroblasts could promote invasiveness of murine mammary carcinoma cells (Matise et al. 2012). Altogether, these studies clearly demonstrate that the tumor-suppressive function of TGF-β relies also on microenvironmental effects.

Importantly, although this point lies outside the scope of this chapter, the mammary gland is particularly suitable for investigating the dual role of TGF-β during carcinogenesis, since many studies based on mouse models of mammary tumors indicate that once transformation has occurred and a tumor has become established, TGF-β actually facilitates tumor progression and subsequent metastatic dissemination, in contrast to its tumor-suppressive effect on the normal and hyperplastic mammary gland (Bottinger et al. 1997a; Forrester et al. 2005; Gorska et al. 2003; Muraoka-Cook et al. 2004; Muraoka-Cook et al. 2006; Muraoka et al. 2002; Muraoka et al. 2003). Tumor-promoter functions of TGF-β are developed in Chap. 7.

6.2.2 The Digestive Tract

The digestive tract is anatomically divided into different parts: the mouth, esophagus, stomach, small intestine, colon, and rectum. The intestinal lumen is lined with a specialized epithelium organized into crypts and villi (van der Flier and Clevers 2009). Stem and proliferating cells are located in the crypts, whereas differentiated cells (enterocytes, entero-endocrine cells, and goblet cells) are located in the villi. Paneth cells constitute the only differentiated cell population settling in the crypts (they do not migrate "upward"). Colorectal adenocarcinoma is the most frequent tumor of the digestive tract. In 1990, Fearon and Vogelstein proposed a model according to which this cancer arises through an ordered sequence of mutations called the adenomacarcinoma sequence (Fearon and Vogelstein 1990). As mentioned earlier in this chapter, the spontaneous development of colorectal tumors in Smad3^{null} mice constitutes strong evidence of the tumor-suppressive role of TGF-β

(Zhu et al. 1998). To explore more precisely the role of TGF-β signaling in cancers of the digestive tract, different cancer predisposition models have been developed:

1. Confirmation of the tumor-suppressive effect of TGF-β signaling on digestive tract tumorigenesis was first obtained in heterozygous Smad4-KO^{het} mice. These mice develop polyps in the proximal gastrointestinal tract, confirming the central role of Smad4 in tumor suppression (Takaku et al. 1999; Xu et al. 2000). Smad4-KOE6sad (sad, serrated adenomas) is a null mutant allele caused by a spontaneous deletion in a splice acceptor site, sufficient to induce serrated adenomas and mixed polyposis (Alberici et al. 2006). Interestingly, knocking down Smad4 in T-lymphocytes (CD4-Cre/Smad4-KO$^{L/L}$) results in epithelial tumors of the oral cavity, stomach, duodenum, rectum, and colon (Kim et al. 2006). This recent observation constitutes new evidence that TGF-β signaling in the microenvironment plays an active role in constraining epithelial tumorigenesis.
2. The best-represented category of digestive tract cancer predisposition models relies on associating TGF-β signaling mutations with mutations that inactivate the adenomatous polyposis coli (APC) gene, which encode an inhibitor of the WNT/β-catenin signaling pathway (van der Flier and Clevers 2009). This pathway is particularly important in the digestive tract, as it maintains the crypt as a "proliferative, undifferentiated, and multipotent" compartment. In contrast, the WNT/β-catenin is not active in the villi, which constitute the "differentiated" compartment. In the absence of WNT ligands, β-catenin binds to a complex containing GSK3β (glycogen synthase kinase-3β), CK1 (casein kinase 1), axin (conductin), and APC. This complex phosphorylates β-catenin, targeting it for ubiquitination and degradation. In the presence of WNT ligands, the receptors Frizzled and LRP (low-density lipoprotein receptor-related protein) are activated, leading to inhibition of the "destruction complex" and stabilization of β-catenin, which can accumulate inside the nucleus and eventually activate transcription of growth-promoting genes. The canonical WNT signal transduction pathway is frequently impaired in human cancers, especially those arising in the digestive tract. As the most common alterations of the WNT/β-catenin pathway in cancer are APC-inactivating and β-catenin-activating mutations, several mouse models have been developed with APC mutations. The different mutants develop small intestinal polyps that are histologically indistinguishable, but the average number of polyps is different according to the mutation: 3 in APC^{1638N} mutants, 30 in APC^{Min} (Min, multiple intestinal neoplasia) mutants, and 300 in APCΔ^{716} mutants. These mouse models are somewhat evocative of human familial adenomatous polyposis (FAP, caused by deletions on chromosome 5q encompassing the APC gene), but in contrast to mice with APC mutations, human FAP patients, who present an increased risk of colorectal cancer, develop polyps mainly in the colon and rectum (not in the small intestine).

   Because TGF-β is impaired in colorectal cancers as a result of mutations affecting the genes encoding the TGF-β receptors and Smad proteins, these mutations have been combined with APC mutations in transgenic mouse models. Whereas mice homozygous for a Smad3-inactivating mutation spontaneously develop colorectal cancers between 18 and 24 weeks of age (Zhu et al. 1998), $APC^{min}/Smad3^{null}$ mice develop tumors much earlier and are moribund by the age of 8 weeks (Sodir et al. 2006). $APC\Delta^{716}/Smad4\text{-}KO\Delta^{Ex1/+}$ mice develop colorectal cancers (Takaku et al. 1998) and permitted the identification of a new type of immature myeloid cell (iMC), expressing matrix metalloproteinases (MMP9 and MMP2) and the CC-chemokine receptor 1 (CCR1), present at the invasion front toward CCL9 (the CCR1 ligand), to facilitate tumor invasion (Kitamura et al. 2007). Inactivation of CCR1 in $APC\Delta^{716}/Smad4\text{-}KO\Delta^{Ex1/+}/CCR1\text{-}KO^{null}$ compound mice inhibits the accumulation of iMCs preventing the tumor invasion. Whereas $Smad4\text{-}KO^{E6sad/+}$ (Hohenstein et al. 2003) develop serrated adenomas and mixed polyposis, $APC^{1638N}/Smad4\text{-}KO^{E6sad/+}$ mice have been shown to develop aggressive colorectal cancers (Alberici et al. 2006). As observed in human tumors, the onset of colorectal tumors in mouse models is systematically associated with Smad4 loss of heterozygosity (LOH), indicating that total loss of Smad4 function is required for transformation, rather than haploid insufficiency. Heterozygous inactivation of Smad2 in a mutated APC background was shown in one study to accelerate tumor progression (Hamamoto et al. 2002) and in another study to have no effect (Takaku et al. 2002). The cause of this discrepancy remains unclear but may be due to the use of different mutant APC alleles. Heterozygous inactivation of TβRI in association with the APC^{Min} mutation is associated with a twofold increase in the number of intestinal tumors and with colorectal cancer onset without loss of heterozygosity at the TβRI locus (Zeng et al. 2009). Thus, in contrast to what is observed upon Smad4 inactivation, haploid insufficiency of TβRI is sufficient to facilitate tumorigenesis. Finally, Villin-Cre-dependent activation of the conditional $T\beta RII\text{-}KO^{L/L}$ conditional allele is associated with colorectal cancer development in an APC^{1638N} background (Munoz et al. 2006). The molecular interaction between the TGF-β signaling and the WNT/β-catenin signaling has been recently described by performing microarray performed with biological material prepared from adenomas from $APC\Delta^{1638/+}/K19\text{-}Cre\text{-}ERT2/Smad4^{L/L}$, which revealed increased levels of β-catenin mRNA and WNT signaling (Freeman et al. 2012).

Altogether, these models show that TGF-β-pathway-inactivating mutations strongly potentiate the capacity of APC mutants to induce colorectal cancers. These mouse models clearly demonstrate that progression of colorectal cancers involves inactivation of TGF-β signaling illustrating its tumor suppressor effect. Notably, by combining APC mutations with TGF-β signaling mutations, it is possible to better mimic human FAP, as the lesions appear in the distal intestine.

3. A third category of digestive tract cancer predisposition models relies on the association of TGF-β signaling mutations with mutations affecting genes other than APC. The ELF protein (embryonic liver fodrin), identified in foregut progenitors, belongs to the β-spectrin family involved in cell polarity (Machnicka

et al. 2012). Double heterozygous Smad4-KO^{het}/ELF-KO^{het} mutants develop colorectal cancers and also gastric and liver tumors (Katuri et al. 2006; Redman et al. 2005; Tang et al. 2005). Activating KRAS mutations are classically associated with the transition from the adenoma to the carcinoma state. LSL-KRAS^{G12D} (LSL, Lox-Stop-Lox) is a conditional (Cre-inducible) allele encoding a constitutively active KRAS mutant (KRAS^{G12D}). Villin-Cre/LSL-KRAS^{G12D}/TβRII-KO$^{L/L}$ mice develop small intestinal and colorectal carcinomas by the age of 20 weeks (Trobridge et al. 2009).

4. A fourth strategy for producing digestive tract cancer predisposition models involves mice with different genetic backgrounds to avoid the lethal inflammatory phenotype observed in mice with compromised TGF-β signaling due to TGF-β or Smad inactivation. To avoid the generalized inflammatory disorders that cause early death of TGF-β1^{null} mice at ~3 weeks of age (Kulkarni et al. 1993; Shull et al. 1992), homozygous inactivation of TGF-β1 was introduced into immunodeficient RAG2^{null} mice, lacking both B and T cells. TGF-β1^{null}/RAG2^{null} double-mutant mice do not develop inflammatory disease, live to adulthood, and develop colorectal cancers by the age of 5 months (Engle et al. 1999). Interestingly, if these mice are made germ-free, they no longer develop colorectal cancer; this indicates that microbe-induced inflammation is crucial to the development of colon cancer (Engle et al. 2002). In the 129/Sv background, all Smad3^{null} mice develop tumors by 6 months of age, whereas in the hybrid 129/Sv;C57BL/6 background, only 30 % of these mice develop tumors (Zhu et al. 1998). Once again, the crucial role of the microenvironment is illustrated.
5. The last approach to developing digestive tract cancer predisposition models involves treating transgenic mice with external agents facilitating colorectal transformation. Indeed, although embedded in the body, the lumen of the digestive tract is in fact in direct contact with the external environment, rendering it more vulnerable to external aggressions. Fatty-acid-binding (FABP)-promoter-mediated Cre recombination makes it possible to target floxed alleles to the colonic crypts (Saam and Gordon 1999). FABP-Cre/TβRII-KO$^{L/L}$ mice displaying targeted inactivation of TβRII in the distal intestine do not develop tumors of the digestive tract but are more sensitive than their "control" littermates to colorectal carcinogenesis induced by azoxymethane (a chemical carcinogen) (Biswas et al. 2004). In a helicobacter-free environment, furthermore, Smad3^{null} mice do not develop colon cancer by the age of 9 months, and infection of these mice induces colon cancer in >50 % of the animals (Maggio-Price et al. 2006). Finally, it has been shown that TGF-β-KO^{het} mice developed more liver and lung tumors after treatment with diethylnitrosamine (DEN) (Tang et al. 1998), whereas transgenic mice overexpressing Smad3 are partially protected against DEN-induced liver carcinogenesis as a result of increased apoptosis (Yang et al. 2006). This again illustrates the crucial role of TGF-β-induced apoptosis in tumor suppression.

6.2.3 The Exocrine Pancreas

The pancreas comprises separate functional units that regulate two major physiological processes: digestion and glucose metabolism [for review, see (Gittes 2009)]. The exocrine pancreas consists of acinar and duct cells. The acinar cells produce digestive enzymes and constitute the bulk of the pancreatic tissue. They are organized into grape-like clusters at the smallest termini of the branching duct system. The ducts, which add mucous and bicarbonates to the enzyme mixture, form a network of increasing size, culminating in main and accessory pancreatic ducts that discharge their content into the duodenum. The endocrine pancreas, consisting of specialized cell types organized into compact islets embedded within the acinar tissue, secretes hormones (glucagon, insulin, pancreatic polypeptide, and somatostatin) into the bloodstream.

Pancreatic ductal adenocarcinoma (PDAC) is a very aggressive neoplasm, always lethal [for review, see (Kern et al. 2011)]. It is the fifth most common cause of death from cancer in the western world. At the time of diagnosis, the median age of PDAC-bearing patients is 65–70 years. PDAC affects the exocrine pancreas and accounts for ~80 % of pancreatic cancers. It arises from precursor lesions with differentiated ductal features. These precursor lesions are divided into three groups: pancreatic intraepithelial neoplasms (PanIN), intraductal papillary mucinous neoplasms (IPMN), and mucinous cystic neoplasms (MCN). Certain genetic alterations are recurrently found in sporadic and familial human PDAC, such as KRAS-activating mutations in over 90 % of cases and Ink4A/Arf- and TP53-inactivating mutations, each in about 50 % of cases [for review, see (Hansel et al. 2003) and (Landi 2009)]. The *SMAD4* gene is located at chromosomal locus 18q21, which is frequently lost in pancreatic cancer, especially in higher-grade lesions (Fukushige et al. 1998; Hahn et al. 1996; Schleger et al. 2000). Introduction of these mutations into mice has made it possible to create more than 20 models, recapitulating the whole spectrum of lesions observed in humans (according to the standard classification described in (Hruban et al. 2001), (Hruban et al. 2006) and http://pathology.jhu.edu/pancreas_panin/). Most of the models developed so far rely on targeted conditional expression in the Lox-Stop-Lox KRAS^{G12D} (LSL-KRAS^{G12D}) knock-in mouse strain (Jackson et al. 2001), which bears a Cre-inducible allele targeting the endogenous KRAS locus with the most common alteration found in pancreatic cancers. Pancreas-specific activation of the LSL-KRAS^{G12D} allele with the Pdx1-Cre (Gu et al. 2002) or Ptf1a/p48-Cre (Kawaguchi et al. 2002) allele induces the development of age-dependent precursor lesions (PanINs lesions in 100 % of animals) and PDAC after 1 year (in about 10 % of mice) (Hingorani et al. 2003). Combining the LSL-KRAS^{G12D} allele with inactivation of the tumor suppressor Ink4A/Arf leads to very aggressive PDAC at a few weeks of age (median survival: approximately 2 months). This is the most robust and aggressive mouse model of PDAC developed to date (Aguirre 2003). This work and the unique opportunity offered by the Cre-lox system have prompted researchers to combine the KRAS^{G12D} allele with many tumor suppressors inactivation and with oncogenic activations observed in humans (Bartholin 2012).

TGF-β signaling mutations combined with KRAS-activating mutations have emerged as very potent inducers of pancreatic cancer in mouse models, illustrating the tumor-suppressive role of TGF-β in this organ. These models provide a genuine opportunity to explore the complex role of TGF-β during carcinogenesis. Since Smad4 deletion is one of the most common genetic alterations in human PDAC, several mouse models have been developed where TGF-β signaling is specifically inactivated in the pancreas. Combined with KRAS activation, homozygous deletion of Smad4 induces cystic lesions, either IPMNs or MCNs (Bardeesy et al. 2006; Izeradjene et al. 2007; Kojima et al. 2007). Smad7 is a repressor of TGF-β signaling. Transgenic mice expressing Smad7 in the pancreas also develop PanINs (Kuang et al. 2006). Combined with TβRII inactivation targeted to the pancreas using the Ptf1a-Cre allele, $KRAS^{G12D}$ induces PDAC with a higher penetrance than observed with Smad4 (Ijichi et al. 2006), indicating that the tumor-suppressive function of TGF-β probably involves Smad4-independent mechanisms. Interestingly, PDAC cells in Ptf1a-Cre/LSL-$Kras^{G12D}$/$T\beta RII^{L/L}$ secrete high levels of several CXC chemokines, which in turn induce in the pancreatic stromal fibroblasts the expression of connective tissue growth factor (CTGF) , a profibrotic and tumor-promoting factor (Ijichi et al. 2011). This observation represents a novel evidence of the complex dialogue existing between the epithelial and the mesenchymal compartment during tumor progression.

Enlarged cells with vacuoles containing other cells have been observed by pathologists in human tumors for over 100 years. These "cells-in-cells" have notably been described in pancreatic cancer (Silverman et al. 1988; Tracey et al. 1984). At first, this cell cannibalism was hypothesized to be a pro-tumor mechanism, killing immune cells or feeding cancer cells upon starvation. More recently, a tumor-suppressive role for "cell-in-cells" process was proposed after the observation that cultured breast cancer cells could invade each other in vitro leading to a cell-death process that was called entosis or homotypic cell cannibalism (HoCC) (Overholtzer et al. 2007). Decreased HoCC has recently been reported in patients with metastasized PDAC (Cano et al. 2012), suggesting that HoCC plays a tumor-suppressive role in human pancreatic cancer. The HoCC decrease in the most aggressive PDACs is due to Nupr1. Nuprlis a stress protein activated at the transcriptional level by TGF-β (Pommier et al. 2012), overexpressed in late stages of PDAC and its metastases (Ito et al. 2005; Su et al. 2001a; b), involved in resistance to gemcitabine (the reference chemotherapy against PDAC) (Giroux 2006), and associated with a poor prognosis in PDAC patients (Hamidi et al. 2012). Nupr1 inhibits HoCC and enhances aggressiveness in PDAC by activating the EMT in response to TGF-β (Cano et al. 2012). Inactivation of Nupr1 in the $KRAS^{G12D}$/Ink4A/Arf^{null} model (Pdx1-Cre/LSL-$KRAS^{G12D}$/Ink4A/Arf-$KO^{L/L}$/$Nupr1^{null}$ mice) results in a less aggressive phenotype characterized by enhancement of cell cannibalism. This validates the view that Nupr1 antagonizes cell cannibalism in vivo within the primary tumor. In Nupr1-depleted cells, unexpectedly, TGF-β stimulates HoCC (Cano et al. 2012). This suggests that HoCC is a tumor-suppressive program activated by TGF-β. The following model can be proposed: in the presence of Nupr1, TGF-β induces EMT and behaves as a tumor promoter; in the absence of Nupr1, TGF-β induces HoCC and behaves as a tumor suppressor.

Transcriptional intermediary factor 1γ (TIF1γ; alias, TRIM33/RFG7/PTC7/ectodermin) belongs to an evolutionarily conserved family of nuclear factors that have been implicated in stem cell pluripotency, embryonic development and tumor suppression (Hatakeyama 2011). Chromosomal breakpoints chromosome on 1p13.1 containing *TIF1γ* gene have been reported in acute megakaryocytic leukemias (Ng et al. 1999), osteochondromas (Sawyer et al. 2002), bronchial large cell carcinomas (Johansson et al. 1994), and childhood papillary thyroid carcinomas (Sawyer et al. 2002). Moreover, abrogation of the closely related *TIF1α* gene in mice caused hepatocellular carcinoma (Khetchoumian et al. 2007). These observations reinforce the idea according to which TIF1γ loss of function could play an active protective role during tumorigenesis. TIF1γ contributes to TGF-β signaling, although its precise functional role is not clear. Some data point toward TIF1γ as a negative regulator of the pathway through its capacity to mono-ubiquinate Smad4 and limit Smad4 nuclear accumulation (Dupont et al. 2005; 2009; Levy et al. 2007). In contrast, other studies have suggested that TIF1γ played an important positive role in transducing TGF-β signaling through its interaction with Smad2 and Smad3 (Bai et al. 2010; Doisne et al. 2009; He et al. 2006). For the reasons above mentioned, e.g., the involvement of TIF1γ in TGF-β signaling through its interaction with the Smad proteins, the location of TIF1γ at a locus that is altered in human tumors and the development of liver tumors in TIF1α knockout mice, a possible role of TIF1γ in pancreatic cancer was assessed. First, TIF1γ expression was reported to be markedly down-regulated in human pancreatic tumors (Vincent et al. 2009). Next, Pdx1-driven TIF1γ inactivation was shown to cooperate with the $KRAS^{G12D}$ oncogene in the mouse pancreas to induce exocrine pancreatic tumors (Vincent et al. 2009; Vincent et al. 2012). The relationship between *Tif1γ* and *Smad4* genes in pancreatic tumors was explored in double-null homozygous mutant mice in $KRAS^{G12D}$ background (Vincent et al. 2012) and revealed a pronounced genetic interaction between Smad4 and TIF1γ inactivation, the Pdx1-Cre; $KRAS^{G12D}$; Smad4-$KO^{L/L}$; TIF1γ-$KO^{L/L}$ (KSSTT) mice exhibiting accelerated tumor initiation (Vincent et al. 2012). The mechanism by which TIF1γ exerts its tumor-suppressive effect remains elusive. The increased tumor initiation observed when TIF1γ and Smad4 are simultaneously inactivated (compared to single gene inactivation) indicates that both genes may independently function as tumor suppressors of KRAS-driven pancreatic tumor. However, whether TIF1γ constrains transformation by modulating TGF-β functions, through mechanisms that are independent of Smad4, remains an open question and constitute a conceivable hypothesis consistent with the phenotype observed in Pdx1-Cre; LSL-$KRAS^{G12D}$; TRβII-$KO^{L/L}$ mice (Ijichi et al. 2006). Indeed, these mice, in which canonical and non-canonical TGF-β pathways are abrogated, present a phenotype resembling the one observed in KSSTT mice. Finally, it cannot be excluded that TIF1γ prevents transformation through mechanisms totally unrelated to TGF-β signaling, resulting from the modulation of general programs affecting broader genomic functions. For instance, TIF1γ and related family members were involved in chromatin remodeling, transcription elongation, and DNA repair (Agricola et al. 2011; Kepkay et al. 2011; Tsai et al. 2010). As briefly evocated in this chapter, after the malignant transformation has occurred by bypassing its anti-proliferative and pro-apoptotic effects, TGF-β acquires oncogenic properties and eventually facilitates tumor progression. In some cases, TGF-β-mediated Smad4 activation may facilitate local invasion and metastatic dissemination by inducing EMT. In malignant tumors that retained a functional Smad4, it is possible that TIF1γ may behave as a tumor suppressor, through its capacity to mono-ubiquitinate Smad4 and limit its nuclear accumulation, thus limiting the aggressiveness of Smad4-positive cancer cells by compromising their capacity to undergo EMT. This hypothesis is supported both by in vitro and in vivo evidence. Indeed, it has been shown that TIF1γ could inhibit TGF-β-induced EMT of human mammary epithelial cells in culture (Hesling et al. 2011). In vivo, even if the double-null mice for TIF1γ and Smad4 develop pancreatic tumors with a shorter latency and a higher penetrance, these tumors present a well-differentiated phenotype contrasting with the poorly differentiated phenotype observed in TIF1γ null mice (Vincent et al. 2012). This phenotype may reflect the capacity of TIF1γ to inhibit Smad4-mediated EMT induced by TGF-β in later stages of tumorigenesis. TIF1γ inactivation cooperates with activated KRAS to promote the onset of poorly differentiated tumors in a Smad4-independent way. However, in vitro and in vivo evidence also supports that the absence of Smad4 could redirect TIF1γ-negative tumors toward a well-differentiated phenotype, consistent with a tumor-suppressive effect of TIF1γ relying on its capacity to inhibit Smad4-dependent EMT in late stage of tumorigenesis. Today, the exact mechanism by which TIF1γ behaves as a tumor suppressor independently of Smad4 in the early phase of transformation is still unclear, and an active tumor- suppressive role of TIF1γ in late stages of tumorigenesis that would constrain Smad4-induced EMT still needs to be demonstrated. The following model can be proposed to illustrate the complex role of TIF1γ in cancer (Fig. 6.2).

6.3 Conclusion

Genetically engineered mouse models bearing mutations found in human tumors have largely contributed to demonstrating in vivo the tumor-suppressive role of TGF-β. They have also provided crucial mechanistic information by revealing the crucial role of TGF-β-induced programs such as apoptosis and senescence in preventing transformation. These models have provided precious information on the capacity of TGF-β to limit progression of pre-neoplastic lesions towards aggressive tumors. By combining TGF-β signaling mutations with other mutations found in human tumors, it has been possible to generate mouse models that better mimic human diseases. These models have also enlightened the crucial role played by the microenvironment during carcinogenesis and have provided powerful systems to explore the origin of cancer cells. Finally, one can envision that these models will represent valuable tools for preclinical studies, an obligate step to develop new anticancer treatments.

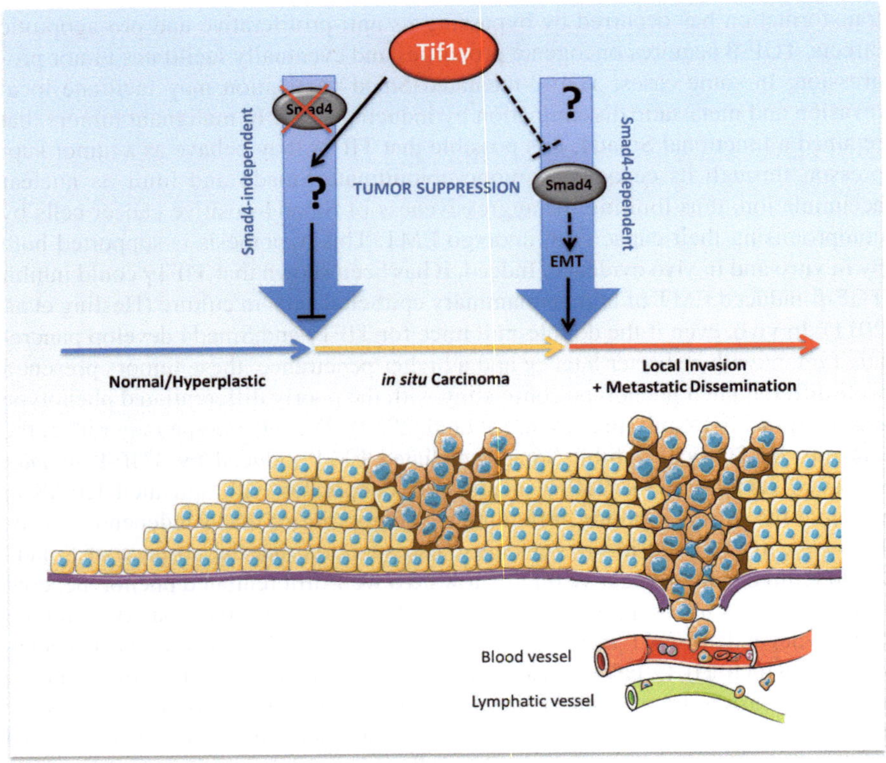

Fig. 6.2 Tumor-suppressive effects of TIF1γ. *Note*: The figures were illustrated using Servier Medical Art

References

Agricola E, Randall RA, Gaarenstroom T, Dupont S, Hill CS (2011) Recruitment of TIF1γ to chromatin via its PHD finger-bromodomain activates its ubiquitin ligase and transcriptional repressor activities. Mol Cell 43:85–96

Aguirre AJ (2003) Activated Kras and Ink4a/Arf deficiency cooperate to produce metastatic pancreatic ductal adenocarcinoma. Genes Dev 17:3112–3126

Alberici P, Jagmohan-Changur S, De Pater E, Van Der Valk M, Smits R, Hohenstein P, Fodde R (2006) Smad4 haploinsufficiency in mouse models for intestinal cancer. Oncogene 25:1841–1851

Amendt C, Schirmacher P, Weber H, Blessing M (1998) Expression of a dominant negative type II TGF-β receptor in mouse skin results in an increase in carcinoma incidence and an acceleration of carcinoma development. Oncogene 17:25–34

Amendt C, Mann A, Schirmacher P, Blessing M (2002) Resistance of keratinocytes to TGF-β-mediated growth restriction and apoptosis induction accelerates re-epithelialization in skin wounds. J Cell Sci 115:2189–2198

Aretz S, Stienen D, Uhlhaas S, Stolte M, Entius MM, Loff S, Back W, Kaufmann A, Keller KM, Blaas SH, Siebert R, Vogt S, Spranger S, Holinski-Feder E, Sunde L, Propping P, Friedl W

(2007) High proportion of large genomic deletions and a genotype phenotype update in 80 unrelated families with juvenile polyposis syndrome. J Med Genet 44:702–709

Ashcroft GS, Yang X, Glick AB, Weinstein M, Letterio JL, Mizel DE, Anzano M, Greenwell-Wild T, Wahl SM, Deng C, Roberts AB (1999) Mice lacking Smad3 show accelerated wound healing and an impaired local inflammatory response. Nat Cell Biol 1:260–266

Bai X, Kim J, Yang Z, Jurynec MJ, Akie TE, Lee J, LeBlanc J, Sessa A, Jiang H, DiBiase A, Zhou Y, Grunwald DJ, Lin S, Cantor AB, Orkin SH, Zon LI (2010) TIF1γ controls erythroid cell fate by regulating transcription elongation. Cell 142:133–143

Bardeesy N, Cheng KH, Berger JH, Chu GC, Pahler J, Olson P, Hezel AF, Horner J, Lauwers GY, Hanahan D, DePinho RA (2006) Smad4 is dispensable for normal pancreas development yet critical in progression and tumor biology of pancreas cancer. Genes Dev 20:3130–3146

Bartholin L (2012) Pancreatic Cancer and the Tumor Microenvironment: Mesenchyme's role in Pancreatic Carcinogenesis. In: PJ Grippo, HG Munshi (Eds.), Pancreatic Cancer and Tumor Microenvironment. Trivandrum (India): Transworld Research Network; p 55–94.

Bhowmick NA, Chytil A, Plieth D, Gorska AE, Dumont N, Shappell S, Washington MK, Neilson EG, Moses HL (2004) TGF-β signaling in fibroblasts modulates the oncogenic potential of adjacent epithelia. Science 303:848–851

Bian Y, Terse A, Du J, Hall B, Molinolo A, Zhang P, Chen W, Flanders KC, Gutkind JS, Wakefield LM, Kulkarni AB (2009) Progressive tumor formation in mice with conditional deletion of TGF-β signaling in head and neck epithelia is associated with activation of the PI3K/Akt pathway. Cancer Res 69:5918–5926

Bierie B, Stover DG, Abel TW, Chytil A, Gorska AE, Aakre M, Forrester E, Yang L, Wagner KU, Moses HL (2008) Transforming growth factor-β regulates mammary carcinoma cell survival and interaction with the adjacent microenvironment. Cancer Res 68:1809–1819

Bierie B, Gorska AE, Stover DG, Moses HL (2009) TGF-β promotes cell death and suppresses lactation during the second stage of mammary involution. J Cell Physiol 219:57–68

Biswas S, Chytil A, Washington K, Romero-Gallo J, Gorska AE, Wirth PS, Gautam S, Moses HL, Grady WM (2004) Transforming growth factor-β receptor type II inactivation promotes the establishment and progression of colon cancer. Cancer Res 64:4687–4692

Biswas S, Trobridge P, Romero-Gallo J, Billheimer D, Myeroff LL, Willson JK, Markowitz SD, Grady WM (2008) Mutational inactivation of TGF-βR2 in microsatellite unstable colon cancer arises from the cooperation of genomic instability and the clonal outgrowth of transforming growth factor-β resistant cells. Genes Chromosomes Cancer 47:95–106

Bornstein S, White R, Malkoski S, Oka M, Han G, Cleaver T, Reh D, Andersen P, Gross N, Olson S, Deng C, Lu SL, Wang XJ (2009) Smad4 loss in mice causes spontaneous head and neck cancer with increased genomic instability and inflammation. J Clin Invest 119:3408–3419

Bottinger EP, Jakubczak JL, Haines DC, Bagnall K, Wakefield LM (1997a) Transgenic mice overexpressing a dominant-negative mutant type II transforming growth factor-β receptor show enhanced tumorigenesis in the mammary gland and lung in response to the carcinogen 7,12-dimethylbenz-[a]-anthracene. Cancer Res 57:5564–5570

Bottinger EP, Jakubczak JL, Roberts IS, Mumy M, Hemmati P, Bagnall K, Merlino G, Wakefield LM (1997b) Expression of a dominant-negative mutant TGF-β type II receptor in transgenic mice reveals essential roles for TGF-β in regulation of growth and differentiation in the exocrine pancreas. EMBO J 16:2621–2633

Boulanger CA, Smith GH (2001) Reducing mammary cancer risk through premature stem cell senescence. Oncogene 20:2264–2272

Boulanger CA, Wagner KU, Smith GH (2005) Parity-induced mouse mammary epithelial cells are pluripotent, self-renewing and sensitive to TGF-β1 expression. Oncogene 24:552–560

Buess M, Terracciano L, Reuter J, Ballabeni P, Boulay JL, Laffer U, Metzger U, Herrmann R, Rochlitz C (2004) Amplification of SKI is a prognostic marker in early colorectal cancer. Neoplasia 6:207–212

Buggiano V, Schere-Levy C, Abe K, Vanzulli S, Piazzon I, Smith GH, Kordon EC (2001) Impairment of mammary lobular development induced by expression of TGF-β1 under the control of WAP promoter does not suppress tumorigenesis in MMTV-infected transgenic mice. Int J Cancer 92:568–576

Callahan R, Smith GH (2000) MMTV-induced mammary tumorigenesis: gene discovery, progression to malignancy and cellular pathways. Oncogene 19:992–1001

Network CGA (2012) Comprehensive molecular characterization of human colon and rectal cancer. Nature 487:330–337

Cano CE, Sandi MJ, Hamidi T, Calvo EL, Turrini O, Bartholin L, Loncle C, Secq V, Garcia S, Lomberk G, Kroemer G, Urrutia R, Iovanna JL (2012) Homotypic cell cannibalism, a cell-death process regulated by the nuclear protein 1, opposes to metastasis in pancreatic cancer. EMBO Mol Med 4(9):964–979

Cheng N, Bhowmick NA, Chytil A, Gorksa AE, Brown KA, Muraoka R, Arteaga CL, Neilson EG, Hayward SW, Moses HL (2005) Loss of TGF-β type II receptor in fibroblasts promotes mammary carcinoma growth and invasion through upregulation of TGF-α-, MSP- and HGF-mediated signaling networks. Oncogene 24:5053–5068

Cui W, Fowlis DJ, Cousins FM, Duffie E, Bryson S, Balmain A, Akhurst RJ (1995) Concerted action of TGF-β1 and its type II receptor in control of epidermal homeostasis in transgenic mice. Genes Dev 9:945–955

Cui W, Fowlis DJ, Bryson S, Duffie E, Ireland H, Balmain A, Akhurst RJ (1996) TGF-β1 inhibits the formation of benign skin tumors, but enhances progression to invasive spindle carcinomas in transgenic mice. Cell 86:531–542

Descargues P, Sil AK, Sano Y, Korchynskyi O, Han G, Owens P, Wang XJ, Karin M (2008) IKKα is a critical coregulator of a Smad4-independent TGF-β Smad2/3 signaling pathway that controls keratinocyte differentiation. Proc Natl Acad Sci USA 105:2487–2492

Dickson MC, Martin JS, Cousins FM, Kulkarni AB, Karlsson S, Akhurst RJ (1995) Defective haematopoiesis and vasculogenesis in transforming growth factor-β1 knock out mice. Development 121:1845–1854

Doisne JM, Bartholin L, Yan KP, Garcia CN, Duarte N, Le Luduec JB, Vincent D, Cyprian F, Horvat B, Martel S, Rimokh R, Losson R, Benlagha K, Marie JC (2009) iNKT cell development is orchestrated by different branches of TGF-β signaling. J Exp Med 206:1365–1378

Dupont S, Zacchigna L, Cordenonsi M, Soligo S, Adorno M, Rugge M, Piccolo S (2005) Germ-layer specification and control of cell growth by Ectodermin, a Smad4 ubiquitin ligase. Cell 121:87–99

Dupont S, Mamidi A, Cordenonsi M, Montagner M, Zacchigna L, Adorno M, Martello G, Stinchfield MJ, Soligo S, Morsut L, Inui M, Moro S, Modena N, Argenton F, Newfeld SJ, Piccolo S (2009) FAM/USP9x, a deubiquitinating enzyme essential for TGF-β signaling, controls Smad4 monoubiquitination. Cell 136:123–135

Engle SJ, Hoying JB, Boivin GP, Ormsby I, Gartside PS, Doetschman T (1999) Transforming growth factor-β1 suppresses nonmetastatic colon cancer at an early stage of tumorigenesis. Cancer Res 59:3379–3386

Engle SJ, Ormsby I, Pawlowski S, Boivin GP, Croft J, Balish E, Doetschman T (2002) Elimination of colon cancer in germ-free transforming growth factor-β1-deficient mice. Cancer Res 62:6362–6366

Fang WB, Jokar I, Chytil A, Moses HL, Abel T, Cheng N (2011) Loss of one TGF-βR2 allele in fibroblasts promotes metastasis in MMTV: polyoma middle T transgenic and transplant mouse models of mammary tumor progression. Clin Exp Metastasis 28:351–366

Fearon ER, Vogelstein B (1990) A genetic model for colorectal tumorigenesis. Cell 61:759–767

Forrester E, Chytil A, Bierie B, Aakre M, Gorska AE, Sharif-Afshar AR, Muller WJ, Moses HL (2005) Effect of conditional knockout of the type II TGF-β eceptor gene in mammary epithelia on mammary gland development and polyomavirus middle T antigen induced tumor formation and metastasis. Cancer Res 65:2296–2302

Fowlis DJ, Cui W, Johnson SA, Balmain A, Akhurst RJ (1996) Altered epidermal cell growth control in vivo by inducible expression of transforming growth factor-β1 in the skin of transgenic mice. Cell Growth Differ 7:679–687

Freeman TJ, Smith JJ, Chen X, Washington MK, Roland JT, Means AL, Eschrich SA, Yeatman TJ, Deane NG, Beauchamp RD (2012) Smad4-mediated signaling inhibits intestinal neoplasia by inhibiting expression of β-catenin. Gastroenterology 142(562–571):e562

Fukuchi M, Nakajima M, Fukai Y, Miyazaki T, Masuda N, Sohda M, Manda R, Tsukada K, Kato H, Kuwano H (2004) Increased expression of c-Ski as a co-repressor in transforming growth factor-β signaling correlates with progression of esophageal squamous cell carcinoma. Int J Cancer 108:818–824

Fukushige S, Furukawa T, Satoh K, Sunamura M, Kobari M, Koizumi M, Horii A (1998) Loss of chromosome 18q is an early event in pancreatic ductal tumorigenesis. Cancer Res 58:4222–4226

Giroux V (2006) p8 is a new target of gemcitabine in pancreatic cancer cells. Clin Cancer Res 12:235–241

Gittes GK (2009) Developmental biology of the pancreas: a comprehensive review. Dev Biol 326:4–35

Go C, Li P, Wang XJ (1999) Blocking transforming growth factor-β signaling in transgenic epidermis accelerates chemical carcinogenesis: a mechanism associated with increased angiogenesis. Cancer Res 59:2861–2868

Go C, He W, Zhong L, Li P, Huang J, Brinkley BR, Wang XJ (2000) Aberrant cell cycle progression contributes to the early-stage accelerated carcinogenesis in transgenic epidermis expressing the dominant negative TGF-βRII. Oncogene 19:3623–3631

Gorska AE, Joseph H, Derynck R, Moses HL, Serra R (1998) Dominant-negative interference of the transforming growth factor-β type II receptor in mammary gland epithelium results in alveolar hyperplasia and differentiation in virgin mice. Cell Growth Differ 9:229–238

Gorska AE, Jensen RA, Shyr Y, Aakre ME, Bhowmick NA, Moses HL (2003) Transgenic mice expressing a dominant-negative mutant type II transforming growth factor-β receptor exhibit impaired mammary development and enhanced mammary tumor formation. Am J Pathol 163:1539–1549

Gu G, Dubauskaite J, Melton DA (2002) Direct evidence for the pancreatic lineage: NGN3+ cells are islet progenitors and are distinct from duct progenitors. Development 129:2447–2457

Guasch G, Schober M, Pasolli HA, Conn EB, Polak L, Fuchs E (2007) Loss of TGF-β signaling destabilizes homeostasis and promotes squamous cell carcinomas in stratified epithelia. Cancer Cell 12:313–327

Hahn SA, Schutte M, Hoque AT, Moskaluk CA, da Costa LT, Rozenblum E, Weinstein CL, Fischer A, Yeo CJ, Hruban RH, Kern SE (1996) DPC4, a candidate tumor suppressor gene at human chromosome 18q21.1.[see comment]. Science 271:350–353

Hamamoto T, Beppu H, Okada H, Kawabata M, Kitamura T, Miyazono K, Kato M (2002) Compound disruption of smad2 accelerates malignant progression of intestinal tumors in apc knockout mice. Cancer Res 62:5955–5961

Hamidi T, Algul H, Cano CE, Sandi MJ, Molejon MI, Riemann M, Calvo EL, Lomberk G, Dagorn JC, Weih F, Urrutia R, Schmid RM, Iovanna JL (2012) Nuclear protein 1 promotes pancreatic cancer development and protects cells from stress by inhibiting apoptosis. J Clin Invest 122:2092–2103

Han G, Lu SL, Li AG, He W, Corless CL, Kulesz-Martin M, Wang XJ (2005) Distinct mechanisms of TGF-β1-mediated epithelial-to-mesenchymal transition and metastasis during skin carcinogenesis. J Clin Invest 115:1714–1723

Hansel DE, Kern SE, Hruban RH (2003) Molecular pathogenesis of pancreatic cancer. Annu Rev Genomics Hum Genet 4:237–256

Hatakeyama S (2011) TRIM proteins and cancer. Nat Rev Cancer 11:792–804

He W, Dorn DC, Erdjument-Bromage H, Tempst P, Moore MA, Massague J (2006) Hematopoiesis controlled by distinct TIF1γ and Smad4 branches of the TGF-β pathway. Cell 125:929–941

Hesling C, Fattet L, Teyre G, Jury D, Gonzalo P, Lopez J, Vanbelle C, Morel AP, Gillet G, Mikaelian I, Rimokh R (2011) Antagonistic regulation of EMT by TIF1γ and Smad4 in mammary epithelial cells. EMBO Rep 12(7):665–672

Hingorani SR, Petricoin EF, Maitra A, Rajapakse V, King C, Jacobetz MA, Ross S, Conrads TP, Veenstra TD, Hitt BA, Kawaguchi Y, Johann D, Liotta LA, Crawford HC, Putt ME, Jacks T, Wright CV, Hruban RH, Lowy AM, Tuveson DA (2003) Preinvasive and invasive ductal pancreatic cancer and its early detection in the mouse. Cancer Cell 4:437–450

Hinshelwood RA, Huschtscha LI, Melki J, Stirzaker C, Abdipranoto A, Vissel B, Ravasi T, Wells CA, Hume DA, Reddel RR, Clark SJ (2007) Concordant epigenetic silencing of transforming growth factor-β signaling pathway genes occurs early in breast carcinogenesis. Cancer Res 67:11517–11527

Hohenstein P, Molenaar L, Elsinga J, Morreau H, van der Klift H, Struijk A, Jagmohan-Changur S, Smits R, van Kranen H, van Ommen GJ, Cornelisse C, Devilee P, Fodde R (2003) Serrated adenomas and mixed polyposis caused by a splice acceptor deletion in the mouse Smad4 gene. Genes Chromosomes Cancer 36:273–282

Howe JR, Roth S, Ringold JC, Summers RW, Jarvinen HJ, Sistonen P, Tomlinson IP, Houlston RS, Bevan S, Mitros FA, Stone EM, Aaltonen LA (1998) Mutations in the SMAD4/DPC4 gene in juvenile polyposis. Science 280:1086–1088

Howe JR, Bair JL, Sayed MG, Anderson ME, Mitros FA, Petersen GM, Velculescu VE, Traverso G, Vogelstein B (2001) Germline mutations of the gene encoding bone morphogenetic protein receptor 1A in juvenile polyposis. Nat Genet 28:184–187

Hruban RH, Adsay NV, Albores-Saavedra J, Compton C, Garrett ES, Goodman SN, Kern SE, Klimstra DS, Kloppel G, Longnecker DS, Luttges J, Offerhaus GJ (2001) Pancreatic intraepithelial neoplasia: a new nomenclature and classification system for pancreatic duct lesions. Am J Surg Pathol 25:579–586

Hruban RH, Rustgi AK, Brentnall TA, Tempero MA, Wright CV, Tuveson DA (2006) Pancreatic cancer in mice and man: the Penn Workshop 2004. Cancer Res 66:14–17

Ijichi H, Chytil A, Gorska AE, Aakre ME, Fujitani Y, Fujitani S, Wright CV, Moses HL (2006) Aggressive pancreatic ductal adenocarcinoma in mice caused by pancreas-specific blockade of transforming growth factor-β signaling in cooperation with active Kras expression. Genes Dev 20:3147–3160

Ijichi H, Chytil A, Gorska AE, Aakre ME, Bierie B, Tada M, Mohri D, Miyabayashi K, Asaoka Y, Maeda S, Ikenoue T, Tateishi K, Wright CV, Koike K, Omata M, Moses HL (2011) Inhibiting Cxcr2 disrupts tumor-stromal interactions and improves survival in a mouse model of pancreatic ductal adenocarcinoma. J Clin Invest 121:4106–4117

Ito Y, Yoshida H, Motoo Y, Iovanna JL, Nakamura Y, Kakudo K, Uruno T, Takamura Y, Miya A, Noguchi S, Kuma K, Miyauchi A (2005) Expression of p8 protein in breast carcinoma; an inverse relationship with apoptosis. Anticancer Res 25:833–837

Izeradjene K, Combs C, Best M, Gopinathan A, Wagner A, Grady WM, Deng CX, Hruban RH, Adsay NV, Tuveson DA, Hingorani SR (2007) Kras(G12D) and Smad4/Dpc4 haploinsufficiency cooperate to induce mucinous cystic neoplasms and invasive adenocarcinoma of the pancreas. Cancer Cell 11:229–243

Jackson EL, Willis N, Mercer K, Bronson RT, Crowley D, Montoya R, Jacks T, Tuveson DA (2001) Analysis of lung tumor initiation and progression using conditional expression of oncogenic K-ras. Genes Dev 15:3243–3248

Jhappan C, Geiser AG, Kordon EC, Bagheri D, Hennighausen L, Roberts AB, Smith GH, Merlino G (1993) Targeting expression of a transforming growth factor-β1 transgene to the pregnant mammary gland inhibits alveolar development and lactation. EMBO J 12:1835–1845

Johansson M, Dietrich C, Mandahl N, Hambraeus G, Johansson L, Clausen PP, Mitelman F, Heim S (1994) Karyotypic characterization of bronchial large cell carcinomas. Int J Cancer 57:463–467

Jones S, Zhang X, Parsons DW, Lin JC, Leary RJ, Angenendt P, Mankoo P, Carter H, Kamiyama H, Jimeno A, Hong SM, Fu B, Lin MT, Calhoun ES, Kamiyama M, Walter K, Nikolskaya T, Nikolsky Y, Hartigan J, Smith DR, Hidalgo M, Leach SD, Klein AP, Jaffee EM, Goggins M, Maitra A, Iacobuzio-Donahue C, Eshleman JR, Kern SE, Hruban RH, Karchin R, Papadopoulos N, Parmigiani G, Vogelstein B, Velculescu VE, Kinzler KW (2008) Core signaling pathways in human pancreatic cancers revealed by global genomic analyses. Science 321:1801–1806

Kang SH, Bang YJ, Im YH, Yang HK, Lee DA, Lee HY, Lee HS, Kim NK, Kim SJ (1999) Transcriptional repression of the transforming growth factor-β type I receptor gene by DNA methylation results in the development of TGF-β resistance in human gastric cancer. Oncogene 18:7280–7286

Katuri V, Tang Y, Li C, Jogunoori W, Deng CX, Rashid A, Sidawy AN, Evans S, Reddy EP, Mishra B, Mishra L (2006) Critical interactions between TGF-β signaling/ELF, and E-cadherin/β -catenin mediated tumor suppression. Oncogene 25:1871–1886

Kawaguchi Y, Cooper B, Gannon M, Ray M, MacDonald RJ, Wright CVE (2002) The role of the transcriptional regulator Ptf1a in converting intestinal to pancreatic progenitors. Nat Genet 32:128–134

Kepkay R, Attwood KM, Ziv Y, Shiloh Y, Dellaire G (2011) KAP1 depletion increases PML nuclear body number in concert with ultrastructural changes in chromatin. Cell Cycle 10:308–322

Kern SE, Shi C, Hruban RH (2011) The complexity of pancreatic ductal cancers and multidimensional strategies for therapeutic targeting. J Pathol 223:295–306

Khetchoumian K, Teletin M, Tisserand J, Mark M, Herquel B, Ignat M, Zucman-Rossi J, Cammas F, Lerouge T, Thibault C, Metzger D, Chambon P, Losson R (2007) Loss of Trim24 (Tif1α) gene function confers oncogenic activity to retinoic acid receptor α. Nat Genet 39:1500–1506

Kim B-G, Li C, Qiao W, Mamura M, Kasperczak B, Anver M, Wolfraim L, Hong S, Mushinski E, Potter M, Kim S-J, Fu X-Y, Deng C, Letterio JJ (2006) Smad4 signalling in T cells is required for suppression of gastrointestinal cancer. Nature 441:1015–1019

Kim SJ, Im YH, Markowitz SD, Bang YJ (2000) Molecular mechanisms of inactivation of TGF-β receptors during carcinogenesis. Cytokine Growth Factor Rev 11:159–168

Kitamura T, Kometani K, Hashida H, Matsunaga A, Miyoshi H, Hosogi H, Aoki M, Oshima M, Hattori M, Takabayashi A, Minato N, Taketo MM (2007) SMAD4-deficient intestinal tumors recruit CCR1+ myeloid cells that promote invasion. Nat Genet 39:467–475

Kojima K, Vickers SM, Adsay NV, Jhala NC, Kim HG, Schoeb TR, Grizzle WE, Klug CA (2007) Inactivation of Smad4 accelerates Kras(G12D)-mediated pancreatic neoplasia. Cancer Res 67:8121–8130

Kordon EC, McKnight RA, Jhappan C, Hennighausen L, Merlino G, Smith GH (1995) Ectopic TGF-β1 expression in the secretory mammary epithelium induces early senescence of the epithelial stem cell population. Dev Biol 168:47–61

Kuang C, Xiao Y, Liu X, Stringfield TM, Zhang S, Wang Z, Chen Y (2006) In vivo disruption of TGF-β signaling by Smad7 leads to premalignant ductal lesions in the pancreas. Proc Natl Acad Sci USA 103:1858–1863

Kulkarni AB, Huh CG, Becker D, Geiser A, Lyght M, Flanders KC, Roberts AB, Sporn MB, Ward JM, Karlsson S (1993) Transforming growth factor-β1 null mutation in mice causes excessive inflammatory response and early death. Proc Natl Acad Sci USA 90:770–774

Kundu SD, Kim IY, Yang T, Doglio L, Lang S, Zhang X, Buttyan R, Kim SJ, Chang J, Cai X, Wang Z, Lee C (2000) Absence of proximal duct apoptosis in the ventral prostate of transgenic mice carrying the C3(1)-TGF-β type II dominant negative receptor. Prostate 43:118–124

Landi S (2009) Genetic predisposition and environmental risk factors to pancreatic cancer: A review of the literature. Mutat Res 681:299–307

Larsson J, Goumans MJ, Sjostrand LJ, van Rooijen MA, Ward D, Leveen P, Xu X, ten Dijke P, Mummery CL, Karlsson S (2001) Abnormal angiogenesis but intact hematopoietic potential in TGF-β type I receptor-deficient mice. EMBO J 20:1663–1673

Letterio JJ, Roberts AB (1998) Regulation of immune responses by TGF-β. Annu Rev Immunol 16:137–161

Levy L, Hill CS (2006) Alterations in components of the TGF-β superfamily signaling pathways in human cancer. Cytokine Growth Factor Rev 17:41–58

Levy L, Howell M, Das D, Harkin S, Episkopou V, Hill CS (2007) Arkadia activates Smad3/Smad4-dependent transcription by triggering signal-induced SnoN degradation. Mol Cell Biol 27:6068–6083

Li W, Qiao W, Chen L, Xu X, Yang X, Li D, Li C, Brodie SG, Meguid MM, Hennighausen L, Deng CX (2003) Squamous cell carcinoma and mammary abscess formation through squamous metaplasia in Smad4/Dpc4 conditional knockout mice. Development 130:6143–6153

Liu X, Alexander V, Vijayachandra K, Bhogte E, Diamond I, Glick A (2001) Conditional epidermal expression of TGF-β1 blocks neonatal lethality but causes a reversible hyperplasia and alopecia. Proc Natl Acad Sci USA 98:9139–9144

Machnicka B, Grochowalska R, Boguslawska DM, Sikorski AF, Lecomte MC (2012) Spectrin-based skeleton as an actor in cell signaling. Cell Mol Life Sci 69:191–201

Maggio-Price L, Treuting P, Zeng W, Tsang M, Bielefeldt-Ohmann H, Iritani BM (2006) Helicobacter infection is required for inflammation and colon cancer in SMAD3-deficient mice. Cancer Res 66:828–838

Markowitz S, Wang J, Myeroff L, Parsons R, Sun L, Lutterbaugh J, Fan RS, Zborowska E, Kinzler KW, Vogelstein B et al (1995) Inactivation of the type II TGF-βreceptor in colon cancer cells with microsatellite instability. Science 268:1336–1338

Massague J (1992) Receptors for the TGF-β family. Cell 69:1067–1070

Massague J (2008) TGF-β in Cancer. Cell 134:215–230

Matise LA, Palmer TD, Ashby WJ, Nashabi A, Chytil A, Aakre M, Pickup MW, Gorska AE, Zijlstra A, Moses HL (2012) Lack of transforming growth factor-β signaling promotes collective cancer cell invasion through tumor-stromal crosstalk. Breast Cancer Res 14:R98

Medrano EE (2003) Repression of TGF-β signaling by the oncogenic protein SKI in human melanomas: consequences for proliferation, survival, and metastasis. Oncogene 22:3123–3129

Mu Y, Gudey SK, Landstrom M (2012) Non-Smad signaling pathways. Cell Tissue Res 347:11–20

Munoz NM, Upton M, Rojas A, Washington MK, Lin L, Chytil A, Sozmen EG, Madison BB, Pozzi A, Moon RT, Moses HL, Grady WM (2006) Transforming growth factor-β receptor type II inactivation induces the malignant transformation of intestinal neoplasms initiated by Apc mutation. Cancer Res 66:9837–9844

Muraoka-Cook RS, Kurokawa H, Koh Y, Forbes JT, Roebuck LR, Barcellos-Hoff MH, Moody SE, Chodosh LA, Arteaga CL (2004) Conditional overexpression of active transforming growth factor-β1in vivo accelerates metastases of transgenic mammary tumors. Cancer Res 64:9002–9011

Muraoka-Cook RS, Shin I, Yi JY, Easterly E, Barcellos-Hoff MH, Yingling JM, Zent R, Arteaga CL (2006) Activated type I TGF-β receptor kinase enhances the survival of mammary epithelial cells and accelerates tumor progression. Oncogene 25:3408–3423

Muraoka RS et al (2002) Blockade of TGF-β inhibits mammary tumor cell viability, migration, and metastases. J Clin Invest 109(12):1551–1559

Muraoka RS, Koh Y, Roebuck LR, Sanders ME, Brantley-Sieders D, Gorska AE, Moses HL, Arteaga CL (2003) Increased malignancy of Neu-induced mammary tumors overexpressing active transforming growth factor-β1. Mol Cell Biol 23:8691–8703

Ng KC, Tan AM, Chong YY, Lau LC, Lou J (1999) Congenital acute megakaryoblastic leukemia (M7) with chromosomal t(1;22)(p13;q13) translocation in a set of identical twins. J Pediatr Hematol Oncol 21:428–430

Nguyen AV, Pollard JW (2000) Transforming growth factor-β3 induces cell death during the first stage of mammary gland involution. Development 127:3107–3118

Nomura M, Li E (1998) Smad2 role in mesoderm formation, left-right patterning and craniofacial development. Nature 393:786–790

Oshima M, Oshima H, Taketo MM (1996) TGF-β receptor type II deficiency results in defects of yolk sac hematopoiesis and vasculogenesis. Dev Biol 179:297–302

Overholtzer M, Mailleux AA, Mouneimne G, Normand G, Schnitt SJ, King RW, Cibas ES, Brugge JS (2007) A nonapoptotic cell death process, entosis, that occurs by cell-in-cell invasion. Cell 131:966–979

Pierce DF Jr, Johnson MD, Matsui Y, Robinson SD, Gold LI, Purchio AF, Daniel CW, Hogan BL, Moses HL (1993) Inhibition of mammary duct development but not alveolar outgrowth during pregnancy in transgenic mice expressing active TGF-β1. Genes Dev 7:2308–2317

Pierce DF Jr, Gorska AE, Chytil A, Meise KS, Page DL, Coffey RJ Jr, Moses HL (1995) Mammary tumor suppression by transforming growth factor-β1 transgene expression. Proc Natl Acad Sci USA 92:4254–4258

Pommier RM, Gout J, Vincent DF, Cano CE, Kaniewski B, Martel S, Rodriguez J, Fourel G, Valcourt U, Marie JC, Iovanna JL, Bartholin L (2012) The human NUPR1/P8 gene is transcriptionally activated by transforming growth factor-β via the Smad signalling pathway. Biochem J 445:285–293

Poser I, Rothhammer T, Dooley S, Weiskirchen R, Bosserhoff AK (2005) Characterization of Sno expression in malignant melanoma. Int J Oncol 26:1411–1417

Proetzel G, Pawlowski SA, Wiles MV, Yin M, Boivin GP, Howles PN, Ding J, Ferguson MW, Doetschman T (1995) Transforming growth factor-β3 is required for secondary palate fusion. Nat Genet 11:409–414

Pu H, Collazo J, Jones E, Gayheart D, Sakamoto S, Vogt A, Mitchell B, Kyprianou N (2009) Dysfunctional transforming growth factor-β receptor II accelerates prostate tumorigenesis in the TRAMP mouse model. Cancer Res 69:7366–7374

Pugh TJ, Weeraratne SD, Archer TC, Pomeranz Krummel DA, Auclair D, Bochicchio J, Carneiro MO, Carter SL, Cibulskis K, Erlich RL, Greulich H, Lawrence MS, Lennon NJ, McKenna A, Meldrim J, Ramos AH, Ross MG, Russ C, Shefler E, Sivachenko A, Sogoloff B, Stojanov P, Tamayo P, Mesirov JP, Amani V, Teider N, Sengupta S, Francois JP, Northcott PA, Taylor MD, Yu F, Crabtree GR, Kautzman AG, Gabriel SB, Getz G, Jager N, Jones DT, Lichter P, Pfister SM, Roberts TM, Meyerson M, Pomeroy SL, Cho YJ (2012) Medulloblastoma exome sequencing uncovers subtype-specific somatic mutations. Nature 488(7409):106–110

Qiao W, Li AG, Owens P, Xu X, Wang XJ, Deng CX (2006) Hair follicle defects and squamous cell carcinoma formation in Smad4 conditional knockout mouse skin. Oncogene 25:207–217

Redman RS, Katuri V, Tang Y, Dillner A, Mishra B, Mishra L (2005) Orofacial and gastrointestinal hyperplasia and neoplasia in smad4+/– and elf+/–/smad4+/– mutant mice. J Oral Pathol Med 34:23–29

Reed JA, Bales E, Xu W, Okan NA, Bandyopadhyay D, Medrano EE (2001) Cytoplasmic localization of the oncogenic protein Ski in human cutaneous melanomas in vivo: functional implications for transforming growth factor-β signaling. Cancer Res 61:8074–8078

Ross SR (2010) Mouse mammary tumor virus molecular biology and oncogenesis. Viruses 2:2000–2012

Saam JR, Gordon JI (1999) Inducible gene knockouts in the small intestinal and colonic epithelium. J Biol Chem 274:38071–38082

Sanford LP, Ormsby I, Gittenberger-de Groot AC, Sariola H, Friedman R, Boivin GP, Cardell EL, Doetschman T (1997) TGF-β2 knockout mice have multiple developmental defects that are non- overlapping with other TGF-β knockout phenotypes. Development 124: 2659–2670

Sawyer JR, Thomas EL, Lukacs JL, Swanson CM, Ding Y, Parham DM, Thomas JR, Nicholas RW (2002) Recurring breakpoints of 1p13 approximately p22 in osteochondroma. Cancer Genet Cytogenet 138:102–106

Schleger C, Arens N, Zentgraf H, Bleyl U, Verbeke C (2000) Identification of frequent chromosomal aberrations in ductal adenocarcinoma of the pancreas by comparative genomic hybridization (CGH). J Pathol 191:27–32

Sellheyer K, Bickenbach JR, Rothnagel JA, Bundman D, Longley MA, Krieg T, Roche NS, Roberts AB, Roop DR (1993) Inhibition of skin development by overexpression of transforming growth factor-β1 in the epidermis of transgenic mice. Proc Natl Acad Sci USA 90:5237–5241

Shull MM, Ormsby I, Kier AB, Pawlowski S, Diebold RJ, Yin M, Allen R, Sidman C, Proetzel G, Calvin D et al (1992) Targeted disruption of the mouse transforming growth factor-β1 gene results in multifocal inflammatory disease. Nature 359:693–699

Siegel PM, Shu W, Cardiff RD, Muller WJ, Massague J (2003) Transforming growth factor-β signaling impairs Neu-induced mammary tumorigenesis while promoting pulmonary metastasis. Proc Natl Acad Sci USA 100:8430–8435

Silverman JF, Dabbs DJ, Finley JL, Geisinger KR (1988) Fine-needle aspiration biopsy of pleomorphic (giant cell) carcinoma of the pancreas. Cytologic, immunocytochemical, and ultrastructural findings. Am J Clin Pathol 89:714–720

Sirard C, de la Pompa JL, Elia A, Itie A, Mirtsos C, Cheung A, Hahn S, Wakeham A, Schwartz L, Kern SE, Rossant J, Mak TW (1998) The tumor suppressor gene Smad4/Dpc4 is required for gastrulation and later for anterior development of the mouse embryo. Genes Dev 12:107–119

Sodir NM, Chen X, Park R, Nickel AE, Conti PS, Moats R, Bading JR, Shibata D, Laird PW (2006) Smad3 deficiency promotes tumorigenesis in the distal colon of ApcMin/+ mice. Cancer Res 66:8430–8438

Stenvers KL, Tursky ML, Harder KW, Kountouri N, Amatayakul-Chantler S, Grail D, Small C, Weinberg RA, Sizeland AM, Zhu HJ (2003) Heart and liver defects and reduced transforming growth factor-β2 sensitivity in transforming growth factor-β type III receptor-deficient embryos. Mol Cell Biol 23:4371–4385

Stephens PJ, Tarpey PS, Davies H, Van Loo P, Greenman C, Wedge DC, Nik-Zainal S, Martin S, Varela I, Bignell GR, Yates LR, Papaemmanuil E, Beare D, Butler A, Cheverton A, Gamble J, Hinton J, Jia M, Jayakumar A, Jones D, Latimer C, Lau KW, McLaren S, McBride DJ, Menzies A, Mudie L, Raine K, Rad R, Chapman MS, Teague J, Easton D, Langerod A, Lee MT, Shen CY, Tee BT, Huimin BW, Broeks A, Vargas AC, Turashvili G, Martens J, Fatima A, Miron P, Chin SF, Thomas G, Boyault S, Mariani O, Lakhani SR, van de Vijver M, van't Veer L, Foekens J, Desmedt C, Sotiriou C, Tutt A, Caldas C, Reis-Filho JS, Aparicio SA, Salomon AV, Borresen-Dale AL, Richardson AL, Campbell PJ, Futreal PA, Stratton MR (2012) The landscape of cancer genes and mutational processes in breast cancer. Nature 486:400–404

Sternlicht MD (2006) Key stages in mammary gland development: the cues that regulate ductal branching morphogenesis. Breast Cancer Res 8:201

Su SB, Motoo Y, Iovanna JL, Berthezene P, Xie MJ, Mouri H, Ohtsubo K, Matsubara F, Sawabu N (2001a) Overexpression of p8 is inversely correlated with apoptosis in pancreatic cancer. Clin Cancer Res 7:1320–1324

Su SB, Motoo Y, Iovanna JL, Xie MJ, Mouri H, Ohtsubo K, Yamaguchi Y, Watanabe H, Okai T, Matsubara F, Sawabu N (2001b) Expression of p8 in human pancreatic cancer. Clin Cancer Res 7:309–313

Subramanian G, Schwarz RE, Higgins L, McEnroe G, Chakravarty S, Dugar S, Reiss M (2004) Targeting endogenous transforming growth factor-β receptor signaling in Smad4-deficient human pancreatic carcinoma cells inhibits their invasive phenotype1. Cancer Res 64:5200–5211

Takaku K, Oshima M, Miyoshi H, Matsui M, Seldin MF, Taketo MM (1998) Intestinal tumorigenesis in compound mutant mice of both Dpc4 (Smad4) and Apc genes. Cell 92:645–656

Takaku K, Miyoshi H, Matsunaga A, Oshima M, Sasaki N, Taketo MM (1999) Gastric and duodenal polyps in Smad4 (Dpc4) knockout mice. Cancer Res 59:6113–6117

Takaku K, Wrana JL, Robertson EJ, Taketo MM (2002) No effects of Smad2 (madh2) null mutation on malignant progression of intestinal polyps in Apc(delta716) knockout mice. Cancer Res 62:4558–4561

Tang B, Bottinger EP, Jakowlew SB, Bagnall KM, Mariano J, Anver MR, Letterio JJ, Wakefield LM (1998) Transforming growth factor-β1is a new form of tumor suppressor with true haploid insufficiency. Nat Med 4:802–807

Tang Y, Katuri V, Srinivasan R, Fogt F, Redman R, Anand G, Said A, Fishbein T, Zasloff M, Reddy EP, Mishra B, Mishra L (2005) Transforming growth factor-β suppresses nonmetastatic colon cancer through Smad4 and adaptor protein ELF at an early stage of tumorigenesis. Cancer Res 65:4228–4237

Tracey KJ, O'Brien MJ, Williams LF, Klibaner M, George PK, Saravis CA, Zamcheck N (1984) Signet ring carcinoma of the pancreas, a rare variant with very high CEA values. Immunohistologic comparison with adenocarcinoma. Dig Dis Sci 29:573–576

Trobridge P, Knoblaugh S, Washington MK, Munoz NM, Tsuchiya KD, Rojas A, Song X, Ulrich CM, Sasazuki T, Shirasawa S (2009) TGF-β receptor inactivation and mutant kras induce intestinal neoplasms in mice via a β-catenin-independent pathway. Gastroenterology 136:1680–1688, e1687

Tsai WW, Wang Z, Yiu TT, Akdemir KC, Xia W, Winter S, Tsai CY, Shi X, Schwarzer D, Plunkett W, Aronow B, Gozani O, Fischle W, Hung MC, Patel DJ, Barton MC (2010) TRIM24 links a non-canonical histone signature to breast cancer. Nature 468:927–932

van der Flier LG, Clevers H (2009) Stem cells, self-renewal, and differentiation in the intestinal epithelium. Annu Rev Physiol 71:241–260

van Hattem WA, Brosens LA, de Leng WW, Morsink FH, Lens S, Carvalho R, Giardiello FM, Offerhaus GJ (2008) Large genomic deletions of Smad4, Bmpr1A and Pten in juvenile polyposis. Gut 57:623–627

Vincent DF, Yan KP, Treilleux I, Gay F, Arfi V, Kaniewski B, Marie JC, Lepinasse F, Martel S, Goddard-Leon S, Iovanna JL, Dubus P, Garcia S, Puisieux A, Rimokh R, Bardeesy N, Scoazec JY, Losson R, Bartholin L (2009) Inactivation of TIF1γ cooperates with Kras to induce cystic tumors of the pancreas. PLoS Genet 5:e1000575

Vincent DF, Gout J, Chuvin N, Arfi V, Pommier RM, Bertolino P, Jonckheere N, Ripoche D, Kaniewski B, Martel S, Langlois JB, Goddard-Leon S, Colombe A, Janier M, Van Seuningen I, Losson R, Valcourt U, Treilleux I, Dubus P, Bardeesy N, Bartholin L (2012) Tif1γ suppresses murine pancreatic tumoral transformation by a Smad4-independent pathway. Am J Pathol 180:2214–2221

Waldrip WR, Bikoff EK, Hoodless PA, Wrana JL, Robertson EJ (1998) Smad2 signaling in extraembryonic tissues determines anterior-posterior polarity of the early mouse embryo. Cell 92:797–808

Wang XJ, Greenhalgh DA, Bickenbach JR, Jiang A, Bundman DS, Krieg T, Derynck R, Roop DR (1997) Expression of a dominant-negative type II transforming growth factor-β receptor in the epidermis of transgenic mice blocks TGF-β-mediated growth inhibition. Proc Natl Acad Sci USA 94:2386–2391

Wang XJ, Liefer KM, Tsai S, O'Malley BW, Roop DR (1999) Development of gene-switch transgenic mice that inducibly express transforming growth factor-β1in the epidermis. Proc Natl Acad Sci USA 96:8483–8488

Weeks BH, He W, Olson KL, Wang XJ (2001) Inducible expression of transforming growth factor-β1 in papillomas causes rapid metastasis. Cancer Res 61:7435–7443

Weinstein M, Yang X, Li C, Xu X, Gotay J, Deng CX (1998) Failure of egg cylinder elongation and mesoderm induction in mouse embryos lacking the tumor suppressor smad2. Proc Natl Acad Sci USA 95:9378–9383

Xu X, Brodie SG, Yang X, Im YH, Parks WT, Chen L, Zhou YX, Weinstein M, Kim SJ, Deng CX (2000) Haploid loss of the tumor suppressor Smad4/Dpc4 initiates gastric polyposis and cancer in mice. Oncogene 19:1868–1874

Yang L, Mao C, Teng Y, Li W, Zhang J, Cheng X, Li X, Han X, Xia Z, Deng H, Yang X (2005) Targeted disruption of Smad4 in mouse epidermis results in failure of hair follicle cycling and formation of skin tumors. Cancer Res 65:8671–8678

Yang L, Huang J, Ren X, Gorska AE, Chytil A, Aakre M, Carbone DP, Matrisian LM, Richmond A, Lin PC, Moses HL (2008) Abrogation of TGF-β signaling in mammary carcinomas recruits Gr-1+CD11b+ myeloid cells that promote metastasis. Cancer Cell 13:23–35

Yang X, Li C, Xu X, Deng C (1998) The tumor suppressor Smad4/Dpc4 is essential for epiblast proliferation and mesoderm induction in mice. Proc Natl Acad Sci USA 95:3667–3672

Yang YA, Tang B, Robinson G, Hennighausen L, Brodie SG, Deng CX, Wakefield LM (2002) Smad3 in the mammary epithelium has a nonredundant role in the induction of apoptosis, but not in the regulation of proliferation or differentiation by transforming growth factor-β. Cell Growth Differ 13:123–130

Yang YA, Zhang GM, Feigenbaum L, Zhang YE (2006) Smad3 reduces susceptibility to hepatocarcinoma by sensitizing hepatocytes to apoptosis through downregulation of Bcl-2. Cancer Cell 9:445–457

Zeng Q, Phukan S, Xu Y, Sadim M, Rosman DS, Pennison M, Liao J, Yang GY, Huang CC, Valle L, Di Cristofano A, de la Chapelle A, Pasche B (2009) TGF-βR1 haploinsufficiency is a potent modifier of colorectal cancer development. Cancer Res 69:678–686

Zhang F, Lundin M, Ristimaki A, Heikkila P, Lundin J, Isola J, Joensuu H, Laiho M (2003) Ski-related novel protein N (SnoN), a negative controller of transforming growth factor-β signaling, is a prognostic marker in estrogen receptor-positive breast carcinomas. Cancer Res 63:5005–5010

Zhu Y, Richardson JA, Parada LF, Graff JM (1998) Smad3 mutant mice develop metastatic colorectal cancer. Cell 94:703–714

Chapter 7
The Multifunctional Roles of TGF-β in Navigating the Metastatic Cascade

Michael K. Wendt and William P. Schiemann

Abstract The role of TGF-β during tumorigenesis is best characterized by the diverse functions this multifunctional cytokine exhibits in early-stage versus late-stage cancers. For instance, during the initial stages of tumorigenesis, TGF-β uniformly acts as a potent tumor suppressor, even in low-grade carcinomas capable of evading the cytostatic activities of TGF-β. However, as carcinoma cells continue to evolve and progress towards more aggressive disease states they typically acquire the ability to surmount the last vestiges of the tumor suppressing activities of TGF-β, ultimately gaining a selective advantage that enables TGF-β to promote their metastatic progression and production of recurrent secondary tumor lesions that are refractory to standard chemotherapies. The molecular, cellular, and microenvironmental mechanisms that permit metastatic carcinoma cells to usurp and commandeer TGF-β for oncogenic activities are highly diverse and remain incompletely understood. Here we review recent advances that provide new insights into how aggressive carcinoma cell populations are selected to respond to the oncogenic activities of TGF-β, focusing specifically on its essential functions coupled to metastatic outgrowth and the acquisition of chemoresistance.

Keywords Chemoresistance • Epithelial-mesenchymal transition • Metastasis • Signal transduction • TGF-β

Abbreviations

ECM	Extracellular matrix
EGF	Epidermal growth factor

M.K. Wendt • W.P. Schiemann (✉)
Case Comprehensive Cancer Center, Case Western Reserve University, Wolstein Research Building, 2103 Cornell Road, Cleveland, OH 44106, USA
e-mail: william.schiemann@case.edu

A. Moustakas and K. Miyazawa (eds.), *TGF-β in Human Disease*,
DOI 10.1007/978-4-431-54409-8_7, © Springer Japan 2013

EMT	Epithelial-mesenchymal transition
ERK	Extracellular signal-regulated kinase
FAK	Focal adhesion kinase
JNK	c-Jun N-terminal kinase
LAP	Latency-associated peptide
MAP kinase	Mitogen-activated protein kinase
MEC	Mammary epithelial cell
MET	Mesenchymal–epithelial transition
NF-κB	Nuclear factor-κB
PI3K	Phosphoinositide-3-kinase
TGF-β	Transforming growth factor-β
TβR-I	TGF-β type I receptor
TβR-II	TGF-β type II receptor
TβR-III	TGF-β type III receptors

7.1 Introduction

The major causes of cancer mortality are the processes of tumor recurrence and secondary tumor metastasis. Indeed, the 5-year survival rates for patients diagnosed with localized versus disseminated disease drops from (1) 99 to 23 % for breast cancers, (2) 90 to 12 % for colorectal cancers, (3) 98 to 15 % for melanomas, (4) 92 to 27 % for ovarian cancers, and (5) 100 to 29 % for prostate cancers (Siegel et al. 2012). Unfortunately, the multistep complexity of metastasis and its associated cascade of gene expression and repression has stymied our understanding of this inefficient and deadly process. Interestingly, recent findings have clearly established the TGF-β pathway as being an important mediator of carcinoma recurrence and metastasis (Taylor et al. 2010), which contrasts dramatically from its ability to suppress cellular transformation by preventing normal cells from proliferating uncontrollably, as well as by inducing their differentiation and apoptosis (Tian et al. 2010). Numerous studies over the last three decades have investigated the mechanisms that underlie these dichotomous functions of TGF-β in normal and malignant cells, a phenomenon that is collectively known as the "TGF-β Paradox." Although a single unifying sequelae operant in eliciting the "TGF-β Paradox" has yet to be defined, a significant body of work has demonstrated that aberrations within the TGF-β signaling system figure prominently in its acquisition of oncogenic activity in developing carcinomas (Rahimi and Leof 2007; Tian and Schiemann 2009; Wendt et al. 2012b). The propagation of TGF-β messages across plasma membranes is initiated by its binding to the TGF-β type II receptor (TβR-II), a Ser/Thr protein kinase that recruits, transphosphorylates, and activates the Ser/Thr protein kinase activity of the TGF-β type I (TβR-I) (Parvani et al. 2011; Tian et al. 2010). Once activated, TβR-I rapidly phosphorylates and activates the latent transcription factors, Smad2 and Smad3, which subsequently accumulate in the nucleus as part of heterocomplexes that contain Smad4 and govern gene expression in a cell- and context-specific manner (Feng and Derynck 2005; Heldin and Moustakas 2012; Massague and Gomis 2006). The Smad2/3 pathway is commonly referred to as "canonical" TGF-β signaling, which is complemented by the ability of TGF-β to activate an ever expanding list of "non-canonical" signaling pathways, including (1) integrins and focal adhesion proteins; (2) PI3K and its effectors, AKT and mTOR; (3) NF-κB and inflammatory mediators; and (4) MAP kinases and their downstream effectors (Parvani et al. 2011; Taylor et al. 2010; Wendt et al. 2009a). Interestingly, a number of studies have demonstrated that genetic or epigenetic disruptions that elicit imbalances between these two core branches of TGF-β signaling represent the crux of the "TGF-β Paradox" and its ability to confer oncogenic functions to TGF-β in developing and progressing carcinomas (Tian and Schiemann 2009; Wendt et al. 2009b). Readers desiring additional information related to specific effector molecules and pathways of the TGF-β signaling system are directed to Chap. 1 herein, as well as to several recent reviews (Parvani et al. 2011; Taylor et al. 2010; Wendt et al. 2009a). Here we review recent findings that depict the function of TGF-β and its effector molecules as potent promoters of carcinoma progression and metastasis.

7.2 Epithelial–Mesenchymal Transition

One cannot address the role of TGF-β in mediating metastasis without first exploring the topic of epithelial–mesenchymal transition (EMT), which at its very core reflects the transdifferentiation of polarized epithelial cells into genetically and phenotypically distinct fibroblastoid-like cells (Kalluri and Weinberg 2009; Taylor et al. 2010; Wendt et al. 2009a). The epithelium is comprised of a monolayer of tightly packed epithelial sheets that not only form the skin but also line the internal cavities (e.g., ductal structures of the mammary gland, lung airways, and gastrointestinal tract), thereby providing a protective barrier to ward off environmental insults. Fully differentiated and polarized epithelial cells also exhibit a variety of specialized secretory, sensory, and glandular functions, many of which are regulated by TGF-β. In stark contrast, the mesenchyme functions to provide structural support for the epithelium, doing so via the production and secretion of the extracellular matrix (ECM), particularly collagens and fibronectins (Hay 2005). Moreover, although epithelial cells are highly confined and immobile, their mesenchymal counterparts are exceedingly mobile and invasive and thus are capable of exiting their associated epithelium and traversing through the underlying basement membrane and ECM (Micalizzi et al. 2010). As a single entity, the process of EMT is insufficient to drive carcinoma metastasis. In fact, EMT programs are a normal physiological process that has been hijacked by carcinoma cells to enable their dissemination to distant locales. In clarifying this issue, Kalluri and Weinberg recently subcategorized EMT into three distinct programs, including (1) Type 1 EMT, which occurs embryonically and developmentally during the formation of the endocardial cushion, neural crest formation, and palate closure and fusion; (2) Type 2 EMT, which occurs during tissue remodeling and repair, as well as during inflammation-driven fibrotic reactions; and (3) Type 3 EMT, which occurs during the metastatic progression of carcinomas (Kalluri and Weinberg 2009). TGF-β is clearly a master regulator of all EMT subtypes and readers desiring a more thorough summary of the role of TGF-β in regulating EMT programs are directed to several comprehensive reviews (Heldin et al. 2012; Taylor et al. 2010; Wendt et al. 2009a). In the succeeding sections, we highlight the mechanisms whereby TGF-β induces EMT reactions, as well as discuss the utility of these events to identify and diagnose metastatic carcinomas.

7.2.1 Mechanisms of EMT Induced by TGF-β

Metastatic carcinoma cells that have emerged from EMT programs induced by TGF-β are endowed with several important characteristics, including heightened invasiveness, resistance to apoptosis, and the appearance of stem-like behaviors (Taylor et al. 2010; Wendt et al. 2009a; Yang and Weinberg 2008). The ability of TGF-β to induce EMT programs requires signaling inputs from both the canonical and noncanonical branches of the TGF-β signaling system, which converge in the nucleus to regulate the expression and activity of a variety of master transcriptional regulators of EMT (Taube et al. 2010). Included in this essential group of EMT transcription factors are members of the Snail (SNAI1 and SNAI2/Slug), ZEB (ZEB1 and ZEB2/SIP1), basic helix-loop-helix (Twist1 and Twist2), Six family of homeobox (Six1), and Forkhead (FOXC2), as well as members of the High Mobility Group proteins (HMGA2), which modify DNA structure to enhance transcription factor binding (Wendt et al. 2012b) (Fig. 7.1, *Mechanisms*). Interestingly, these transcription factors play essential roles during the performance of developmental Type 1 EMT programs, as well as during the initiation of oncogenic Type 3 EMT programs, suggesting that the inappropriate reactivation of embryonic EMT programs underlies the acquisition of metastatic phenotypes in response to TGF-β-driven EMT (Micalizzi et al. 2010). For instance, the basic helix-loop-helix transcription factor Twist1 plays a critical role in regulating developmental cell fate determination and differentiation reactions (Leptin 1991). However, consistent with the notion that metastatic carcinoma cells hijack physiological EMT programs during their systemic dissemination (Kang and Massague 2004), Twist1 expression is dramatically upregulated in metastatic breast cancer cells, a reaction that is regulated by TGF-β and essential for their metastasis in mice (Yang et al. 2004). Moreover, heterologous expression of Twist1 also empowers mammary epithelial cells with stem-like characteristics and increases their tumor-initiating capacity (Mani et al. 2008), thereby introducing the concept that EMT plays a number of diverse roles during carcinoma metastasis ranging from their increased invasiveness to their enhanced survival to their heightened chemoresistance (Fig. 7.1, *Consequences*). We recently added to this growing list of EMT functions by demonstrating that recombinant expression of Twist1 was sufficient to initiate the

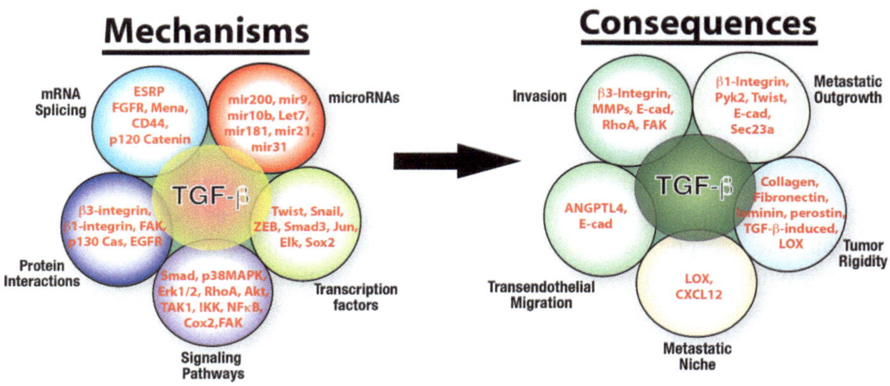

Fig. 7.1 The mechanisms and consequences of oncogenic TGF-β signaling during tumor progression and metastasis. TGF-β is a multifunctional cytokine that suppresses tumor formation by inhibiting cell proliferation and by stimulating cell differentiation or apoptosis. During carcinoma development and progression, TGF-β function is readily converted to that of a tumor promoter, leading to enhanced carcinoma cell EMT, invasion, and metastasis. Several of the established mechanisms of oncogenic TGF-β signaling are shown (*left*). The consequences of these events in relation to specific steps of the metastatic cascade are shown (*right*). At present, the precise "cause and effect" link between specific mechanisms and their associated consequences is expanding at a rapid rate; however, understanding how the consequences and pathophysiological readouts of oncogenic TGF-β signaling relate to multiple mechanistic events remains a daunting challenge to basic and translational researchers. See text for details

pulmonary outgrowth of disseminated breast cancer cells that otherwise succumb to systemic dormancy (Wendt et al. 2011b) (Fig. 7.1, *Consequences*). Along these lines, recent findings indicate that transcription factors coupled to the induction of EMT programs work in concert both directly and indirectly to bring about EMT reactions. For instance, Casas and colleagues (Casas et al. 2011) recently demonstrated that the ability of Twist1 to repress E-cadherin expression during the EMT and metastasis of breast cancer cells was absolutely dependent upon its stimulation of Snail2 (Slug) (Fig. 7.1, *Mechanisms*). Similar to the response of Twist1 to TGF-β, members of the Snail family of zinc finger transcription factors (SNAI1-3) are also induced in cells stimulated by TGF-β (Horiguchi et al. 2008), leading to the direct transcriptional repression of a variety of epithelial genes, including E-cadherin (Hajra et al. 2002) (Fig. 7.1, *Consequences*). Finally, TGF-β readily stimulates the expression of high mobility group A2 (HMGA2), which promotes EMT programs by upregulating the expression of Snail1, Snail2, and Twist, which collectively converge in downregulating that of E-cadherin (Thuault et al. 2006) (Fig. 7.1). Future studies need to define spatiotemporal regulation of these EMT transcription factors by TGF-β in developing and progressing carcinomas and, more importantly, establish the specific steps in the metastatic cascade that are dependent upon these transcription factors.

7.2.2 *Defining the Metastatic EMT Morphology*

To date, EMT programs have been predominantly studied in the context of traditional 2D-culture systems, whose artificial nature imposes significant cellular and genetic adaptations necessary for cells to thrive in this biomechanically rigid microenvironment. Indeed, the prominent feature of transitioning epithelial cells in 2D-cultures is their formation of actin-rich filopodia and actin stress fibers as they acquire fibroblastoid-like morphologies. Teleologically, these morphological features are in keeping with the acquisition of motile, invasive, and metastatic phenotypes by carcinoma cells. Unfortunately, these actin-based architectures are an artifact of propagating mesenchymal cells on rigid 2D-cultures and, in fact, are not observed in mesenchymal cells in vivo (Hay 2005). Similarly, the advent of 3D-organotypic culture systems and the propagation of post-EMT cells in these systems have also called into question the precise morphology exhibited by fully transitioned carcinoma cells in vivo, particularly in tumor microenvironments that possess elevated expression and activity of TGF-βs. Indeed, we recently established that applying breast cancer cells to 3D-organotypic cultures results in the formation of intricate branched structures that readily assume dense multicellular spherical structures in response to TGF-β (Wendt et al. 2010, 2011a, b, 2012a). Importantly, the formation of these "invasospheres" transpires through an EMT-dependent mechanism and reflects the ability of TGF-β to promote the invasion and metastasis of these same breast cancer cells in mice (Wendt et al. 2010, 2012a). Collectively, these findings suggest that the histological association of sarcomatoid morphologies with EMT phenotypes may be misleading and misguiding when attempting to identify post-EMT carcinoma cells in vivo. In an effort to alleviate the current controversies surrounding EMT phenotypes in vivo, future studies clearly need to identify the morphologies and features of post-EMT cells at specific steps of the metastatic cascade, as well as to determine their ability to predict the overall survival and clinical outcome of patients with metastatic disease.

7.2.3 *The Diagnostic Value of EMT*

The crosstalk and linkages between the aforementioned EMT transcription factors have greatly increased our understanding of how TGF-β regulates epithelial plasticity in normal and malignant cells. However, while the list of "master regulators" of EMT programs continues to expand at an alarming rate (Fig. 7.1; *Mechanisms*), the utility of these transcription factors to function as diagnostic or predictive biomarkers of metastatic progression, as well as novel therapeutic targets to alleviate carcinoma dissemination is compromised at an equally alarming rate. Indeed, single cell analysis of circulating tumor cells has shown that these cells simultaneously express epithelial and mesenchymal markers, suggesting the existence of metastable and dynamic EMT phenotypes in carcinoma cells that have escaped the primary tumor

(Armstrong et al. 2011). Given the difficulty to pathologically identify EMT in vivo and the plasticity of EMT programs in highly malignant cells, the overall clinical utility and effectiveness of detecting and targeting of post-EMT carcinoma cells harboring mesenchymal characteristics remains to be determined. For instance, EMT and its counterpart, mesenchymal–epithelial transition (MET), are transient processes that likely take place numerous times during the evolution of a tumor and its dissemination to vital organs. Indeed, several recent studies that examined the plasticity of a variety of breast and other carcinoma cell progression series indicate that the maintenance of a perpetual mesenchymal-state limits the overall tumorigenicity and aggressiveness of fully transitioned cells in response to TGF-β (Korpal et al. 2011; Wendt et al. 2011a, b). As such, evolving the capacity to undergo successive cycles of EMT:MET selects for carcinoma cells that possess the highest metastatic potential. Moving forward, future studies need to identify and validate novel biomarkers capable of differentiating primary, secondary, and tertiary EMT states, which may provide novel diagnostic and therapeutic regimens to alleviate metastatic disease.

7.3 Emerging Paradigms in TGF-β-Mediated Gene Regulation

7.3.1 Diversity in Smad Signaling

The specific set of genes targeted by canonical TGF-β signaling varies greatly in a cell- and context-specific manner. For example, the expression of a particular gene may be induced by TGF-β in fibroblasts, whose proliferation is stimulated by TGF-β; however, this same gene may be potently downregulated by TGF-β in epithelial cells, whose proliferation is inhibited by TGF-β. This diversity in TGF-β signaling has historically been believed to reflect the differential ability of TGF-β to couple to its canonical and noncanonical effectors in distinct cells types, as well as the differential influence of heterologous signals that impinge upon TGF-β effector molecules. However, recent findings have demonstrated an important alternative mechanism, namely the differential and physical interactions between Smad3 with specific "master" transcription factors, including Oct4 in embryonic stem cells (ESCs), Myod1 in myotubes, and PU.1 in pro-B cells (Mullen et al. 2011), thereby enabling Smad3 and TGF-β to direct cell-fate outcomes during tissue morphogenesis and remodeling (Fig. 7.1, *Mechanisms*). Canonical TGF-β signaling is also impacted by the direct phosphorylation of Smad2/3 by a variety of Ser/Thr protein kinases, including members of the CDK and MAP kinase families, GSK-3β, PKC, CK1γ2, and several others (Hayashida et al. 2003; Matsuura et al. 2005; Matsuzaki 2011; Wrighton et al. 2009) (Fig. 7.1, *Mechanisms*). Secondary phosphorylation of Smad2/3 by these protein kinases can alter the subcellular localization, stability, and transcriptional activity of these canonical TGF-β effectors, particularly their regulation of the expression of metastasis-associated genes (Hayashida et al. 2003). Along these lines, the amplified activation of oncogenic signaling systems (e.g., receptor tyrosine kinases, Ras, PI3K, etc.), which modulate the activation status of Smad2/3-directed protein kinases, further dysregulates the repertoire of canonical TGF-β signaling and enables developing carcinoma cells to evade the tumor suppressing activities of TGF-β (Fig. 7.1, *Mechanisms*).

7.3.2 Regulation of MicroRNAs

In addition to traditional modes of gene expression regulated by canonical and noncanonical TGF-β signaling systems, recent studies have established TGF-β as a potent regulator of microRNAs in developing carcinomas, particularly during their acquisition of metastatic phenotypes. The linkage of microRNAs to oncogenic TGF-β signaling was initially described by its regulation of the miR-200 family of microRNAs, which function in suppressing the expression of the EMT transcription factors, ZEB1 and ZEB2 (SIP1) (Gregory et al. 2008; Korpal et al. 2008) (Fig. 7.1, *Mechanisms*). During EMT programs and metastatic progression, TGF-β promotes the downregulation of miR-200 family members, which stabilizes expression of ZEB1 and ZEB2 and enables their binding to and repression of the *Cdh1* (E-cadherin) promoter, resulting in a loss of E-cadherin expression in transiting carcinoma cells (Fig. 7.1, *Consequences*). Intervening studies have since expanded the repertoire of gene transcripts that are targeted and regulated by the miR-200 family, which are remarkably correlated with those targeted and regulated by TGF-β (Howe et al. 2011; Schliekelman et al. 2011). Interestingly, a significant portion of genes targeted by members of the miR-200 family are coupled to the production of proteins operant in producing or sensing alterations in the tumor microenvironment. Included in this growing list of microenvironmental sensors are the ECM molecules fibronectins, collagens, and laminins, as well as cell surface receptors that connect the extracellular and intracellular milieu, such as integrins and growth factor receptors (Fig. 7.1, *Consequences*). Importantly, we recently demonstrated that aberrant expression of these factors underlies the initiation of the "TGF-β Paradox" and promotes the acquisition of oncogenic TGF-β signaling in developing and progressing carcinomas (Galliher and Schiemann 2006, 2007; Galliher-Beckley and Schiemann 2008; Taylor et al. 2011; Wendt and Schiemann 2009; Wendt et al. 2009b). Additionally, recent findings implicate a powerful feed-forward signaling loop between TGF-β and miR-200 family members in driving oncogenic EMT and carcinoma metastasis (Gregory et al. 2011). Collectively, these findings establish miR-200 family members as tumor suppressors capable of preventing oncogenic TGF-β signaling and its induction of EMT in normal epithelial cells (Schliekelman et al. 2011). Surprisingly, two recent studies demonstrated that highly metastatic 4T1 breast cancer cells are more epithelial-like as compared to their isogenic and nonmetastatic 4T07 counterparts (Korpal et al. 2011; Wendt et al. 2011a, b). Amongst the differences between these two cell types is the reestablishment of miR-200 in

4T1 cells, an event that aids in the production and secretion of metastasis-promoting proteins necessary for the outgrowth of macroscopic metastases (Korpal et al. 2011) (Fig. 7.1, *Consequences*).

Besides regulating the expression of miR-200 family members, TGF-β also dictates the activities of microRNAs by promoting their processing. For instance, Smad2/3 interacts physically with DROSHA, which enhances the conversion of primary miR-21 transcripts into their pre-miR-21 counterparts in a Smad4-independent manner (Davis et al. 2008) (Fig. 7.1, *Mechanisms*). The coupling of TGF-β to EMT programs also proceeds via its ability to upregulate the expression of miRs 21 and 31, resulting in the downregulation of Tiam1 (Cottonham et al. 2010; Zavadil et al. 2007), and miR-155, leading to the repression of RhoA expression (Kong et al. 2008) (Fig. 7.1, *Consequences*). Finally, we have identified miR-181a as a TGF-β-inducible "metastamir" that enhances breast cancer metastasis by facilitating the downregulation of the pro-apoptotic molecule, Bim, thereby conferring miR-181a-expressing carcinoma cells resistance to anoikis (Taylor et al. 2013) (Fig. 7.1, *Consequences*).

7.3.3 Alternative Splicing

The splicing of mRNAs is a post-transcriptional modification that functions to dramatically increase the diversity of gene products produced from a static genome. In fact, transcription of the vast majority of the genes in the genome (~90 %) gives rise to two or more protein products due to the process of alternative splicing. Recently, TGF-β has been suggested to play a major role in mRNA splicing, particularly in developing carcinomas through its ability to suppress the expression of epithelial splicing regulatory proteins (ESRPs), which are a class of RNA-binding proteins that govern the splicing of epithelial gene transcripts (Horiguchi et al. 2012; Warzecha et al. 2010). Indeed, the ability of distinct protein variants to impact cancer development, EMT, and metastasis is rapidly coming to the forefront of cancer biology, including that regulated by TGF-β. For instance, the initiation of alternative gene splicing programs that transpire during EMT reactions driven by TGF-β results in part from its ability to downregulate the expression of ESRP2 via a ZEB1/δEF1- and ZEB2/SIP1-dependent manner (Horiguchi et al. 2012) (Fig. 7.1, *Mechanisms*). Perhaps the best studied example of differential gene splicing governed by TGF-β relates to its regulation of the four member fibroblast growth factor receptor (FGFR) family, of which FGFRs 1–3 are subject to alternative splicing that greatly enhances isoform diversity and function (Fig. 7.1, *Mechanisms*). TGF-β regulates two major splicing events in FGFRs. First, alternative splicing of the extracellular Ig-like domain I (α-loop) results in either the inclusion or the exclusion of the first Ig-like domain. Importantly, exclusion of Ig-like domain I drastically increases the affinity of the FGFR to bind members of the FGF ligand family, which currently comprises 22 individual cytokines (Jain and Turner 2012; Wesche et al. 2011). Second, alternative splicing of the distal portion of Ig-like domain III dictates the specificity of

FGFRs to bind specific FGF ligands (Holzmann et al. 2012; Jain and Turner 2012; Werner et al. 1992). Inclusion of exon 8 results in the production of FGFR-IIIb isoforms capable of binding to FGF7 (KGF) and FGF10, while inclusion of exon 9 results in the production of FGFR-IIIc isoforms that selectively bind FGF2 and FGF4 (Holzmann et al. 2012; Jain and Turner 2012; Werner et al. 1992). Interestingly, FGFR-IIIb and -IIIc isoforms are preferentially expressed in epithelial and mesenchymal cells, respectively, and the induction of EMT programs by TGF-β is sufficient to switch epithelial FGFR-IIIb expression to that of its -IIIc counterpart in post-EMT cells (Shirakihara et al. 2011) (Fig. 7.1, *Mechanisms*). It should be noted, however, that the metastatic cells that exhibit epithelial morphologies and characteristics retain FGFR-IIIc expression (Chaffer et al. 2006), suggesting the presence of a metastable phenotype in metastatic cells that have emerged from successive rounds of EMT:MET. Future studies need to further address the consequences of FGFR splicing during TGF-β-driven EMT and metastasis, particularly during the latter stages of metastatic progression and emergence from micrometastatic dormancy (Korpal et al. 2011; Wendt et al. 2011a, b), and as a potential predictive biomarker and therapeutic target for recurrent metastatic tumors.

7.4 TGF-β and the Metastatic Cascade

7.4.1 Primary Tumor Local Invasion

The initial step of the metastatic cascade requires tumor cells to exit the confines of a primary tumor and invade into the surrounding tissue. TGF-β plays a critical role in stimulating tumor cell invasion through its ability to upregulate the expression and activity of extracellular proteases charged with degrading tumor ECM and supporting basement membrane. For instance, TGF-β robustly induces the expression of several matrix metalloproteinases (MMPs), including MMPs 2, 3, 7, 13, MT1, and especially MMP-9 (Wendt et al. 2009a) (Fig. 7.1, *Consequences*). Although the signaling mechanisms that couple TGF-β to MMP expression remain to be determined definitively, recent studies have implicated the noncanonical effector, TAK1 and its downstream targets as mediators of MMP-9 expression stimulated by TGF-β (Safina et al. 2008) (Fig. 7.1, *Mechanisms*). In general, genetic depletion or pharmacologic inactivation of MMPs drastically decreases the ability of TGF-β to stimulate the invasion of metastatic carcinoma cells (Wendt et al. 2009a). Besides their ability to degrade ECM during carcinoma invasion, MMPs also cleave LAP (latency-associated peptide) that permits the liberation of biologically active TGF-β from inactive ECM depots, thereby initiating a positive feed-forward loop coupled to elevated autocrine TGF-β signaling (Rifkin 2005; Taylor et al. 2010).

TGF-β also enhances carcinoma invasiveness by altering their production of and response to ECM molecules within the developing tumor microenvironment (Stover et al. 2007). Indeed, when stimulated by TGF-β, metastatic carcinoma cells elevate their production of a wide array of ECM proteins, including collagens (Ignotz and Massague 1986), fibronectin (Ignotz and Massague 1986), periostin (Wen et al. 2010), and laminins (Giannelli et al. 2005; Kumar et al. 1995), as well as a newly described protein known as TGF-β-induced [TGFBI; (Nummela et al. 2012)] that limits cell adhesion to structural ECM proteins (Fig. 7.1, *Consequences*). Integrins are heterodimeric transmembrane receptors that function as conduits capable of linking the ECM to intracellular signaling and cytoskeletal systems (Hall 2009; Zutter 2007). Principle players involved in TGF-β-mediated response to ECM proteins are αvβ3 integrins, whose expression is robustly induced by TGF-β via a Smad2/3-independent mechanism (Galliher and Schiemann 2006; Pechkovsky et al. 2008). Amplified expression of αvβ3 integrin results in its indirect interaction with TβR-II, a reaction mediated by focal adhesion kinase (FAK); (Wendt and Schiemann 2009). Importantly, the incorporation of αvβ3 integrin within activated TGF-β receptor complexes enables Src to phosphorylate TβR-II at Tyr284 (Galliher and Schiemann 2007). This phosphorylation event facilitates the binding of Grb2 and ShcA to TβR-II leading to amplified noncanonical TGF-β signaling through MAP kinases (Galliher et al. 2006; Galliher and Schiemann 2007; Galliher-Beckley and Schiemann 2008). Additionally, αvβ3 integrin:TβR-II complexes couple TGF-β to FAK (Wendt and Schiemann 2009), p130Cas (Wendt et al. 2009b), and Pyk2 (Wendt et al. 2012a), which further enhance the coupling of TGF-β to several noncanonical effector molecules, including NF-κB (Neil and Schiemann 2008), Cox-2 (Neil et al. 2008), and PGE2 (Tian and Schiemann 2010). In fact, pharmacologic or genetic inactivation of FAK effectively blocks the invasion and metastasis of breast cancer cells (Wendt and Schiemann 2009), while similar approaches against the FAK target, p130Cas, only inhibit breast cancer invasion and primary tumor dissemination without affecting the overall extent of pulmonary metastasis (Wendt et al. 2009b). Thus, these findings point to the existence of novel rate-limiting steps during initiation of the metastatic cascade in response to TGF-β.

Finally, β3 integrin also governs oncogenic TGF-β signaling by interacting physically with EGFR, a reaction that enables fibronectin to transactivate EGFR in the absence of EGF ligands (Balanis et al. 2011). Along these lines, we showed that FAK serves as a bridge that links EGFR with TβR-II in post-EMT cells, leading to their acquisition of invasive phenotypes when stimulated with EGF (Wendt et al. 2010). Finally, β3 integrin binds FGF and the co-expression of both molecules is highly predictive for lung cancer recurrence (Massabeau et al. 2009; Mori et al. 2008). Collectively, these studies establish the unique ability of integrins, particularly β3 integrin, to influence both the mechanisms of TGF-β signaling and the metastatic consequences of these events (Fig. 7.1). Moreover, these findings highlight the ability of oncogenic TGF-β signaling to cooperate with and activate a cascade of aberrant growth factor signaling events that coalesce to promote several aspects of the metastatic cascade.

7.4.2 Establishment of the Metastatic Niche

The pathophysiology of TGF-β in developing and progressing carcinomas in many respects reflects changes in its expression and bioavailability within the tumor microenvironments, as well as changes within the biomechanics of the tumor microenvironment. For instance, we recently established that TGF-β upregulates the expression of lysyl oxidase (LOX) (Taylor et al. 2011), a secreted copper amine oxidase that enzymatically crosslinks collagen to elastin (Erler et al. 2009; Erler and Giaccia 2006). The outcome of these LOX-mediated crosslinking reactions produces hydrogen peroxide as a by-product that serves as a novel second messenger for TGF-β (Taylor et al. 2011), as well as generates the rigid, palpable characteristics of carcinomas as compared to their surrounding normal tissues (Butcher et al. 2009) (Fig. 7.1, *Consequences*). Collectively, these events drive the oncogenic activities of TGF-β, including its stimulation of EMT, invasive, and metastatic phenotypes (Taylor et al. 2010, 2011) in part by amplifying integrin-mediated signaling events (Levental et al. 2009) and establishing the premetastatic niche (Erler et al. 2009). The latter activity reflects the ability of crosslinked collagens to serve as homing and docking sites during the recruitment of myeloid progenitors to future metastatic sites. Importantly, the recruitment of myeloid progenitors and formation of a premetastatic niche at peripheral sites is controlled in part by TGF-β via its regulation of stromal cell derived factor-1 [SDF-1; (Yang et al. 2008)] (Fig. 7.1, *Consequences*). Taken together, these studies suggest a model wherein aberrant TGF-β signaling upregulates primary tumor expression of LOX, while simultaneously downregulating that of SDF-1, leading to the homing of myeloid progenitors and premetastatic niche formation at peripheral sites of high SDF-1 expression, such as the lungs, brain, and bone marrow.

7.4.3 Transendothelial Cell Migration

Upon exiting the primary tumor, carcinoma cells must intravasate and extravasate the blood or lymphatic vessels before they can invade into a secondary tumor site (Nguyen et al. 2009). The processes of intravasation and extravasation are reminiscent of those employed by immune cells, whose ability to perform these tasks remain incompletely understood. Not surprisingly, even less is known mechanistically with regard to how metastatic cells emulate these events. Recently, TGF-β was shown to induce angiopoietin-like 4 (ANGPTL4) expression, which enables breast cancer cells to disrupt the vasculature and transit through the pulmonary endothelium (Padua et al. 2008) (Fig. 7.1, *Consequences*). Along these lines, the downregulated expression of E-cadherin in post-EMT prostate cancer cells facilitates their migration through endothelial cell layers (Drake et al. 2009) (Fig. 7.1, *Consequences*). Taken together, these findings highlight two consequences of TGF-β signaling that are operant in facilitating transendothelial migration of carcinoma cells during their dissemination. Future studies need to bolster our understanding of transendothelial cell migration and its regulation by TGF-β, as well as to determine the therapeutic potential of targeting the effectors operant in this process as a novel means to prevent carcinoma cell seeding at secondary tumor sites.

7.4.4 Secondary Tumor Outgrowth

Following their dissemination and invasion into secondary tumor sites, metastatic carcinoma cells must surmount the tendency to undergo systemic dormancy, which may represent the most significant obstacle of the metastatic cascade (Brackstone et al. 2007). Indeed, estimates indicate that even small tumors (~1 cm) can release more than one million cells each day into the blood stream (Chiang and Massague 2008; Talmadge and Fidler 2010), signifying that the metastatic cascade is a highly inefficient process that is subject to two major limitations, namely the inability of disseminated cells to (1) escape the lethality of vascular shear stress and (2) circumvent metastatic dormancy. With respect to the latter phenomenon, it is important to remember that metastatic cells typically exit biomechanically rigid primary tumors and subsequently take up residence in new organs that are noticeably more compliant (Butcher et al. 2009). Indeed, using 3D-organotypic cultures that recapitulate rigid primary tumor and compliant pulmonary microenvironments, we observed that a compliant microenvironment could restore some of the tumor suppressing activities of TGF-β to highly metastatic breast cancer cells (Allington et al. 2009; Taylor et al. 2011), indicating that the extent to which TGF-β exhibits oncogenic activities is directly proportional to the degree of microenvironmental rigidity. Interestingly, metastatic cells that have completed an EMT program readily circumvent the suppressive activities of compliant microenvironments and instead utilize TGF-β as a means to promote their outgrowth (Wendt et al. 2011a, b, 2012a). Mechanistically, increased tissue compliance inhibits the coupling of TGF-β to Smad4 (Korpal et al. 2009; Wendt et al. 2011a, 2012a) (Fig. 7.1, *Mechanisms*), which prevents the expression of Pyk2 necessary to mediate escape from metastatic dormancy (Wendt et al. 2012a) (Fig. 7.1, *Consequences*). These events are also regulated by the reciprocal expression patterns of E-cadherin and β1 integrin, such that dormant cells are typically E-cadherin high, β1 integrin low, while outgrowth proficient cells are typically β1 integrin high, E-cadherin low (Barkan et al. 2008, 2010; Shibue and Weinberg 2009; Wendt et al. 2011a, 2012a). Importantly, we determined that EMT programs represent the molecular switch between metastatic dormancy and proficiency due to the downregulation of E-cadherin expression, which interacts in a heterotypic manner with the extracellular domain of β1 integrin (Whittard et al. 2002), downregulating its expression and that of Pyk2, collectively preventing the initiation of proliferative programs (Wendt et al. 2011a, b, 2012a). Likewise, the expression and activity of FAK has also been implicated in the initiation of metastatic outgrowth (Shibue and Weinberg 2009), and as such, FAK and Pyk2 represent essential TGF-β effectors that link altered integrin expression and ECM biomechanics with the micrometastatic machinery operant in driving secondary tumor formation (Fig. 7.1, *Consequences*).

7.5 Conclusions

In years past, cancer biologists have analyzed tumor development and progression primarily using a "tumoricentric" approach—i.e., focusing on studies that determine the molecular and genetic basis of how tumors are initiated, progress, and ultimately impact their surroundings. New studies emphasizing translational and tissue-based research approaches have ushered in a new arena of cancer biology aimed at understanding how clinical regimens and targeted therapies impact the aforementioned processes of tumor development and metastatic progression. A primary focus of these analyses deals with the development of chemoresistance, which typically reflects three distinct subtypes of acquired resistance: (1) development of secondary mutations within the drug target; (2) compensatory activation and/or amplification of additional growth promoting pathways; and (3) acquisition of EMT phenotypes following drug treatment (Baum et al. 2008; Sharma et al. 2010; Singh and Settleman 2010). With respect to the third mechanism, recent studies suggest that chemoresistant cells do in fact display a mesenchymal phenotype indicative of EMT programs. Thus, the principal question moving forward relates to whether chemoresistant carcinoma cells are actually derived from a low-frequency resident population that was selected out of the bulk tumor, or whether these chemoresistant variants emerged from their performance of an EMT program? Although additional experimentation is clearly warranted, the chemoresistance of metastatic carcinoma cells has been associated with the acquisition of EMT programs (Creighton et al. 2009; Farmer et al. 2009; Sharma et al. 2010; Taube et al. 2010), particularly that driven by TGF-β (Mani et al. 2008; Shipitsin et al. 2007). Likewise, TGF-β has also been observed to drive an EMT program coupled to radioinsensitivity (Barcellos-Hoff and Akhurst 2009). Collectively, these findings suggest that co-administering TGF-β antagonists with targeted chemotherapies may provide novel inroads to reverse this consequence of EMT programs, and in doing so, may sensitize late-stage carcinoma cells not only to newly developed targeted chemotherapies but also to traditional genotoxic and cytotoxic agents. Future studies need to address this hypothesis and determine the most effective drug and dosing combinations needed to eradicate these recalcitrant tumor subpopulations. Likewise, future studies need to continue elucidating the molecular mechanisms whereby TGF-β drives carcinoma development and metastatic progression, paying particular attention at determining how the flux through the TGF-β signaling system changes at distinct steps of the metastatic cascade, and at defining novel TGF-β-based biomarkers capable of predicting clinical outcomes, directing treatment regimens, and serving as novel drug targets.

Acknowledgments Members of the Schiemann Laboratory are thanked for critical reading of the manuscript. W.P.S. was supported in part by grants from the National Institutes of Health (CA129359), the Department of Defense (BC084561), and pilot funding from the Case Comprehensive Cancer Center (P30 CA043703), while M.K.W. was supported by the National Institutes of Health (CA166140).

References

Allington TM, Galliher-Beckley AJ, Schiemann WP (2009) Activated Abl kinase inhibits oncogenic transforming growth factor-β signaling and tumorigenesis in mammary tumors. FASEB J 23:4231–4243

Armstrong AJ, Marengo MS, Oltean S et al (2011) Circulating tumor cells from patients with advanced prostate and breast cancer display both epithelial and mesenchymal markers. Mol Cancer Res 9:997–1007

Balanis NYM, Wendt MK, Schiemann WP, Carlin CR (2011) A conformationally flexible region of the EGF receptor regulates formation of membrane protrusions by controlling selective binding to fibronectin receptors. Mol Biol Cell 22:4288–4301

Barcellos-Hoff MH, Akhurst RJ (2009) Transforming growth factor-β in breast cancer: too much, too late. Breast Cancer Res 11:202

Barkan D, Kleinman H, Simmons JL et al (2008) Inhibition of metastatic outgrowth from single dormant tumor cells by targeting the cytoskeleton. Cancer Res 68:6241–6250

Barkan D, El Touny LH, Michalowski AM et al (2010) Metastatic growth from dormant cells induced by a Col-I-enriched fibrotic environment. Cancer Res 70:5706–5716

Baum B, Settleman J, Quinlan MP (2008) Transitions between epithelial and mesenchymal states in development and disease. Semin Cell Dev Biol 19:294–308

Brackstone M, Townson JL, Chambers AF (2007) Tumour dormancy in breast cancer: an update. Breast Cancer Res 9:208

Butcher DT, Alliston T, Weaver VM (2009) A tense situation: forcing tumour progression. Nat Rev Cancer 9:108–122

Casas E, Kim J, Bendesky A, Ohno-Machado L, Wolfe CJ, Yang J (2011) Snail2 is an essential mediator of Twist1-induced epithelial mesenchymal transition and metastasis. Cancer Res 71:245–254

Chaffer CL, Brennan JP, Slavin JL, Blick T, Thompson EW, Williams ED (2006) Mesenchymal-to-epithelial transition facilitates bladder cancer metastasis: role of fibroblast growth factor receptor-2. Cancer Res 66:11271–11278

Chiang AC, Massague J (2008) Molecular basis of metastasis. N Engl J Med 359:2814–2823

Cottonham CL, Kaneko S, Xu L (2010) miR-21 and miR-31 converge on TIAM1 to regulate migration and invasion of colon carcinoma cells. J Biol Chem 285:35293–35302

Creighton CJ, Li X, Landis M et al (2009) Residual breast cancers after conventional therapy display mesenchymal as well as tumor-initiating features. Proc Natl Acad Sci USA 106:13820–13825

Davis BN, Hilyard AC, Lagna G, Hata A (2008) SMAD proteins control DROSHA-mediated microRNA maturation. Nature 454:56–61

Drake JM, Strohbehn G, Bair TB, Moreland JG, Henry MD (2009) ZEB1 enhances transendothelial migration and represses the epithelial phenotype of prostate cancer cells. Mol Biol Cell 20:2207–2217

Erler JT, Giaccia AJ (2006) Lysyl oxidase mediates hypoxic control of metastasis. Cancer Res 66:10238–10241

Erler JT, Bennewith KL, Cox TR et al (2009) Hypoxia-induced lysyl oxidase is a critical mediator of bone marrow cell recruitment to form the premetastatic niche. Cancer Cell 15:35–44

Farmer P, Bonnefoi H, Anderle P et al (2009) A stroma-related gene signature predicts resistance to neoadjuvant chemotherapy in breast cancer. Nat Med 15:68–74

Feng XH, Derynck R (2005) Specificity and versatility in TGF-β signaling through Smads. Annu Rev Cell Dev Biol 21:659–693

Galliher AJ, Schiemann WP (2006) β3 integrin and Src facilitate transforming growth factor-β mediated induction of epithelial-mesenchymal transition in mammary epithelial cells. Breast Cancer Res 8:R42

Galliher AJ, Schiemann WP (2007) Src phosphorylates Tyr284 in TGF-β type II receptor and regulates TGF-β stimulation of p38 MAPK during breast cancer cell proliferation and invasion. Cancer Res 67:3752–3758
Galliher AJ, Neil JR, Schiemann WP (2006) Role of transforming growth factor-β in cancer progression. Future Oncol 2:743–763
Galliher-Beckley AJ, Schiemann WP (2008) Grb2 binding to Tyr284 in TβR-II is essential for mammary tumor growth and metastasis stimulated by TGF-β. Carcinogenesis 29:244–251
Giannelli G, Bergamini C, Fransvea E, Sgarra C, Antonaci S (2005) Laminin-5 with transforming growth factor-β1 induces epithelial to mesenchymal transition in hepatocellular carcinoma. Gastroenterology 129:1375–1383
Gregory PA, Bert AG, Paterson EL et al (2008) The miR-200 family and miR-205 regulate epithelial to mesenchymal transition by targeting ZEB1 and SIP1. Nat Cell Biol 10:593–601
Gregory PA, Bracken CP, Smith E et al (2011) An autocrine TGF-β/ZEB/miR-200 signaling network regulates establishment and maintenance of epithelial-mesenchymal transition. Mol Biol Cell 22:1686–1698
Hajra KM, Chen DY, Fearon ER (2002) The SLUG zinc-finger protein represses E-cadherin in breast cancer. Cancer Res 62:1613–1618
Hall A (2009) The cytoskeleton and cancer. Cancer Metastasis Rev 28:5–14
Hay ED (2005) The mesenchymal cell, its role in the embryo, and the remarkable signaling mechanisms that create it. Dev Dyn 233:706–720
Hayashida T, Decaestecker M, Schnaper HW (2003) Cross-talk between ERK MAP kinase and Smad signaling pathways enhances TGF-β-dependent responses in human mesangial cells. FASEB J 17:1576–1578
Heldin CH, Moustakas A (2012) Role of Smads in TGFβ signaling. Cell Tissue Res 347:21–36
Heldin CH, Vanlandewijck M, Moustakas A (2012) Regulation of EMT by TGFβ in cancer. FEBS Lett 586:1959–1970
Holzmann K, Grunt T, Heinzle C et al (2012) Alternative splicing of fibroblast growth factor receptor IgIII loops in cancer. J Nucleic Acids 2012:950508
Horiguchi K, Shirakihara T, Nakano A, Imamura T, Miyazono K, Saitoh M (2008) Role of Ras signaling in the induction of Snail by TGF-β. J Biol Chem 284:245–253
Horiguchi K, Sakamoto K, Koinuma D et al (2012) TGF-β drives epithelial-mesenchymal transition through δEF1-mediated downregulation of ESRP. Oncogene 31:3190–3201
Howe EN, Cochrane DR, Richer JK (2011) Targets of miR-200c mediate suppression of cell motility and anoikis resistance. Breast Cancer Res 13:R45
Ignotz RA, Massague J (1986) Transforming growth factor-β stimulates the expression of fibronectin and collagen and their incorporation into the extracellular matrix. J Biol Chem 261:4337–4345
Jain VK, Turner NC (2012) Challenges and opportunities in the targeting of fibroblast growth factor receptors in breast cancer. Breast Cancer Res 14:208
Kalluri R, Weinberg RA (2009) The basics of epithelial-mesenchymal transition. J Clin Invest 119:1420–1428
Kang Y, Massague J (2004) Epithelial-mesenchymal transitions: twist in development and metastasis. Cell 118:277–279
Kong W, Yang H, He L et al (2008) MicroRNA-155 is regulated by the transforming growth factor β/Smad pathway and contributes to epithelial cell plasticity by targeting RhoA. Mol Cell Biol 28:6773–6784
Korpal M, Lee ES, Hu G, Kang Y (2008) The miR-200 family inhibits epithelial-mesenchymal transition and cancer cell migration by direct targeting of E-cadherin transcriptional repressors ZEB1 and ZEB2. J Biol Chem 283:14910–14914
Korpal M, Yan J, Lu X, Xu S, Lerit DA, Kang Y (2009) Imaging transforming growth factor-β signaling dynamics and therapeutic response in breast cancer bone metastasis. Nat Med 15:960–966
Korpal M, Ell BJ, Buffa FM et al (2011) Direct targeting of Sec23a by miR-200 s influences cancer cell secretome and promotes metastatic colonization. Nat Med 17:1101–1108

Kumar NM, Sigurdson SL, Sheppard D, Lwebuga-Mukasa JS (1995) Differential modulation of integrin receptors and extracellular matrix laminin by transforming growth factor-β 1 in rat alveolar epithelial cells. Exp Cell Res 221:385–394
Leptin M (1991) Twist and snail as positive and negative regulators during Drosophila mesoderm development. Genes Dev 5:1568–1576
Levental KR, Yu H, Kass L et al (2009) Matrix crosslinking forces tumor progression by enhancing integrin signaling. Cell 139:891–906
Mani SA, Guo W, Liao MJ et al (2008) The epithelial-mesenchymal transition generates cells with properties of stem cells. Cell 133:704–715
Massabeau C, Rouquette I, Lauwers-Cances V et al (2009) Basic fibroblast growth factor-2/β3 integrin expression profile: Signature of local progression after chemoradiotherapy for patients with locally advanced non-small-cell lung cancer. Int J Radiat Oncol Biol Phys 75:696–702
Massague J, Gomis RR (2006) The logic of TGFβ signaling. FEBS Lett 580:2811–2820
Matsuura I, Wang G, He D, Liu F (2005) Identification and characterization of ERK MAP kinase phosphorylation sites in Smad3. Biochemistry 44:12546–12553
Matsuzaki K (2011) Smad phosphoisoform signaling specificity: the right place at the right time. Carcinogenesis 32:1578–1588
Micalizzi DS, Farabaugh SM, Ford HL (2010) Epithelial-mesenchymal transition in cancer: parallels between normal development and tumor progression. J Mammary Gland Biol Neoplasia 15:117–134
Mori S, Wu CY, Yamaji S et al (2008) Direct binding of integrin αvβ3 to FGF1 plays a role in FGF1 signaling. J Biol Chem 283:18066–18075
Mullen AC, Orlando DA, Newman JJ et al (2011) Master transcription factors determine cell-type-specific responses to TGF-β signaling. Cell 147:565–576
Neil JR, Schiemann WP (2008) Altered TAB1:IκB kinase interaction promotes transforming growth factor β-mediated nuclear factor-κB activation during breast cancer progression. Cancer Res 68:1462–1470
Neil JR, Johnson KM, Nemenoff RA, Schiemann WP (2008) COX-2 inactivates Smad signaling and enhances EMT stimulated by TGF-β through a PGE2-dependent mechanism. Carcinogenesis 29:2227–2235
Nguyen DX, Bos PD, Massague J (2009) Metastasis: from dissemination to organ-specific colonization. Nat Rev Cancer 9:274–284
Nummela P, Lammi J, Soikkeli J, Saksela O, Laakkonen P, Holtta E (2012) Transforming growth factor β-induced (TGFBI) is an anti-adhesive protein regulating the invasive growth of melanoma cells. Am J Pathol 180:1663–1674
Padua D, Zhang XH, Wang Q et al (2008) TGFβ primes breast tumors for lung metastasis seeding through angiopoietin-like 4. Cell 133:66–77
Parvani JG, Taylor MA, Schiemann WP (2011) Noncanonical TGF-β signaling during mammary tumorigenesis. J Mammary Gland Biol Neoplasia 16:127–146
Pechkovsky DV, Scaffidi AK, Hackett TL et al (2008) Transforming growth factor β1 induces αvβ3 integrin expression in human lung fibroblasts via a β3 integrin-, c-Src-, and p38 MAPK-dependent pathway. J Biol Chem 283:12898–12908
Rahimi RA, Leof EB (2007) TGF-β signaling: a tale of two responses. J Cell Biochem 102:593–608
Rifkin DB (2005) Latent transforming growth factor-β (TGF-β) binding proteins: Orchestrators of TGF-β availability. J Biol Chem 280:7409–7412
Safina A, Ren MQ, Vandette E, Bakin AV (2008) TAK1 is required for TGF-β1-mediated regulation of matrix metalloproteinase-9 and metastasis. Oncogene 27:1198–1207
Schliekelman MJ, Gibbons DL, Faca VM et al (2011) Targets of the tumor suppressor miR-200 in regulation of the epithelial-mesenchymal transition in cancer. Cancer Res 71:7670–7682
Sharma SV, Lee DY, Li B et al (2010) A chromatin-mediated reversible drug-tolerant state in cancer cell subpopulations. Cell 141:69–80
Shibue T, Weinberg RA (2009) Integrin β1-focal adhesion kinase signaling directs the proliferation of metastatic cancer cells disseminated in the lungs. Proc Natl Acad Sci USA 106:10290–10295

Shipitsin M, Campbell LL, Argani P et al (2007) Molecular definition of breast tumor heterogeneity. Cancer Cell 11:259–273

Shirakihara T, Horiguchi K, Miyazawa K et al (2011) TGF-β regulates isoform switching of FGF receptors and epithelial-mesenchymal transition. EMBO J 30:783–795

Siegel R, Naishadham D, Jemal A (2012) Cancer statistics, 2012. CA Cancer J Clin 62:10–29

Singh A, Settleman J (2010) EMT, cancer stem cells and drug resistance: An emerging axis of evil in the war on cancer. Oncogene 29:4741–4751

Stover DG, Bierie B, Moses HL (2007) A delicate balance: TGF-β and the tumor microenvironment. J Cell Biochem 101:851–861

Talmadge JE, Fidler IJ (2010) The biology of cancer metastasis: historical perspective. Cancer Res 70:5649–5669

Taube JH, Herschkowitz JI, Komurov K et al (2010) Core epithelial-to-mesenchymal transition interactome gene-expression signature is associated with claudin-low and metaplastic breast cancer subtypes. Proc Natl Acad Sci USA 107:15449–15454

Taylor MA, Parvani JG, Schiemann WP (2010) The pathophysiology of epithelial-mesenchymal transition induced by transforming growth factor-β in normal and malignant mammary epithelial cells. J Mammary Gland Biol Neoplasia 15:169–190

Taylor MA, Amin JD, Kirschmann DA, Schiemann WP (2011) Lysyl oxidase contributes to mechanotransduction-mediated regulation of transforming growth factor-β signaling in breast cancer cells. Neoplasia 13:406–418

Taylor MA, Sossey-Alaoui K, Thompson CL, Danielpour D, Schiemann WP (2013) TGF-β upregulates miR-181a expression to promote breast cancer metastasis. J Clin Invest 123:150–163

Thuault S, Valcourt U, Petersen M, Manfioletti G, Heldin CH, Moustakas A (2006) Transforming growth factor-β employs HMGA2 to elicit epithelial-mesenchymal transition. J Cell Biol 174:175–183

Tian M, Schiemann WP (2009) The TGF-β paradox in human cancer: an update. Future Oncol 5:259–271

Tian M, Schiemann WP (2010) PGE2 receptor EP2 mediates the antagonistic effect of COX-2 on TGF-β signaling during mammary tumorigenesis. FASEB J 24:1105–1116

Tian M, Neil JR, Schiemann WP (2010) Transforming growth factor-β and the hallmarks of cancer. Cell Signal 23:951–962

Warzecha CC, Jiang P, Amirikian K et al (2010) An ESRP-regulated splicing programme is abrogated during the epithelial-mesenchymal transition. EMBO J 29:3286–3300

Wen W, Chau E, Jackson-Boeters L, Elliott C, Daley TD, Hamilton DW (2010) TGF-β1 and FAK regulate periostin expression in PDL fibroblasts. J Dent Res 89:1439–1443

Wendt M, Schiemann W (2009) Therapeutic targeting of the focal adhesion complex prevents oncogenic TGF-β signaling and metastasis. Breast Cancer Res 11:R68

Wendt MK, Allington TM, Schiemann WP (2009a) Mechanisms of the epithelial-mesenchymal transition by TGF-β. Future Oncol 5:1145–1168

Wendt MK, Smith JA, Schiemann WP (2009b) p130Cas is required for mammary tumor growth and transforming growth factor-β-mediated metastasis through regulation of Smad2/3 activity. J Biol Chem 284:34145–34156

Wendt MK, Smith JA, Schiemann WP (2010) Transforming growth factor-β-induced epithelial-mesenchymal transition facilitates epidermal growth factor-dependent breast cancer progression. Oncogene 29:6485–6498

Wendt MK, Molter J, Flask, CA, Schiemann, WP (2011a) In vivo dual substrate bioluminesent imaging. J Vis Exp 56:4288–4301

Wendt MK, Taylor MA, Schiemann BJ, Schiemann WP (2011b) Down-regulation of epithelial cadherin is required to initiate metastatic outgrowth of breast cancer. Mol Biol Cell 22:2423–2435

Wendt MK, Schiemann BJ, Parvani JG, Lee YH, Kang Y, Schiemann WP (2012a) TGF-β stimulates Pyk2 expression as part of an epithelial-mesenchymal transition program required for metastatic outgrowth of breast cancer. Oncogene Jun 28: doi 10.1038/onc.2012.230

Wendt MK, Tian M, Schiemann WP (2012b) Deconstructing the mechanisms and consequences of TGF-β-induced EMT during cancer progression. Cell Tissue Res 347:85–101
Werner S, Duan DS, de Vries C, Peters KG, Johnson DE, Williams LT (1992) Differential splicing in the extracellular region of fibroblast growth factor receptor 1 generates receptor variants with different ligand-binding specificities. Mol Cell Biol 12:82–88
Wesche J, Haglund K, Haugsten EM (2011) Fibroblast growth factors and their receptors in cancer. Biochem J 437:199–213
Whittard JD, Craig SE, Mould AP et al (2002) E-cadherin is a ligand for integrin α2β1. Matrix Biol 21:525–532
Wrighton KH, Lin X, Feng XH (2009) Phospho-control of TGF-β superfamily signaling. Cell Res 19:8–20
Yang J, Weinberg RA (2008) Epithelial-mesenchymal transition: at the crossroads of development and tumor metastasis. Dev Cell 14:818–829
Yang J, Mani SA, Donaher JL et al (2004) Twist, a master regulator of morphogenesis, plays an essential role in tumor metastasis. Cell 117:927–939
Yang L, Huang J, Ren X et al (2008) Abrogation of TGFβ signaling in mammary carcinomas recruits Gr-1^{+}CD11b^{+} myeloid cells that promote metastasis. Cancer Cell 13:23–35
Zavadil J, Narasimhan M, Blumenberg M, Schneider RJ (2007) Transforming growth factor-β and microRNA:mRNA regulatory networks in epithelial plasticity. Cells Tissues Organs 185:157–161
Zutter MM (2007) Integrin-mediated adhesion: tipping the balance between chemosensitivity and chemoresistance. Adv Exp Med Biol 608:87–100

Chapter 8
TGF-β Signaling in Leukemogenesis

Kazuhito Naka and Atsushi Hirao

Abstract Transforming growth factor-β (TGF-β) signaling plays diverse roles in regulating cell proliferation, differentiation, and "stemness" in a cell context-dependent manner. In the first section of this chapter, we outline the genetic changes that attenuate TGF-β signaling in patients with acute myelogenous leukemia (AML), or the blast crisis phase of chronic myelogenous leukemia (CML), or pediatric acute lymphoblastic leukemia (ALL). In the second section, we discuss recent advances in stem cell research indicating that TGF-β signaling does not always suppress leukemogenesis. In fact, TGF-β signaling paradoxically sustains the survival and resistance to therapy of the CML stem cells that are responsible for disease recurrence in CML patients treated with tyrosine kinase inhibitors (TKIs). We examine evidence implicating TGF-β and its downstream effectors Akt and FOXO in the in vivo maintenance of the self-renewal ability of both TKI-resistant CML stem cells and the normal hematopoietic stem cells (HSCs) from which they are derived. Increased knowledge of the complex effects of TGF-β signaling may lead to improved diagnostic and therapeutic tools that can benefit leukemia patients.

Keywords TGF-β • Smad • FOXO • CML stem cells • TKI-resistance

K. Naka (✉)
Exploratory Project on Cancer Stem Cells, Cancer Research Institute,
Kanazawa University, Kakuma-machi, Kanazawa 920-1192, Japan
e-mail: kazunaka@staff.kanazawa-u.ac.jp

A. Hirao
Division of Molecular Genetics, Cancer Research Institute, Kanazawa University,
Kakuma-machi, Kanazawa 920-1192, Japan
e-mail: ahirao@staff.kanazawa-u.ac.jp

A. Moustakas and K. Miyazawa (eds.), *TGF-β in Human Disease*,
DOI 10.1007/978-4-431-54409-8_8, © Springer Japan 2013

Abbreviations

ALL	Acute lymphoblastic leukemia
AML	Acute myelogenous leukemia
ANLL	Acute nonlymphoblastic leukemia
APL	Acute promyelocytic leukemia
ATRA	All-*trans* retinoic acid
BMP	Bone morphogenetic protein
Bmpr1a	BMP receptor type 1A
CCyR	Complete cytogenetic response
CML	Chronic myelogenous leukemia
cPML	Cytoplasmic isoform of PML
FAB	French–American–British
GFAP	Glial fibrillary acidic protein
HSC	Hematopoietic stem cell
KIR	Killer immunoglobulin-like receptor
L-GMP	Leukemic-granulocyte-macrophage progenitor
LRC	Lipid raft clustering
LSK	Lineage$^-$Sca-1$^+$c-Kit$^+$
MEF	Mouse embryonic fibroblast
MPN	Myeloproliferative neoplasm
NK	Natural killer
PI3K	Phosphatidylinositol-3-OH kinase
PML	Promyelocytic leukemia
RARα	Retinoic acid receptor α
ROS	Reactive oxygen species
SARA	Smad anchor for receptor activation
SCL	Stem cell leukemia
SOD	Superoxide dismutase
TGF-β	Transforming growth factor-β
TKI	Tyrosine kinase inhibitor
WHO	World Health Organization

8.1 Genetic Alterations Affecting the TGF-β-Smad Pathway in Leukemia Patients

The transforming growth factor-β (TGF-β) family of highly conserved and highly homologous secreted polypeptides (Massagué et al. 1990) was described in Chap. 1. Briefly, TGF-β molecules bind to two heteromeric cell surface receptors with serine/threonine kinase activity: TGF-β receptor type I (TGF-βRI) (also known as activin receptor-like kinase 5; ALK5) and TGF-β receptor type II (TGF-βRII) (Massagué 2000). Binding of TGF-β to TGF-βRII triggers the recruitment and activation of TGF-βRI, which phosphorylates downstream Smad proteins (Heldin et al. 1997). That TGF-β signaling is involved in leukemogenesis has been known (Blank and Karlsson 2011; Dong and Blobe 2006; Kim and Letterio 2003; Larsson and Karlsson 2005). It has been known that TGF-β signaling pathway is altered in hematological malignancies such as acute myelogenous leukemia (AML) and final fatal phase of chronic myelogenous leukemia (CML) termed "blast crisis." AML is a heterogeneous clonal disease characterized by rapid accumulation of abnormal immature white blood cells in the bone marrow and peripheral blood. Although chronic phase CML is a relatively benign myeloproliferative neoplasm (MPN), CML patients in blast crisis reveal rapid proliferation of immature leukemic blasts which are resistant to therapy. In addition, genetic changes that attenuate TGF-β signaling also drive the excessive cell proliferation that characterizes acute lymphoblastic leukemia (ALL) originated from T-cell and B-cell lineage.

In this section of this chapter, we examine this assumption in the context of major chromosomal alterations that affect TGF-β signaling in human leukemia patients.

8.1.1 Acute Myelogenous Leukemia

8.1.1.1 t(8;21)(q22;q22)-Positive AML

The traditional French–American–British (FAB) classification system used to diagnose AML patients in twentieth century was based on cellular morphology and cytochemistry. Starting in 2001, a new World Health Organization (WHO) classification system that takes into account cell morphology, patient and clinical factors, and genetic and chromosomal alterations has been used to provide a scientific basis for AML diagnosis (Jaffe et al. 2001; Swerdlow et al. 2008). The chromosomal translocation t(8;21)(q21.3;q22.12), which generates the AML1–ETO fusion oncoprotein, is a causal genetic change in cases of "immature" type AML (defined as M0 in the FAB system). The normal AML1 protein, also known as Runx1, CBFA2 or PEBP2αB, is a member of the *runt* family of transcription factors. In mammalian embryos, AML1 is critical for the development of definitive hematopoiesis. In adults, AML1 is dispensable for the maintenance of normal hematopoietic stem cells (HSCs) but required for the maturation of megakaryocytes and the differentiation of T- and B-lymphoblasts (Ichikawa et al. 2004). The normal ETO protein, also known as MTG8, is involved in transcriptional regulation because it recruits co-repressors such as N-CoR, mSin3A, and SMRT. In the AML1–ETO fusion oncoprotein, a critical lysine residue is acetylated by the transcriptional coactivator p300, an event implicated in AML initiation (Wang et al. 2011). Importantly, AML1–ETO interacts with Smad3 and inhibits its ability to bind to DNA, suppressing TGF-β-Smad signaling (Jakubowiak et al. 2000) (Fig. 8.1). These results support the hypothesis that AML1–ETO drives AML leukemogenesis because it abrogates TGF-β-mediated tumor-suppressive signaling.

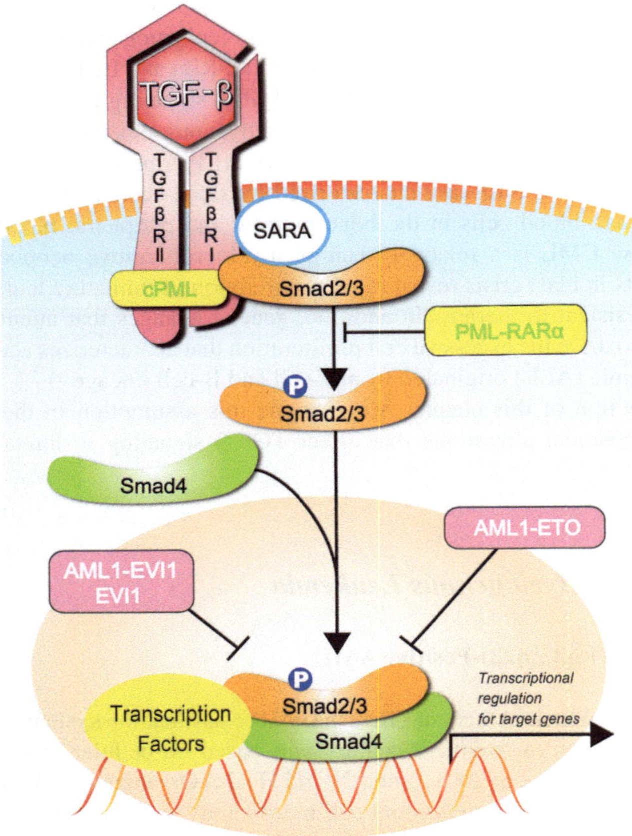

Fig. 8.1 The role of TGF-β signaling in human leukemogenesis. TGF-β signaling is transduced by a cytoplasmic complex containing TGF-β type I receptor (TGF-βRI), TGF-β type II receptor (TGF-βRII), Smad2/3, SARA and cPML. Binding of TGF-β to TGF-βRII triggers the activation of TGF-βRI, which phosphorylates downstream Smad2/3 proteins. The phosphorylated Smad2/3 molecules interact with Smad4, which regulates the expression of genes able to suppress leukemogenesis in the nucleus. The PML–RARα fusion oncoprotein, which causes APL, inhibits cPML function and blocks tumor-suppressive TGF-β signaling. Similarly, the AML1–ETO oncoprotein expressed in AML patients and the AML1–EVI1 oncoprotein or normal EVI1 protein expressed in blast crisis CML patients interact with Smad3 proteins, interfering with their DNA-binding capacity and abrogating TGF-β-mediated growth arrest

8.1.1.2 t(15;17)(q22;q21)-Positive Acute Promyelocytic Leukemia

Acute promyelocytic leukemia (APL, defined as M3 in the FAB classification) is a subtype of AML characterized by the accumulation of abnormal granulated promyelocytes in the bone marrow and peripheral blood. The causal t(15;17)(q24.1;q21.2) translocation generates an oncoprotein (PML–RARα) in which the promyelocytic leukemia (PML) protein is fused to the retinoic acid receptor α (RARα). The normal

PML protein, which is essential for the formation of a dot-like nuclear structure called the "PML nuclear body," has been implicated in transcription, maintenance of genomic integrity, antiviral responses, apoptosis induction, tumor suppression, and stemness. The normal RARα protein belongs to the nuclear receptor superfamily and transduces all-*trans* retinoic acid (ATRA)-mediated transcriptional signaling. APL patients treated with ATRA show PML–RARα degradation and enhanced differentiation of APL cells. Mouse embryonic fibroblasts (MEFs) obtained from *Pml*-deficient mice are resistant to TGF-β-dependent growth arrest, suggesting that PML is involved in the activation of the TGF-β pathway (Lin et al. 2004). Indeed, Lin et al. (2004) reported that a cytoplasmic isoform of PML (cPML) is crucial for TGF-β signaling. The cPML protein participates in the complex containing TGF-βRI, TGF-βRII, Smad2/3, and the "Smad anchor for receptor activation" (SARA) proteins, thereby promoting TGF-β signal transduction. The expression of PML–RARα inhibits cPML function and blocks TGF-β signaling (Fig. 8.1). In *Pml*-deficient MEFs, restoration of cPML function restores normal TGF-β-dependent growth arrest, cellular senescence, and apoptosis. Treatment of APL cell line, NM4, with ATRA increased the interaction between cPML and Smad2/3 and improved their TGF-β responses, implicating PML–RARα-mediated inhibition of TGF-β signaling in APL leukemogenesis (Lin et al. 2004).

8.1.2 Blast Crisis CML

Human CML is a myeloproliferative neoplasm (MPN) that occurs in three phases: (1) an initial chronic phase, (2) an accelerated phase, and (3) finally a blast crisis phase (Ren 2005). The majority of CML patients suffer from the t(9;22)(q34;q11.2) translocation that generates the Philadelphia chromosome and the *BCR–ABL* fusion oncogene, which encodes a constitutively active tyrosine kinase (de Klein et al. 1982; Rowley 1973). Because chronic phase CML cells can still produce functionally normal mature blood cells similar to those in healthy individuals (Fialkow et al. 1977), newly diagnosed CML patients exhibit a massive expansion of myeloid cells that extends for 5–15 years (chronic phase). The acquisition of additional genetic changes and/or epigenetic alterations during this period drives the progression of the disease to the accelerated phase. In the last and lethal blast crisis phase, cells of the myeloid, B-lymphoid, or bi-phenotypic myeloid, and lymphoid mixed lineage are observed (Calabretta and Perrotti 2004).

Blast crisis CML is associated with expression of the AML1–EVI1 fusion oncoprotein generated by the t(3;21)(q26.2;q22.12) translocation. The normal EVI1 protein encodes a zinc-finger transcription factor known to regulate HSC maintenance (Goyama et al. 2008; Yuasa et al. 2005). Excessively high levels of the normal EVI1 protein alone can also cause blast crisis CML (Mitani et al. 1994; Ogawa et al. 1996). Notably, both the normal EVI1 protein and the AML1–EVI1 fusion oncoprotein interact with Smad3 (Fig. 8.1). This association attenuates the DNA-binding activity of Smad3 and thus inhibits the TGF-β signaling pathway (Kurokawa et al. 1998a, b). These findings indicate that loss of TGF-β signaling is implicated in the progression of CML from the chronic phase to blast crisis and that this interference with TGF-β signaling is due to the interaction of Smad3 with EVI1 or AML1–EVI1.

8.1.3 T-Cell ALL

Letterio and colleagues have examined Smad2/3 proteins in a spectrum of patients with various forms of childhood ALL including T-cell ALL, pre-B-cell ALL, and acute nonlymphoblastic leukemia (ANLL) (Wolfraim et al. 2004). Although Smad2 protein is present at normal levels in leukemic cells of all ALL subtypes examined, Smad3 protein is low in leukemic cells from T-cell ALL patients but not in leukemic cells from pre-B cell ALL or ANLL (Wolfraim et al. 2004). However, Smad3 mRNA has been present at normal levels in leukemic cells from all T-cell ALL patients. Furthermore, no evidence for a deletion or point mutation in any region of the all nine exon of the Smad3 gene (*MADH3*) has been found in any T-cell ALL patients. Nevertheless, although the molecular mechanism responsible for the deficiency of Smad3 protein in these patients has yet to be elucidated, it is tempting to speculate that it might underlie the leukemogenesis of pediatric T-cell ALL. In mice, the loss of a single *Smad3* allele impairs the growth-suppressive effects of TGF-β on normal T-cells (Wolfraim et al. 2004). Moreover, mice harboring a heterozygous deletion of *Smad3* plus homozygous loss of $p27^{Kip1}$ spontaneously develop T-cell leukemia (Wolfraim et al. 2004). These results suggest that a combination of attenuated TGF-β signaling due to *Smad3* deficiency and the homozygous inactivation of $p27^{Kip1}$ might cause pediatric ALL. Indeed, deletions and germline mutations of the *CDKN1B* gene encoding the $p27^{Kip1}$ protein are frequent in pediatric ALL patients (Komuro et al. 1999; Takeuchi et al. 2002).

8.1.4 B-Cell ALL

The t(12;21)(p13.2;q22.12) translocation generating the fusion oncoprotein TEL–AML1 (also known as ETV6–RUNX1) is found in patients with childhood B-cell precursor ALL. Like AML1–EVI1 in AML patients, TEL–AML1 binds to Smad3 and compromises its ability to activate transcription of its target genes (Ford et al. 2009). Murine B-cell progenitor cells expressing TEL–AML1 proliferate more slowly than parental cells but are more resistant to TGF-β-mediated growth suppression (Ford et al. 2009). Importantly, mice expressing a *TEL–AML1* transgene show a greatly expanded number of pre-pro-B cells (Ford et al. 2009). Thus, whereas expression of TEL–AML1 alone is insufficient for the development of full-blown leukemia, this oncoprotein can promote the proliferation of preleukemic cells.

Section summary: The available evidence supports a tumor-suppressive function for TGF-β signaling, at least in differentiated leukemia cells.

8.2 Role of TGF-β-FOXO Signaling in CML Stem Cells

Although TGF-β mediates cytostatic signaling in normal and/or premalignant epithelial cells, it has been clarified that TGF-β promotes their own proliferation, invasion, and metastatic behavior in advanced malignant cells in epithelial-derived tumors (Ikushima and Miyazono 2010; Massagué 2008, 2012; Wakefield and Roberts 2002). In contrast to the case of solid tumor, it has been supposed that suppression of TGF-β signaling pathway is a cause of hematopoietic malignancy, including AML, ALL, and blast crisis CML as described in Sect. 8.1. However, TGF-β signaling and its triggering of the downstream Akt–FOXO pathway paradoxically regulates the self-renewal of CML stem cells and has been linked with their therapeutic resistance. In CML patients with recurrent disease, the reappearance of leukemia is due to the generation of a large number of differentiated CML cells from a rare population of therapy-resistant CML stem cells. Because CML stem cells originate from normal HSCs, it has been suggested that HSCs and CML stem cells share the same molecular mechanisms for regulating the maintenance of their self-renewal ability. In this section of this chapter, we examine the roles of TGF-β and Akt–FOXO signaling in CML stem cells and normal HSCs.

8.2.1 *Shared Hallmarks of HSCs and CML Stem Cells*

Several lines of evidence implicate normal HSCs as the cell of origin for CML stem cells expressing *BCR–ABL*. (1) Chronic phase CML cells can give rise to multiple lineages of functionally normal mature blood cells (Fialkow et al. 1977), indicating that these CML cells retain normal pluripotent differentiation capacity. (2) The transplantation of murine HSCs engineered to express the human *BCR–ABL* gene induces CML-like MPN in recipient mice (Daley et al. 1990; Elefanty and Cory 1992; Gishizky et al. 1993; Kelliher et al. 1990; Pear et al. 1998). In contrast, the transplantation of murine committed hematopoietic progenitor cells expressing *BCR–ABL* does not induce CML (Huntly et al. 2004). (3) CML stem cells are functionally defined by their ability to induce CML once transplanted into recipient mice. In mice that develop CML-like disease due to retroviral transduction of the *BCR–ABL* gene, CML stem cells can be purified from a rare Lineage^{-}Sca-1^{+}c-Kit^{+} (LSK) population (i.e., cells that bear the marker profile of normal HSCs) (Hu et al. 2006; Ito et al. 2008; Zhao et al. 2007, 2009). (4) In an inducible *BCR–ABL* transgenic mouse model in which expression of the *BCR–ABL* gene is controlled by a Tet-regulated 3′-enhancer of the murine *SCL* (stem cell leukemia) gene (Huettner et al. 2003; Koschmieder et al. 2005), the CML stem cells driving disease are restricted to populations with the phenotype of long-term HSCs (i.e., CD34^{-}Flt3^{-}LSK or CD150^{+}Flt3^{-}CD48^{-}LSK) (Reynaud et al. 2011; Schemionek et al. 2010; Zhang 2012). (5) In chronic phase CML patients, the CML stem cells responsible for the disease can be isolated from a cell population expressing surface markers associated with primitive hematopoietic cells (Holyoake et al. 1999, 2001; Pellicano et al. 2009). These observations collectively bolster the hypothesis that HSCs and CML stem cells regulate their self-renewal ability using the same mechanisms. Indeed, several signaling pathways, including those involving Wnt-β-catenin (Scheller et al. 2006; Zhao et al. 2007), PML (Ito et al. 2008), and Hedgehog (Dierks et al. 2008; Zhao et al. 2009), have been implicated in the regulation of stem cell fate during both normal hematopoiesis and CML leukemogenesis. Some researchers have taken these similarities in properties between HSCs and CML stem cells as evidence that CML is a "stem cell disease" (Naka et al. 2010a) and therefore contend that elucidating mechanisms in HSCs will lead to a better understanding of how CML stem cells are perpetuated.

8.2.2 *Akt–FOXO Signaling in HSCs*

HSCs top the hierarchy of the normal hematopoietic system and are defined by both their ability to reproduce themselves, a property known as self-renewal, and their capacity to give rise to all mature hematopoietic cell lineages. Because HSCs have to sustain hematopoiesis for the life of an individual, the maintenance of HSC quiescence in vivo is crucial for homeostasis. HSC quiescence is adopted within defined "bone marrow niches," which provide a comfortable microenvironment where HSCs can remain in the non-dividing, G0 phase of the cell cycle (Suda et al. 2005). Quiescent HSCs generate only very low levels of reactive oxygen species (ROS), allowing these cells to escape any DNA damage that might be caused by oxidative stress. Genomic integrity is thought to be crucial for the preservation of HSC self-renewal capacity, and it is speculated that dysfunction of genomic integrity causes generation of the Philadelphia chromosome and thereby *BCR–ABL* fusion oncogene.

The self-renewal of normal HSCs also depends on signaling involving the phosphatidylinositol-3-OH kinase (PI3K)–Akt pathway and its downstream targets FOXO1, FOXO3a, FOXO4, and FOXO6, which are members of the Forkhead O (FOXO) subfamily of transcription factors (Greer and Brunet 2005). In the absence of stimulation by growth factors or insulin, FOXOs are present in an active state in a cell's nucleus and freely induce their transcriptional targets. When a growth factor or insulin binds to the appropriate cell surface receptor, PI3K–Akt signaling is activated, and activated Akt directly phosphorylates FOXOs, resulting in their nuclear exclusion and subsequent degradation in the cytoplasm.

FOXO deficiencies have profound effects on HSCs. For example, Tothova et al. reported a marked decrease in HSCs in mice with triple conditional deletions of the *Foxo1*, *Foxo3a*, and *Foxo4* genes (Tothova et al. 2007). The HSCs from these triple *Foxo*-deficient mice not only exhibited a self-renewal defect but also showed elevated ROS. Notably, the expression of superoxide dismutase (*SOD*) genes was impaired in triple *Foxo*-deficient HSCs. In two other studies, Foxo3a was found to be solely responsible for the maintenance of the HSC pool (Miyamoto et al. 2007;

Yalcin et al. 2008). *Foxo3a*$^{-/-}$ HSCs showed reduced expression of the Foxo target genes *SOD2* and *catalase* and exhibited increased ROS, defective self-renewal, and decreased expression of the CDK inhibitors p27^{Kip1} and p57^{Kip2} (Miyamoto et al. 2007; Yalcin et al. 2008). Importantly, p57^{Kip2} and p27^{Kip1} are required for the maintenance and quiescence of HSCs (Matsumoto et al. 2011; Zou et al. 2011). Thus, the FOXO factors are responsible for regulating HSC quiescence, ROS levels and maintenance of self-renewal ability.

The Akt–FOXO pathway in quiescent HSCs within a bone marrow niche is regulated by lipid raft clustering (LRC) (Yamazaki et al. 2009). Lipid raft microdomains, which are patches in the plasma membrane enriched in cholesterol, glycosphingolipids, and glycolipids, control cytokine signaling in murine HSCs. Yamazaki et al. demonstrated that HSCs freshly isolated from mouse bone marrow lacked LRC. Notably, LRC-induced cytokine signaling is essential for the cell cycle reentry of HSCs. Conversely, inhibition of LRC attenuates cytokine signaling, which results in repression of Akt activity followed by Foxo activation. Therefore, the status of the lipid rafts within HSCs in a bone marrow niche fine-tunes Akt–Foxo signaling and regulates HSC quiescence.

8.2.3 TGF-β Signaling in HSCs

Numerous in vitro studies have indicated that TGF-β signaling is involved in regulating mouse and human HSCs (Larsson and Karlsson 2005). Yamazaki et al. clarified that TGF-β signaling controls HSC self-renewal ability by inhibiting LRC and thereby attenuating cytokine signaling (Yamazaki et al. 2009). The inhibition of LRC by TGF-β in HSCs suppressed Akt and induced nuclear localization of Foxo3a. These results link TGF-β signaling with the LRC–PI3K–Akt–Foxo signaling pathway that regulates HSC quiescence in vivo.

While it had been hoped that a genetic approach might reveal the precise role of TGF-β in modulating HSCs in vivo, it has been difficult to carry out the required experiments due to the early lethality and massive inflammatory response in null mutant mice disrupted in the *TGF-β1* (Kulkarni et al. 1993; Shull et al. 1992), *TGF-βRI* (Larsson et al. 2001), or *TGF-βRII* (Oshima et al. 1996) genes. To overcome these difficulties, conditional knockout mice with hematopoietic cell-specific loss of the *TGF-βRI* (*ALK5*) or *TGF-βRII* genes were established using the Cre/LoxP system. Interestingly, while conditional mutants lacking *TGF-βRI* (*ALK5*) had no significant defects in the quiescence and maintenance of HSC pool (Larsson et al. 2003, 2005), *TGF-βRII*-deficient conditional mutants exhibited obvious abnormalities in HSCs (Yamazaki et al. 2011). Similarly, HSCs of conditional *Smad4*-deficient mice showed a significant reduction in self-renewal capacity following transplantation into normal recipient mice (Karlsson et al. 2007). Interestingly, although Smad4 plays an integral role in both the TGF-β and bone morphogenetic protein (BMP) pathways, studies of conditional mutant mice disrupted in the *Smad1* and *Smad5* genes revealed that canonical BMP signaling is dispensable for HSC maintenance (Singbrant et al. 2010). Thus, it is the TGF-β-Smad pathway that is required for HSC maintenance in vivo (Yamazaki et al. 2011). However, conditional disruption of the *BMP receptor type 1A* (*Bmpr1a*) gene in mice results in an increased number of osteoblasts, which expand the size of the HSC bone marrow niches. This expansion in turn facilitates the enlargement of the HSC pool, indicating that BMP signaling is indirectly involved in the control of HSC numbers (Zhang et al. 2003).

HSCs themselves produce TGF-β in a latent form but are not able to process it into its active structure. Yamazaki et al. have shown that bone marrow glial cells expressing glial fibrillary acidic protein (GFAP) regulate the maintenance of quiescent HSCs by activating their latent TGF-β (Yamazaki et al. 2011). These GFAP-expressing glial cells have been identified as non-myelinating Schwann cells that ensheath sympathetic nerves in the bone marrow. Importantly, when the number of active TGF-β-producing cells is reduced via autonomic nerve denervation, HSCs are missing from the bone marrow (Yamazaki et al. 2011). Taken together, these results indicate that the signaling of TGF-β family members within the bone marrow niche has profound regulatory effects on HSC properties.

8.2.4 Opposing Roles of FOXOs in CML Non-stem Cells vs. CML Stem Cells

Given that CML stem cells originate from normal HSCs, it has been of great interest to determine whether CML stem cells conserve the regulatory mechanisms found in normal HSCs. Studies using CML cell lines have shown that, in CML non-stem cells, activated FOXOs can induce apoptosis or cell cycle arrest (Essafi et al. 2005; Ghaffari et al. 2003; Komatsu et al. 2003). The BCR–ABL fusion protein in CML cells drives strong Akt activation that leads to nuclear export of Foxo3a and forceful repression of Foxo3a functions (Essafi et al. 2005; Ghaffari et al. 2003; Komatsu et al. 2003). When TKIs block BCR–ABL and suppress Akt activity, Foxo3a remains activated and triggers apoptosis or cell cycle arrest (Fig. 8.2, left). Based on these findings, it has been believed that FOXOs are suppressors of CML leukemogenesis. However, Komatsu et al. insightfully reported several years ago that Foxo3a is involved in the acquisition of dormancy in CML cell lines exposed to imatinib (Komatsu et al. 2003). Indeed, when we investigated whether Foxo3a was similarly implicated in the maintenance of CML stem cells using a mouse CML model, we found that, unlike CML non-stem cells, CML stem cells exhibit decreased Akt phosphorylation and increased nuclear localization of Foxo3a in vivo (Naka et al. 2010b). In line with this observation, Lee et al. reported that Foxo3a was localized in the nucleus of a primitive $CD150^{+}CD48^{-}CD41^{-}LSK$ population in a *Pten-deficient* MPN mouse model (Lee et al. 2010). Thus, it appears that the maintenance of the rare CML stem cell population is controlled by different regulators than those governing the majority of CML non-stem cell population.

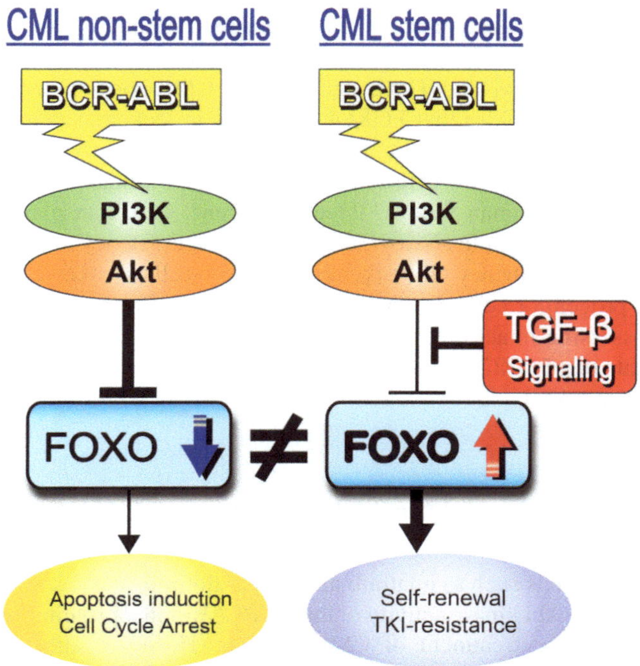

Fig. 8.2 Opposing roles of the TGF-β-FOXO axis in CML stem cells vs. CML non-stem cells. FOXO transcription factor is generally thought to suppress tumorigenesis because activated FOXO induces the apoptosis or cell cycle arrest. In CML non-stem cells (*left*), FOXO functions are suppressed by BCR–ABL, which drives the activation of PI3K–Akt signaling pathway. In CML non-stem cells treated with TKIs, BCR–ABL activity is blocked, Akt activity is suppressed, and FOXOs remain activated. Differentiated leukemic cells are thus reduced in number. Paradoxically, in CML stem cells (*right*), the TGF-β-FOXO axis is tumor-promoting. BCR–ABL-triggered activation of the PI3K–Akt signaling pathway is inhibited by TGF-β signaling. As a consequence, FOXOs remain active and drive the self-renewal and TKI resistance of CML stem cells. Because BCR–ABL does not function in CML stem cells, TKIs have no effect on them and cannot eradicate them. Thus, it is possible that the therapeutic suppression of the TGF-β-FOXO axis might be a novel strategy for the eradication of CML stem cells and thus the inhibition of leukemia recurrence that is frequently lethal

To clarify the functional role of Foxo3a in CML stem cells, we transplanted *Foxo3a*$^{+/+}$ and *Foxo3a*$^{-/-}$ CML stem cells into recipient mice (Naka et al. 2010b). Both groups of recipients died of CML-like MPN with similar kinetics, indicating that Foxo3a is dispensable for the initiation of CML. However, serial transplantation of *Foxo3a*$^{+/+}$ and *Foxo3a*$^{-/-}$ CML stem cells revealed that the ability of CML stem cells to perpetuate disease in vivo is significantly decreased by *Foxo3a* deficiency. Moreover, recipients that received *Foxo3a*$^{+/+}$ CML stem cells developed both ALL and CML, demonstrating that these CML stem cells are able to generate malignancies in multiple hematopoietic lineages. In contrast, recipients transplanted with *Foxo3a*$^{-/-}$ CML stem cells developed neither ALL nor CML. Thus, Foxo3a sustains the long-term maintenance of multi-lineage leukemia-initiating potential in CML stem cells. These findings underpin the paradoxical concept that FOXOs play opposite roles in the survival of CML stem cells vs. CML non-stem cells (Fig. 8.2).

The counterintuitive idea that FOXOs are active in leukemia stem cells can be extended to AML. In AML patients, Foxo signaling is activated in primitive $CD34^{+}lineage^{low}$ bone marrow cells (Sykes et al. 2011). Similarly, in a mouse AML model harboring the human *MLL-AF9* leukemic gene, mice transplanted with leukemia cells triply deficient for *Foxo1/3/4* display a longer latency to disease coupled with a decreased frequency of leukemia-initiating cells (leukemic-granulocyte-macrophage progenitor cells; L-GMP) compared to mice transplanted with leukemia cells possessing intact Foxo alleles (Sykes et al. 2011). Taken together, these results indicate that FOXOs are involved in the maintenance of leukemic stem cells in both CML and AML.

8.2.5 FOXO is Crucial for TKI Resistance in CML Stem Cells

The discovery of TKIs such as imatinib mesylate, nilotinib, and dasatinib that could directly inhibit the Abl kinase activity of the BCR–ABL oncoprotein was a dramatic breakthrough for the treatment of CML patients. However, although 95 % of imatinib-treated chronic phase CML patients survive for over 5 years (Druker et al. 2006), resistance to or intolerance of these drugs inevitably develops (Bhatia et al. 2003; Corbin et al. 2011; Cortes et al. 2004; Goldman and Gordon 2006; Savona and Talpaz 2008). Recent studies have demonstrated that TKIs are potent killers of differentiated leukemic cells, which are proliferating, but cannot deplete primitive CML stem cells, which are quiescent (Copland et al. 2006; Graham et al. 2002; Jorgensen et al. 2007; Konig et al. 2008a, b, c). It is possible that quiescent CML stem cells remaining after a first round of TKI treatment acquire mutations of the *BCR–ABL* gene itself. Such mutations would be manifested as TKI resistance in CML patients treated with a second round of TKI therapy (Gorre et al. 2001; Roumiantsev et al. 2002). Thus, many oncologists believe that TKI inhibitors coupled with novel therapeutics capable of eradicating quiescent CML stem cells may prevent disease recurrence and greatly benefit CML patients (Kavalerchik et al. 2008).

Because FOXOs play an essential role in the long-term maintenance of CML stem cells, we examined whether FOXOs also contribute to their TKI resistance. Although mice transplanted with $Foxo3a^{+/+}$ or $Foxo3a^{-/-}$ CML stem cells showed similar rates of lethality, the survival of mice that received *Foxo*-deficient CML stem cells and were treated with imatinib was significantly prolonged (Naka et al. 2010b). Furthermore, after imatinib administration, the number of residual CML stem cells present in mice transplanted with $Foxo3a^{-/-}$ CML stem cells was less than in mice transplanted with $Foxo3a^{+/+}$ CML stem cells. Interestingly, Muschen and colleagues have shown that the *BCL-6* proto-oncogene, which is a downstream effector of FOXO, is essential for both the maintenance of CML stem cells and the TKI resistance of ALL cells positive for the Philadelphia chromosome (Duy et al. 2011;

Hurtz et al. 2011). Therefore, therapeutics targeting FOXOs might help to eradicate CML stem cells in CML patients and render TKI-resistant cells in Ph⁺ALL patients susceptible to treatment.

8.2.6 *TGF-β Regulates FOXO Activity and the Self-Renewal Ability of CML Stem Cells*

Because TGF-β signaling regulates the Akt–FOXO pathway in normal HSCs, we investigated the impact of the TGF-β inhibitor Ly364947 on CML stem cell functions in vivo (Naka et al. 2010b). The administration of Ly364947 alone to CML-affected mice leads to Akt activation and decreased nuclear Foxo3a in CML stem cells, demonstrating that TGF-β is a critical regulator of Akt and Foxo in these cells in vivo. Notably, although administration of Ly364947 alone does not extend the survival of CML-affected mice, the administration of Ly364947 combined with imatinib treatment significantly reduces recipient lethality, CML stem cell frequency, and CML infiltration in the lung (Naka et al. 2010b). These findings indicate that the TGF-β-FOXO axis plays a critical role in the maintenance of TKI-resistant CML stem cells.

The sources of TGF-β that could support CML stem cell survival in vivo are under intense investigation. For normal HSCs, the activation of latent TGF-β is mediated by GFAP-expressing glial cells in the bone marrow (Yamazaki et al. 2011), making these cells important candidates for the source of TGF-β also maintaining the CML stem cell niche. In addition, Yokota et al. have demonstrated that osteoclasts are involved in the maintenance of dormant CML cells in the bone marrow through their release of TGF-β (Yokota et al. 2010). Recently, it was reported that an individual's genotype at the *killer immunoglobulin-like receptor* (*KIR*) genes is predictive of a complete cytogenetic response (CCyR) and survival rate in CML patients who are treated with imatinib (Marin et al. 2012). KIRs are expressed by natural killer (NK) cells, which are cells of the innate immune response. The physiological functions of NK cells, which include cytotoxicity and cytokine production, are controlled by a balance of inhibitory and activatory signals from cell surface receptors such as the KIRs. Significantly, NK cells are increased in number in CML patients treated with TKIs (Kreutzman et al. 2010; Yong et al. 2009). Marin et al. found that patients carrying the activatory receptor *KIR2DS1* have a significantly lower probability of achieving CCyR within 2 years of TKI treatment compared with patients who are *KIR2DS1*-negative (Marin et al. 2012). Importantly, NK cells expressing *KIR2DS1* secrete TGF-β (Ghio et al. 2009), implicating these cells as a possible source of the TGF-β driving the TKI resistance of CML stem cells.

Because TGF-β signaling is responsible for the maintenance of TKI-resistant CML stem cells in vivo, the suppression of TGF-β signaling so as to eradicate CML stem cells is a plausible therapeutic strategy. Although TGF-β-Foxo signaling is essential for the maintenance of both normal HSCs and CML stem cells, inhibition of TGF-β has a much more dramatic effect on CML stem cells than on normal HSCs. Notably, in the presence of imatinib, the treatment in vitro of primitive CD34^{+} human CML cells with TGF-β inhibitors attenuates their viability (Moller et al. 2007; Naka et al. 2010b). These results suggest that judicious inhibition of TGF-β-FOXO signaling may lead to efficient eradication of residual CML stem cells.

Section Summary: Because the activation of TGF-β signaling may be implicated in CML leukemogenesis, the application of TGF-β inhibitors to the treatment of CML patients is a promising therapeutic avenue. If detrimental effects on normal hematopoietic cells can be avoided, it is possible that the inhibition of TGF-β signaling might specifically deplete CML stem cells in patients with chronic phase CML and that a combination of a TGF-β inhibitor with TKIs might be effective in removing residual CML stem cells that survive therapy with TKI alone.

8.3 Conclusion

In this chapter, we have reviewed the multifaceted role of TGF-β signaling in leukemogenesis. As investigators have believed for many years, TGF-β signaling in normal HSCs helps to suppress the rise of several types of differentiated leukemia cells, including those present in AML, blast crisis CML, and pediatric ALL. Therefore, it is important to elucidate how the chromosomal translocations associated with these disorders generate oncoproteins that can inhibit TGF-β-mediated growth suppression. It is also important to recognize that TGF-β-FOXO signaling has the opposite effect on CML stem cells and is responsible for their survival and TKI resistance. Moreover, like normal HSCs, CML stem cells are quiescent, stress-resistant, and maintain their self-renewal ability within a niche. Thus, for the successful development of new small molecule drugs capable of eliminating CML stem cells, it may be advantageous to target the TGF-β signaling network (including elements of the Akt–FOXO pathway) and the niche cells that support it. The ultimate goal for this field is to develop novel therapeutics that can eradicate leukemia stem cells, thereby forestalling disease recurrence and providing concrete benefits for patients.

References

Bhatia R, Holtz M, Niu N et al (2003) Persistence of malignant hematopoietic progenitors in chronic myelogenous leukemia patients in complete cytogenetic remission following imatinib mesylate treatment. Blood 101:4701–4707. doi:10.1182/blood-2002-09-2780

Blank U, Karlsson S (2011) The role of Smad signaling in hematopoiesis and translational hematology. Leukemia 25:1379–1388. doi:10.1038/leu.2011.95

Calabretta B, Perrotti D (2004) The biology of CML blast crisis. Blood 103:4010–4022. doi:10.1182/blood-2003-12-4111

Copland M, Hamilton A, Elrick LJ et al (2006) Dasatinib (BMS-354825) targets an earlier progenitor population than imatinib in primary CML but does not eliminate the quiescent fraction. Blood 107:4532–4539. doi:10.1182/blood-2005-07-2947

Corbin AS, Agarwal A, Loriaux M, Cortes J, Deininger MW, Druker BJ (2011) Human chronic myeloid leukemia stem cells are insensitive to imatinib despite inhibition of BCR-ABL activity. J Clin Invest 121:396–409. doi:10.1172/JCI35721
Cortes J, O'brien S, Kantarjian H (2004) Discontinuation of imatinib therapy after achieving a molecular response. Blood 104:2204–2205. doi:10.1182/blood-2004-04-1335
Daley GQ, Van Etten RA, Baltimore D (1990) Induction of chronic myelogenous leukemia in mice by the P210bcr/abl gene of the Philadelphia chromosome. Science 247:824–830
de Klein A, Van Kessel AG, Grosveld G et al (1982) A cellular oncogene is translocated to the Philadelphia chromosome in chronic myelocytic leukaemia. Nature 300:765–767
Dierks C, Beigi R, Guo GR et al (2008) Expansion of Bcr-Abl-positive leukemic stem cells is dependent on Hedgehog pathway activation. Cancer Cell 14:238–249. doi:10.1016/j.ccr.2008.08.003
Dong M, Blobe GC (2006) Role of transforming growth factor-β in hematologic malignancies. Blood 107:4589–4596. doi:10.1182/blood-2005-10-4169
Druker BJ, Guilhot F, O'brien SG et al (2006) Five-year follow-up of patients receiving imatinib for chronic myeloid leukemia. N Engl J Med 355:2408–2417. doi:10.1056/NEJMoa062867
Duy C, Hurtz C, Shojaee S et al (2011) BCL6 enables Ph^+ acute lymphoblastic leukaemia cells to survive BCR-ABL1 kinase inhibition. Nature 473:384–388. doi:10.1038/nature09883
Elefanty AG, Cory S (1992) Hematologic disease induced in BALB/c mice by a bcr-abl retrovirus is influenced by the infection conditions. Mol Cell Biol 12:1755–1763
Essafi A, Fernandez De Mattos S, Hassen YA et al (2005) Direct transcriptional regulation of Bim by FoxO3a mediates STI571-induced apoptosis in Bcr-Abl-expressing cells. Oncogene 24:2317–2329. doi:10.1038/sj.onc.1208421
Fialkow PJ, Jacobson RJ, Papayannopoulou T (1977) Chronic myelocytic leukemia: clonal origin in a stem cell common to the granulocyte, erythrocyte, platelet and monocyte/macrophage. Am J Med 63:125–130
Ford AM, Palmi C, Bueno C et al (2009) The TEL-AML1 leukemia fusion gene dysregulates the TGF-β pathway in early B lineage progenitor cells. J Clin Invest 119:826–836. doi:10.1172/JCI36428
Ghaffari S, Jagani Z, Kitidis C, Lodish HF, Khosravi-Far R (2003) Cytokines and BCR-ABL mediate suppression of TRAIL-induced apoptosis through inhibition of forkhead FOXO3a transcription factor. Proc Natl Acad Sci USA 100:6523–6528. doi:10.1073/pnas.0731871100
Ghio M, Contini P, Negrini S, Boero S, Musso A, Poggi A (2009) Soluble HLA-I-mediated secretion of TGF-β1 by human NK cells and consequent down-regulation of anti-tumor cytolytic activity. Eur J Immunol 39:3459–3468. doi:10.1002/eji.200939728
Gishizky ML, Johnson-White J, Witte ON (1993) Efficient transplantation of BCR-ABL-induced chronic myelogenous leukemia-like syndrome in mice. Proc Natl Acad Sci USA 90:3755–3759
Goldman J, Gordon M (2006) Why do chronic myelogenous leukemia stem cells survive allogeneic stem cell transplantation or imatinib: does it really matter? Leuk Lymphoma 47:1–7. doi:10.1080/10428190500407996
Gorre ME, Mohammed M, Ellwood K et al (2001) Clinical resistance to STI-571 cancer therapy caused by BCR-ABL gene mutation or amplification. Science 293:876–880. doi:10.1126/science.1062538
Goyama S, Yamamoto G, Shimabe M et al (2008) Evi-1 is a critical regulator for hematopoietic stem cells and transformed leukemic cells. Cell Stem Cell 3:207–220. doi:10.1016/j.stem.2008.06.002
Graham SM, Jorgensen HG, Allan E et al (2002) Primitive, quiescent, Philadelphia-positive stem cells from patients with chronic myeloid leukemia are insensitive to STI571 in vitro. Blood 99:319–325. doi:10.1182/blood.V99.1.319
Greer EL, Brunet A (2005) FOXO transcription factors at the interface between longevity and tumor suppression. Oncogene 24:7410–7425. doi:10.1038/sj.onc.1209086

Heldin CH, Miyazono K, Ten Dijke P (1997) TGF-β signalling from cell membrane to nucleus through SMAD proteins. Nature 390:465–471. doi:10.1038/37284

Holyoake T, Jiang X, Eaves C, Eaves A (1999) Isolation of a highly quiescent subpopulation of primitive leukemic cells in chronic myeloid leukemia. Blood 94:2056–2064

Holyoake TL, Jiang X, Jorgensen HG et al (2001) Primitive quiescent leukemic cells from patients with chronic myeloid leukemia spontaneously initiate factor-independent growth in vitro in association with up-regulation of expression of interleukin-3. Blood 97:720–728. doi:10.1182/blood.V97.3.720

Hu Y, Swerdlow S, Duffy TM, Weinmann R, Lee FY, Li S (2006) Targeting multiple kinase pathways in leukemic progenitors and stem cells is essential for improved treatment of Ph^+ leukemia in mice. Proc Natl Acad Sci USA 103:16870–16875. doi:10.1073/pnas.0606509103

Huettner CS, Koschmieder S, Iwasaki H et al (2003) Inducible expression of BCR/ABL using human CD34 regulatory elements results in a megakaryocytic myeloproliferative syndrome. Blood 102:3363–3370. doi:10.1182/blood-2003-03-0768

Huntly BJ, Shigematsu H, Deguchi K et al (2004) MOZ-TIF2, but not BCR-ABL, confers properties of leukemic stem cells to committed murine hematopoietic progenitors. Cancer Cell 6:587–596. doi:10.1016/j.ccr.2004.10.015

Hurtz C, Hatzi K, Cerchietti L et al (2011) BCL6-mediated repression of p53 is critical for leukemia stem cell survival in chronic myeloid leukemia. J Exp Med 208:2163–2174. doi:10.1084/jem.20110304

Ichikawa M, Asai T, Saito T et al (2004) AML-1 is required for megakaryocytic maturation and lymphocytic differentiation, but not for maintenance of hematopoietic stem cells in adult hematopoiesis. Nat Med 10:299–304. doi:10.1038/nm997

Ikushima H, Miyazono K (2010) TGFβ signalling: a complex web in cancer progression. Nat Rev Cancer 10:415–424. doi:10.1038/nrc2853

Ito K, Bernardi R, Morotti A et al (2008) PML targeting eradicates quiescent leukaemia-initiating cells. Nature 453:1072–1078. doi:10.1038/nature07016

Jaffe ES, Harris NL, Stein H, Vardiman JW et al (2001) World Health Organization classification of tumours. Pathology and genetics of tumours of haematopoietic and lymphoid tissues. IARC, Lyon

Jakubowiak A, Pouponnot C, Berguido F et al (2000) Inhibition of the transforming growth factor β1 signaling pathway by the AML1/ETO leukemia-associated fusion protein. J Biol Chem 275:40282–40287. doi:10.1074/jbc.C000485200

Jorgensen HG, Allan EK, Jordanides NE, Mountford JC, Holyoake TL (2007) Nilotinib exerts equipotent antiproliferative effects to imatinib and does not induce apoptosis in CD34+ CML cells. Blood 109:4016–4019. doi:10.1182/blood-2006-11-057521

Karlsson G, Blank U, Moody JL et al (2007) Smad4 is critical for self-renewal of hematopoietic stem cells. J Exp Med 204:467–474. doi:10.1084/jem.20060465

Kavalerchik E, Goff D, Jamieson CH (2008) Chronic myeloid leukemia stem cells. J Clin Oncol 26:2911–2915. doi:10.1200/JCO.2008.17.5745

Kelliher MA, Mclaughlin J, Witte ON, Rosenberg N (1990) Induction of a chronic myelogenous leukemia-like syndrome in mice with v-abl and BCR/ABL. Proc Natl Acad Sci USA 87:6649–6653

Kim SJ, Letterio J (2003) Transforming growth factor-β signaling in normal and malignant hematopoiesis. Leukemia 17:1731–1737. doi:10.1038/sj.leu.2403069

Komatsu N, Watanabe T, Uchida M et al (2003) A member of Forkhead transcription factor FKHRL1 is a downstream effector of STI571-induced cell cycle arrest in BCR-ABL-expressing cells. J Biol Chem 278:6411–6419. doi:10.1074/jbc.M211562200

Komuro H, Valentine MB, Rubnitz JE et al (1999) p27KIP1 deletions in childhood acute lymphoblastic leukemia. Neoplasia 1:253–261

Konig H, Copland M, Chu S, Jove R, Holyoake TL, Bhatia R (2008a) Effects of dasatinib on SRC kinase activity and downstream intracellular signaling in primitive chronic myelogenous leukemia hematopoietic cells. Cancer Res 68:9624–9633. doi:10.1158/0008-5472.CAN-08-1131

Konig H, Holtz M, Modi H et al (2008b) Enhanced BCR-ABL kinase inhibition does not result in increased inhibition of downstream signaling pathways or increased growth suppression in CML progenitors. Leukemia 22:748–755. doi:10.1038/sj.leu.2405086

Konig H, Holyoake TL, Bhatia R (2008c) Effective and selective inhibition of chronic myeloid leukemia primitive hematopoietic progenitors by the dual Src/Abl kinase inhibitor SKI-606. Blood 111:2329–2338. doi:10.1182/blood-2007-05-092056

Koschmieder S, Gottgens B, Zhang P et al (2005) Inducible chronic phase of myeloid leukemia with expansion of hematopoietic stem cells in a transgenic model of BCR-ABL leukemogenesis. Blood 105:324–334. doi:10.1182/blood-2003-12-4369

Kreutzman A, Juvonen V, Kairisto V et al (2010) Mono/oligoclonal T and NK cells are common in chronic myeloid leukemia patients at diagnosis and expand during dasatinib therapy. Blood 116:772–782. doi:10.1182/blood-2009-12-256800

Kulkarni AB, Huh CG, Becker D et al (1993) Transforming growth factor β 1 null mutation in mice causes excessive inflammatory response and early death. Proc Natl Acad Sci USA 90:770–774

Kurokawa M, Mitani K, Imai Y, Ogawa S, Yazaki Y, Hirai H (1998a) The t(3;21) fusion product, AML1/Evi-1, interacts with Smad3 and blocks transforming growth factor-β-mediated growth inhibition of myeloid cells. Blood 92:4003–4012

Kurokawa M, Mitani K, Irie K et al (1998b) The oncoprotein Evi-1 represses TGF-β signalling by inhibiting Smad3. Nature 394:92–96. doi:10.1038/27945

Larsson J, Karlsson S (2005) The role of Smad signaling in hematopoiesis. Oncogene 24:5676–5692. doi:10.1038/sj.onc.1208920

Larsson J, Goumans MJ, Sjostrand LJ et al (2001) Abnormal angiogenesis but intact hematopoietic potential in TGF-β type I receptor-deficient mice. EMBO J 20:1663–1673. doi:10.1093/emboj/20.7.1663

Larsson J, Blank U, Helgadottir H et al (2003) TGF-β signaling-deficient hematopoietic stem cells have normal self-renewal and regenerative ability in vivo despite increased proliferative capacity in vitro. Blood 102:3129–3135. doi:10.1182/blood-2003-04-1300

Larsson J, Blank U, Klintman J, Magnusson M, Karlsson S (2005) Quiescence of hematopoietic stem cells and maintenance of the stem cell pool is not dependent on TGF-β signaling in vivo. Exp Hematol 33:592–596. doi:10.1016/j.exphem.2005.02.003

Lee JY, Nakada D, Yilmaz OH et al (2010) mTOR activation induces tumor suppressors that inhibit leukemogenesis and deplete hematopoietic stem cells after Pten deletion. Cell Stem Cell 7:593–605. doi:10.1016/j.stem.2010.09.015

Lin HK, Bergmann S, Pandolfi PP (2004) Cytoplasmic PML function in TGF-β signalling. Nature 431:205–211. doi:10.1038/nature02783

Marin D, Gabriel IH, Ahmad S et al (2012) KIR2DS1 genotype predicts for complete cytogenetic response and survival in newly diagnosed chronic myeloid leukemia patients treated with imatinib. Leukemia 26:296–302. doi:10.1038/leu.2011.180

Massagué J (2000) How cells read TGF-β signals. Nat Rev Mol Cell Biol 1:169–178. doi:10.1038/35043051

Massagué J (2008) TGFβ in cancer. Cell 134:215–230. doi:10.1016/j.cell.2008.07.001

Massagué J (2012) TGFβ signalling in context. Nat Rev Mol Cell Biol 13:616–630. doi:10.1038/nrm3434

Massagué J, Cheifetz S, Boyd FT, Andres JL (1990) TGF-β receptors and TGF-β binding proteoglycans: recent progress in identifying their functional properties. Ann N Y Acad Sci 593:59–72

Matsumoto A, Takeishi S, Kanie T et al (2011) p57 is required for quiescence and maintenance of adult hematopoietic stem cells. Cell Stem Cell 9:262–271. doi:10.1016/j.stem.2011.06.014

Mitani K, Ogawa S, Tanaka T et al (1994) Generation of the AML1-EVI-1 fusion gene in the t(3;21)(q26;q22) causes blastic crisis in chronic myelocytic leukemia. EMBO J 13:504–510

Miyamoto K, Araki KY, Naka K et al (2007) Foxo3a is essential for maintenance of the hematopoietic stem cell pool. Cell Stem Cell 1:101–112. doi:10.1016/j.stem.2007.02.001

Moller GM, Frost V, Melo JV, Chantry A (2007) Upregulation of the TGFβ signalling pathway by Bcr-Abl: implications for haemopoietic cell growth and chronic myeloid leukaemia. FEBS Lett 581:1329–1334. doi:10.1016/j.febslet.2007.02.048

Naka K, Hoshii T, Hirao A (2010a) Novel therapeutic approach to eradicate tyrosine kinase inhibitor resistant chronic myeloid leukemia stem cells. Cancer Sci 101:1577–1581. doi:10.1111/j.1349-7006.2010.01584.x

Naka K, Hoshii T, Muraguchi T et al (2010b) TGF-β-FOXO signalling maintains leukaemia-initiating cells in chronic myeloid leukaemia. Nature 463:676–680. doi:10.1038/nature08734

Ogawa S, Kurokawa M, Tanaka T et al (1996) Increased Evi-1 expression is frequently observed in blastic crisis of chronic myelocytic leukemia. Leukemia 10:788–794

Oshima M, Oshima H, Taketo MM (1996) TGF-β receptor type II deficiency results in defects of yolk sac hematopoiesis and vasculogenesis. Dev Biol 179:297–302. doi:10.1006/dbio.1996.0259

Pear WS, Miller JP, Xu L et al (1998) Efficient and rapid induction of a chronic myelogenous leukemia-like myeloproliferative disease in mice receiving P210 bcr/abl-transduced bone marrow. Blood 92:3780–3792

Pellicano F, Cilloni D, Helgason GV et al (2009) FOXO transcription factor activity is partially retained in quiescent CML stem cells and induced by tyrosine kinase inhibitors in CML progenitor cells. Blood. doi:10.1182/blood-2009-06-226621

Ren R (2005) Mechanisms of BCR-ABL in the pathogenesis of chronic myelogenous leukaemia. Nat Rev Cancer 5:172–183. doi:10.1038/nrc1567

Reynaud D, Pietras E, Barry-Holson K et al (2011) IL-6 controls leukemic multipotent progenitor cell fate and contributes to chronic myelogenous leukemia development. Cancer Cell 20:661–673. doi:10.1016/j.ccr.2011.10.012

Roumiantsev S, Shah NP, Gorre ME et al (2002) Clinical resistance to the kinase inhibitor STI-571 in chronic myeloid leukemia by mutation of Tyr-253 in the Abl kinase domain P-loop. Proc Natl Acad Sci USA 99:10700–10705. doi:10.1073/pnas.162140299

Rowley JD (1973) Letter: a new consistent chromosomal abnormality in chronic myelogenous leukaemia identified by quinacrine fluorescence and Giemsa staining. Nature 243:290–293

Savona M, Talpaz M (2008) Getting to the stem of chronic myeloid leukaemia. Nat Rev Cancer 8:341–350. doi:10.1038/nrc2368

Scheller M, Huelsken J, Rosenbauer F et al (2006) Hematopoietic stem cell and multilineage defects generated by constitutive β-catenin activation. Nat Immunol 7:1037–1047. doi:10.1038/ni1387

Schemionek M, Elling C, Steidl U et al (2010) BCR-ABL enhances differentiation of long-term repopulating hematopoietic stem cells. Blood 115:3185–3195. doi:10.1182/blood-2009-04-215376

Shull MM, Ormsby I, Kier AB et al (1992) Targeted disruption of the mouse transforming growth factor-β 1 gene results in multifocal inflammatory disease. Nature 359:693–699. doi:10.1038/359693a0

Singbrant S, Karlsson G, Ehinger M et al (2010) Canonical BMP signaling is dispensable for hematopoietic stem cell function in both adult and fetal liver hematopoiesis, but essential to preserve colon architecture. Blood 115:4689–4698. doi:10.1182/blood-2009-05-220988

Suda T, Arai F, Hirao A (2005) Hematopoietic stem cells and their niche. Trends Immunol 26:426–433. doi:10.1016/j.it.2005.06.006

Swerdlow SH, Campo E, Harris NL et al (2008) WHO classification of tumours of haematopoietic and lymphoid tissues. IARC, Lyon

Sykes SM, Lane SW, Bullinger L et al (2011) AKT/FOXO signaling enforces reversible differentiation blockade in myeloid leukemias. Cell 146:697–708. doi:10.1016/j.cell.2011.07.032

Takeuchi C, Takeuchi S, Ikezoe T, Bartram CR, Taguchi H, Koeffler HP (2002) Germline mutation of the p27/Kip1 gene in childhood acute lymphoblastic leukemia. Leukemia 16:956–958. doi:10.1038/sj.leu.2402408

Tothova Z, Kollipara R, Huntly BJ et al (2007) FoxOs are critical mediators of hematopoietic stem cell resistance to physiologic oxidative stress. Cell 128:325–339. doi:10.1016/j.cell.2007.01.003
Wakefield LM, Roberts AB (2002) TGF-β signaling: positive and negative effects on tumorigenesis. Curr Opin Genet Dev 12:22–29
Wang L, Gural A, Sun XJ et al (2011) The leukemogenicity of AML1-ETO is dependent on site-specific lysine acetylation. Science 333:765–769. doi:10.1126/science.1201662
Wolfraim LA, Fernandez TM, Mamura M et al (2004) Loss of Smad3 in acute T-cell lymphoblastic leukemia. N Engl J Med 351:552–559. doi:10.1056/NEJMoa031197
Yalcin S, Zhang X, Luciano JP et al (2008) Foxo3 is essential for the regulation of ataxia telangiectasia mutated and oxidative stress-mediated homeostasis of hematopoietic stem cells. J Biol Chem 283:25692–25705. doi:10.1074/jbc.M800517200
Yamazaki S, Iwama A, Takayanagi SI, Eto K, Ema H, Nakauchi H (2009) TGF-β as a candidate bone marrow niche signal to induce hematopoietic stem cell hibernation. Blood 113:1250–1256. doi:10.1182/blood-2008-04-146480
Yamazaki S, Ema H, Karlsson G et al (2011) Nonmyelinating Schwann cells maintain hematopoietic stem cell hibernation in the bone marrow niche. Cell 147:1146–1158. doi:10.1016/j.cell.2011.09.053
Yokota A, Kimura S, Tanaka R et al (2010) Osteoclasts are involved in the maintenance of dormant leukemic cells. Leuk Res 34:793–799. doi:10.1016/j.leukres.2009.08.034
Yong AS, Keyvanfar K, Hensel N et al (2009) Primitive quiescent $CD34^+$ cells in chronic myeloid leukemia are targeted by in vitro expanded natural killer cells, which are functionally enhanced by bortezomib. Blood 113:875–882. doi:10.1182/blood-2008-05-158253
Yuasa H, Oike Y, Iwama A et al (2005) Oncogenic transcription factor Evi1 regulates hematopoietic stem cell proliferation through GATA-2 expression. EMBO J 24:1976–1987. doi:10.1038/sj.emboj.7600679
Zhang B (2012) Altered mircroenvironmental regulation of leukemic and normal stem cells in chronic myelogenous leukemia. Cancer Cell 21:577–592. doi:10.1016/j.ccr.2012.02.018
Zhang J, Niu C, Ye L et al (2003) Identification of the haematopoietic stem cell niche and control of the niche size. Nature 425:836–841. doi:10.1038/nature02041
Zhao C, Blum J, Chen A et al (2007) Loss of β-catenin impairs the renewal of normal and CML stem cells in vivo. Cancer Cell 12:528–541. doi:10.1016/j.ccr.2007.11.003
Zhao C, Chen A, Jamieson CH et al (2009) Hedgehog signalling is essential for maintenance of cancer stem cells in myeloid leukaemia. Nature 458:776–779. doi:10.1038/nature07737
Zou P, Yoshihara H, Hosokawa K et al (2011) $p57^{Kip2}$ and $p27^{Kip1}$ cooperate to maintain hematopoietic stem cell quiescence through interactions with Hsc70. Cell Stem Cell 9:247–261. doi:10.1016/j.stem.2011.07.003

Chapter 9
TGF-β in Skin Cancer and Fibrosis

Gangwen Han, Zheyi Han, and Xiao-Jing Wang

Abstract Among all three isoforms of the TGF-β ligand, TGF-β1 is the predominant isoform in the skin. In normal skin, canonical TGF-β signaling components, i.e., TGF-β receptors and signaling Smads, are broadly and highly expressed, whereas TGF-β ligands are expressed at very low levels. These expression patterns determine that the TGF-β signaling input to the skin is low under normal conditions but high once TGF-β ligands are upregulated under disease conditions. TGF-β1 is a potent growth inhibitor of epidermal keratinocytes, which dictates its tumor suppressive effect in early stages of skin cancer. However, cancer cells lose TGF-β-induced growth inhibition at late stages, and TGF-β-induced angiogenesis and skin inflammation create an environment favorable for skin cancer progression and metastasis. In fibrotic skin diseases, TGF-β plays a key role in activating fibroblast proliferation and stimulating the production of extracellular matrix proteins. Also, TGF-β affects many aspects of skin wound healing and in turn influences cutaneous scarring after skin damage. Fully understanding the mechanisms of TGF-β in the pathogenesis of skin cancer and fibrotic diseases will help design novel strategies in treating skin diseases.

Keywords Cutaneous scarring • Smad • Squamous cell carcinoma (SCC) • Systemic sclerosis (SSc) • TGF-β • Wound healing

9.1 Introduction

Transforming growth factor-β (TGF-β) is a ubiquitous and multifunctional cytokine that regulates a variety of events in cells and tissues, including cell proliferation, differentiation, migration, angiogenesis, inflammation, and immune surveillance.

G. Han • Z. Han • X.-J. Wang (✉)
Department of Pathology, University of Colorado Denver,
Anschutz Medical Campus, Aurora, CO 80045, USA
e-mail: xj.wang@ucdenver.edu

A. Moustakas and K. Miyazawa (eds.), *TGF-β in Human Disease*,
DOI 10.1007/978-4-431-54409-8_9, © Springer Japan 2013

The function of TGF-β is largely dependent on its targeted cells and is influenced by other molecules and signaling. Thus, the in vivo role of TGF-β is complicated and context dependent. The skin is histologically divided into two parts: the epidermis and the dermis. Keratinocytes in the epidermis and fibroblasts in the dermis represent two major cell types showing opposite effects in response to TGF-β: TGF-β inhibits keratinocytes but promotes fibroblast proliferation (Anzano et al. 1982; Coffey et al. 1988; Shipley et al. 1986). Among the three TGF-β isoforms, TGF-β1 is the predominant isoform in most tissue types, including the skin (Frank et al. 1996; Quan et al. 2002; Wang 2001), and is expressed at a very low level in normal skin but elevated in skin diseases and cancer (Han et al. 2005; Quan et al. 2002; Querfeld et al. 1999). The role of TGF-β in the pathogenesis of skin cancer and fibrotic diseases has been extensively studied. In addition to the direct effect of TGF-β on keratinocytes and fibroblasts, TGF-β-mediated inflammation, angiogenesis, and regulation of immune cells have additional impacts on the development of skin cancer and fibrosis (Li et al. 2006; Martin 1997; Seifert and Mrowietz 2009; Wei et al. 2011). Particularly, recent studies on the receptors and Smads, components of TGF-β signaling in transgenic/knockout mouse models further advanced our understanding of the roles of TGF-β signaling in skin cancer and fibrotic diseases (Han and Wang 2011; Lakos et al. 2004; Sonnylal et al. 2007; Wei et al. 2011). This chapter will focus on the role of TGF-β signaling in the pathogenesis of non-melanoma skin cancer and skin fibrotic disorders, the latter including systemic sclerosis (SSc), hypertrophic scarring, and keloid scarring. Because hypertrophic and keloid scarring are consequences of wound healing affected by TGF-β during the healing process, the role of TGF-β signaling in cutaneous wound healing is also discussed.

9.2 TGF-β Signaling in Non-melanoma Skin Cancer

Non-melanoma skin squamous cell carcinoma (SCC) is the most common human cancer developed from epidermal keratinocytes. Genetic mutations and continuous proliferation of mutant keratinocytes are the prerequisites for SCC development. Studies indicate that angiogenesis, inflammation, and alterations in immune response in the stroma significantly influence the development of skin cancer and metastasis. Many molecules and signaling pathways have been implicated in the pathogenesis of SCC. The role of TGF-β and its downstream signaling components have been extensively investigated in SCC. The TGF-β signaling pathway consists of three components: (1) Ligands, i.e., TGF-β1, 2, and 3. (2) Secreted ligands function through two serine/threonine kinase receptors, TGF-β receptor type I (TGF-βRI) and type II (TGF-βRII), both of which are necessary for signal transduction. TGF-βRII receptor directly binds to ligands and interacts with TGF-βRI. (3) Intracellular downstream mediators, i.e., Smads, are phosphorylated and activated by TGF-βRI. To date, many components of TGF-β signaling have been studied for their differential roles in normal keratinocytes, SCC cells and tumor stromal cells. This chapter will give a brief review over these differential roles of TGF-β signaling in skin carcinogenesis.

9.2.1 *TGF-β Inhibits Tumor Formation of* ras *Transformed Keratinocytes*

TGF-β was initially identified as a growth factor stimulating the growth of normal rat kidney fibroblasts (Anzano et al. 1982; Sporn 1999, 2006). Soon, however, TGF-β was found to inhibit mink lung epithelial cell growth (Tucker et al. 1984). Thus, the diversity of TGF-β's functions on different cell types was quickly recognized (Roberts et al. 1985). In particular, TGF-β inhibits the growth of normal human keratinocytes but can lose its growth inhibitory responses as it happens in the skin cancer cell line SCC-25 due to the possibility that tumor cells lack TGF-β receptors (Coffey et al. 1988; Shipley et al. 1986). However, tumor development in normal keratinocytes requires an initial oncogenic mutation. The *Ras* mutation in keratinocytes is considered an important initiating factor for papilloma development in mouse skin (Roop et al. 1986) and was identified in some human skin cancers (Dlugosz et al. 2002). Thus, *Ras*-transfected keratinocytes have been used to study the role of TGF-β signaling in skin carcinogenesis. In vitro studies showed that *v-ras*^{Ha} transfected normal mouse keratinocytes increased basal expression and secretion of TGF-β1, and grafting *v-ras*^{Ha} transfected normal keratinocytes onto nude mice formed well-differentiated papillomas with increased expression of TGF-β1 protein in the basal and spinous layers of papillomas (Glick et al. 1991, 1994). However, *v-ras*^{Ha} transfected TGF-β1 null mouse keratinocytes grafted onto nude mice progressed rapidly to multifocal SCC. Tumor proliferation was also elevated in grafts initiated from *v-ras*^{Ha} transfected TGF-β1 null keratinocytes compared to cells with *v-ras*^{Ha} transfected wild-type keratinocytes. These studies indicated that TGF-β provides a tumor-suppressing function in skin carcinogenesis (Glick et al. 1991, 1994). Similarly, transient inactivation of TGF-β signaling by infecting with a dominant-negative TGF-βRII causes chromosome instability. These phenomena, observed in keratinocytes with TGF-β1 deletion or expressing dominant-negative TGF-βRII, could be suppressed by exogenous TGF-β1 (Glick et al. 1999). In the TGF-β signaling pathway, Smad3 is a critical factor mediating TGF-β signaling to its target genes (Millet and Zhang 2007). Smad3 null keratinocytes transduced with the *v-ras*^{Ha} gene lost their growth inhibitory response to TGF-β. Nude mice with *v-ras*^{Ha}-transduced Smad3 null keratinocyte grafts developed more papillomas and progressed to SCC with a higher frequency than nude mice with *v-ras*^{Ha}-transduced Smad3 wild-type keratinocyte grafts (Vijayachandra et al. 2003). Therefore, data based on in vitro models of keratinocyte transformation studies demonstrated that TGF-β signaling inhibits keratinocyte proliferation and defects in TGF-β signaling accelerate tumor progression in multistage mouse carcinogenesis (Glick et al. 1999). HaCaT is a spontaneously immortalized normal human keratinocyte cell line (Boukamp et al. 1988). The cell line keeps most characteristics of normal keratinocytes and remains non-tumorigenic after long-term culture and passage. HaCaT cell became tumorigenic after transducing with a *ras* oncogene (Fusenig and Boukamp 1998; Gold et al. 2000), Within established *ras-transfected* HaCaT cells, benign and malignant clones have been characterized based on their behaviors in forming tumor after transplantation into nude mice (Fusenig and Boukamp 1998). HaCaT-*ras* cells with stable transduction of dominant-negative TGF-βRII lose differentiation and form metastatic SCC in vivo (Ganapathy et al. 2010). In contrast to the growth inhibitory effect of TGF-β on normal keratinocytes, HaCaT-*ras* cells with high TGF-β1 production in cultured conditions have increased tumorigenesis in vivo, and transducing TGF-β1 or TGF-β2 to the benign HaCaT-*ras* clone increases tumor formation, invasion, and metastasis in nude mice (Davies et al. 2012). Furthermore, the HaCaT-*ras* cells producing more TGF-β still retain the response to TGF-β-induced growth inhibition in vitro (Davies et al. 2012), suggesting that the growth inhibitory effect of TGF-β1 is insufficient for tumor suppression in vivo.

9.2.2 In Vivo Expression of TGF-β1 in the Skin Inhibits Benign Tumor Formation and Promotes Malignant Conversion in Skin Carcinogenesis

The two-stage chemical carcinogenesis model is a well-established experimental system to study the mechanism of skin tumor development in vivo (Filler et al. 2007; Yuspa 1986). In this model, skin tumor is first initiated by exposing the skin to carcinogens such as the mutagen dimethylbenzanthracene (DMBA), which can cause mutation in the oncogene *Ha-ras* (Fujiki et al. 1989); second, tumor formation is promoted by treatments of a tumor promoter, such as 12-*O*-tetradecanoylphorbol-13-acetate (TPA). Some benign tumors (primarily papillomas) can convert to malignant SCCs after acquiring additional spontaneous genetic mutations (Akhurst and Balmain 1999). In wild-type mice, TGF-β1 mRNA expression is quickly induced by TPA in suprabasal cells of the epidermis (Akhurst et al. 1988). Chemically induced papillomas and carcinomas have elevated levels of TGF-β1 mRNA in the keratinocyte compartment (Fowlis et al. 1992; Patamalai et al. 1994). To investigate the role of TGF-β in skin homeostasis and carcinogenesis in vivo, keratinocyte-specific TGF-β transgenic mice were developed. Transgenic mouse models that target TGF-β1 to epidermal keratinocytes using different keratin promoters demonstrate that functions of TGF-β1 in skin development and epidermal keratinocyte proliferation depending on the location of TGF-β1 expression in different epidermal layers. When TGF-β1 was targeted to suprabasal layers of the epidermis driven by keratin 1 (K1) promoter (K1.TGF-β1), transgenic mice died at early neonatal stages (Sellheyer et al. 1993) due to the inhibition of keratinocyte proliferation. In contrast, mice overexpressing TGF-β1 driven by keratin 6 or keratin 10 promoters (K6.TGF-β1 or K10.TGF-β mice) lived to adulthood with no significant histological changes other than resistance to TPA-induced epidermal hyperplasia (Cui et al. 1995; Fowlis et al. 1996). However, study of long-term chemical carcinogenesis on K6.TGF-β1 and K10.TGF-β1 transgenic mice led to the conclusion that TGF-β1 has a biphasic effect on skin carcinogenesis: inhibiting tumorigenesis at an early stage but promoting malignant tumor conversion and rapid metastasis at a later stage (Cui et al. 1996). Using gene-switch-TGF-β1 transgenic mice to induce TGF-β

1 transgene expression in the skin at specific stages of chemical carcinogenesis, we found that TGF-β1 transgene induction at early stage inhibited tumor formation but TGF-β1 induction in papillomas significantly promoted tumor conversion from benign to malignant with increased metastasis (Weeks et al. 2001). Moreover, TGF-β transgenic papillomas exhibited down-regulation of TGF-β receptors and their signal transducer Smads, loss of the invasion suppressor E-cadherin/catenin complex in the cell membrane, elevated expression of matrix metalloproteinases and increased angiogenesis. Thus, although down regulation of TGF-β signaling components in tumor epithelia abolishes TGF-β-induced tumor cell growth inhibition, tumor stroma, which has intact TGF-β signaling components, respond to increased TGF-β ligand to promote tumor metastasis (Weeks et al. 2001).

9.2.3 In Vivo Abrogation of TGF-β Receptors in Skin Promotes Tumor Development and Metastasis

Mice with germline TGF-βRII deletion are embryonic lethal around 10.5 days of gestation (Oshima et al. 1996). Interruption of TGF-βRII function by either overexpressing dominant-negative TGF-βRII (ΔβRII) receptors or conditional deletion of TGF-βRII in specific epidermis has been utilized to study the role of TGF-β receptors in skin homeostasis and carcinogenesis (Amendt et al. 1998; Go et al. 1999; Guasch et al. 2007). Transgenic mice expressing ΔβRII in suprabasal keratinocytes driven by truncated mouse loricrin promoter (ML.ΔβRII) develop hyperproliferative skin conditions in the early neonatal stage but normalize in adulthood. In vitro cultured primary keratinocytes from these mice demonstrated resistance to exogenous TGF-β1-induced growth inhibition (Wang et al. 1997). When these transgenic mice were subjected to the two-stage chemical carcinogenesis study, they developed tumors earlier with a larger number of benign papillomas and higher frequencies of malignant cancer formation and metastasis compared to wild-type mice (Go et al. 1999). Analysis showed that ML.ΔβRII tumors had increased expressions of vascular endothelial growth factor (VEGF), a pro-angiogenesis factor, and decreased expression of thrombospondin-1, an angiogenesis inhibitor. Both factors are attributed to the increased angiogenesis found in tumors and malignant carcinoma progression and metastasis. Increased angiogenesis correlated with elevated endogenous TGF-β1 in ML.ΔβRII tumors (Go et al. 1999). Similarly, when ΔβRII was expressed in basal keratinocytes in mouse skin targeted by the keratin 5 promoter, the transgenic mice also have accelerated skin tumor development and malignancy transformation in chemical carcinogenesis (Amendt et al. 1998) compared to wild-type mice. Moreover, keratinocyte-specific TGF-βRII knockout mice in which deletion of TGF-βRII in stratified epithelia driven by the keratin 14 promoter results in spontaneous SCC formation in the transitional area between the stratified squamous epithelium of the skin and the mucosal epithelium, e.g. the anal and genital regions but not in mouse back skin (Guasch et al. 2007). Further analyses revealed that in comparison with mouse back skin, the epithelium of the transitional zone in the anus naturally showed enhanced *Ras*-MAPK signaling, locally increased inflammation and aberrant differentiation, which favor to tumorigenesis in the anal and genital regions. In aged mice with TGF-βRII deletion, reduced apoptosis and high epidermal proliferation due to blocking of TGF-β signaling eventually caused anal and genital SCC development. Tumor formation in these regions did not happen in young mice when enhanced proliferation, as a result of TGF-βRII deletion, was balanced with increased apoptosis (Guasch et al. 2007). Furthermore, when DMBA alone was applied to TGF-βRII knockout mice skin or grafting of *Ha-Ras*-infected TGF-βRII knockout keratinocytes onto nude mice, mice quickly developed invasive/metastatic SCC (Guasch et al. 2007). Therefore, the loss of TGF-β function in epidermal keratinocytes promotes skin carcinogenesis, but development of malignant tumors requires additional initiation factors (Amendt et al. 1998; Go et al. 1999; Guasch et al. 2007). Human malignant skin cancer frequently exhibits overexpression of TGF-β1 but reduced expression of TGF-βRII. To better understand how this combination affects cancer prognosis, we studied skin carcinogenesis using the chemical carcinogenesis protocol in inducible TGF-β transgenic mice with and without functional TGF-β receptors. In this study, TGF-β1/ΔβRII compound mice were generated; these mice stably express ΔβRII, which inhibits cell binding to the TGF-β ligand, but TGF-β1 can be induced at specific stages of carcinogenesis. When TGF-β1 expression was induced in the papilloma stage, mice developed malignant SCC at a high frequency, with more metastasis to the lymph nodes and lung. Overall, 30 % of tumors in mice overexpressing TGF-β1 exhibited spindle cell carcinoma (SPCC). TGF-β1 induction in the papilloma with inactivation of TGF-βRII function in the TGF-β/ΔβRII compound mice also developed more malignant SCC and metastasis. However, SPCC was rarely observed in TGF-β1/ΔβRII mice. Consistently, TGF-β1-induced SCCs exhibited loss of E-cadherin, an EMT marker. This phenomenon was not common in ΔβRII or TGF-β1/ΔβRII compound mice. Further analysis showed that TGF-β1-mediated SCCs had significant levels of Hey1 and Jag1, while upregulation of Erk1, JNK1, and RhoA/Rac1 was also observed in all TGF-β1 and ΔβRII mice. Therefore, TGF-β1-induced EMT and metastasis can be uncoupled. TGF-β1 mediates EMT via the activation of Notch signaling and requires intact TGF-β signaling components, whereas increased angiogenesis and matrix metalloproteinase (MMP) caused by TGF-β-induced MAPK and Rho/Rac activation via a Smad-independent mechanism largely contribute to cancer metastasis (Han et al. 2005). This study highlights an important fact that EMT cannot always be used to predict the metastatic potential.

9.2.4 Differential Roles of TGF-β Signaling Smads in Skin Cancer

The discovery and study of Smad family members mediating TGF-β signaling has significantly improved the understanding of the TGF-β signaling pathway's multiple biological functions. Eight isoforms of Smads have been found in mammals and are generally classified into three subtypes: R-Smads (Smad1, 2, 3, 5, 8), Co-Smad (Smad4), and I-Smads (Smad6, 7) (Derynck and Zhang 1996; Massague 1996).

Germline deletions of Smad1, Smad2, Smad4, Smad5, and Smad7 in mice result in embryonic death (Chang et al. 1999; Heyer et al. 1999; Kleiter et al. 2010; Nomura and Li 1998; Sirard et al. 1998; Tremblay et al. 2001; Waldrip et al. 1998; Weinstein et al. 1998; Yang et al. 1998). Smad3 knockout mice survive to adulthood (Datto et al. 1999; Yang et al. 1999; Zhu et al. 1998). To overcome embryonic lethality caused by knocking out individual Smads, conditional knockout mice targeting Smad deletion to keratinocytes were generated to study the role of Smads in skin carcinogenesis (Bornstein et al. 2007). So far, the potential roles of TGF-β signaling Smads, i.e., Smad2, Smad3, and Smad4, in skin carcinogenesis have been investigated, but the role of other Smads in skin tumors remains to be clarified. Thus, the current discussion will be focused on the role of TGF-β signaling Smads in skin cancer.

9.2.4.1 The Role of Smad2 in Skin Carcinogenesis

Loss or reduction of Smad2 expression occurs in several human SCCs, including genital SCCs, head and neck SCCs and cervical SCCs (Han and Wang 2011). Immunostaining has shown Smad2 protein reduction/loss in 70 % of human skin cancers. Smad2 loss is especially high in poorly differentiated SCCs (Hoot et al. 2008). Although Smad2 mutation was not frequent in human SCCs (Han and Wang 2011), we found that approximately 67 % of poorly differentiated human skin SCCs have loss of heterozygosity (LOH) at the Smad2 locus, possibly due to accumulated ultraviolet radiation-induced genomic damage (Hoot et al. 2008). Since germline homozygous Smad2 deletion in mice results in embryonic lethality (Heyer et al. 1999; Nomura and Li 1998; Waldrip et al. 1998; Weinstein et al. 1998), the role of Smad2 in skin carcinogenesis was first investigated in Smad2 heterozygous mice (Smad2+/−) because these mice lived a normal lifespan without spontaneous tumor development. Smad2+/− mice developed a greater number of less-differentiated tumors compared to wild-type control mice when exposed to the two-stage chemical carcinogenesis protocol (Tannehill-Gregg et al. 2004). In our established conditional Smad2 knockout mice that target Smad2 deletions to keratinocytes using the keratin 14 (K14) promoter, we found Smad2 deletion did not develop spontaneous skin tumors throughout 18 months of observation (Hoot et al. 2008). Moreover, even in the presence of a DMBA-induced *H-ras* mutation, a genetic alteration mimicking early stage human skin cancer, mice with epidermal Smad2 deletion still failed to develop skin tumors without TPA promotion. Thus, Smad2 loss alone is insufficient to promote initiated cells to develop into cancer. Similar to the observation in Smad2+/− mice (Tannehill-Gregg et al. 2004), when the two-stage chemical carcinogenesis protocol was applied to mice with epidermal Smad2 loss, mice developed skin tumors early with increased numbers, and accelerated malignant conversion compared to wild-type mice (Hoot et al. 2008). Histological analysis showed that Smad2 loss induced papillomas to undergo early EMT, and about 25 % malignant tumors in conditional Smad2 knockout mice were classified as SPCC, which are rare in wild-type mice. Moreover, Smad2 deficient tumors have significantly increased angiogenesis in the stroma. Mechanism analyses revealed that levels of TGF-β1, TGF-β signaling, Smad3/Smad4, and VEGF were similar between

Smad2 deletion tumors and wild-type tumors. Proliferation and apoptosis in tumors have a very limited effect from Smad2 deletion. However, Smad2 deletion significantly increased Snail and hepatocyte growth factor (HGF) expression in keratinocytes and tumor epithelial cells. Further analysis identified that Smad2 loss causes Snail and HGF upregulation via loss of Smad2-mediated transcriptional repression and enhanced Smad3/4-mediated transactivation for Snail and HGF in keratinocytes and tumor cells (Hoot et al. 2008; Hoot et al. 2010). These results provide potential new targets for treating skin cancer.

9.2.4.2 The Role of Smad3 in Skin Carcinogenesis

Smad3 acts as a major mediator for the TGF-β signaling pathway (Millet and Zhang 2007; Nakao et al. 1997). With respect to the role of Smad3 in SCC, Smad3 null keratinocytes transduced with *v-ras*^{Ha} exhibited significant reduction of TGF-β-induced cell growth arrest and increased tumorigenesis after grafting onto nude mice. Overexpression of Smad3 in wild-type keratinocytes by infection with adenoviral-Smad3 induced keratinocyte growth arrest and senescence. All of the changes observed in Smad3 null keratinocytes cannot be amended by transfecting Smad2 and Smad4 (Bae et al. 2009; Vijayachandra et al. 2003). In contrast to the findings in Smad3 null keratinocytes and keratinocyte grafting bioassays, a carcinogenesis study in Smad3 knockout mice has shown that Smad3 is required for tumor formation in the two-stage carcinogenesis model. Germline Smad3 knockout mice did not develop spontaneous skin tumors after long-term observation (Datto et al. 1999; Yang et al. 1999; Zhu et al. 1998). Interestingly, both homozygous and heterozygous Smad3 knockout mice have shown resistance to chemically induced skin carcinogenesis (Li et al. 2004a; Tannehill-Gregg et al. 2004). Furthermore, Smad3 knockout tumors show reduced cell proliferation and inflammation, but increased apoptosis in response to TPA treatment (Li et al. 2004a). The observed phenomena may be attributed to Smad3 deletion-mediated blocking of TGF-β signaling, evidenced by the reduction of TGF-β-induced activator protein-1 family members and TGFα in TPA treated Smad3 null cells and tissues. Moreover, tumor tissues exhibited reduced leukocyte infiltration, particularly a reduction of tumor-associated macrophage infiltration. Consistently, the pro-inflammatory cytokine, IL-1β, and the monocyte/macrophage-attracting chemokine, MCP-1, are significantly reduced in Smad3 null tumors compared to wild-type tumors (Li et al. 2004a). Therefore, TGF-β mediated inflammation appears to require Smad3, which could explain the difference between Smad3-mediated tumor suppressive effects inside the keratinocytes and Smad3-mediated oncogenic effects on the tumor stroma.

9.2.4.3 The Role of Smad4 in Skin Carcinogenesis

To avoid embryonic lethality caused by germline deletion of Smad4 (Sirard et al. 1998; Yang et al. 1998), mouse models with Smad4 deletion specifically targeted at keratinocytes have been established (Bornstein et al. 2009; Li et al. 2003;

Qiao et al. 2006; Yang et al. 2005). Smad4 deficient skin develops progressive alopecia with hair follicle degeneration without affecting keratinocyte differentiation (Owens et al. 2008; Yang et al. 2005). Molecular mechanisms for this pathological alteration are attributed to the loss of Smad4 that blocked Smad1/5 mediated desmoglein-4 transcription (Owens et al. 2008). When Smad4 was deleted in epidermal keratinocytes, mammary epithelium or head and neck epithelium, all mice developed spontaneous SCC (Bornstein et al. 2009; Li et al. 2003; Yang et al. 2005). These findings further confirmed that Smad4 mutation acts as an oncogenic factor (Hahn et al. 1996; Thiagalingam et al. 1996). Tumors in Smad4 knockout mice have reductions in phosphatase and tensin homolog (PTEN) and $p21^{CIP/WAF}$, but upregulation of AKT, increased cell proliferation and nuclear accumulation of cyclin D1 (Qiao et al. 2006). In the established Smad4/PTEN double knockout mice, skin tumor formation was significantly accelerated compared to Smad4 deletion alone (Qiao et al. 2006; Yang et al. 2005). When Smad4 and PTEN genes were simultaneously deleted in epithelial cells of the upper digestive tract, forestomach tumor formation was also accelerated (Teng et al. 2006). Thus, Smad4 and PTEN act synergistically to regulate epidermal proliferation and differentiation. Loss of both genes may contribute to the initiation and promotion of tumorigenesis (Teng et al. 2006; Yang et al. 2005). The tumor suppressive role of Smad4 is also shown by the deletion of Smad4 in oral epithelia, which resulted in spontaneous head and neck SCC with downregulation of Fanc/Brca pathway genes, which increased genomic instability and inflammation (Bornstein et al. 2009). Thus, there are several critical differences between Smad2 and Smad4 deletion in the epithelium. Smad2 deletion in the epidermis promoted EMT at quite an early stage and caused significant increase in SPCC formation in chemical carcinogen-initiated skin cancer, but Smad2 loss itself is insufficient to initiate tumorigenesis. Increased angiogenesis observed in chemically induced Smad2−/−tumors was caused by directly upregulating HGF through recruiting Smad4 onto the SBE site of the HGF promoter (Hoot et al. 2010). Smad4 deletion in epithelial cells could initiate spontaneous carcinoma development by downregulation of PTEN or Fanc/Brca genes. EMT phenomena or SPCC was seldom observed in the Smad4 deletion tumors. Furthermore, Smad4 deleted but not Smad2 deleted tumors have increased expression of TGF-β1 and VEGF, which contributes to increased angiogenesis and inflammation (Bornstein et al. 2009; Owens et al. 2010; Qiao et al. 2006; Yang et al. 1998).

9.3 TGF-β and Systemic Sclerosis

SSc, also known as scleroderma, is a chronic and fatal autoimmune disease characterized by excessive extracellular matrix accumulation in connective tissues. SSc affects the skin and internal organs, including the lungs, gastrointestinal tract, and heart. Clinically, SSc is classified into two major types: diffuse SSc, characterized as cutaneous scleroderma with at least one internal organ involvement, and limited cutaneous scleroderma, characterized as limited skin scleroderma without internal organ involvement. The major pathologic feature of SSc is excess collagen deposition in the extracellular matrix. Although the mechanism of skin fibrosis remains unknown, abnormal dermal fibroblast activation and chronic inflammation have been implicated in the pathogenesis of SSc. TGF-β has been shown to promote fibroblast proliferation (Wynn 2007), stimulate the production of collagen by fibroblasts (Bettinger et al. 1996; Wu et al. 2012), and induce inflammation in the skin (Li et al. 2004b). Therefore, accumulated evidence indicates that TGF-β signaling plays a key pathogenic role in SSc.

9.3.1 Increased TGF-β Expression in SSc Lesions

Histologically, fibrosis is a cardinal feature of SSc, but vasculopathy and inflammatory infiltration may also be observed in the early stages of SSc (Beyer et al. 2012; Gabrielli et al. 2009; Varga and Abraham 2007). Fibrosis is characterized by excessive collagen formation in connective tissue, produced by activated fibroblasts. Since TGF-β is shown to promote fibroblast proliferation and enhance collagen production in normal human fibroblasts, TGF-β has been implicated in the pathogenesis of fibrotic disorders (Varga and Jimenez 1986). SSc patients have high levels of plasma TGF-β, and fibroblasts derived from SSc patients display elevated levels of TGF-β mRNA expression (Higley et al. 1994; Vuorio et al. 1991). Abnormal expression of TGF-β is shown to be involved in SSc skin lesion formation (Gay et al. 1992; Gruschwitz et al. 1990; Rudnicka et al. 1994). Increased TGF-β1 was also found in endothelial cells at the early stages of SSc skin (Gabrielli et al. 1993). For the three isoforms of TGF-β (1, 2, and 3) in mammals, the mRNA for each has been detected in inflamed areas of either localized or diffused SSc, but not in sclerotic or healthy skin. However, TGF-β1 and TGF-β2 proteins were confirmed in the inflammatory skin of patients, whereas TGF-β3 protein appears to be present in the dermis of both patients and healthy controls, indicating a reduced specificity of TGF-β3 in the pathogenesis of SSc (Querfeld et al. 1999). Furthermore, TGF-β2 mRNA was found to be co-localized with pro-α1(I) collagen expression around dermal blood vessels in the inflammatory stage of SSc, but no expression of TGF-β2 or pro-α1(I) collagen was found in the dermis of fibrotic stage patients, suggesting that TGF-β2 might be more important in the pathogenesis of early SSc (Kulozik et al. 1990). In several established SSc animal models, increased TGF-β acts as a key mediator in fibrosis formation, and the reduction of TGF-β expression by several therapeutic strategies shows antifibrotic effects (Artlett 2010; Batteux et al. 2011; Iwamoto et al. 2011; Rosenbloom et al. 2010; Yamamoto 2010). These experiments further verified the role of TGF-β in the pathogenesis of SSc.

9.3.2 Upregulation of TGF-β Receptors in SSc Fibroblasts

It was also reported that alteration of TGF-β receptors occurred in SSc fibroblasts. Increased TGF-βRI and TGF-βRII mRNAs were found in cultured fibroblasts derived from SSc patients compared to healthy control fibroblasts. Consistently, overexpression of either TGF-βRI or TGF-βRII significantly increases α2 (I) collagen promoter activity in transient transfection assays in dermal fibroblasts (Kawakami et al. 1998). Furthermore, both TGF-βRI and TGF-βRII increased in dermal fibroblasts and inflammatory cells around vascular blood in localized scleroderma and SSc lesional skin in comparison with healthy controls (Kubo et al. 2001, 2002). These studies suggest that overexpression of TGF-βRI and TGF-βRII in dermal fibroblasts results in autocrine TGF-β signaling, which plays a pathological role in SSc development. However, these findings have not been fully verified by other studies. One study reported that SSc derived fibroblasts have high levels of TGF-β RI expression, but 30 % decreased TGF-βRII expression compared to normal healthy skin derived fibroblasts (Pannu et al. 2004). Furthermore, normal fibroblasts infected with adenoviral-TGF-βRI but not adenoviral-TGF-βRII demonstrated increased basal level of type-I collagen expression. They concluded that increasing the ratio of TGF-βRI to TGF-βRII contributed to collagen formation involved in the pathogenesis of SSc (Pannu et al. 2004). Perturbation of TGF-βRI and TGF-βRII in the pathogenesis of fibrosis has also been seen in animal models of SSc. In the induced TGF-βRI (βRI^{CA}/Cre-ER) mouse models, postnatal induction of TGF-βRI in fibroblasts resulted in multiple organ fibrosis mimicking features of SSc in humans (Sonnylal et al. 2007). When constitutively active TGF-βRI was induced in fibroblasts in βRI^{CA}/Cre-ER transgenic mice, the mice developed generalized dermal fibrosis in their skin and a small-vessel vasculopathy with both smooth muscle and endothelial abnormalities in the lungs and kidneys (Sonnylal et al. 2007). Additionally, overexpression of dominant-negative TGF-βRII in fibroblasts in TGF-βRIIΔk-fib mice also caused lung and skin fibrosis as endogenous TGF-βRI was activated in transgenic fibroblasts (Denton et al. 2003, 2005; Hoyles et al. 2008).

9.3.3 Abnormal Expression of Smads in SSc

Type I collagen proteins, which consist of glycine- and proline-rich two α1(I) chains (*COL1A1*) and one α2(I) chain (*COL1A2*), have been recognized as major excess protein in SSc lesion (Jinnin 2010). *COL1A2* genes contain Smad binding sites at their promoters. Transient overexpression of Smad3 and Smad4, but not Smad1 or Smad2, causes trans-activation of the human *COL1A2* promoter in normal human dermal fibroblasts (Chen et al. 1999; Dennler et al. 1998; Zhang et al. 2000). In vitro studies demonstrated that scleroderma-derived fibroblasts had increased *COL1A2* promoter activity with high levels of Smad3 binding to the promoter compared to normal control fibroblasts (Jinnin et al. 2006). Compared to fibroblasts derived from healthy adult volunteers, increased Smad3 expression is found in fibroblasts from scleroderma patients (Mori et al. 2003). Since subcutaneous injection of bleomycin into mouse skin for 4 weeks results in fibrosis in the skin and lung, bleomycin-induced fibrosis has been extensively used to study the pathogenesis of SSc (Yamamoto 2010). When the bleomycin-inducing sclerosis protocol was applied to Smad3 knockout mice, inflammatory infiltration and TGF-β expression in the dermis at an early time point were similar in Smad3 knockout and wild-type mice.

However, by day 28 after injection, lesional skin from Smad3 knockout mice exhibited attenuated fibrosis, lower synthesis and accumulation of collagen, and reduced collagen gene transcription in situ, compared to wild-type mice (Lakos et al. 2004). These data support the idea that Smad3 mediates TGF-β-induced fibrosis in SSc thus targeting Smad3 signaling could at least in part ameliorate scleroderma. Smad7, as an antagonist of the TGF-β signaling pathway, has been reported to inhibit α1 (I) collagen production in normal human fibroblasts and decrease α-smooth muscle actin expression in hypertrophic scar-derived fibroblasts (Kopp et al. 2005). Reduced Smad7 functions, due to either reduced expression or functional defect, have been observed in SSc. In one study, SSc lesions or SSc derived fibroblasts have reduced Smad7 expression but increased Smad3 expression in comparison with that of normal skin or fibroblasts, indicating that Smad7 deficiency and Smad3 upregulation attribute to upregulated TGF-β signaling in scleroderma (Dong et al. 2002). In a separate study, SSc lesions and SSc derived fibroblasts were found to express high levels of Smad7 and TGF-βRI, and TGF-β signaling Smads remained unchanged compared to normal controls (Asano et al. 2004). Further analysis showed that the inhibitory effect of Smad7 on human α2(I) collagen promoter activity was completely impaired in SSc fibroblasts, suggesting that increased Smad7 in SSc fibroblasts may be caused by autocrine TGF-β signaling (Asano et al. 2004). The discrepancy of Smad7 expression levels in SSc skin and fibroblasts has also been reported by other studies (Kreuter et al. 2006; Zhu et al. 2012) and may be due to different sample sources or different stages of SSc (Asano et al. 2004). Recent reports indicate that the Smad1 pathway may also be involved in the pathogenesis of SSc. The first evidence came from a study on forced TGF-βRI expression-induced fibrosis, which upregulates collagen production and CCN2 (CTGF) via the TGF-βRI/Smad1 and ERK1/2 pathways (Pannu et al. 2007). The same group later showed that total and phosphorylated Smad1 levels were significantly elevated in SSc skin biopsy samples and cultured SSc fibroblasts and correlated with elevated CCN2 protein levels. Furthermore, this group identified that Smad1 was a direct activator of the CCN2 gene through binding the GC-motif in the CCN2 promoter. SSc fibroblasts exhibited elevated CCN2 promoter activity, which is correlated with the level of increased Smad1. Knocking down Smad1 in SSc fibroblasts reduced production of CCN2 and collagen (Pannu et al. 2008). Therefore, it is likely that Smad1 and Smad3 play complementary roles in the pathogenesis of SSc.

9.3.4 TGF-β Regulates Other Fibrotic Factors Implicated in SSc

In addition to TGF-β, many cytokines, chemokines, and growth factors have been implicated in the pathogenesis of SSc. However, TGF-β seems to still be a master regulator in SSc as evidenced by its crosstalk with other molecules during SSc fibrosis. The platelet derived growth factor (PDGF) pathway has long been recognized as another key player in fibrotic diseases. PDGF regulates many aspects of fibrotic diseases including promoting myofibroblast proliferation and stimulating production of inflammatory cytokines and ECM. In SSc, TGF-β enhances the function of PDGF by upregulating expression of PDGF receptors in fibroblasts (Bonner 2004). c-Abelson (c-Abl), a non-receptor tyrosine kinase associated with chronic myelogenous leukemia (CML), is produced by TGF-β-activated fibroblasts and is also involved in TGF-β-induced fibrotic processes (Bhattacharyya et al. 2009; Rosenbloom et al. 2010). Inhibition of c-Abl by dasatinib or nilotinib has shown antifibrotic effects (Beyer et al. 2012). CCN2 (CTGF), early growth response gene I (EGR1), and Endothelin 1 (ET-1) are all induced by TGF-β and upregulated in SSc skin. All have been shown to mediate TGF-β-induced fibrosis. Targeting CCN2, EGR1 (e.g., imatinib) or ET-1 (e.g., Bosentan) may become new therapeutic strategies for treating fibrotic diseases (Bhattacharyya et al. 2012).

9.4 TGF-β and Cutaneous Hypertrophic and Keloid Scarring

Hypertrophic scarring and keloid scarring are frequently observed in clinical practice. Hypertrophic scarring occurs when the raised scar is limited to within the margin of the original injury, while keloid scarring is defined as a continually growing scar extending beyond the original injury. Scarring is usually the outcome of normal wound healing and can regress spontaneously after injury; however, hypertrophic scars caused by acute injuries such as severe burns usually result in life-long physical disability. Keloid scarring has been shown to have a strong genetic predisposition and rarely regresses (Tuan and Nichter 1998). Clinically, hypertrophic scarring and keloid scarring cause the patient physical and psychological distress and can significantly affect the patient's quality of life (Shih and Bayat 2010). Causally, both result from overgrowth of fibrous tissue characterized by excess production of ECM and collagen formation after skin damage due to injury or surgery (Seifert and Mrowietz 2009; Tziotzios et al. 2012). Many cytokines and growth factors are proposed to be involved in scar formation; TGF-β and its signaling pathway, which influences all aspects of wound healing, play key roles in regulating scar development (Seifert and Mrowietz 2009).

9.4.1 TGF-β and Scar Free Wounds

Previous studies have demonstrated that fetal wounds heal without scar formation in humans and various experimental animal models (Bullard et al. 2003). Compared to adult wounds, fetal wounds are characterized by fine and reticular collagen deposition with less cross-linking, less inflammation, higher hyaluronic acid concentration, and a greater ratio of collagen type III to type I (Namazi et al. 2011). Multiple differences in molecules and gene expression have been identified between fetal wounds and adult wounds (Larson et al. 2010; Namazi et al. 2011). Among them, the role of TGF-β has been extensively studied. The expression of TGF-β1 and TGF-β2 is increased in adult wounds while levels of TGF-β1 and TGF-β2 are low in fetal wounding mouse models (Nath et al. 1994; Sullivan et al. 1995; Whitby and Ferguson 1991). Lack of TGF-β1 has also been observed in fetal platelets (Olutoye et al. 1996). To verify the pathological role of TGF-β1 and TGF-β2 in scar formation, TGF-β1 and TGF-β2 neutralizing antibodies were used in adult rat wounds, which resulted in reduced monocyte and macrophage profile, neovascularization, fibronectin, collagen III and collagen I deposition in the early stages of wound healing. Only given a single antibody to either TGF-β1 or TGF-β2 did the wound show less or no relieved scarring (Shah et al. 1995). In addition, some studies indicated that scar-less wounds have significantly increased levels of TGF-β3, while TGF-β1 levels remained unchanged (Eslami et al. 2009; Hsu et al. 2001). Administration of the TGF-β3 peptide to the wounded area exhibited an antifibrotic effect (Shah et al. 1995). Those studies clearly demonstrate isoform-specific differences of TGF-β playing different roles in wound healing and cutaneous scarring. Although several other reports have shown increased TGF-β1 and TGF-β2 in fetal wounds after acute wounding, these increases clear rapidly. This transient induction of TGF-β1 and TGF-β2 in fetal wounds might not be enough to induce inflammation (Cowin et al. 2001; Martin et al. 1993; Soo et al. 2003). Moreover, recent studies indicate that fetal fibroblasts do not exhibit TGF-β1-induced collagen production when compared with their mature counterparts (Rolfe et al. 2007). Thus, the difference in response to TGF-β stimuli between fetal fibroblasts and mature fibroblasts may also contribute to scar-free fetal wounds.

9.4.2 *TGF-β and Cutaneous Scarring*

The mechanism of keloid formation is still poorly understood, and no effective therapy for keloids exists. As studies have shown that TGF-β has a potent role in stimulating the proliferation of fibroblasts and causing inflammation, activation of TGF-β signaling has been postulated to play a central role in the pathogenesis of keloids. Compared to normal skin, keloid-derived fibroblasts have high expression of TGF-β1 and TGF-β2 proteins, but expression of TGF-β3 remains unchanged (Lee et al. 1999). Both TGF-β1 and VEGF expressions were increased in keloid fibroblasts and the upregulation of VEGF expression was dependent on TGF-β (Fujiwara et al. 2005). Furthermore, upregulation of TGF-β1, TGF-βRI, and TGF-βRII has been shown in burn-induced hypertrophic scar tissue in vivo (Ghahary et al. 1993; Schmid et al. 1998; Zhang et al. 1995). In addition, high levels of TGF-β signaling Smads including Smad2, 3, 4, TGF-βRI and TGF-βRII proteins were reported in keloid and hypertrophic scar tissue, and keloid-derived fibroblasts (Phan et al. 2005;

Tsujita-Kyutoku et al. 2005). In contrast, inhibitory Smads including Smad6 and Smad7 have been shown to decrease in keloid tissue (Yu et al. 2006). In vitro studies showed that TGF-β1 increases collagen production by keloid fibroblasts and promotes proliferation and migration of keloid fibroblasts (Bettinger et al. 1996; Wu et al. 2012). These results indicated the pathogenic role of TGF-β signaling in cutaneous scar formation (Mrowietz and Seifert 2009; Seifert and Mrowietz 2009). Ideally, accelerated wound healing without scars or with minimal scarring is the aim for perfect regeneration in adult skin injuries. TGF-β is produced by many cellular components involved in wound healing, including keratinocytes, platelets, and macrophages, and therefore production of TGF-β increases after injury (Singer and Clark 1999). However, the controversy surrounding the role of TGF-β in wound healing has existed for a long time and is still not resolved. Analysis identified that TGF-β3 in serum plays a crucial role in mediating keratinocyte migration and inhibiting dermal fibroblast migration, implying TGF-β3 is key factor initiating re-epithelialization at early stages of wound healing as keratinocytes are exposed to serum due to skin damage (Bandyopadhyay et al. 2006; Henry et al. 2003). In contrast, transgenic mice overexpressing the TGF-β1 transgene in keratinocytes exhibited delayed healing after burn injury or excisional wounding due to delayed re-epithelialization and significant skin inflammation (Chan et al. 2002; Wang et al. 2006; Yang et al. 2001). Consistently, TGF-β1 knockout mice showed accelerated re-epithelialization during incisional wound repair in comparison with wild-type mice (Koch et al. 2000), and transgenic mice overexpressing a dominant-negative TGF-βRII receptor or keratinocyte-specific deletion of TGF-βRII receptor also exhibited accelerated re-epithelialization in skin wounds (Amendt et al. 2002; Guasch et al. 2007). Smad3 knockout mice, in which TGF-β signaling is partially abolished, exhibited accelerated wound healing, featuring increased keratinocyte proliferation and migration, and reduced monocyte infiltration (Ashcroft et al. 1999). Overexpression of Smad2 in basal epidermal keratinocytes delayed wound healing due to a defect in basal keratinocyte migration (Hosokawa et al. 2005). These studies indicate that both ectopically and endogenously expressed TGF-β1 inhibits re-epithelialization and activation of TGF-β signaling in skin wounds might promote scar formation. Thus, a better wound healing outcome may be achieved by selectively blocking the negative effects of TGF-β1 (Massague 1999). Recently, we reported that transgenic mice undergoing temporal induction of Smad7, an antagonist of the TGF-β signaling pathway, in the epidermis during wound healing exhibited accelerated wound healing characterized by increased re-epithelialization and reduced inflammation, angiogenesis and the production of type I collagen in wound stroma. The mechanism involved in blocking TGF-β-induced epithelial proliferation arrest and inhibiting NFκB signaling in the epidermis by Smad7 (Han et al. 2011; Hong et al. 2007). These findings highlight the importance of future studies to evaluate whether temporal Smad7 application in skin wounds can be used as a therapeutic strategy for impaired wound healing related to defects in keratinocyte migration, excessive inflammation, and scarring.

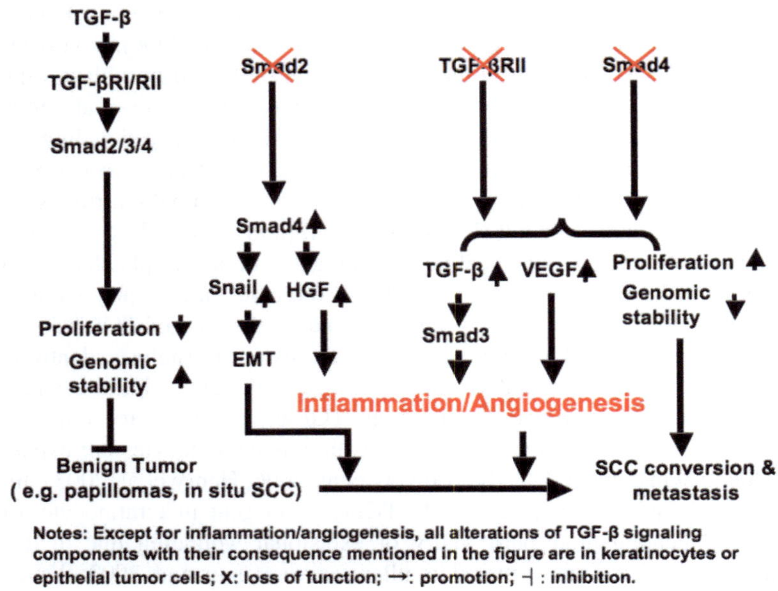

Fig. 9.1 Schematic summarizing the role of TGF-β signaling in skin carcinogenesis. TGF-β expression in skin suppresses tumor formation associated with inhibiting proliferation and maintaining genomic stability of keratinocytes. During tumor development, TGF-β-induced inflammation and angiogenesis accelerate malignant SCC conversion and metastasis. Smad2 loss in epithelia causes increased Snail and HGF by recruiting Smad4 to the promoter of Snail and HGF in tumor cells. These molecular changes lead to EMT and increasing angiogenesis, which promote SCC formation. TGF-βRII or Smad4-deficient cells express high level of TGF-β but cell itself loses the response to TGF-β-induced tumor suppression, and additional TGF-β increase stromal angiogenesis and inflammation, all of which accelerate SCC development and metastasis. Smad3 loss in skin abrogates TGF-β-induced inflammation during skin carcinogenesis

9.5 Concluding Remarks

Skin cancer and fibrotic disease are common skin diseases lacking efficient therapies. TGF-β has been implicated in the pathogenesis of both due to its differential regulation of epidermal keratinocytes, fibroblasts, and other dermal cell components. The role of TGF-β signaling in skin carcinogenesis is schematically summarized in Fig. 9.1. TGF-β first acts as a tumor suppressor in carcinogenesis as TGF-β inhibits normal keratinocyte proliferation and maintains genomic stability. However, TGF-β has also been demonstrated to play a crucial role in promoting malignant skin cancer conversion and cancer metastasis. Dominant-negative TGF-βRII transgenic mice and Smad4 deletion tumors have also increased TGF-β expression associated with tumor development and metastasis. Although SCC tumor cells express high levels of TGF-β, cancer cells lost their response to TGF-β-induced growth inhibition due to the deficient component of TGF-β signaling in tumor cells.

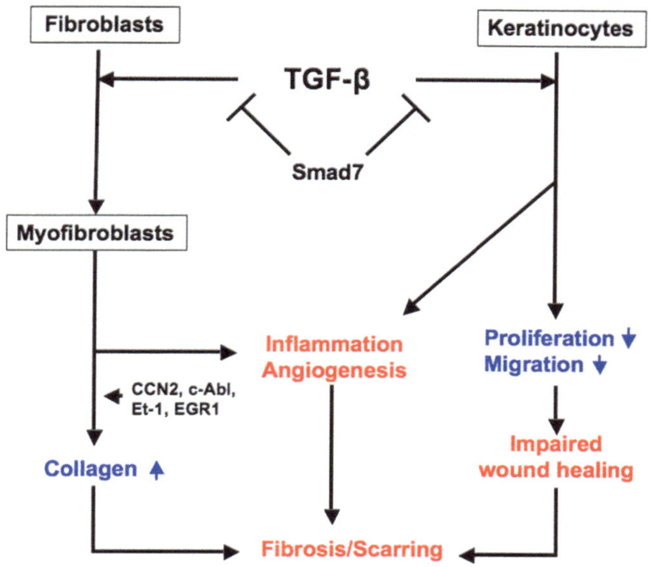

Fig. 9.2 Schematic summarizing the role of TGF-β signaling in fibrotic diseases. TGF-β acts a key player in mediating fibroblast activation to differentiate into myofibroblasts and stimulates the production of collagen via Smad3 and Smad1/Erk pathways. TGF-β also synergizes with other fibrotic factors such as CCN2, c-Ab1, Et-1, EGR1 in promoting collagen production. Excess collagen production and TGF-β-induced inflammation and angiogenesis contribute to skin fibrosis. In response to skin damage, TGF-β directly stimulates inflammation and angiogenesis in the stroma and inhibits keratinocyte migration via Smad2/3 signaling, which result in delayed wound healing and skin scarring. Overexpression of Smad7 or deletion of Smad3 reduces skin fibrosis/skin scarring and promotes wound healing through anti-inflammation, accelerating keratinocyte migration, and reducing collagen production

However, TGF-β-induced angiogenesis and inflammation in tumor stroma further accelerate tumor development and metastasis. Therefore, TGF-β contributes to either tumor inhibition or promotion depending on tumor stages. Figure 9.2 summarizes the role of TGF-β signaling in fibrotic diseases and cutaneous wound healing. In skin fibrotic diseases, it is well established that TGF-β plays a crucial role in fibrosis formation due to its ability to activate fibroblasts and stimulate collagen production in the extracellular matrix. Activation of TGF-β receptors, especially TGF-βRI, and abnormal expression of Smads have been documented in skin fibrosis. Based on wound healing studies in the fetus, the isoforms of TGF-β subtypes may have different roles in cutaneous scar formation. TGF-β1 is a negative factor in cutaneous wound healing and promotes skin scarring due to its action in fibroblast activation and induction of inflammation. Blocking TGF-β signaling may lead to novel strategies for treating impaired wound and skin fibrotic diseases in humans.

Acknowledgments The original work from the Wang laboratory is supported by NIH grants CA79998, CA87849 and AR061792. The authors thank Pamela Garl for carefully proofreading this chapter.

References

Akhurst RJ, Balmain A (1999) Genetic events and the role of TGF β in epithelial tumour progression. J Pathol 187:82–90
Akhurst RJ, Fee F, Balmain A (1988) Localized production of TGF-β mRNA in tumour promoter-stimulated mouse epidermis. Nature 331:363–365
Amendt C, Schirmacher P, Weber H, Blessing M (1998) Expression of a dominant negative type II TGF-β receptor in mouse skin results in an increase in carcinoma incidence and an acceleration of carcinoma development. Oncogene 17:25–34
Amendt C, Mann A, Schirmacher P, Blessing M (2002) Resistance of keratinocytes to TGFβ-mediated growth restriction and apoptosis induction accelerates re-epithelialization in skin wounds. J Cell Sci 115:2189–2198
Anzano MA, Roberts AB, Meyers CA, Komoriya A, Lamb LC, Smith JM, Sporn MB (1982) Synergistic interaction of two classes of transforming growth factors from murine sarcoma cells. Cancer Res 42:4776–4778
Artlett CM (2010) Animal models of scleroderma: fresh insights. Curr Opin Rheumatol 22:677–682
Asano Y, Ihn H, Yamane K, Kubo M, Tamaki K (2004) Impaired Smad7-Smurf-mediated negative regulation of TGF-β signaling in scleroderma fibroblasts. J Clin Invest 113:253–264
Ashcroft GS, Yang X, Glick AB, Weinstein M, Letterio JL, Mizel DE, Anzano M, Greenwell-Wild T, Wahl SM, Deng C, Roberts AB (1999) Mice lacking Smad3 show accelerated wound healing and an impaired local inflammatory response. Nat Cell Biol 1:260–266
Bae DS, Blazanin N, Licata M, Lee J, Glick AB (2009) Tumor suppressor and oncogene actions of TGFβ1 occur early in skin carcinogenesis and are mediated by Smad3. Mol Carcinog 48:441–453
Bandyopadhyay B, Fan J, Guan S, Li Y, Chen M, Woodley DT, Li W (2006) A "traffic control" role for TGFβ3: orchestrating dermal and epidermal cell motility during wound healing. J Cell Biol 172:1093–1105
Batteux F, Kavian N, Servettaz A (2011) New insights on chemically induced animal models of systemic sclerosis. Curr Opin Rheumatol 23:511–518
Bettinger DA, Yager DR, Diegelmann RF, Cohen IK (1996) The effect of TGF-β on keloid fibroblast proliferation and collagen synthesis. Plast Reconstr Surg 98:827–833
Beyer C, Distler O, Distler JH (2012) Innovative antifibrotic therapies in systemic sclerosis. Curr Opin Rheumatol 24:274–280
Bhattacharyya S, Ishida W, Wu M, Wilkes M, Mori Y, Hinchcliff M, Leof E, Varga J (2009) A non-Smad mechanism of fibroblast activation by transforming growth factor-β via c-Abl and Egr-1: selective modulation by imatinib mesylate. Oncogene 28:1285–1297
Bhattacharyya S, Wei J, Varga J (2012) Understanding fibrosis in systemic sclerosis: shifting paradigms, emerging opportunities. Nat Rev Rheumatol 8:42–54
Bonner JC (2004) Regulation of PDGF and its receptors in fibrotic diseases. Cytokine Growth Factor Rev 15:255–273
Bornstein S, Hoot K, Han GW, Lu SL, Wang XJ (2007) Distinct roles of individual Smads in skin carcinogenesis. Mol Carcinog 46:660–664
Bornstein S, White R, Malkoski S, Oka M, Han G, Cleaver T, Reh D, Andersen P, Gross N, Olson S, Deng C, Lu SL, Wang XJ (2009) Smad4 loss in mice causes spontaneous head and neck cancer with increased genomic instability and inflammation. J Clin Invest 119:3408–3419
Boukamp P, Petrussevska RT, Breitkreutz D, Hornung J, Markham A, Fusenig NE (1988) Normal keratinization in a spontaneously immortalized aneuploid human keratinocyte cell line. J Cell Biol 106:761–771
Bullard KM, Longaker MT, Lorenz HP (2003) Fetal wound healing: current biology. World J Surg 27:54–61
Chan T, Ghahary A, Demare J, Yang L, Iwashina T, Scott PG, Tredget EE (2002) Development, characterization, and wound healing of the keratin 14 promoted transforming growth factor-β1 transgenic mouse. Wound Repair Regen 10:177–187

Chang H, Huylebroeck D, Verschueren K, Guo Q, Matzuk MM, Zwijsen A (1999) Smad5 knockout mice die at mid-gestation due to multiple embryonic and extraembryonic defects. Development 126:1631–1642

Chen SJ, Yuan W, Mori Y, Levenson A, Trojanowska M, Varga J (1999) Stimulation of type I collagen transcription in human skin fibroblasts by TGF-β: involvement of Smad 3. J Invest Dermatol 112:49–57

Coffey RJ Jr, Sipes NJ, Bascom CC, Graves-Deal R, Pennington CY, Weissman BE, Moses HL (1988) Growth modulation of mouse keratinocytes by transforming growth factors. Cancer Res 48:1596–1602

Cowin AJ, Holmes TM, Brosnan P, Ferguson MW (2001) Expression of TGF-β and its receptors in murine fetal and adult dermal wounds. Eur J Dermatol 11:424–431

Cui W, Fowlis DJ, Cousins FM, Duffie E, Bryson S, Balmain A, Akhurst RJ (1995) Concerted action of TGF-β 1 and its type II receptor in control of epidermal homeostasis in transgenic mice. Genes Dev 9:945–955

Cui W, Fowlis DJ, Bryson S, Duffie E, Ireland H, Balmain A, Akhurst RJ (1996) TGFβ1 inhibits the formation of benign skin tumors, but enhances progression to invasive spindle carcinomas in transgenic mice. Cell 86:531–542

Datto MB, Frederick JP, Pan L, Borton AJ, Zhuang Y, Wang XF (1999) Targeted disruption of Smad3 reveals an essential role in transforming growth factor β-mediated signal transduction. Mol Cell Biol 19:2495–2504

Davies M, Prime SS, Eveson JW, Price N, Ganapathy A, D'Mello A, Paterson IC (2012) Transforming growth factor-β enhances invasion and metastasis in Ras-transfected human malignant epidermal keratinocytes. Int J Exp Pathol 93:148–156

Dennler S, Itoh S, Vivien D, ten Dijke P, Huet S, Gauthier JM (1998) Direct binding of Smad3 and Smad4 to critical TGF β-inducible elements in the promoter of human plasminogen activator inhibitor-type 1 gene. EMBO J 17:3091–3100

Denton CP, Zheng B, Evans LA, Shi-wen X, Ong VH, Fisher I, Lazaridis K, Abraham DJ, Black CM, de Crombrugghe B (2003) Fibroblast-specific expression of a kinase-deficient type II transforming growth factor β (TGFβ) receptor leads to paradoxical activation of TGFβ signaling pathways with fibrosis in transgenic mice. J Biol Chem 278:25109–25119

Denton CP, Lindahl GE, Khan K, Shiwen X, Ong VH, Gaspar NJ, Lazaridis K, Edwards DR, Leask A, Eastwood M, Leoni P, Renzoni EA, Bou Gharios G, Abraham DJ, Black CM (2005) Activation of key profibrotic mechanisms in transgenic fibroblasts expressing kinase-deficient type II transforming growth factor-{β} receptor (T{β}RII{Δ}k). J Biol Chem 280:16053–16065

Derynck R, Zhang Y (1996) Intracellular signalling: the mad way to do it. Curr Biol 6:1226–1229

Dlugosz A, Merlino G, Yuspa SH (2002) Progress in cutaneous cancer research. J Investig Dermatol Symp Proc 7:17–26

Dong C, Zhu S, Wang T, Yoon W, Li Z, Alvarez RJ, ten Dijke P, White B, Wigley FM, Goldschmidt-Clermont PJ (2002) Deficient Smad7 expression: a putative molecular defect in scleroderma. Proc Natl Acad Sci USA 99:3908–3913

Eslami A, Gallant-Behm CL, Hart DA, Wiebe C, Honardoust D, Gardner H, Hakkinen L, Larjava HS (2009) Expression of integrin αvβ6 and TGF-β in scarless vs scar-forming wound healing. J Histochem Cytochem 57:543–557

Filler RB, Roberts SJ, Girardi M (2007) Cutaneous two-stage chemical carcinogenesis. CSH Protoc 2007: pdb prot4837

Fowlis DJ, Flanders KC, Duffie E, Balmain A, Akhurst RJ (1992) Discordant transforming growth factor β 1 RNA and protein localization during chemical carcinogenesis of the skin. Cell Growth Differ 3:81–91

Fowlis DJ, Cui W, Johnson SA, Balmain A, Akhurst RJ (1996) Altered epidermal cell growth control in vivo by inducible expression of transforming growth factor β 1 in the skin of transgenic mice. Cell Growth Differ 7:679–687

Frank S, Madlener M, Werner S (1996) Transforming growth factors β1, β2, and β3 and their receptors are differentially regulated during normal and impaired wound healing. J Biol Chem 271:10188–10193

Fujiki H, Suganuma M, Yoshizawa S, Kanazawa H, Sugimura T, Manam S, Kahn SM, Jiang W, Hoshina S, Weinstein IB (1989) Codon 61 mutations in the c-Harvey-ras gene in mouse skin tumors induced by 7,12-dimethylbenz[a]anthracene plus okadaic acid class tumor promoters. Mol Carcinog 2:184–187

Fujiwara M, Muragaki Y, Ooshima A (2005) Upregulation of transforming growth factor-β1 and vascular endothelial growth factor in cultured keloid fibroblasts: relevance to angiogenic activity. Arch Dermatol Res 297:161–169

Fusenig NE, Boukamp P (1998) Multiple stages and genetic alterations in immortalization, malignant transformation, and tumor progression of human skin keratinocytes. Mol Carcinog 23: 144–158

Gabrielli A, Di Loreto C, Taborro R, Candela M, Sambo P, Nitti C, Danieli MG, DeLustro F, Dasch JR, Danieli G (1993) Immunohistochemical localization of intracellular and extracellular associated TGF β in the skin of patients with systemic sclerosis (scleroderma) and primary Raynaud's phenomenon. Clin Immunol Immunopathol 68:340–349

Gabrielli A, Avvedimento EV, Krieg T (2009) Scleroderma. N Engl J Med 360:1989–2003

Ganapathy A, Paterson IC, Prime SS, Eveson JW, Pring M, Price N, Threadgold SP, Davies M (2010) TGF-β inhibits metastasis in late stage human squamous cell carcinoma of the skin by a mechanism that does not involve Id1. Cancer Lett 298:107–118

Gay S, Trabandt A, Moreland LW, Gay RE (1992) Growth factors, extracellular matrix, and oncogenes in scleroderma. Arthritis Rheum 35:304–310

Ghahary A, Shen YJ, Scott PG, Gong Y, Tredget EE (1993) Enhanced expression of mRNA for transforming growth factor-β, type I and type III procollagen in human post-burn hypertrophic scar tissues. J Lab Clin Med 122:465–473

Glick AB, Sporn MB, Yuspa SH (1991) Altered regulation of TGF-β 1 and TGF-α in primary keratinocytes and papillomas expressing v-Ha-ras. Mol Carcinog 4:210–219

Glick AB, Lee MM, Darwiche N, Kulkarni AB, Karlsson S, Yuspa SH (1994) Targeted deletion of the TGF-β 1 gene causes rapid progression to squamous cell carcinoma. Genes Dev 8: 2429–2440

Glick A, Popescu N, Alexander V, Ueno H, Bottinger E, Yuspa SH (1999) Defects in transforming growth factor-β signaling cooperate with a Ras oncogene to cause rapid aneuploidy and malignant transformation of mouse keratinocytes. Proc Natl Acad Sci USA 96:14949–14954

Go C, Li P, Wang XJ (1999) Blocking transforming growth factor β signaling in transgenic epidermis accelerates chemical carcinogenesis: a mechanism associated with increased angiogenesis. Cancer Res 59:2861–2868

Gold LI, Jussila T, Fusenig NE, Stenback F (2000) TGF-β isoforms are differentially expressed in increasing malignant grades of HaCaT keratinocytes, suggesting separate roles in skin carcinogenesis. J Pathol 190:579–588

Gruschwitz M, Muller PU, Sepp N, Hofer E, Fontana A, Wick G (1990) Transcription and expression of transforming growth factor type β in the skin of progressive systemic sclerosis: a mediator of fibrosis? J Invest Dermatol 94:197–203

Guasch G, Schober M, Pasolli HA, Conn EB, Polak L, Fuchs E (2007) Loss of TGFβ signaling destabilizes homeostasis and promotes squamous cell carcinomas in stratified epithelia. Cancer Cell 12:313–327

Hahn SA, Schutte M, Hoque AT, Moskaluk CA, da Costa LT, Rozenblum E, Weinstein CL, Fischer A, Yeo CJ, Hruban RH, Kern SE (1996) DPC4, a candidate tumor suppressor gene at human chromosome 18q21.1. Science 271:350–353

Han G, Wang XJ (2011) Roles of TGFβ signaling Smads in squamous cell carcinoma. Cell Biosci 1:41

Han G, Lu SL, Li AG, He W, Corless CL, Kulesz-Martin M, Wang XJ (2005) Distinct mechanisms of TGF-β1-mediated epithelial-to-mesenchymal transition and metastasis during skin carcinogenesis. J Clin Invest 115:1714–1723

Han G, Li F, Ten Dijke P, Wang XJ (2011) Temporal smad7 transgene induction in mouse epidermis accelerates skin wound healing. Am J Pathol 179:1768–1779

Henry G, Li W, Garner W, Woodley DT (2003) Migration of human keratinocytes in plasma and serum and wound re-epithelialisation. Lancet 361:574–576

Heyer J, Escalante-Alcalde D, Lia M, Boettinger E, Edelmann W, Stewart CL, Kucherlapati R (1999) Postgastrulation Smad2-deficient embryos show defects in embryo turning and anterior morphogenesis. Proc Natl Acad Sci USA 96:12595–12600

Higley H, Persichitte K, Chu S, Waegell W, Vancheeswaran R, Black C (1994) Immunocytochemical localization and serologic detection of transforming growth factor β 1. Association with type I procollagen and inflammatory cell markers in diffuse and limited systemic sclerosis, morphea, and Raynaud's phenomenon. Arthritis Rheum 37:278–288

Hong S, Lim S, Li AG, Lee C, Lee YS, Lee EK, Park SH, Wang XJ, Kim SJ (2007) Smad7 binds to the adaptors TAB2 and TAB3 to block recruitment of the kinase TAK1 to the adaptor TRAF2. Nat Immunol 8:504–513

Hoot KE, Lighthall J, Han G, Lu SL, Li A, Ju W, Kulesz-Martin M, Bottinger E, Wang XJ (2008) Keratinocyte-specific Smad2 ablation results in increased epithelial-mesenchymal transition during skin cancer formation and progression. J Clin Invest 118:2722–2732

Hoot KE, Oka M, Han G, Bottinger E, Zhang Q, Wang XJ (2010) HGF upregulation contributes to angiogenesis in mice with keratinocyte-specific Smad2 deletion. J Clin Invest 120:3606–3616

Hosokawa R, Urata MM, Ito Y, Bringas P Jr, Chai Y (2005) Functional significance of Smad2 in regulating basal keratinocyte migration during wound healing. J Invest Dermatol 125: 1302–1309

Hoyles RK, Khan K, Shiwen X, Howat SL, Lindahl GE, Leoni P, du Bois RM, Wells AU, Black CM, Abraham DJ, Denton CP (2008) Fibroblast-specific perturbation of transforming growth factor β signaling provides insight into potential pathogenic mechanisms of scleroderma-associated lung fibrosis: exaggerated response to alveolar epithelial injury in a novel mouse model. Arthritis Rheum 58:1175–1188

Hsu M, Peled ZM, Chin GS, Liu W, Longaker MT (2001) Ontogeny of expression of transforming growth factor-β 1 (TGF-β 1), TGF-β 3, and TGF-β receptors I and II in fetal rat fibroblasts and skin. Plast Reconstr Surg 107:1787–1794; discussion 1795–1786

Iwamoto N, Distler JH, Distler O (2011) Tyrosine kinase inhibitors in the treatment of systemic sclerosis: from animal models to clinical trials. Curr Rheumatol Rep 13:21–27

Jinnin M (2010) Mechanisms of skin fibrosis in systemic sclerosis. J Dermatol 37:11–25

Jinnin M, Ihn H, Mimura Y, Asano Y, Tamaki K (2006) Potential regulatory elements of the constitutive up-regulated α2(I) collagen gene in scleroderma dermal fibroblasts. Biochem Biophys Res Commun 343:904–909

Kawakami T, Ihn H, Xu W, Smith E, LeRoy C, Trojanowska M (1998) Increased expression of TGF-β receptors by scleroderma fibroblasts: evidence for contribution of autocrine TGF-β signaling to scleroderma phenotype. J Invest Dermatol 110:47–51

Kleiter I, Song J, Lukas D, Hasan M, Neumann B, Croxford AL, Pedre X, Hovelmeyer N, Yogev N, Mildner A, Prinz M, Wiese E, Reifenberg K, Bittner S, Wiendl H, Steinman L, Becker C, Bogdahn U, Neurath MF, Steinbrecher A, Waisman A (2010) Smad7 in T cells drives T helper 1 responses in multiple sclerosis and experimental autoimmune encephalomyelitis. Brain 133:1067–1081

Koch RM, Roche NS, Parks WT, Ashcroft GS, Letterio JJ, Roberts AB (2000) Incisional wound healing in transforming growth factor-β1 null mice. Wound Repair Regen 8:179–191

Kopp J, Preis E, Said H, Hafemann B, Wickert L, Gressner AM, Pallua N, Dooley S (2005) Abrogation of transforming growth factor-β signaling by SMAD7 inhibits collagen gel contraction of human dermal fibroblasts. J Biol Chem 280:21570–21576

Kreuter A, Hyun J, Skrygan M, Sommer A, Tomi NS, Breuckmann F, Altmeyer P, Gambichler T (2006) Ultraviolet A1 phototherapy decreases inhibitory SMAD7 gene expression in localized scleroderma. Arch Dermatol Res 298:265–272

Kubo M, Ihn H, Yamane K, Tamaki K (2001) Up-regulated expression of transforming growth factor β receptors in dermal fibroblasts in skin sections from patients with localized scleroderma. Arthritis Rheum 44:731–734

Kubo M, Ihn H, Yamane K, Tamaki K (2002) Upregulated expression of transforming growth factor-β receptors in dermal fibroblasts of skin sections from patients with systemic sclerosis. J Rheumatol 29:2558–2564

Kulozik M, Hogg A, Lankat-Buttgereit B, Krieg T (1990) Co-localization of transforming growth factor β 2 with α 1(I) procollagen mRNA in tissue sections of patients with systemic sclerosis. J Clin Invest 86:917–922

Lakos G, Takagawa S, Chen SJ, Ferreira AM, Han G, Masuda K, Wang XJ, DiPietro LA, Varga J (2004) Targeted disruption of TGF-β/Smad3 signaling modulates skin fibrosis in a mouse model of scleroderma. Am J Pathol 165:203–217

Larson BJ, Longaker MT, Lorenz HP (2010) Scarless fetal wound healing: a basic science review. Plast Reconstr Surg 126:1172–1180

Lee TY, Chin GS, Kim WJ, Chau D, Gittes GK, Longaker MT (1999) Expression of transforming growth factor β 1, 2, and 3 proteins in keloids. Ann Plast Surg 43:179–184

Li W, Qiao W, Chen L, Xu X, Yang X, Li D, Li C, Brodie SG, Meguid MM, Hennighausen L, Deng CX (2003) Squamous cell carcinoma and mammary abscess formation through squamous metaplasia in Smad4/Dpc4 conditional knockout mice. Development 130:6143–6153

Li AG, Lu SL, Zhang MX, Deng C, Wang XJ (2004a) Smad3 knockout mice exhibit a resistance to skin chemical carcinogenesis. Cancer Res 64:7836–7845

Li AG, Wang D, Feng XH, Wang XJ (2004b) Latent TGFβ1 overexpression in keratinocytes results in a severe psoriasis-like skin disorder. EMBO J 23:1770–1781

Li AG, Lu SL, Han G, Hoot KE, Wang XJ (2006) Role of TGFβ in skin inflammation and carcinogenesis. Mol Carcinog 45:389–396

Martin P (1997) Wound healing—aiming for perfect skin regeneration. Science 276:75–81

Martin P, Dickson MC, Millan FA, Akhurst RJ (1993) Rapid induction and clearance of TGFβ 1 is an early response to wounding in the mouse embryo. Dev Genet 14:225–238

Massague J (1996) TGFβ signaling: receptors, transducers, and Mad proteins. Cell 85:947–950

Massague J (1999) Wounding Smad. Nat Cell Biol 1:E117–E119

Millet C, Zhang YE (2007) Roles of Smad3 in TGF-β signaling during carcinogenesis. Crit Rev Eukaryot Gene Expr 17:281–293

Mori Y, Chen SJ, Varga J (2003) Expression and regulation of intracellular SMAD signaling in scleroderma skin fibroblasts. Arthritis Rheum 48:1964–1978

Mrowietz U, Seifert O (2009) Keloid scarring: new treatments ahead. Actas Dermosifiliogr 100(Suppl 2):75–83

Nakao A, Imamura T, Souchelnytskyi S, Kawabata M, Ishisaki A, Oeda E, Tamaki K, Hanai J, Heldin CH, Miyazono K, ten Dijke P (1997) TGF-β receptor-mediated signalling through Smad2, Smad3 and Smad4. EMBO J 16:5353–5362

Namazi MR, Fallahzadeh MK, Schwartz RA (2011) Strategies for prevention of scars: what can we learn from fetal skin? Int J Dermatol 50:85–93

Nath RK, LaRegina M, Markham H, Ksander GA, Weeks PM (1994) The expression of transforming growth factor type β in fetal and adult rabbit skin wounds. J Pediatr Surg 29:416–421

Nomura M, Li E (1998) Smad2 role in mesoderm formation, left-right patterning and craniofacial development. Nature 393:786–790

Olutoye OO, Yager DR, Cohen IK, Diegelmann RF (1996) Lower cytokine release by fetal porcine platelets: a possible explanation for reduced inflammation after fetal wounding. J Pediatr Surg 31:91–95

Oshima M, Oshima H, Taketo MM (1996) TGF-β receptor type II deficiency results in defects of yolk sac hematopoiesis and vasculogenesis. Dev Biol 179:297–302

Owens P, Bazzi H, Engelking E, Han G, Christiano AM, Wang XJ (2008) Smad4-dependent desmoglein-4 expression contributes to hair follicle integrity. Dev Biol 322:156–166

Owens P, Engelking E, Han G, Haeger SM, Wang XJ (2010) Epidermal Smad4 deletion results in aberrant wound healing. Am J Pathol 176:122–133

Pannu J, Gore-Hyer E, Yamanaka M, Smith EA, Rubinchik S, Dong JY, Jablonska S, Blaszczyk M, Trojanowska M (2004) An increased transforming growth factor β receptor type I:type II ratio contributes to elevated collagen protein synthesis that is resistant to inhibition via a kinase-deficient transforming growth factor β receptor type II in scleroderma. Arthritis Rheum 50:1566–1577

Pannu J, Nakerakanti S, Smith E, ten Dijke P, Trojanowska M (2007) Transforming growth factor-β receptor type I-dependent fibrogenic gene program is mediated via activation of Smad1 and ERK1/2 pathways. J Biol Chem 282:10405–10413
Pannu J, Asano Y, Nakerakanti S, Smith E, Jablonska S, Blaszczyk M, ten Dijke P, Trojanowska M (2008) Smad1 pathway is activated in systemic sclerosis fibroblasts and is targeted by imatinib mesylate. Arthritis Rheum 58:2528–2537
Patamalai B, Burow DL, Gimenez-Conti I, Zenklusen JC, Conti CJ, Klein-Szanto AJ, Fischer SM (1994) Altered expression of transforming growth factor-β 1 mRNA and protein in mouse skin carcinogenesis. Mol Carcinog 9:220–229
Phan TT, Lim IJ, Aalami O, Lorget F, Khoo A, Tan EK, Mukhopadhyay A, Longaker MT (2005) Smad3 signalling plays an important role in keloid pathogenesis via epithelial-mesenchymal interactions. J Pathol 207:232–242
Qiao W, Li AG, Owens P, Xu X, Wang XJ, Deng CX (2006) Hair follicle defects and squamous cell carcinoma formation in Smad4 conditional knockout mouse skin. Oncogene 25:207–217
Quan T, He T, Kang S, Voorhees JJ, Fisher GJ (2002) Ultraviolet irradiation alters transforming growth factor β/smad pathway in human skin in vivo. J Invest Dermatol 119:499–506
Querfeld C, Eckes B, Huerkamp C, Krieg T, Sollberg S (1999) Expression of TGF-β 1, -β 2 and -β 3 in localized and systemic scleroderma. J Dermatol Sci 21:13–22
Roberts AB, Anzano MA, Wakefield LM, Roche NS, Stern DF, Sporn MB (1985) Type β transforming growth factor: a bifunctional regulator of cellular growth. Proc Natl Acad Sci USA 82:119–123
Rolfe KJ, Richardson J, Vigor C, Irvine LM, Grobbelaar AO, Linge C (2007) A role for TGF-β1-induced cellular responses during wound healing of the non-scarring early human fetus? J Invest Dermatol 127:2656–2667
Roop DR, Lowy DR, Tambourin PE, Strickland J, Harper JR, Balaschak M, Spangler EF, Yuspa SH (1986) An activated Harvey ras oncogene produces benign tumours on mouse epidermal tissue. Nature 323:822–824
Rosenbloom J, Castro SV, Jimenez SA (2010) Narrative review: fibrotic diseases: cellular and molecular mechanisms and novel therapies. Ann Intern Med 152:159–166
Rudnicka L, Varga J, Christiano AM, Iozzo RV, Jimenez SA, Uitto J (1994) Elevated expression of type VII collagen in the skin of patients with systemic sclerosis. Regulation by transforming growth factor-β. J Clin Invest 93:1709–1715
Schmid P, Itin P, Cherry G, Bi C, Cox DA (1998) Enhanced expression of transforming growth factor-β type I and type II receptors in wound granulation tissue and hypertrophic scar. Am J Pathol 152:485–493
Seifert O, Mrowietz U (2009) Keloid scarring: bench and bedside. Arch Dermatol Res 301:259–272
Sellheyer K, Bickenbach JR, Rothnagel JA, Bundman D, Longley MA, Krieg T, Roche NS, Roberts AB, Roop DR (1993) Inhibition of skin development by overexpression of transforming growth factor β 1 in the epidermis of transgenic mice. Proc Natl Acad Sci USA 90:5237–5241
Shah M, Foreman DM, Ferguson MW (1995) Neutralisation of TGF-β 1 and TGF-β 2 or exogenous addition of TGF-β 3 to cutaneous rat wounds reduces scarring. J Cell Sci 108(Pt 3): 985–1002
Shih B, Bayat A (2010) Genetics of keloid scarring. Arch Dermatol Res 302:319–339
Shipley GD, Pittelkow MR, Wille JJ Jr, Scott RE, Moses HL (1986) Reversible inhibition of normal human prokeratinocyte proliferation by type β transforming growth factor-growth inhibitor in serum-free medium. Cancer Res 46:2068–2071
Singer AJ, Clark RA (1999) Cutaneous wound healing. N Engl J Med 341:738–746
Sirard C, de la Pompa JL, Elia A, Itie A, Mirtsos C, Cheung A, Hahn S, Wakeham A, Schwartz L, Kern SE, Rossant J, Mak TW (1998) The tumor suppressor gene Smad4/Dpc4 is required for gastrulation and later for anterior development of the mouse embryo. Genes Dev 12:107–119
Sonnylal S, Denton CP, Zheng B, Keene DR, He R, Adams HP, Vanpelt CS, Geng YJ, Deng JM, Behringer RR, de Crombrugghe B (2007) Postnatal induction of transforming growth factor β signaling in fibroblasts of mice recapitulates clinical, histologic, and biochemical features of scleroderma. Arthritis Rheum 56:334–344

Soo C, Beanes SR, Hu FY, Zhang X, Dang C, Chang G, Wang Y, Nishimura I, Freymiller E, Longaker MT, Lorenz HP, Ting K (2003) Ontogenetic transition in fetal wound transforming growth factor-β regulation correlates with collagen organization. Am J Pathol 163:2459–2476

Sporn MB (1999) TGF-β: 20 years and counting. Microbes Infect 1:1251–1253

Sporn MB (2006) The early history of TGF-β, and a brief glimpse of its future. Cytokine Growth Factor Rev 17:3–7

Sullivan KM, Lorenz HP, Meuli M, Lin RY, Adzick NS (1995) A model of scarless human fetal wound repair is deficient in transforming growth factor β. J Pediatr Surg 30:198–202; discussion 202–193

Tannehill-Gregg SH, Kusewitt DF, Rosol TJ, Weinstein M (2004) The roles of Smad2 and Smad3 in the development of chemically induced skin tumors in mice. Vet Pathol 41:278–282

Teng Y, Sun AN, Pan XC, Yang G, Yang LL, Wang MR, Yang X (2006) Synergistic function of Smad4 and PTEN in suppressing forestomach squamous cell carcinoma in the mouse. Cancer Res 66:6972–6981

Thiagalingam S, Lengauer C, Leach FS, Schutte M, Hahn SA, Overhauser J, Willson JK, Markowitz S, Hamilton SR, Kern SE, Kinzler KW, Vogelstein B (1996) Evaluation of candidate tumour suppressor genes on chromosome 18 in colorectal cancers. Nat Genet 13:343–346

Tremblay KD, Dunn NR, Robertson EJ (2001) Mouse embryos lacking Smad1 signals display defects in extra-embryonic tissues and germ cell formation. Development 128:3609–3621

Tsujita-Kyutoku M, Uehara N, Matsuoka Y, Kyutoku S, Ogawa Y, Tsubura A (2005) Comparison of transforming growth factor-β/Smad signaling between normal dermal fibroblasts and fibroblasts derived from central and peripheral areas of keloid lesions. In Vivo 19:959–963

Tuan TL, Nichter LS (1998) The molecular basis of keloid and hypertrophic scar formation. Mol Med Today 4:19–24

Tucker RF, Shipley GD, Moses HL, Holley RW (1984) Growth inhibitor from BSC-1 cells closely related to platelet type β transforming growth factor. Science 226:705–707

Tziotzios C, Profyris C, Sterling J (2012) Cutaneous scarring: pathophysiology, molecular mechanisms, and scar reduction therapeutics Part II. Strategies to reduce scar formation after dermatologic procedures. J Am Acad Dermatol 66:13–24

Varga J, Abraham D (2007) Systemic sclerosis: a prototypic multisystem fibrotic disorder. J Clin Invest 117:557–567

Varga J, Jimenez SA (1986) Stimulation of normal human fibroblast collagen production and processing by transforming growth factor-β. Biochem Biophys Res Commun 138:974–980

Vijayachandra K, Lee J, Glick AB (2003) Smad3 regulates senescence and malignant conversion in a mouse multistage skin carcinogenesis model. Cancer Res 63:3447–3452

Vuorio T, Kahari VM, Black C, Vuorio E (1991) Expression of osteonectin, decorin, and transforming growth factor-β 1 genes in fibroblasts cultured from patients with systemic sclerosis and morphea. J Rheumatol 18:247–251

Waldrip WR, Bikoff EK, Hoodless PA, Wrana JL, Robertson EJ (1998) Smad2 signaling in extraembryonic tissues determines anterior-posterior polarity of the early mouse embryo. Cell 92:797–808

Wang XJ (2001) Role of TGFβ signaling in skin carcinogenesis. Microsc Res Tech 52:420–429

Wang XJ, Greenhalgh DA, Bickenbach JR, Jiang A, Bundman DS, Krieg T, Derynck R, Roop DR (1997) Expression of a dominant-negative type II transforming growth factor β (TGF-β) receptor in the epidermis of transgenic mice blocks TGF-β-mediated growth inhibition. Proc Natl Acad Sci USA 94:2386–2391

Wang XJ, Han G, Owens P, Siddiqui Y, Li AG (2006) Role of TGF β-mediated inflammation in cutaneous wound healing. J Investig Dermatol Symp Proc 11:112–117

Weeks BH, He W, Olson KL, Wang XJ (2001) Inducible expression of transforming growth factor β1 in papillomas causes rapid metastasis. Cancer Res 61:7435–7443

Wei J, Bhattacharyya S, Tourtellotte WG, Varga J (2011) Fibrosis in systemic sclerosis: emerging concepts and implications for targeted therapy. Autoimmun Rev 10:267–275

Weinstein M, Yang X, Li C, Xu X, Gotay J, Deng CX (1998) Failure of egg cylinder elongation and mesoderm induction in mouse embryos lacking the tumor suppressor smad2. Proc Natl Acad Sci USA 95:9378–9383

Whitby DJ, Ferguson MW (1991) Immunohistochemical localization of growth factors in fetal wound healing. Dev Biol 147:207–215
Wu CS, Wu PH, Fang AH, Lan CC (2012) FK506 inhibits the enhancing effects of transforming growth factor-β1 on collagen expression and TGF-β/Smad signaling in keloid fibroblasts: implication for new therapeutic approach. Br J Dermatol 167(3):532–541
Wynn TA (2007) Common and unique mechanisms regulate fibrosis in various fibroproliferative diseases. J Clin Invest 117:524–529
Yamamoto T (2010) Animal model of systemic sclerosis. J Dermatol 37:26–41
Yang X, Li C, Xu X, Deng C (1998) The tumor suppressor SMAD4/DPC4 is essential for epiblast proliferation and mesoderm induction in mice. Proc Natl Acad Sci USA 95:3667–3672
Yang X, Letterio JJ, Lechleider RJ, Chen L, Hayman R, Gu H, Roberts AB, Deng C (1999) Targeted disruption of SMAD3 results in impaired mucosal immunity and diminished T cell responsiveness to TGF-β. EMBO J 18:1280–1291
Yang L, Chan T, Demare J, Iwashina T, Ghahary A, Scott PG, Tredget EE (2001) Healing of burn wounds in transgenic mice overexpressing transforming growth factor-β 1 in the epidermis. Am J Pathol 159:2147–2157
Yang L, Mao C, Teng Y, Li W, Zhang J, Cheng X, Li X, Han X, Xia Z, Deng H, Yang X (2005) Targeted disruption of Smad4 in mouse epidermis results in failure of hair follicle cycling and formation of skin tumors. Cancer Res 65:8671–8678
Yu H, Bock O, Bayat A, Ferguson MW, Mrowietz U (2006) Decreased expression of inhibitory SMAD6 and SMAD7 in keloid scarring. J Plastic Reconstr Aesthet Surg 59:221–229
Yuspa SH (1986) Cutaneous chemical carcinogenesis. J Am Acad Dermatol 15:1031–1044
Zhang K, Garner W, Cohen L, Rodriguez J, Phan S (1995) Increased types I and III collagen and transforming growth factor-β 1 mRNA and protein in hypertrophic burn scar. J Invest Dermatol 104:750–754
Zhang W, Ou J, Inagaki Y, Greenwel P, Ramirez F (2000) Synergistic cooperation between Sp1 and Smad3/Smad4 mediates transforming growth factor β1 stimulation of α 2(I)-collagen (COL1A2) transcription. J Biol Chem 275:39237–39245
Zhu Y, Richardson JA, Parada LF, Graff JM (1998) Smad3 mutant mice develop metastatic colorectal cancer. Cell 94:703–714
Zhu H, Li Y, Qu S, Luo H, Zhou Y, Wang Y, Zhao H, You Y, Xiao X, Zuo X (2012) MicroRNA expression abnormalities in limited cutaneous scleroderma and diffuse cutaneous scleroderma. J Clin Immunol 32:514–522

Chapter 10
The Role of TGF-β in Cutaneous Melanoma Biology

Delphine Javelaud and Alain Mauviel

Abstract Ample evidence indicates that transforming growth factor-β (TGF-β) signaling plays a major role at various stages of carcinogenesis. While it may represent a tumor suppressor pathway at early stages of cancer progression, essentially due to its cytostatic activity, it has also become clear that TGF-β may act as a potent tumor promoter via both autocrine and paracrine mechanisms: TGF-β enhances tumor cell migration and homing to various metastatic sites, allows tumor escape from the immune system and promotes peri-tumoral vasculogenesis. This chapter reviews the current literature on the implication of TGF-β signaling in melanoma.

Keywords Invasion • Melanoma • Metastasis • TGF-β

10.1 Introduction

Melanomas arise from the transformation of melanocytes, cells that derive from melanoblasts in the neural crest that migrate and differentiate before localizing essentially in the hair follicles and in the epidermis. They produce melanin and are responsible for the pigmentation of hair and skin. Premalignant melanocytic lesions include clinicopathologic entities such as lentigo, nevi, or dysplastic nevi (Chin et al. 2006) Melanoma is the most lethal skin cancer and its incidence increases faster than any other malignancy. Although early stage melanoma patients can be treated successfully by surgical resection of the primary tumor, the prognosis for metastatic disease is dismal with an overall 5-year mortality rate of 90 %: melanomas are highly aggressive tumors that exhibit strong metastatic potential together

D. Javelaud (✉) • A. Mauviel
Institut Curie, Team "TGF-β and Oncogenesis", INSERM U1021, CNRS UMR 3347, University Center, Building 110, Orsay, France
e-mail: delphine.javelaud@curie.fr; alain.mauviel@curie.fr

A. Moustakas and K. Miyazawa (eds.), *TGF-β in Human Disease*,
DOI 10.1007/978-4-431-54409-8_10, © Springer Japan 2013

with high resistance to common anticancer treatments, including chemotherapy, interleukin-2-based immunotherapy, and a combination of chemotherapy and immunotherapy.

The key malignant transition is thought to occur from radial growth phase (RGP) to vertical growth phase (VGP) of primary melanoma lesions, which results in expansive infiltration of the dermis and its constituent structures, such as blood and lymphatic vessels, presaging distant metastasis.

Genetic alterations occur early during melanoma development and contribute to melanoma development. They include inactivation of the *INK4a/ARF* melanoma susceptibility locus, activating mutation of *NRAS*, *BRAF*, and *cKIT* oncogenes, alteration/loss of *PTEN*, *p53* mutations and *CDK4* mutations [see Table 10.1 and (Dutton-Regester and Hayward 2012; Flaherty et al. 2012; Tsao et al. 2012)].

Targeted therapeutic strategies have emerged over the recent years, accompanied with new hopes to cure patients with advanced melanoma. For example, Vemurafenib (Plexxikon/Roche) or GSK2118436 (GlaxoSmithKline) are highly selective and very potent mutated BRAF inhibitors. BRAF is the most commonly activated oncogene in melanoma, with approximately 50 % of advanced melanomas harboring such mutations. About 80–90 % of the BRAF mutations found in melanoma consist of an activating substitution of glutamate for valine at position 600 of the kinase domain within the BRAF amino-acid sequence ($BRAF^{V600E}$). In a randomized phase III trial comparing Vemurafenib to the commonly used chemotherapeutic drug, Dacarbazin, Vemurafenib treatment led to a decrease of 63 % for relative risk of death and to a significant increase in overall survival (+6 months). However, some patients presented resistance to this BRAF inhibitor and new devastating progression of the disease. Identification of drug resistance mechanisms developed by tumors and rapid preclinical and clinical testing of strategies to overcome such resistance are under investigation. Ongoing trials are combining BRAF and MEK inhibitors with the goal of delaying or overcoming resistance as a result of reactivation of the MAPK pathway [reviewed in (Flaherty et al. 2012), ASCO 2012, http://chicago2012.asco.org/Abstracts.aspx]. Another drug, Ipilimumab, a blocking antibody targeting CTLA4 (Cytotoxic T lymphocyte-associated antigen4), a protein involved in the differentiation of immunosuppressive regulatory T cells, has been tested in stage IV melanoma patients. Initial results indicate a clear increase in overall survival (ASCO 2012, http://chicago2012.asco.org/Abstracts.aspx).

Given the heterogeneity of genetic alterations found in melanoma patients and the weak success of actual therapeutic treatments coupled with exacerbated resistance mechanisms, further efforts have to be realized in developing targeted therapies. In this context, numerous preclinical studies have provided convincing evidence that TGF-β and its cognate signaling cascade play a major role in the progression of malignant melanoma to metastasis and could therefore represent a relevant target for preventing melanoma development. This chapter summarizes the literature on the pleiotropic activities of TGF-β in cutaneous melanoma biology (Fig. 10.1).

Table 10.1 Selected genetic alterations in malignant melanoma thought to be involved in melanomagenesis

Gene	Functions/pathways affected	Alteration in melanoma (frequency)
Signaling factors		
BRAF	Oncogene/MAPK signaling	Point mutation (50 %)
NRAS	Oncogene/MAPK signaling	Point mutation (20 %)
MEK	Oncogene/MAPK signaling	Point mutation (1–2 %)
KIT	Oncogene/MAPK and PI3K signaling	Point mutation (<1 %)
ERBB4	Oncogene/PI3K signaling	Point mutation (15–20 %)
PTK2B	Oncogene/MAPK signaling	Point mutation (2.5 %)
NEDD9	Melanoma metastasis gene	Amplification (55–60 %)
AKT1, AKT2, AKT3	Oncogene/PI3K signaling	Point mutation (<1 %) or amplification (25 % for AKT3)
PTEN	Tumor supressor/PI3K signaling repression	Point mutation (50–60 %)/ hemizygous deletion (50–60 %) homozygous deletion (10 %)
CTNNB1	Cell adhesion/transcriptional co-activator	Point mutation (30 %)
EGFR	Oncogene/β-catenin and MAPK signaling	Amplification associated with polysomy 7 (80 %)
MET	Oncogene/promoter of cellular invasion	Amplification (47 %)
Cell cycle and apoptosis regulators		
CDK4	Promoter of cell proliferation	Point mutation or amplification (5 %)
CDKN2A/INK4A	Tumor supressor-cell cycle inhibitor	Point mutation or deletion (30 %)
CCND1	Promoter of cell proliferation	Amplification (10 %)
TP53	Tumor supressor- cyle inhibitor/ apoptosis inducer	Point mutation (5 %)
APAF1	Apoptosis inducer	Silenced/promoter inactivation (40 %)
Transcription factors		
MITF	Melanocyte-specific lineage transcription factor	Amplification (20 %)
ETV1	Oncogene/MAPK signaling and MITF induction	Amplification (10–20 %)
MYC	Oncogene/Promoter of cell proliferation	Amplification (20 %)

From Flaherty et al. (2012), Meyle and Guldberg (2009), Miller and Mihm (2006)

10.2 TGF-β Signaling Molecules in Melanoma

Increased expression and secretion of the different TGF-β isoforms in melanoma cell lines when compared with normal melanocytes has been reported by several studies (Albino et al. 1991; Krasagakis et al. 1994; Rodeck et al. 1991; 1994).

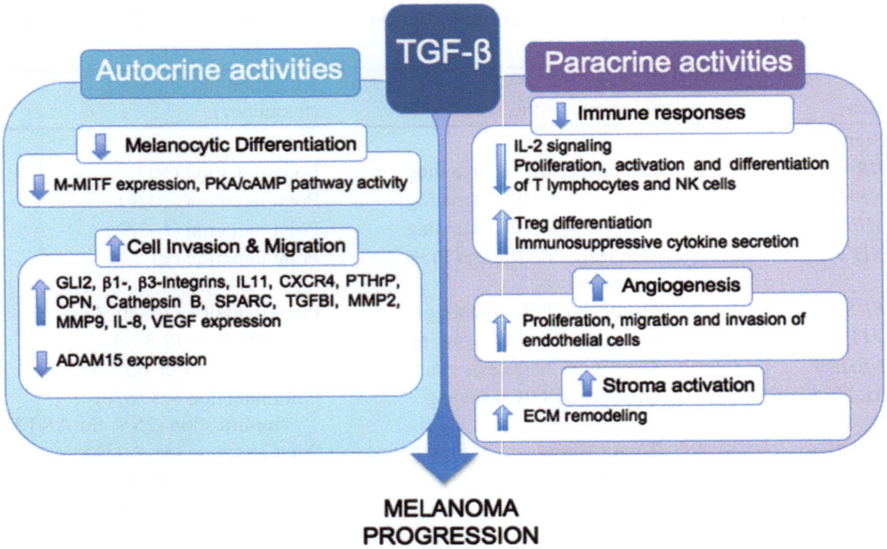

Fig. 10.1 Multiple roles of TGF-β in promoting melanoma development

In situ, TGF-β1 is secreted by normal melanocytes and melanomas at various stages, while TGF-β2 and TGF-β3 are not expressed in normal melanocytes but only heterogeneously in nevi and melanoma. Expression of TGF-β2 and TGF-β3 seems to appear early in melanoma progression and to increase with tumor progression (Van Belle et al. 1996). A correlation between TGF-β2 expression levels and tumor thickness has also been reported (Reed et al. 1994). Similarly, increased TGF-β1 and TGF-β2 plasma levels are observed at later stages of tumor development, while no significant differences have been reported between those of healthy patients and those from patients with primary or locally invasive melanoma (Krasagakis et al. 1998). Thus despite some discrepancies, all these studies point toward an increase in TGF-β expression levels that correlates with tumor progression.

Interestingly, TGF-β treatment of melanoma cells induces expression of TGF-β1 and TGF-β2 in a positive amplification loop. Moreover, patients with MMP-1 positive metastases had significantly shorter disease-free survival compared to patients with MMP-1 negative metastases (median 11.2 vs. 17.0 months, $p=0.0383$) (Nikkola et al. 2002) and MMP-1 secreted by melanoma cell was shown to be involved in the activation of TGF-β from its latent form (Iida and McCarthy 2007).

Expression of TGF-βII receptor (TβRII) mRNA is much more heterogeneously distributed in primary melanomas compared with benign melanocytic nevi. Melanoma progression appeared not to be associated with a complete loss of *TβRII* gene expression, since all skin metastases revealed clearly detectable although heterogeneous levels of TβRII mRNA expression (Schmid et al. 1995). Melastatin (TRPM1), a member of the transient receptor potential (TRPM) cation channel family, is robustly expressed in benign and dysplastic nevi and in melanomas in situ, but it is only variably expressed in invasive melanomas. Melastatin expression levels show widespread downregulation in melanoma metastases and are inversely correlated with metastatic potential and good prognosis in melanoma (Duncan et al. 1998; 2001). Intron 6 of the *TRPM1* gene hosts the gene for miR-211, a microRNA whose expression is restricted to the melanocyte lineage (Gaur et al. 2007). miR-211 and melastatin share the same promoter and are expressed coordinately in melanocytes and melanomas. It has been described that TβRII is one of the direct targets of miR-211 (Levy et al. 2010), suggesting that epigenetic events may control the expression and activity of the TGF-β in melanoma cells during disease progression.

CD105/endoglin, a co-receptor associated with TβRII that modulates the activity of the latter, is heterogeneously expressed in small clusters of tumor cells within primary melanomas and in intradermal nevi and metastatic melanoma lesions (Altomonte et al. 1996; Pardali et al. 2011).

Several studies have described a functional Smad signaling in melanoma. Basal phosphorylation of Smad3 is detectable in vitro by western blot analysis of whole melanoma cell extracts, demonstrating a constitutive activation of the TGF-β receptors. Furthermore, a pan-TGF-β antibody was shown to inhibit basal Smad3/Smad4-dependent transcriptional activity in melanoma cells (Javelaud et al. 2007; Rodeck et al. 1999). Also, exogenous TGF-β was shown to induce further a Smad3/Smad4 transcriptional response, demonstrating the functionality of the entire signaling cascade (Javelaud et al. 2007; 2008; Rodeck et al. 1999). Frequent nuclear phosphorylation of Smad2 has been detected in clinical specimens of benign and malignant cutaneous melanocytic neoplasms (Lo and Witte 2008), suggesting that the TGF-β pathway is constitutively activated in melanomas. Yet, another study showed that Smad2 phosphorylation correlated with low tumor thickness but not with overall survival and development of metastases suggesting that it would not be a useful prognostic marker (Mnich et al. 2007). This indicates the need for further studies to characterize both the phosphorylation and cellular localization of Smad2, Smad3, and Smad4 in clinical samples to determine their exact activation state throughout the various stages of melanoma progression. Interestingly, it has also been shown that phosphorylation of Smad2 in melanoma cells is associated with Smad2 phosphorylation in neighboring keratinocytes, indicating that melanoma-derived TGF-β affects the microenvironment and adjacent cells in a paracrine fashion (Mnich et al. 2007). Genome-wide expression analysis of nearly a hundred human melanoma cell lines has identified two groups with very distinct gene expression profiles: the first one is characterized by high expression of neural crest and melanocytic differentiation markers, the other one is characterized by the expression of a number of genes associated with a more aggressive phenotype, whose concomitant expression is reminiscent of a TGF-β signature (Hoek et al. 2006). These two groups present very distinct behaviors: the melanocytic differentiation signature is associated with strong proliferation but weak invasive capacity, while the TGF-β signature characterizes highly invasive melanoma cell lines with low proliferation rate (Hoek et al. 2006).

10.3 Resistance to TGF-β Induced Cell Cycle Arrest in Melanoma Cells: A Complex, Somewhat Unresolved Issue

TGF-β exerts potent anti-proliferative activity on normal melanocytes but has weak or no effect on melanoma cells. During tumor progression, melanoma cells acquire a growing resistance to TGF-β-dependent growth inhibition (Krasagakis et al. 1999). The latter may be explained by attenuation or inhibition of the TGF-β pathway in these cancer cells. For example, CD105 expression is correlated with cell sensitivity to TGF-β treatment, suggesting that expression of this co-receptor may modulate the cellular response to TGF-β (Altomonte et al. 1996). Recently, Yilmaz et al. described that TGF-β induction of DLX-2 (Distal-less homeobox) attenuates growth suppressive TGF-β signaling in a negative feedback loop (Yilmaz et al. 2011). Expression of filamin, a cytoskeletal actin–binding protein, has also been suggested to repress TGF-β signaling by direct interaction with Smad2 and Smad4 (Sasaki et al. 2001). SKI and SNON oncoproteins are direct repressors of Smad2/3 transcriptional activity and are highly expressed in melanoma cells lines as well in melanoma samples (Javelaud et al. 2011; Reed et al. 2001). In some melanoma cell lines, downregulation of SKI expression using antisense SKI vectors or shRNA restored TGF-β-mediated growth inhibition through $p21^{CIP1}$ induction and inhibited subcutaneous tumor growth in xenograft experiments (Chen et al. 2009; Reed et al. 2001). MIA (melanoma inhibitory activity) is a secreted protein expressed in melanomas but not in melanocytes and is associated with tumor progression in vivo (Bosserhoff et al. 1997). Rothhammer et al. have shown that MIA induces SKI and SNO expression, leading to reduced Smad2/3 expression in melanoma cells (Rothhammer and Bosserhoff 2006).

Despite overexpression of different proteins known to repress TGF-β signaling, mechanisms of resistance to TGF-β-induced cell growth arrest are likely more complicated than due to a single inhibitory protein. Autocrine activation of the pathway in melanoma cell has been documented in a variety of settings (Lo and Witte 2008; Rodeck et al. 1999). High phosphorylation levels of Smad3 and Smad2 linker domain phosphorylation have been described in both melanoma cell lines and clinical melanoma specimens (Cohen-Solal et al. 2011). The latter phosphorylation event specifically inhibits the expression of the cell cycle inhibitors $p21^{CIP1}$ and p15, and resulting cell growth arrest by TGF-β, but does not alter the induction of other TGF-β target genes such as PAI-1 (Cohen-Solal et al. 2011). At this stage, the mechanisms responsible for Smad2 linker domain phosphorylation are not fully understood and several studies have implicated high levels of SKI, GSK3, CDK, and MAPK (p38, JNK and ERK) in the phosphorylation of Smad2/3 linker domains (Cohen-Solal et al. 2011; Lin et al. 2010). In particular, melanomas are characterized by hyperactivity of the MEK/ERK cascade, often induced by activating mutations of the BRAF and NRAS oncogenes (Chin et al. 2006; Flaherty et al. 2012; Tsao et al. 2012). ERK activation may contribute to the phosphorylation of Smad2/3 linker domains and elicit TGF-β-induced growth arrest resistance.

Shellman et al. have shown that forced overexpression of NRAS in primary melanoma cell lines counteracts TGF-β-induced growth inhibition by preventing the accumulation of hypophosphorylated Rb protein (Shellman et al. 2000).

Another important protein that may be implicated in the loss of growth inhibition by TGF-β is Id2, a positive regulator of cell growth and plays a critical role prompting G1/S cell cycle progression. Id2 inhibits CDK inhibitors expression by interfering with bHLH, ETS and PAX transcription factor activity and growth suppressing activity of pRb by direct interaction. Id2 is negatively regulated by TGF-β, yet Schlegel et al. have shown a differential inhibition of Id2 by TGF-β in melanoma cells, depending on whether they are of the proliferative or invasive type (Schlegel et al. 2009).

During melanoma initiation and progression, a number of genetic and epigenetic alterations occur, including amplification of c-MYC or cyclin D1 overexpression, hyperactivation of CDK4 due to loss of the CDK inhibitor p16 function or activating mutations in the CDK4 gene, alterations in the activity of CDK inhibitors such as $p21^{CIP1}$, p27, or p15. All of them may participate in a cellular activation state whose endpoint is an attenuation of the TGF-β anti-proliferative effects.

PAX3, a paired-box transcription factor involved in melanocyte differentiation and survival, could also participate in resistance to TGF-β-induced cell growth arrest. Its forced overexpression in melanoma cells attenuates TGF-β anti-proliferative activity and TGF-β represses PAX3 expression (Yang et al. 2008).

10.4 Promotion of Melanoma Aggressiveness by Autocrine TGF-β Signaling

Melanoma cells secrete high levels of TGF-β ligands and most of them are largely insensitive to their anti-proliferative activity. Autocrine TGF-β is also involved in a number of additional cellular processes, as it can act on tumor cells and their microenvironment by autocrine and paracrine mechanisms that contribute to tumor development and aggressiveness.

Cell autonomous effects of TGF-β in epithelial tumor cells have been long identified. In particular, this pathway is involved in the promotion of epithelial to mesenchymal transition (EMT), remodeling of extracellular matrix, cell migration, and cell survival (Tian et al. 2011). EMT takes place during essential processes of embryogenesis such as gastrulation. Coordinated modifications in gene expression promote cell migration through changes in the expression of cell–cell adhesion molecules of the cadherin family. Consequently, epithelial cells acquire a fibroblast-like phenotype (Davies et al. 2002). This phenomenon is also observed during the metastatic development of epithelial tumors. Melanoma derives from the neuroectoderm, yet melanoma cells undergo gene expression changes and acquire a fibroblast-like phenotype similar to what is observed in epithelial tumor. This melanoma-specific phenomenon may be qualified a pseudo-EMT.

In melanoma, TGF-β has been shown to promote tumor invasion via induction of β1- and β3-integrins and inhibition of *CDH1*, promoting pseudo-EMT (Janji et al. 1999). There is also strong evidence for a direct role of autocrine TGF-β signaling in melanoma development and progression. For example, overexpression of the inhibitory Smad7 in melanoma cells reduces their capacity to form colonies in an anchorage-independent manner, dramatically inhibits the secretion of matrix metalloproteinases MMP-2 and MMP-9 and their capacity to invade Matrigel, and delays tumor growth in a subcutaneous injection model in *nude* mice (Javelaud et al. 2005). Furthermore, inhibition of constitutive TGF-β activity by Smad7 overexpression dramatically reduced experimental bone metastasis development in mice, by preventing TGF-β induction of genes known to mediate osteolytic breast cancer bone metastases, such as IL-11, PTHrP, CXCR4, and OPN (Javelaud et al. 2007). In a model of in vivo human skin grafting onto mice, it has been observed that overexpression of Smad7 in melanoma cells blocks invasion with melanoma cells being kept in close proximity to the dermal-epidermal junction (DiVito et al. 2010). Smad7 was found to increase N-cadherin expression in melanoma cells, thus promoting homotypic interactions between melanoma cells and dermal fibroblasts. Smad7 overexpression also inhibited β-catenin T41/S45 phosphorylation, resulting in the stabilization of β-catenin (DiVito et al. 2010). Of note, increased cytoplasmic/nuclear β-catenin expression has been linked with good prognosis and increased survival of melanoma patients (Chien et al. 2009).

Treatment of melanoma cell lines with SD-208, a highly specific small molecule inhibitor of TβRI, blocks in vitro TGF-β induction of Smad3 phosphorylation, TGF-β induced Smad3/Smad4–driven transcription, as well as invasion through Matrigel. SD-208 was also shown to reduce the development of experimental osteolytic bone metastases following intracardiac inoculation of tumor cells in *nude* mice (Mohammad et al. 2011). SD208 was also found to reduce the growth of established bone metastases in the same model. Together with the data obtained with Smad7 overexpression, these experiments demonstrate that inhibition of constitutive activation of TGF-β signaling may represent a valuable therapeutic option to prevent melanoma progression to metastases and inhibit the growth and dissemination of existing metastases.

We have identified the transcription factor GLI2 as a direct transcriptional target of TGF-β signaling in a variety of normal and transformed cell lines, including melanoma (Dennler et al. 2007; 2009). GLI2 is known to be a mediator of the Sonic Hedgehog pathway and numerous studies have established its role in cancer development (Javelaud et al. 2012). In melanoma cells, we found that GLI2 regulates the invasive capacity via downregulation of E-cadherin, a protein that plays a critical role in maintaining melanocyte interactions with epidermal keratinocytes, and upregulation of metalloproteinases, MMP2 and MMP9, that contribute to basement membrane invasion by tumor cells (Alexaki et al. 2010). GLI2 has also been shown to induce the expression of proteins known to participate in the metastatic process, such as WNT5a (Dissanayake et al. 2008; Reddy et al. 2001) and PTHrP (Sterling et al. 2006).

Numerous studies described modulation of gene expression by the TGF-β pathway, participating in melanoma development. For example, TGF-β induces expression of integrins, proteases such as Cathepsin B, and that of matrix remodeling proteins such as SPARC involved in the promotion of cellular invasion and migration, and represses the expression of ADAM15 (A Disintegrin And Metalloproteinase) in cooperation with IFN-γ (Ungerer et al. 2010). ADAM15 is expressed in melanocytes and endothelial cells of benign nevi and melanoma tissues and is downregulated in melanoma metastases compared to primary melanoma lesions (Ungerer et al. 2010). It seems to have tumor suppressor function in melanoma, as tail vein injection of recombinant disintegrin domain of ADAM15 together with mouse melanoma cells reduces the formation of lung metastases in mice (Trochon-Joseph et al. 2004). Furthermore, overexpression of ADAM15 has been shown to reduce migration and invasive capacity of melanoma cells in vitro, probably driven by induction of α5β1 integrin expression and increased adhesion to fibronectin (Chen et al. 2008; Ungerer et al. 2010).

TGFBI (transforming growth factor-β induced) is a TGF-β1-inducible ECM protein that plays a role in the invasive growth of melanoma cells (Nummela et al. 2012). It is highly expressed during melanoma VGP and in metastatic cell lines, has anti-adhesive properties and promotes the growth of subcutaneous tumor xenografts in mice. TGFBI localizes at tumor edges, together with fibrillar fibronectin/tenascin-C/periostin structures that characteristically surround melanoma cells at the invasive front. Both in tumors generated in nude mice and in human melanoma metastases.

Genome wide expression analysis of a large series of human melanoma cell lines revealed distinct groups characterized by specific invasive and proliferative behaviors. On the one end, highly proliferative cell lines that express neural crest and melanocytic differentiation markers, such as MITF, TYR, and MLANA, were found, while another group was composed of cell lines that grow slower than the former, are more invasive, and display high expression of a set of genes controlled by the TGF-β pathway, including TGFBI, SERPINE1, and CTGF, to cite a few (Hoek et al. 2006). These extreme phenotypes were indistinguishable by their BRAF or NRAS mutation status. It thus appeared from this broad study that cell autonomous TGF-β activation is associated with a characteristic TGF-β gene signature together with an invasive cell behavior. The genetic context may still contribute to TGF-β-induced invasive behavior of melanoma cells, as it has been shown that forced TGF-β pathway activation in immortalized melanocytes via overexpression of a constitutively activated TβRI only promotes invasion into the dermis of organotypic human skin cultures in the presence of mutated $BRAF^{V600E}$ and PTEN deficiency (Lo and Witte 2008). This was also associated with increased levels and activities of MMP-2 and MMP-9.

One important point to keep in mind is that human melanoma tumors are, in essence, highly heterogeneous and present both invasive and proliferative regions simultaneously. These behaviors are described as reversible and may be under the control of signals originating in the tumor microenvironment (Hoek et al. 2008).

Both primary and metastatic melanomas contain a heterogeneous mix of more or less differentiated cells. Strikingly actively disseminating cells are more uniform with low levels of pigment and high levels of TGF-β2 expression (Pinner et al. 2009). Several lines of evidences have suggested that TGF-β ligands negatively control melanocyte differentiation. Melanocytes in the epidermis rarely proliferate and produce melanin pigments that protect the skin from deleterious ultraviolet light (UV), while melanocytes in the hair follicles repeatedly proliferate and differentiate for hair pigmentation in every hair cycle. The hair follicle repeats its cyclic regeneration and regression with alternating phases, anagen (growing phase), catagen (regressing phase), and telogen (resting phase) to regrow new hair (Fuchs 2007). The fates of follicular melanocytes, including their proliferation, differentiation, and death all happen in synchronization with hair cycle progression (Foitzik et al. 2000; Nishimura et al. 2002). The catagen phase is an apoptosis-driven process accompanied by terminal differentiation, proteolysis, and matrix remodeling. TGF-β plays an important role in catagen regulation as an inhibitor of keratinocyte proliferation and as an inducer of apoptosis (Foitzik et al. 2000; Soma et al. 2003; Tumbar et al. 2004). Melanocyte stem cells reside with keratinocyte stem cells in the hair follicle bulge enriched in active TGF-β. Melanocyte stem cell quiescence is preceded by activation of TGF-β signaling, loss of Ki67 expression, downregulation of melanogenic gene expression, and dramatic morphologic changes from dendritic shape into a slender, oval shape with shrinkage resulting from an increased nuclear/cytoplasmic ratio (Nishimura et al. 2010). Upregulation of TGF-β1/2 expression in the niche area and phospho-Smad2 expression were detected prior to the morphologic changes, downregulation of melanogenic genes, and loss of Ki67 expression by melanocyte stem cells. Activation of TGF-β signaling plays dual roles in melanocyte stem cell maintenance: (a), through inhibition of stem cell differentiation in the stem cell niche and (b), via induction of stem cell quiescence that requires BCL2 expression for cell survival (Nishimura et al. 2010). TGF-β may also prevent melanocyte maturation through a direct repression of M-MITF expression (Kim et al. 2004; Pierrat et al. 2012). M-MITF is a master transcription factor of the melanocytic lineage that controls cell survival, migration, and differentiation. This factor directly activates expression of melanogenic genes such as those encoding tyrosinase, TRP1, and TRP2. TGF-β inhibits M-MITF promoter activity by two different mechanisms: repression of CREB activity through an inhibition of cAMP/PKA signaling and induction of GLI2, the latter directly binding to the MITF promoter to repress transcription by mechanisms that remain to be identified (Pierrat et al. 2012). Proteolysis of tyrosinase has also been involved in melanogenesis inhibition by TGF-β (Martinez-Esparza et al. 1997).

TGF-β also induces dedifferentiation of melanoma cells and promotes cell migration and invasion, thereby promoting metastasis. This dedifferentiation is reversible, transient, and dynamic and probably dependent upon hypoxia and the tumor microenvironment (Hoek et al. 2008; Pinner et al. 2009).

10.5 The Role of Tumor-Derived TGF-β on the Microenvironment

Paracrine effects of TGF-β on surrounding cells in the tumor microenvironment may be advantageous for melanoma cells. Two main roles are to be pointed out. Suppressive effects of TGF-β on the immune system may allow tumor cells to escape from immune surveillance, and pro-angiogenic properties of TGF-β could support nutrition of the tumor and facilitate metastasis. In addition, stimulation of stromal cells by TGF-β leads to increased production of reciprocally paracrine-acting growth factors. Also, melanoma cells can modulate their surrounding stroma for their own benefits through the paracrine activity of TGF-β. Stimulation of production of ECM proteins such as collagen, fibronectin, tenascin, and α2-integrin, by stromal fibroblasts provides a scaffold for melanoma cells to adhere and migrate, leading to increased survival and metastasis formation (Berking et al. 2001). Finally, modulation of the expression of proteases and their inhibitors by TGF-β likely facilitates invasion by means of stromal remodeling.

Aberrant TGF-β signaling in mice results in large vascular and endothelial defects, demonstrating the central role of TGF-β in embryonic vascular morphogenesis and in the establishment and maintenance of vessel wall integrity (Goumans et al. 2003). TGF-β can regulate the initiation and development of new blood vessels as well as the maturation of newly formed vessels. These processes involve not only stimulation of proliferation, migration, and invasion of endothelial cells during initiation of angiogenesis but also increase in endothelial cell adhesion, basement membrane deposition and recruitment of pericytes, and vascular smooth muscle cells during vessel maturation (Tian et al. 2011). In melanoma cells, TGF-β enhances angiogenesis by activating the expression of proangiogenic factors such as Interleukin (IL)-8 and VEGF (Liu et al. 2005).

TGF-β exerts immunoregulatory functions by inhibiting the proliferation, activation, and differentiation of lymphocytes, as well as the natural killer and dendritic cell functions, all these events leading to the suppression of the antitumor immune response (Arteaga et al. 1993; Letterio and Roberts 1998; Wrzesinski et al. 2007). Noteworthy, transplantation of murine bone marrow expressing a dominant-negative TβRII in a syngeneic mouse model of melanoma leads to the generation of mature leukocytes capable of a potent antitumor response (Shah et al. 2002).

TGF-β also exerts repressory effects on IL-2 activity, the main activator of immune response: TGF-β inhibits the phosphorylation and activation of components of the JAK/STAT cascade downstream of the IL-2 receptor and expression of IL-2 targets genes such as c-MYC and Cyclin D2 (Bright et al. 1997; Nelson et al. 2003). TGF-β is also an important differentiation factor for regulatory T cells (Foxp3+ Treg), a subset of strongly immunosuppressive T cells often found in tumors, including melanoma.

MCP-1 (monocyte chemoattractant protein-1), a potent chemokine that recruits macrophages, is expressed by malignant but not normal melanocytes (Nesbit et al. 2001). Monocytes and macrophages infiltrate melanomas and produce factors promoting tumor development and invasion, including immunosuppressive cytokines and angiogenic factors (Dirkx et al. 2006). IL-10 is also expressed in primary and metastatic melanoma lesions (Dummer et al. 1996) and elevated levels of IL-10 have been measured in sera of patients with advanced melanoma that correlate with tumor progression (Moretti et al. 2001; Nemunaitis et al. 2001). TGF-β increases both MCP-1 and IL-10 expression in melanoma cells, by means of a crosstalk between the Smad, AKT, and MAPK/ERK pathways. Thus, TGF-β enhances MCP-1 mediated monocyte/macrophage migration and IL-10 immunosuppressive functions (Diaz-Valdes et al. 2011).

10.6 Other TGF-β Family Members in Melanoma

While the above paragraphs have focused on TGF-β *stricto sensu*, other members of the TGF-β family of growth factors are also expressed and/or contribute to melanoma development. A brief overview of the cognate literature is provided below.

It has been shown that Activin inhibits the proliferation and induces apoptosis of primary melanocytes. Expression of Activin and its receptors has been detected in melanoma cell lines that show a response to exogenous Activin. Yet those cell lines were found to be resistant to Activin's anti-proliferative activity (Stove et al. 2004).

Melanoma cells, unlike normal melanocytes, express and secrete high level of Follistatin. Follistatin is a secreted protein that binds to activin and blocks activin-induced signaling (Stove et al. 2004). It was suggested that early during melanoma progression, secretion of Follistatin may protect against the cytostatic function of activin while, on the other hand, melanoma cells at a later stage may use activin, either autocrine or paracrine, to maintain a state of dedifferentiation and create a microenvironment supportive of melanoma growth.

Expression of multiple BMPs (bone morphogenetic proteins) including BMP-2, -4, -7, is upregulated in melanomas compared to nevi (Rothhammer et al. 2005), and expression of BMP-4 and BMP-7 was found to increase during disease progression. BMP-4 was found to be implicated in the control of the migratory and invasive properties of melanoma cells, as inhibition of BMP signaling by means of overexpression of the BMP inhibitor chordin, or by specific downregulation of BMP4 expression, resulted in a strong reduction of both invasive and migratory capacities of melanoma cells (Rothhammer et al. 2005), possibly via reduced MMP-1, 2, 3, and 9 expression (Rothhammer et al. 2008). Dido1 (Death inducer-obliterator 1) was recently identified as a direct and specific BMP target gene. This protein was first described as an early apoptosis regulator protein and as a tumor suppressor gene in hematological myeloid neoplasm (Jilaveanu et al. 2009). Upon apoptosis induction, Dido-1 is translocated into the nucleus and activates pro-caspase expression (Garcia-Domingo et al. 2003). It is highly expressed in the nucleus of melanoma cell lines derived from metastases (Braig and Bosserhoff 2013). Dido1 expression induced by BMP-4 signaling regulates anchorage independent growth, apoptosis resistance as well as the migratory and invasive capacities of melanoma cells (Braig and Bosserhoff 2013). In melanocytes, BMP-4 inhibits melanogenesis by blocking Tyrosinase expression (Yaar et al. 2006). Accordingly, overexpression of Noggin, a BMP antagonist, in the epithelium of hair follicles in mice, induces a darker coat color compared to wild-type mice (Sharov et al. 2005). Inversely, BMP-2 promotes melanogenesis by inducing the expression of Tyrosinase (Bilodeau et al. 2001). In astrocytoma cells, BMP-2 also inhibits the expression of HGF (hepatocyte growth factor) (Chattopadhyay et al. 2004), a potent mitogenic, motogenic, and morphogenic cytokine for melanoma cells (Noonan et al. 2003).

While BMP7 is overexpressed in melanoma tissues, it may exert antitumor activities. For instance, it has been described that BMP7 inhibits the proliferation of both normal and malignant melanocytes, an effect that is alleviated by concomitant overexpression of Noggin (Hsu et al. 2008). Noggin expression is also correlated with an induction of Nodal and VEGF expression in a subset of but not all melanoma cells lines. Induction of these growth-promoting factors may participate in the restoration of proliferation (Hsu et al. 2008). Another study only found a minor effect of BMP7 on melanoma proliferation while potently inhibiting of both the migratory and invasive capacities of melanoma cells (Na et al. 2009).

Nodal expression is restricted to embryonic tissues, epithelial stem cells, and cancer cells. Nodal is absent in normal skin and only rarely detected in poorly invasive radial growth phase melanomas. Yet, it is observed in up to 60 % of cases of vertical growth phase primary melanoma lesions and melanoma metastases. Its expression correlates with melanoma progression and experimentally, its depletion in melanoma cells decreases anchorage-independent growth and plasticity, concomitant with a marked abrogation of tumorigenicity (Topczewska et al. 2006; Yu et al. 2010).

Cripto-1, a Nodal co-receptor, is also expressed in melanomas and contributes to invasion and proliferation of melanoma cells (De Luca et al. 2011). In addition to mediating Nodal signaling, Cripto-1 modulates signaling of other TGF-β family ligands, including GDF-1, GDF-3, Activin, and TGF-β.

MIC-1 (macrophage inhibitory cytokine-1)/GDF-15 (growth differentiation factor-15) is expressed in a high proportion of melanoma cell lines compared to normal melanocytes and was detected in all examined metastatic melanoma biopsies (Boyle et al. 2009). GDF-15 levels are also increased five- to sixfold in sera from patients with an advanced melanoma compared to healthy donors (Huh et al. 2010). Affymetrix analysis of microdissected fresh frozen melanocytic nevi and melanoma samples showed upregulated expression of GDF-15 in primary and metastatic melanomas compared to melanocytic nevi (Mauerer et al. 2011). Interestingly, progression-free survival in melanoma patients with low GDF-15 staining was significantly higher compared to patients with high GDF-15 expression (Mauerer et al. 2011). Inhibition of GDF-15 expression with a specific shRNA strongly reduced tumor growth in an in vivo mouse xenograft model (Boyle et al. 2009; Huh et al. 2010). Inhibition of GDF-15 was found to delay melanoma tumor vascular development, subsequently affecting tumor cell proliferation and apoptosis. Melanoma cells secrete GDF-15 that promotes directional blood vessel development, in combination with VEGF (Huh et al. 2010). Expression of GDF-15 in melanoma cells is directly controlled by the transcription factor M-MITF (Boyle et al. 2009).

GDF-3/ Vgr-2 expression is also correlated with the metastatic capacity of melanoma cell lines and its overexpression promotes the growth of implanted melanoma tumors in a syngeneic mouse model. Moreover, GDF-3 expression is accompanied by an increased expression of CD24/CD44, markers of melanoma stem cell-like cells/melanoma-initiating cells (Ehira et al. 2010).

Together, these studies demonstrate the complexity of the TGF-β ligand/signaling network in melanoma and warrants further investigations.

10.7 Therapeutic Targeting of TGF-β to Fight Melanoma

There is broad experimental evidence for the potential benefit of targeting TGF-β signaling for cancer treatment. Different strategies have been developed to inhibit the TGF-β pathway: inhibition of TGF-β ligands expression with antisense molecules, blocking of ligands with ligand traps, and interference with receptor signaling by means of small molecules. These have been extensively reviewed over the last few years (Connolly et al. 2012; Hawinkels and Ten Dijke 2011; Wrzesinski et al. 2007).

For melanoma treatment, several anti-TGF-β therapies have been studied and they clearly present promising results. Antisense molecule targeting mRNA encoding the TGF-β2, called Trabedersen or AP12009, has been tested in clinical phase I/ II study in patients with advanced pancreatic, malignant melanoma, or colorectal cancer. Traberdersen is safe and well tolerated and has shown promising efficacy. For instance, enrolled melanoma patients ($n = 14$) had reached a median overall survival of 9.3 months at the time of publication (Oettle et al. 2012).

Because TGF-β is abundantly secreted by tumor cells or by their microenvironment during disease progression, monoclonal TGF-β neutralizing antibodies and soluble TβRII or TβRIII have been elaborated to trap TGF-β ligands. In preclinical studies, this strategy has shown encouraging results in particular for breast cancer treatment (Connolly et al. 2012). An anti-TGF-β monoclonal neutralizing antibody, GC1008 (Fresolimumab) is under evaluation in a clinical phase I/II study in patients with advanced melanoma. In 2008, 5/22 patients with advanced melanoma achieved stable disease or partial responses. The most frequently reported drug-related side effects were skin rash/lesions including eruptive non-malignant keratoacanthomas and squamous cell carcinomas (Morris et al. 2008).

A number of small molecule inhibitors have been developed that target TGF-β receptor kinases. The majority of TβRI/ALK5 inhibitors also block the related activin and Nodal receptors ACVR1B/ALK4 and ACVR1C/ALK7, but with reduced affinity (Fu et al. 2008; Inman et al. 2002; Tojo et al. 2005). Several studies have illustrated the power of these inhibitors in the cancer treatment. As an example, in a mouse model of melanoma derived-bone metastasis, we have demonstrated that treatment with SD-208, a TβRI inhibitor, significantly reduces the incidence and size of osteolytic lesions in mice with established bone metastases compared to vehicle-treated mice (Mohammad et al. 2011). Unfortunately, a major drawback of these pharmacologic compounds is their high cardiotoxicity in humans, which has taken most of them out of clinical trials (Orphanos et al. 2009).

As discussed above, TGF-β plays a central role in allowing tumor immune escape. Some new antitumor strategies therefore consist in a combination of anti-TGF-β treatment and a vaccine, as blockade of TGF-β activity markedly enhances immunotherapy in preclinical models (Penafuerte and Galipeau 2008). It has been shown that TGF-β inhibits GM-CSF-induced maturation of bone marrow-derived dendritic cells, as well as expression of MHC class II and co-stimulatory molecules. Also, GM-CSF-secreting autologous immune vaccines have shown very good response in terms of immune stimulation and survival duration. The efficacy of these vaccines has been improved with a combination of TGF-β inhibition by means of TGF-β2 anti-sense oligonucleotides or specific shRNA targeting the proprotein convertase furin to prevent the conversion of latent TGF-β into an active growth factor (Olivares et al. 2011; Senzer et al. 2012).

High doses of IL-2 are a somewhat conventional treatment for metastatic melanoma in the USA despite limited response. Interestingly, nanoscale liposomal polymeric gel (nanolipogels) delivery of a TGF-β inhibitor together with IL-2 in a sustained fashion to the tumor microenvironment was found to significantly delay tumor growth and increase survival in tumor-bearing mice, accompanied with an increase in NK and TIL (tumor infiltrating lymphocytes) activity (Park et al. 2012).

10.8 Concluding Remarks

There is broad evidence underlying the involvement of TGF-β signaling in melanoma progression and metastasis development. TGF-β acts not only on the tumor cells but also on their microenvironment including stromal fibroblasts, endothelial and immune cells. Preclinical studies have reported very potent effects of anti-TGF-β approaches for cancer therapy. However, the effects of anti-TGF-β drugs have been less robust than hoped for in the clinical setting, one reason being their strong cardiotoxic activity. As numerous studies have exemplified the crosstalks between MEK/ERK and TGF-β pathways in epithelial cancers, it will be interesting to study the effect of TGF-β inhibitors in combination with the new melanoma drugs specifically targeting mutant BRAF or MEK. One may hope that combinatorial therapies will allow lower dosage to be used to circumvent drug toxicity and increase the clinical efficacy of TGF-β inhibitors.

References

Albino AP, Davis BM, Nanus DM (1991) Induction of growth factor RNA expression in human malignant melanoma: markers of transformation. Cancer Res 51:4815–4820

Alexaki VI, Javelaud D, Van Kempen LC et al (2010) GLI2-mediated melanoma invasion and metastasis. J Natl Cancer Inst 102:1148–1159

Altomonte M, Montagner R, Fonsatti E et al (1996) Expression and structural features of endoglin (CD105), a transforming growth factor β1 and β3 binding protein, in human melanoma. Br J Cancer 74:1586–1591

Arteaga CL, Hurd SD, Winnier AR et al (1993) Anti-transforming growth factor (TGF)-β antibodies inhibit breast cancer cell tumorigenicity and increase mouse spleen natural killer cell activity. Implications for a possible role of tumor cell/host TGF-β interactions in human breast cancer progression. J Clin Invest 92:2569–2576

Berking C, Takemoto R, Schaider H et al (2001) Transforming growth factor-β1 increases survival of human melanoma through stroma remodeling. Cancer Res 61:8306–8316

Bilodeau ML, Greulich JD, Hullinger RL et al (2001) BMP-2 stimulates tyrosinase gene expression and melanogenesis in differentiated melanocytes. Pigment Cell Res 14:328–336

Bosserhoff AK, Kaufmann M, Kaluza B et al (1997) Melanoma-inhibiting activity, a novel serum marker for progression of malignant melanoma. Cancer Res 57:3149–3153

Boyle GM, Pedley J, Martyn AC et al (2009) Macrophage inhibitory cytokine-1 is overexpressed in malignant melanoma and is associated with tumorigenicity. J Invest Dermatol 129:383–391

Braig S, Bosserhoff AK (2013) Death inducer-obliterator 1 (Dido1) is a BMP target gene and promotes BMP-induced melanoma progression. Oncogene 32(7):837–848

Bright JJ, Kerr LD, Sriram S (1997) TGF-β inhibits IL-2-induced tyrosine phosphorylation and activation of Jak-1 and Stat 5 in T lymphocytes. J Immunol 159:175–183

Chattopadhyay N, T-Felt Hansen J, Godbole MM, Brown EM (2004) Transforming growth factor β receptor family ligands inhibit hepatocyte growth factor synthesis and secretion from astrocytoma cells. Brain Res Mol Brain Res 121:146–150

Chen Q, Meng LH, Zhu CH et al (2008) ADAM15 suppresses cell motility by driving integrin α5β1 cell surface expression via Erk inactivation. Int J Biochem Cell Biol 40:2164–2173

Chen D, Lin Q, Box N et al (2009) SKI knockdown inhibits human melanoma tumor growth in vivo. Pigment Cell Melanoma Res 22:761–772

Chien AJ, Moore EC, Lonsdorf AS et al (2009) Activated Wnt/β-catenin signaling in melanoma is associated with decreased proliferation in patient tumors and a murine melanoma model. Proc Natl Acad Sci USA 106:1193–1198

Chin L, Garraway LA, Fisher DE (2006) Malignant melanoma: genetics and therapeutics in the genomic era. Genes Dev 20:2149–2182

Cohen-Solal KA, Merrigan KT, Chan JL et al (2011) Constitutive Smad linker phosphorylation in melanoma: a mechanism of resistance to transforming growth factor-β-mediated growth inhibition. Pigment Cell Melanoma Res 24:512–524

Connolly EC, Freimuth J, Akhurst RJ (2012) Complexities of TGF-β targeted cancer therapy. Int J Biol Sci 8:964–978

Davies H, Bignell GR, Cox C et al (2002) Mutations of the BRAF gene in human cancer. Nature 417:949–954

De Luca A, Lamura L, Strizzi L et al (2011) Expression and functional role of CRIPTO-1 in cutaneous melanoma. Br J Cancer 105:1030–1038

Dennler S, Andre J, Alexaki I et al (2007) Induction of sonic hedgehog mediators by transforming growth factor-β: Smad3-dependent activation of Gli2 and Gli1 expression in vitro and in vivo. Cancer Res 67:6981–6986

Dennler S, Andre J, Verrecchia F, Mauviel A (2009) Cloning of the human GLI2 Promoter: transcriptional activation by transforming growth factor-β via SMAD3/β-catenin cooperation. J Biol Chem 284:31523–31531

Diaz-Valdes N, Basagoiti M, Dotor J et al (2011) Induction of monocyte chemoattractant protein-1 and interleukin-10 by TGFβ1 in melanoma enhances tumor infiltration and immunosuppression. Cancer Res 71:812–821

Dirkx AE, Oude Egbrink MG et al (2006) Monocyte/macrophage infiltration in tumors: modulators of angiogenesis. J Leukoc Biol 80:1183–1196

Dissanayake SK, Olkhanud PB, O'Connell MP et al (2008) Wnt5A regulates expression of tumor-associated antigens in melanoma via changes in signal transducers and activators of transcription 3 phosphorylation. Cancer Res 68:10205–10214

DiVito KA, Trabosh VA, Chen YS et al (2010) Smad7 restricts melanoma invasion by restoring N-cadherin expression and establishing heterotypic cell-cell interactions in vivo. Pigment Cell Melanoma Res 23:795–808

Dummer W, Bastian BC, Ernst N et al (1996) Interleukin-10 production in malignant melanoma: preferential detection of IL-10-secreting tumor cells in metastatic lesions. Int J Cancer 66:607–610

Duncan LM, Deeds J, Hunter J et al (1998) Down-regulation of the novel gene melastatin correlates with potential for melanoma metastasis. Cancer Res 58:1515–1520

Duncan LM, Deeds J, Cronin FE et al (2001) Melastatin expression and prognosis in cutaneous malignant melanoma. J Clin Oncol 19:568–576

Dutton-Regester K, Hayward NK (2012) Reviewing the somatic genetics of melanoma: from current to future analytical approaches. Pigment Cell Melanoma Res 25:144–154

Ehira N, Oshiumi H, Matsumoto M et al (2010) An embryo-specific expressing TGF-β family protein, growth-differentiation factor 3 (GDF3), augments progression of B16 melanoma. J Exp Clin Cancer Res 29:135

Flaherty KT, Hodi FS, Fisher DE (2012) From genes to drugs: targeted strategies for melanoma. Nat Rev Cancer 12:349–361

Foitzik K, Lindner G, Mueller-Roever S et al (2000) Control of murine hair follicle regression (catagen) by TGF-β1 in vivo. FASEB J 14:752–760

Fu K, Corbley MJ, Sun L, Friedman JE, Shan F, Papadatos JL et al (2008) SM16, an orally active TGF-β type I receptor inhibitor prevents myofibroblast induction and vascular fibrosis in the rat carotid injury model. Arterioscler Thromb Vasc Biol 28:665–671

Fuchs E (2007) Scratching the surface of skin development. Nature 445:834–842

Garcia-Domingo D, Ramirez D, Gonzalez de Buitrago G, Martinez AC (2003) Death inducer-obliterator 1 triggers apoptosis after nuclear translocation and caspase upregulation. Mol Cell Biol 23:3216–3225

Gaur A, Jewell DA, Liang Y et al (2007) Characterization of microRNA expression levels and their biological correlates in human cancer cell lines. Cancer Res 67:2456–2468

Goumans MJ, Lebrin F, Valdimarsdottir G (2003) Controlling the angiogenic switch: a balance between two distinct TGF-β receptor signaling pathways. Trends Cardiovasc Med 13:301–307

Hawinkels LJ, Ten Dijke P (2011) Exploring anti-TGF-β therapies in cancer and fibrosis. Growth Factors 29:140–152

Hoek KS, Schlegel NC, Brafford P et al (2006) Metastatic potential of melanomas defined by specific gene expression profiles with no BRAF signature. Pigment Cell Res 19:290–302

Hoek KS, Eichhoff OM, Schlegel NC et al (2008) In vivo switching of human melanoma cells between proliferative and invasive states. Cancer Res 68:650–656

Hsu MY, Rovinsky SA, Lai CY et al (2008) Aggressive melanoma cells escape from BMP7-mediated autocrine growth inhibition through coordinated Noggin upregulation. Lab Invest 88:842–855

Huh SJ, Chung CY, Sharma A, Robertson GP (2010) Macrophage inhibitory cytokine-1 regulates melanoma vascular development. Am J Pathol 176:2948–2957

Iida J, McCarthy JB (2007) Expression of collagenase-1 (MMP-1) promotes melanoma growth through the generation of active transforming growth factor-β. Melanoma Res 17:205–213

Inman GJ, Nicolas FJ, Callahan JF et al (2002) SB-431542 is a potent and specific inhibitor of transforming growth factor-β superfamily type I activin receptor-like kinase (ALK) receptors ALK4, ALK5, and ALK7. Mol Pharmacol 62:65–74

Janji B, Melchior C, Gouon V et al (1999) Autocrine TGF-β-regulated expression of adhesion receptors and integrin-linked kinase in HT-144 melanoma cells correlates with their metastatic phenotype. Int J Cancer 83:255–262

Javelaud D, Delmas V, Moller M et al (2005) Stable overexpression of Smad7 in human melanoma cells inhibits their tumorigenicity in vitro and in vivo. Oncogene 24:7624–7629

Javelaud D, Mohammad KS, McKenna CR et al (2007) Stable overexpression of Smad7 in human melanoma cells impairs bone metastasis. Cancer Res 67:2317–2324

Javelaud D, Alexaki VI, Mauviel A (2008) Transforming growth factor-β in cutaneous melanoma. Pigment Cell Melanoma Res 21:123–132

Javelaud D, van Kempen L, Alexaki VI et al (2011) Efficient TGF-β/SMAD signaling in human melanoma cells associated with high c-SKI/SnoN expression. Mol Cancer 10:2

Javelaud D, Pierrat MJ, Mauviel A (2012) Crosstalk between TGF-β and hedgehog signaling in cancer. FEBS Lett 586:2016–2025
Jilaveanu LB, Aziz SA, Kluger HM (2009) Chemotherapy and biologic therapies for melanoma: do they work? Clin Dermatol 27:614–625
Kim DS, Park SH, Park KC (2004) Transforming growth factor-β1 decreases melanin synthesis via delayed extracellular signal-regulated kinase activation. Int J Biochem Cell Biol 36:1482–1491
Krasagakis K, Garbe C, Schrier PI, Orfanos CE (1994) Paracrine and autocrine regulation of human melanocyte and melanoma cell growth by transforming growth factor β in vitro. Anticancer Res 14:2565–2571
Krasagakis K, Tholke D, Farthmann B et al (1998) Elevated plasma levels of transforming growth factor (TGF)-β1 and TGF-β2 in patients with disseminated malignant melanoma. Br J Cancer 77:1492–1494
Krasagakis K, Kruger-Krasagakes S, Fimmel S et al (1999) Desensitization of melanoma cells to autocrine TGF-β isoforms. J Cell Physiol 178:179–187
Letterio JJ, Roberts AB (1998) Regulation of immune responses by TGF-β. Annu Rev Immunol 16:137–161
Levy C, Khaled M, Iliopoulos D et al (2010) Intronic miR-211 assumes the tumor suppressive function of its host gene in melanoma. Mol Cell 40:841–849
Lin Q, Chen D, Timchenko NA, Medrano EE (2010) SKI promotes Smad3 linker phosphorylations associated with the tumor-promoting trait of TGFβ. Cell Cycle 9:1684–1689
Liu G, Zhang F, Lee J, Dong Z (2005) Selective induction of interleukin-8 expression in metastatic melanoma cells by transforming growth factor-β 1. Cytokine 31:241–249
Lo RS, Witte ON (2008) Transforming growth factor-β activation promotes genetic context-dependent invasion of immortalized melanocytes. Cancer Res 68:4248–4257
Martinez-Esparza M, Jimenez-Cervantes C, Beermann F et al (1997) Transforming growth factor-β1 inhibits basal melanogenesis in B16/F10 mouse melanoma cells by increasing the rate of degradation of tyrosinase and tyrosinase-related protein-1. J Biol Chem 272:3967–3972
Mauerer A, Roesch A, Hafner C et al (2011) Identification of new genes associated with melanoma. Exp Dermatol 20:502–507
Meyle KD, Guldberg P (2009) Genetic risk factors for melanoma. Hum Genet 126:499–510
Miller AJ, Mihm MC Jr (2006) Melanoma. N Engl J Med 355:51–65
Mnich CD, Hoek KS, Oberholzer PA et al (2007) Reduced pSmad2 immunodetection correlates with increased primary melanoma thickness. Melanoma Res 17:131–136
Mohammad KS, Javelaud D, Fournier PG et al (2011) TGF-β-RI kinase inhibitor SD-208 reduces the development and progression of melanoma bone metastases. Cancer Res 71:175–184
Moretti S, Chiarugi A, Semplici F et al (2001) Serum imbalance of cytokines in melanoma patients. Melanoma Res 11:395–399
Morris JCGIS, Tan AR, Lawrence DP, Olencki TE et al (2008) Phase I/II study of GC1008: A human anti-transforming growth factor-β (TGFβ) monoclonal antibody (MAb) in patients with advanced malignant melanoma (MM) or renal cell carcinoma (RCC). J Clin Oncol 26:9028
Na YR, Seok SH, Kim DJ et al (2009) Bone morphogenetic protein 7 induces mesenchymal-to-epithelial transition in melanoma cells, leading to inhibition of metastasis. Cancer Sci 100:2218–2225
Nelson BH, Martyak TP, Thompson LJ et al (2003) Uncoupling of promitogenic and antiapoptotic functions of IL-2 by Smad-dependent TGF-β signaling. J Immunol 170:5563–5570
Nemunaitis J, Fong T, Shabe P et al (2001) Comparison of serum interleukin-10 (IL-10) levels between normal volunteers and patients with advanced melanoma. Cancer Invest 19:239–247
Nesbit M, Schaider H, Miller TH, Herlyn M (2001) Low-level monocyte chemoattractant protein-1 stimulation of monocytes leads to tumor formation in nontumorigenic melanoma cells. J Immunol 166:6483–6490
Nikkola J, Vihinen P, Vlaykova T et al (2002) High expression levels of collagenase-1 and stromelysin-1 correlate with shorter disease-free survival in human metastatic melanoma. Int J Cancer 97:432–438

Nishimura EK, Jordan SA, Oshima H et al (2002) Dominant role of the niche in melanocyte stem-cell fate determination. Nature 416:854–860

Nishimura EK, Suzuki M, Igras V et al (2010) Key roles for transforming growth factor β in melanocyte stem cell maintenance. Cell Stem Cell 6:130–140

Noonan FP, Dudek J, Merlino G, De Fabo EC (2003) Animal models of melanoma: an HGF/SF transgenic mouse model may facilitate experimental access to UV initiating events. Pigment Cell Res 16:16–25

Nummela P, Lammi J, Soikkeli J et al (2012) Transforming growth factor β-induced (TGFBI) is an anti-adhesive protein regulating the invasive growth of melanoma cells. Am J Pathol 180:1663–1674

Oettle H, Seufferlein T, Luger T et al (2012) Final results of a phase I/II study in patients with pancreatic cancer, malignant melanoma, and colorectal carcinoma with trabedersen. J Clin Oncol 30:4034

Olivares J, Kumar P, Yu Y et al (2011) Phase I trial of TGF-β 2 antisense GM-CSF gene-modified autologous tumor cell (TAG) vaccine. Clin Cancer Res 17:183–192

Orphanos GS, Ioannidis GN, Ardavanis AG (2009) Cardiotoxicity induced by tyrosine kinase inhibitors. Acta Oncol 48:964–970

Pardali E, van der Schaft DW, Wiercinska E et al (2011) Critical role of endoglin in tumor cell plasticity of Ewing sarcoma and melanoma. Oncogene 30:334–345

Park J, Wrzesinski SH, Stern E et al (2012) Combination delivery of TGF-β inhibitor and IL-2 by nanoscale liposomal polymeric gels enhances tumour immunotherapy. Nat Mater 11(10):895–905

Penafuerte C, Galipeau J (2008) TGF β secreted by B16 melanoma antagonizes cancer gene immunotherapy bystander effect. Cancer Immunol Immunother 57:1197–1206

Pierrat MJ, Marsaud V, Mauviel A, Javelaud D (2012) Expression of microphthalmia-associated transcription factor (MITF), which is critical for melanoma progression, is inhibited by both transcription factor GLI2 and transforming growth factor-β. J Biol Chem 287:17996–18004

Pinner S, Jordan P, Sharrock K et al (2009) Intravital imaging reveals transient changes in pigment production and Brn2 expression during metastatic melanoma dissemination. Cancer Res 69:7969–7977

Reddy S, Andl T, Bagasra A et al (2001) Characterization of Wnt gene expression in developing and postnatal hair follicles and identification of Wnt5a as a target of Sonic hedgehog in hair follicle morphogenesis. Mech Dev 107:69–82

Reed JA, McNutt NS, Prieto VG, Albino AP (1994) Expression of transforming growth factor-β 2 in malignant melanoma correlates with the depth of tumor invasion. Implications for tumor progression. Am J Pathol 145:97–104

Reed JA, Bales E, Xu W et al (2001) Cytoplasmic localization of the oncogenic protein Ski in human cutaneous melanomas in vivo: functional implications for transforming growth factor β signaling. Cancer Res 61:8074–8078

Rodeck U, Melber K, Kath R et al (1991) Constitutive expression of multiple growth factor genes by melanoma cells but not normal melanocytes. J Invest Dermatol 97:20–26

Rodeck U, Bossler A, Graeven U et al (1994) Transforming growth factor β production and responsiveness in normal human melanocytes and melanoma cells. Cancer Res 54:575–581

Rodeck U, Nishiyama T, Mauviel A (1999) Independent regulation of growth and SMAD-mediated transcription by transforming growth factor β in human melanoma cells. Cancer Res 59:547–550

Rothhammer T, Bosserhoff AK (2006) Influence of melanoma inhibitory activity on transforming growth factor-β signaling in malignant melanoma. Melanoma Res 16:309–316

Rothhammer T, Poser I, Soncin F et al (2005) Bone morphogenic proteins are overexpressed in malignant melanoma and promote cell invasion and migration. Cancer Res 65:448–456

Rothhammer T, Braig S, Bosserhoff AK (2008) Bone morphogenetic proteins induce expression of metalloproteinases in melanoma cells and fibroblasts. Eur J Cancer 44:2526–2534

Sasaki A, Masuda Y, Ohta Y et al (2001) Filamin associates with Smads and regulates transforming growth factor-β signaling. J Biol Chem 276:17871–17877

Schlegel NC, Eichhoff OM, Hemmi S et al (2009) Id2 suppression of p15 counters TGF-β-mediated growth inhibition of melanoma cells. Pigment Cell Melanoma Res 22:445–453

Schmid P, Itin P, Rufli T (1995) In situ analysis of transforming growth factor-β s (TGF-β 1, TGF-β 2, TGF-β 3), and TGF-β type II receptor expression in malignant melanoma. Carcinogenesis 16:1499–1503

Senzer N, Barve M, Kuhn J et al (2012) Phase I trial of "bi-shRNAi(furin)/GMCSF DNA/autologous tumor cell" vaccine (FANG) in advanced cancer. Mol Ther 20:679–686

Shah AH, Tabayoyong WB, Kundu SD et al (2002) Suppression of tumor metastasis by blockade of transforming growth factor β signaling in bone marrow cells through a retroviral-mediated gene therapy in mice. Cancer Res 62:7135–7138

Sharov AA, Fessing M, Atoyan R et al (2005) Bone morphogenetic protein (BMP) signaling controls hair pigmentation by means of cross-talk with the melanocortin receptor-1 pathway. Proc Natl Acad Sci USA 102:93–98

Shellman YG, Chapman JT, Fujita M et al (2000) Expression of activated N-ras in a primary melanoma cell line counteracts growth inhibition by transforming growth factor-β. J Invest Dermatol 114:1200–1204

Soma T, Dohrmann CE, Hibino T, Raftery LA (2003) Profile of transforming growth factor-β responses during the murine hair cycle. J Invest Dermatol 121:969–975

Sterling JA, Oyajobi BO, Grubbs B et al (2006) The hedgehog signaling molecule Gli2 induces parathyroid hormone-related peptide expression and osteolysis in metastatic human breast cancer cells. Cancer Res 66:7548–7553

Stove C, Vanrobaeys F, Devreese B et al (2004) Melanoma cells secrete follistatin, an antagonist of activin-mediated growth inhibition. Oncogene 23:5330–5339

Tian M, Neil JR, Schiemann WP (2011) Transforming growth factor-β and the hallmarks of cancer. Cell Signal 23:951–962

Tojo M, Hamashima Y, Hanyu A et al (2005) The ALK-5 inhibitor A-83-01 inhibits Smad signaling and epithelial-to-mesenchymal transition by transforming growth factor-β. Cancer Sci 96:791–800

Topczewska JM, Postovit LM, Margaryan NV et al (2006) Embryonic and tumorigenic pathways converge via Nodal signaling: role in melanoma aggressiveness. Nat Med 12:925–932

Trochon-Joseph V, Martel-Renoir D, Mir LM et al (2004) Evidence of antiangiogenic and anti-metastatic activities of the recombinant disintegrin domain of metargidin. Cancer Res 64:2062–2069

Tsao H, Chin L, Garraway LA, Fisher DE (2012) Melanoma: from mutations to medicine. Genes Dev 26:1131–1155

Tumbar T, Guasch G, Greco V et al (2004) Defining the epithelial stem cell niche in skin. Science 303:359–363

Ungerer C, Doberstein K, Burger C et al (2010) ADAM15 expression is downregulated in melanoma metastasis compared to primary melanoma. Biochem Biophys Res Commun 401:363–369

Van Belle P, Rodeck U, Nuamah I et al (1996) Melanoma-associated expression of transforming growth factor-β isoforms. Am J Pathol 148:1887–1894

Wrzesinski SH, Wan YY, Flavell RA (2007) Transforming growth factor-β and the immune response: implications for anticancer therapy. Clin Cancer Res 13:5262–5270

Yaar M, Wu C, Park HY et al (2006) Bone morphogenetic protein-4, a novel modulator of melanogenesis. J Biol Chem 281:25307–25314

Yang G, Li Y, Nishimura EK et al (2008) Inhibition of PAX3 by TGF-β modulates melanocyte viability. Mol Cell 32:554–563

Yilmaz M, Maass D, Tiwari N et al (2011) Transcription factor Dlx2 protects from TGFβ-induced cell-cycle arrest and apoptosis. EMBO J 30:4489–4499

Yu L, Harms PW, Pouryazdanparast P et al (2010) Expression of the embryonic morphogen Nodal in cutaneous melanocytic lesions. Mod Pathol 23:1209–1214

Chapter 11
The Transforming Growth Factor-Beta (TGF-β) in Liver Fibrosis

Isabel Fabregat and Patricia Sancho

Abstract Liver fibrosis is the final consequence of many chronic liver injuries that later develop in cirrhosis and hepatocellular carcinoma (HCC), which are leading causes of morbidity and mortality worldwide. The transforming growth factor-beta (TGF-β) represents a key cytokine that increases in liver in its activated form upon damage and triggers important cellular events during any progression stage of the disease. TGF-β mediates activation of hepatic stellate cells (HSCs) to myofibroblasts and induces cell death and epithelial mesenchymal transition (EMT) of hepatocytes. Both processes may facilitate extracellular matrix (ECM) deposition and scar formation. Regulatory T cells, important negative regulators of inflammation, depend on TGF-β for terminal differentiation, indicating its impact in the inflammatory response. Oxidative stress plays an essential role in mediating liver fibrosis, and recent studies demonstrate that TGF-β contributes to the reactive oxygen species (ROS) production and oxidative damage. Indeed, the active implication of TGF-β signaling in the progression of liver fibrosis makes this cytokine an attractive therapeutic target. In addition to the increasing number of compounds aimed at direct inhibition of the TGF-β pathway, the recent discovery of new downstream molecules with crucial roles in liver fibrosis development, such as NADPH oxidases, is opening the therapeutic perspectives.

Keywords Cell death • Chronic injury • Epithelial mesenchymal transition (EMT) • Hepatic stellate cell (HSC) • Hepatocyte • Inflammation • Myofibroblast • NADPH oxidases (NOX) • ROS

I. Fabregat (✉)
Bellvitge Biomedical Research Institute (IDIBELL), Barcelona, Spain
e-mail: ifabregat@idibell.cat

P. Sancho
Spanish National Cancer Research Center (CNIO), Madrid, Spain

A. Moustakas and K. Miyazawa (eds.), *TGF-β in Human Disease*,
DOI 10.1007/978-4-431-54409-8_11,

Abbreviations

ALK5	Activin receptor-like kinase 5
CLD	Chronic liver disease
ECM	Extracellular matrix
EMT	Epithelial mesenchymal transition
HCC	Hepatocellular carcinoma
HCV	Hepatitis C virus
HSCs	Hepatic stellate cells
NASH	Non-alcoholic steatohepatitis
NOX	NADPH oxidase
ROS	Reactive oxygen species

11.1 Introduction

Liver fibrosis is the final consequence of many chronic liver injuries (Brenner 2009) that later develop in cirrhosis and hepatocellular carcinoma (HCC), which are leading causes of morbidity and mortality worldwide. The main etiologies of chronic liver diseases in industrialized countries include chronic hepatitis C virus (HCV) infection, alcohol, and non-alcoholic steatohepatitis (NASH). Regardless of the etiology, all the chronic liver diseases follow a common course: from middle inflammation, to more severe inflammation, to fibrosis and finally to cirrhosis. A complex and multistep process is involved in the progression to a chronic liver injury, which is evidenced by intracellular signal transduction changes, alteration in cell–cell and cell–extracellular matrix contacts and a drastic transdifferentiation of different cell types. During many years, research has been focused on the dissection of these pathways to develop new therapeutic approaches.

One of the cytokines whose levels increase in any kind of chronic liver disease (CLD) is the transforming growth factor-beta (TGF-β), which triggers important cellular events related to fibrogenesis and repair (Dooley and ten Dijke 2012; Hayashi and Sakai 2012). Most liver cells are sensitive to TGF-β, inducing both the canonical Smad-mediated and the non-canonical Smad-independent downstream signals. During the development of fibrosis, hepatic stellate cells (HSCs) respond to TGF-β moving to a myofibroblast phenotype, which in turn produces the higher deposition of extracellular matrix (ECM) proteins. TGF-β also plays essential roles during the inflammatory process linked to liver fibrosis, since it mediates the terminal differentiation of regulatory T cells, important negative regulators of inflammation. TGF-β induces cell death and epithelial mesenchymal transition of hepatocytes, and recent evidences indicate that this process might also contribute to the ECM deposition and scar formation. Activation of liver sinusoidal endothelial cells and neoangiogenesis is also partially facilitated by TGF-β. Finally, TGF-β contributes to the reactive oxygen species (ROS) production and it is well known that oxidative stress plays an essential role in mediating liver fibrosis.

While the TGF-β role as "master" cytokine in chronic liver diseases is very clear, the complexity of the underlying response in cells and in the organ leading to the drastic changes observed is currently not fully understood. In this chapter, we will update the knowledge about the essential role of TGF-β in liver fibrosis, the proposed molecular mechanisms that mediate its actions, as well as new therapeutic approaches to inhibit its signaling.

11.2 Animal Models for the Study of Liver Fibrosis

During the last years, several mouse models of experimental fibrosis have been used for the study of the pathogenesis and molecular mechanisms associated with the diverse human pathologies leading to liver fibrosis. Among them, chemically induced fibrosis with hepatotoxic agents has been extensively investigated: thioacetamide, dimethylnitrosamine (DMN), and, most importantly, carbon tetrachloride (CCl_4), cause centrilobular parenchymal injury and fibrosis. These agents are processed by the cytochrome P-450 in hepatocytes, which releases damaging products causing massive hepatocyte cell death (Constandinou et al. 2005). In addition, concanavalin A is commonly used as a model for human chronic hepatitis since it triggers immune system-mediated fibrosis with similar histological characteristics (Louis et al. 2000). Finally, bile duct ligation constitutes a very used model for cholestatic fibrosis, triggering extrahepatic biliary atresia and primary sclerosing cholangitis (Constandinou et al. 2005). Most importantly, knockout mice have become a powerful strategy for the last years to study the molecular mechanisms of fibrosis, focusing on the contribution of one or more genes to the pathogenesis of the disease (Hayashi and Sakai 2011). These knockout mice, in addition, have helped to establish new genetic models of liver fibrosis such as the $Mdr2^{-/-}$ mouse, which develops spontaneous sclerosing cholangitis (Fickert et al. 2004).

Studies with transgenic mice that overexpress TGF-β have demonstrated that this cytokine alone is sufficient to induce fibrosis, independently of the primary cause of the disease. The hepatic expression of TGF-β induces upregulation of pro-collagen I and pro-collagen III mRNAs in the hepatic tissue, and deposition of extracellular matrix in the sinusoid (Kanzler et al. 2001). Similar results were obtained in a conditional tetracycline-regulated expression of TGF-β1 in liver of transgenic mice, where fibrosis progressed to an intermediary state (Ueberham et al. 2003). In both cases, activation of HSC was observed. Inversely, in experimental models of liver fibrosis, the fibrogenic process can be attenuated simply by blockade of TGF-β signaling (Ueno et al. 2000). These results together point out to the relevant role played by TGF-β in liver fibrogenesis.

11.3 Effects of TGF-β in Liver Cells: Relevance in Liver Fibrosis

The response to chronic liver injury involves different cell types and undergoes different phases (Dooley and ten Dijke 2012). Initially, liver injury induces epithelial cell stress, which causes cell death either through necrosis or apoptosis. Death-mediated signals and necrotic cells induce a strong inflammation and wound-healing response, as well as activation of HSC. These events may conduct to liver regeneration and repair, but acute setting addresses a fibrogenic process. Below we will discuss the role of TGF-β in these processes.

11.3.1 Role of TGF-β in the Activation of Hepatic Stellate Cells to Myofibroblasts

Regulation of extracellular matrix accumulation in acute and chronic liver injuries involves different mechanisms, but HSCs appear to be the principal effecter in all cases (Friedman 2010). The HSCs are the major storage site of retinoids in the body and are present in the space of Disse in close contact with hepatocytes and the sinusoid. When the HSC is activated, it loses its retinoid content, increases proliferation and motility, expresses new markers, such as smooth muscle actin, and produces ECM proteins. In the normal liver, sinusoidal endothelial cells and Kupffer cells (macrophages) contain relatively high levels of TGF-β mRNA, whereas HSCs express little amounts of TGF-β. However, in response to pro-fibrogenic stimuli HSCs express the three different isoforms of TGF-β and contribute to the development of fibrosis through both autocrine and paracrine loops of TGF-β-stimulated collagen production (Inagaki et al. 2005). The main cell responsible for the fibrosis is the myofibroblast, which produces the fibrous scar found in all chronic liver diseases. Different lines of evidence support the hypothesis that one of the main sources of these myofibroblasts are the quiescent HSC that become activated in response to TGF-β (Fig. 11.1): (1) downregulation of TGF-β expression in liver, by using adenoviruses or genetically modified animals, reveal a failure in HSC activation and fibrogenesis (Hellerbrand et al. 1999; Kanzler et al. 1999); (2) in vitro experiments reveal that HSCs are responsive to TGF-β treatment and transduce a signal that may play important roles in fibrogenesis (Dooley et al. 2000); (3) gene transfer of Smad7, the member of inhibitory Smads, inhibits experimental fibrogenesis, which is coincident with arrested transdifferentiation of primary cultured HSCs to myofibroblasts (Dooley et al. 2003). The response of HSCs to TGF-β, leading to, e.g., induction of α2 (I) collagen expression, is mediated by phosphorylation of Smad2 and Smad3 and subsequent nuclear translocation of a Smad-containing complex (Dooley et al. 2001). Maximal expression of collagen type I in activated HSCs requires Smad3 in vivo and in culture (Schnabl et al. 2001). Interestingly, Smad3 is not necessary for HSC activation as assessed by alpha-SMA expression, but is necessary for inhibition

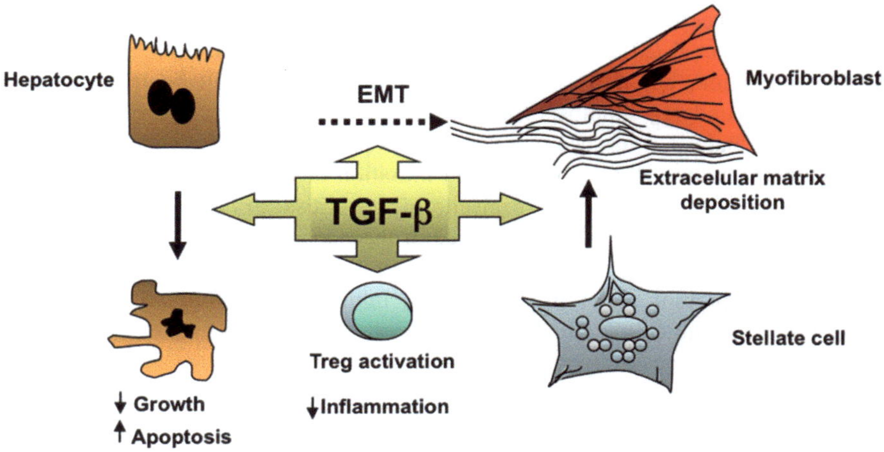

Fig. 11.1 Effects of TGF-β in liver and inflammatory cells. *Effects that may counteract liver fibrosis*: (1) TGF-β triggers activation of hepatic stellate cells to myofibroblats, which are considered the main producers of extracellular matrix proteins. (2) TGF-β induces growth inhibition and cell death of hepatocytes, which impair liver regeneration. (3) The hepatocytes that survive to the inhibitory signals may respond to TGF-β undergoing epithelial mesenchymal transition (EMT). Although controversial, different reports indicate that this process exists and would facilitate ECM deposition and scar formation. *Effects that may counteract liver fibrosis*: regulatory T cells, important negative regulators of inflammation, depend on TGF-β for terminal differentiation, which would have beneficial consequences impairing the fibrotic process

of proliferation of HSCs, which is TGF-β-dependent, and is required for TGF-β1-mediated Smad-containing DNA-binding complex formation in cultured HSCs. These data indicate that HSCs are responsive to TGF-β treatment and transduce a signal that may play an important role in liver fibrogenesis. Myofibroblasts display decreased availability of surface receptors for TGF-β, which could be based on autocrine stimulation. However, lack of activated Smad complexes with DNA-binding activity and absence of α2 (I) collagen transcription inhibition by latency-associated peptide (LAP)/anti-TGF-β antibody raise the possibility of TGF-β signaling independent receptor downregulation in myofibroblasts (Dooley et al. 2000).

11.3.2 Role of TGF-β on Hepatocytes: Relevance in Liver Fibrosis

Chronic liver injuries are characterized by persistent hepatocyte damage and death, induced by chemical toxicity, metabolic overload, viral/microbial infections, etc., which cause metabolic deregulation and oxidative stress. Several modes of cell death have been classified in the damaged liver, including apoptosis and necrosis.

It is now fully accepted that hepatocyte death is critical for hepatic fibrosis (Brenner 2009; Nikolaou et al. 2012). It appears that the primary response to injury would be liver regeneration, but if it is blocked, the default mode will be liver fibrosis. If hepatocytes undergo apoptosis without compensatory proliferation, fibrosis again would result. Indeed, it has been proven that apoptosis and phagocytosis of hepatocytes directly induce HSC activation and initiation of fibrosis (Jiang et al. 2010), and hepatocyte apoptotic bodies during chronic hepatitis C infection amplify stellate cell activation. TGF-β1 might be involved in the impairment of liver regeneration and in amplifying hepatocyte apoptosis (Fig. 11.1). Indeed, TGF-β is an important regulatory suppressor factor in hepatocytes, inhibiting proliferation (Carr et al. 1986) and inducing cell death (Oberhammer et al. 1992). The increase in TGF-β levels in the first stages of liver fibrogenesis may be responsible for an imbalance in the proliferative and survival signals in hepatocytes, contributing to the failure in liver regeneration.

However, paradoxically, in addition to its suppressor effects, TGF-β also induces anti-apoptotic signals in proliferating hepatocytes and hepatoma cells (Valdes et al. 2004; Caja et al. 2007), through activation of the epidermal growth receptor (EGFR) pathway (Murillo et al. 2005). Cells that survive to TGF-β-induced apoptotic signals undergo epithelial mesenchymal transition (EMT) (Gotzmann et al. 2002; Valdes et al. 2002; Caja et al. 2007; 2011; Kaimori et al. 2007), a physiological process during embryogenesis, in which an epithelial cell loses expression of adhesion molecules, such as E-cadherin, and other components responsible for cell polarity. Instead, they express mesenchymal components of cytoskeleton and acquire motility and scattering properties (Thiery et al. 2009). Certain evidences indicate that a crosstalk exists between the genetic programs that control TGF-β-induced growth arrest/apoptosis and those that regulate EMT. Indeed, once hepatocytes undergo EMT they become resistant to TGF-β-induced apoptosis (Valdes et al. 2002), a process in which transcription factors of the Snail family, repressors of the E-cadherin gene, are involved (Franco et al. 2010). A closely related phenotypic conversion is also detected in some models of fibrosis and may be associated with disease progression (Lopez-Novoa and Nieto 2009). In the case of the liver, the role of EMT from hepatocytes to myofibroblasts is perhaps the most intriguing and controversial of recent hypotheses on the origin mechanisms of liver fibrosis (Wells 2011). Strong evidences indicate that hepatocytes from transgenic animals overexpressing Snail (a master gene involved in EMT through its capacity to repress E-cadherin gene, among others) fully undergo EMT (Franco et al. 2010) and might propagate liver fibrosis progression (Rowe et al. 2011). However, under normal genetic background, data from different experimental approaches in animals and humans show controversy. Some reports support a role for EMT in epithelial cells in the liver that might transform into myofibroblasts (Zeisberg et al. 2007; Dooley et al. 2008), whereas others show no evidence of EMT in models of hepatic fibrosis (Taura et al. 2010; Chu et al. 2011). Further experiments are required to fully conclude that TGF-β plays a role in transdifferentiation of hepatocytes to myofibroblasts through EMT processes (Fig. 11.1).

11.4 Crosstalk Between TGF-β and Inflammatory Signals

Inflammation plays an essential role in the development of liver fibrosis. When a chronic injury takes place, a large infiltration of mononuclear cells, which include macrophages, lymphocytes, eosinophils, and plasma cells, occur. Mobilization of lymphocytes produces lymphokines that activate macrophages, which, in turn, stimulate lymphocytes, fibroblasts, and other inflammatory cells, thus setting the stage for persistence of an inflammatory response (Wynn and Barron 2010). Furthermore, macrophages produce pro-fibrotic mediators, including TGF-β1 and PDGF, and control extracellular matrix turnover by regulating the balance of various matrix metalloproteases and tissue inhibitors of metalloproteases. Examples of knockout mice that are resistant to fibrosis because they have less inflammation include those with gene deletions of TNF-α or Toll-like receptor 4 (TLR4), among others (Kitamura et al. 2002; Seki et al. 2007).

Crosstalk between TGF-β and inflammatory signals occurs at different levels. On one side, from studies in different tissues including the liver, TGF-β is believed to play an important role in the regulation of the immune system. Indeed, it activates the differentiation of regulatory T cells (Treg) (Hammerich et al. 2011), a unique subset of $CD4^+$ T-helper cells that control effector T-cell responses to prevent autoimmune reactions. Activated Treg produce the anti-inflammatory cytokine IL-10, which would have beneficial effects in a pro-fibrotic process (Fig. 11.1). However, on the other side, perturbation of TGF-β signaling by pro-inflammatory cytokines in liver cells contributes to both fibrogenesis and carcinogenesis (fibro-carcinogenesis). Smad proteins have intermediate linker regions between conserved Mad homology (MH) 1 and MH2 domains. TGF-β type I receptor and pro-inflammatory cytokine-activated kinases differentially phosphorylate Smad2 and Smad3 to create phosphoisoforms that are phosphorylated at the COOH-terminal (C), linker (L), or both (L/C) regions (Matsuzaki 2009). TGF-β and pro-inflammatory cytokines synergistically enhance collagen synthesis by activated hepatic stellate cells via pSmad2L/C and pSmad3L/C pathways. During chronic liver disease progression, pre-neoplastic hepatocytes persistently affected by TGF-β together with pro-inflammatory cytokines come to exhibit the same carcinogenic (mitogenic) pSmad3L and fibrogenic pSmad2L/C signaling as do myofibroblasts, thereby accelerating liver fibrosis while increasing risk of HCC (Matsuzaki 2009). c-Jun N-terminal kinase (JNK) activated by pro-inflammatory cytokines is mediating this perturbed hepatocytic TGF-β signaling (Yoshida et al. 2005). Under normal conditions, to avoid unlimited extracellular matrix deposition, Smad7 induced by TGF-β negatively regulates its pro-fibrogenic response. In the presence of pro-inflammatory cytokines and activation of the JNK and MAPKs pathways, Smad7 cannot be induced by the pSmad3L pathway (Yoshida and Matsuzaki 2012). Another example of modulation of TGF-β signals by pro-inflammatory cytokines comes from studies in the TLR4-chimeric mice (Seki et al. 2007). In quiescent HSCs, TLR4 activation not only upregulates chemokine secretion and induces chemotaxis of Kupffer cells, but also downregulates the TGF-β pseudoreceptor Bambi, to sensitize HSCs to TGF-β-induced signals and allow unrestricted activation by Kupffer cells. Clinical relevance of the crosstalk between TLR4 and the TGF-β/Bambi signaling has been demonstrated in studies of liver fibrosis progression in hepatitis C and hypercholesterolemic patients (Guo et al. 2009; Teratani et al. 2012). Finally, there is evidence that Th2 cytokines cooperate with TGF-β to induce fibrosis (Wynn 2008). IL-13 activates the production of latent TGF-β in macrophages and upregulates the expression of proteins that cleave the Latent Association Protein (LAP), which contributes to the release of active TGF-β (Lee et al. 2001).

11.5 Reactive Oxygen Species in Liver Fibrosis: Connection with the TGF-β Pathway

ROS, including H_2O_2, OH·, and O_2^-, are critical intermediates in both the normal physiology and pathological conditions of liver cells. When the equilibrium between ROS generation and the antioxidant defense of the cell is disrupted, it results in an oxidative stress process (Sies and Cadenas 1985). As commented above, fibrosis has been well documented in many chronic liver diseases, usually beginning with an inflammatory phase which progresses to fibrosis after chronic oxidative stress (Diesen and Kuo 2010). ROS play a central role in the development of liver fibrosis/cirrhosis by both alcohol and hepatitis virus core proteins (Perlemuter et al. 2003; Dionisio et al. 2009). In addition, oxidative stress markers have been detected in the serum of and biopsy samples from liver cirrhosis patients and in experimental liver fibrosis/cirrhosis animals (Yadav et al. 2002; Pawlak et al. 2008). Moreover, in liver biopsies, areas of fibrosis were localized to areas with increased 4-hydroxy-2′-nonenal (4-HNE), a marker of lipid peroxidation (MacDonald et al. 2001; Seki et al. 2005).

In relation to TGF-β, ROS play a complex role promoting fibrosis progression. On one side, they constitute a commonly known downstream effector implicated in TGF-β signaling (Liu and Gaston Pravia 2010). On the other side, ROS may also promote fibrosis activating latent TGF-β through either LAP direct oxidation and subsequent release of the cytokine (Pociask et al. 2004) or via MMP activation (Wang et al. 2005). Indeed, LAP/TGF-β1 complex has been proposed to function as an oxidative stress sensor (Jobling et al. 2006). Finally, ROS can also stimulate the expression and secretion of TGF-β in a positive feedback loop in many types of cells, including hepatic stellate cells and hepatocytes (Proell et al. 2007; Boudreau et al. 2009).

11.5.1 Subcellular Sources of ROS in Liver Fibrosis

The primary cellular sources of oxidative stress during the inflammatory phase of liver fibrosis are mainly neutrophils, Kupffer cells, and, specially, hepatocytes. Although for many years the mitochondria have been considered as the major source for ROS in the living cells, we have to consider two additional ROS-producing systems playing determinant roles in the liver pathophysiology, such as the P450 system in hepatocytes and the NADPH oxidases (NOX) proteins in different liver cells.

Mitochondria play a central role for ROS production in the liver, since hepatocytes contain hundreds of these organelles and the mitochondrial electron transport is disrupted in a great number of pathophysiological circumstances, resulting in increased electron leak (Murphy 2009). Indeed, several reports have suggested a central role for mitochondrial ROS in hepatic toxicity in models of hepatic cholestasis (Graf et al. 2002; Fang et al. 2004) and alcoholic disease (Kukielka et al. 1994; Zhu et al. 2012). Most importantly, several reports have shown that direct treatment with TGF-β induces a prolonged mitochondrial ROS production in rat hepatocytes (Albright et al. 2003; Herrera et al. 2004). This fact can be attributed to its capacity of downregulating the expression of several antioxidant enzymes, such as glutaredoxin, catalase, superoxide dismutase, and glutathione peroxidase (GPx) (Franklin et al. 2003; Herrera et al. 2004).

CYP2E1, the hepatocytic member of the cytochrome P450 oxidase system, is involved in the metabolism of xenobiotics in the body. Most drugs and hepatotoxins are detoxified by CYP2E1, which can generate ROS as a byproduct of the oxidative reaction. Both in vitro experiments and animal studies in vivo have demonstrated that CYP2E1 is an important source of ROS in alcohol-induced liver injury, and its expression is inducible by alcohol (Zhu et al. 2012). Importantly, it has been reported that TGF-β enhances hepatocyte toxicity in cells overexpressing CYP2E1 upon ethanol exposure (Zhuge and Cederbaum 2006).

Other main source of ROS implicated in TGF-β signaling and fibrosis is the NOX family of NADPH oxidases. This family has been discovered for homology to $gp91^{phox}$, the phagocytic oxidase. Nowadays, the NOX family includes seven different members NOX1 to NOX5, DUOX1, and DUOX2 (Bedard and Krause 2007) whose main function is active ROS production. NOX proteins have been previously related to fibrosis in several organs such as lung (Hecker et al. 2009), pancreas (Masamune et al. 2008), kidney (Sedeek et al. 2010), and heart (Cucoranu et al. 2005). In the liver, several reports have demonstrated a key role for NOX proteins in the progression of hepatic fibrosis (De Minicis et al. 2010; Cui et al. 2011; Paik et al. 2011; Jiang et al. 2012; Sancho et al. 2012). The isoforms expressed by the different resident populations of the liver are mainly NOX1, NOX2, and NOX4 (Paik et al. 2011).

11.5.2 Implication of ROS in the Molecular Mechanisms Mediating Liver Fibrosis

One of the most studied mechanisms of fibrogenesis actually influenced by ROS is myofibroblast activation. In the liver, stellate cell transdifferentiation into myofibroblast is inhibited by antioxidants (Foo et al. 2011; Abhilash et al. 2012). Indeed,

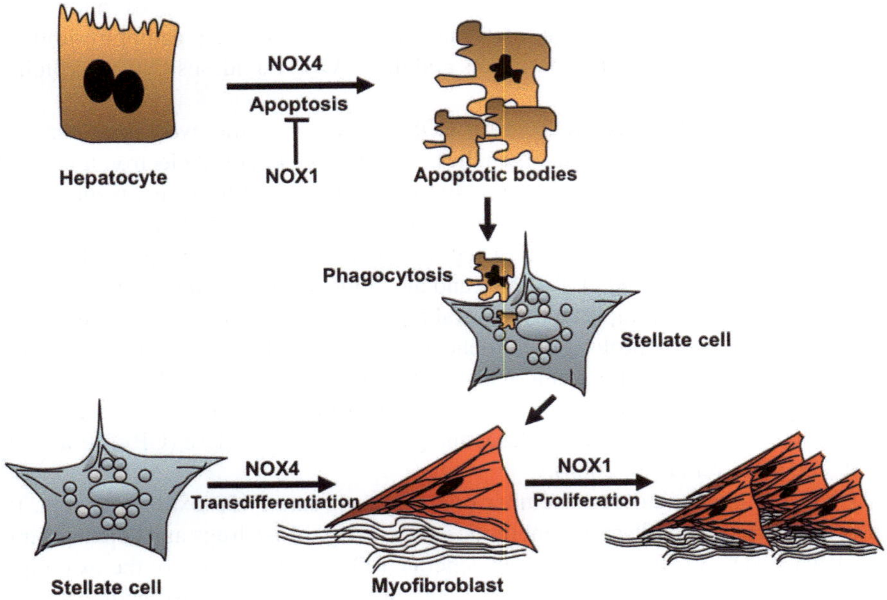

Fig. 11.2 NOX proteins play crucial different roles during liver fibrosis development. Opposite functions of NOX1 and NOX4 in hepatocytes: NOX1 protects cells from pro-apoptotic stimuli and mediates proliferation, while NOX4 promotes TGF-β-induced cell death. Afterwards, stellate cells can phagocytose the resulting apoptotic bodies, which functions as a triggering signal for activation. Primed stellate cells can also suffer transdifferentiation into myofibroblasts in response to TGF-β, a process where NOX4 plays a determinant role controlling the acquisition of the activated phenotype. Finally, and once fully activated, NOX1 favors myofibroblast proliferation, contributing to fibrosis development

NOX4 downstream TGF-β has been described as the main mediator for myofibroblast activation in different organs such as heart (Cucoranu et al. 2005), lung (Hecker et al. 2009), kidney (Bondi et al. 2010), and diseased prostatic stroma (Sampson et al. 2011). Equivalently, it has been demonstrated in cultured HSC that TGF-β-induced transdifferentiation is accompanied by NOX4-derived ROS (Proell et al. 2007), which can be a useful target for therapeutic approaches (Ikeda et al. 2011). Very recently, two different reports have described a key role for NOX4 in hepatic myofibroblasts activation downstream TGF-β (Jiang et al. 2012; Sancho et al. 2012) both in vivo and in vitro (Fig. 11.2). In these works, HSC activation was attenuated either by NOX4 downregulation or in a Nox4$^{-/-}$ genetic background, and, importantly, the myofibroblast-activated state could be also reversed by NOX4 downregulation (Sancho et al. 2012). However, the role of NOX proteins in liver fibrogenesis is not only circumscribed to NOX4. Thus, studies performed in Nox1$^{-/-}$, Nox2$^{-/-}$, or p47phox$^{-/-}$ mice have pointed out the importance of NOX1 and NOX2 in fibrosis development (De Minicis et al. 2010; Jiang et al. 2010; Cui et al. 2011; Paik et al. 2011). Concretely, NOX1 promotes myofibroblast proliferation by PTEN inactivation to positively regulate an Akt/FOXO4/p27 signaling pathway (Cui et al. 2011). Indeed, NOX1 seems to mediate the pro-fibrogenic effects exclusively in endogenous liver cells, while NOX2 could be implicated in both endogenous liver cells and bone marrow-derived cells (Paik et al. 2011), possibly acting in the process of phagocytosis of dead hepatocytes (Jiang et al. 2010) (Fig. 11.2).

Finally, promotion of hepatocyte apoptosis constitutes also a crucial mechanism influenced by TGF-β-induced ROS. In fact, and as mentioned before, TGF-β induces apoptosis through ROS that is derived from both mitochondria and NOX activity (Herrera et al. 2004). Indeed, pretreatment with antioxidants block apoptosis (Sanchez et al. 1996; Herrera et al. 2001). Recently, it has been described that hepatocytes express different members of the NOX family, mainly NOX1, NOX2, and NOX4 (Murillo et al. 2007), which play opposite roles in the control of hepatocyte survival and death. Indeed, NOX4 is necessary to mediate apoptosis induced by TGF-β (Carmona-Cuenca et al. 2008; Caja et al. 2009), but the pro-apoptotic effect of the cytokine can be attenuated when NOX1 is active (Sancho et al. 2009; Sancho and Fabregat 2011; Ortiz et al. 2012) (Fig. 11.2). In addition, $Nox4^{-/-}$ hepatocytes are also resistant to apoptosis induction by other stimuli, such as FasL and TNF-α/actinomycin D (Jiang et al. 2012). In addition, NOX1 activity might further contribute to the inflammatory process promoting COX-2 expression and prostaglandin synthesis in hepatocytes (Sancho et al. 2011). Interestingly, dual NOX4/NOX1 pharmacological inhibition with GKT137831 is able to diminish both the apparition of fibrogenic markers and hepatocyte apoptosis in vivo upon bile duct ligation (Jiang et al. 2012), reinforcing the relevant role of NOX1 and NOX4 in liver fibrosis and opening new perspectives for its treatment.

11.6 TGF-β Pathway Inhibitors as a Promising Therapy in Liver Fibrosis

During the last years, after the role of TGF-β signaling in cancer and other pathologies, including fibrosis, became established, a great effort has been made in order to develop different approaches to inhibit TGF-β pathway. Thus, the number of possible compounds used either in preclinical or clinical studies related to fibrosis is continuously growing, thanks to previous experiences in other pathologies. The different strategies to block the TGF-β pathway can be classified as: (1) ligand traps, which include blocking antibodies and inhibitory peptides; (2) antisense oligos; (3) receptor kinase inhibitors; (4) Smad inhibitors; and (5) indirect inhibitors (Table 11.1). However, the list of compounds tested for liver fibrosis is rather reduced when compared with all the available inhibitors, since clinical efforts have concentrated for the last few years in blocking the underlying pathology specific for each type of fibrosis.

One of the most studied strategies for inhibiting the TGF-β pathway related to liver fibrosis is the ligand trapping, either by soluble receptors or inhibitory peptides. Indeed, several studies have demonstrated the antifibrotic potential of a

Table 11.1 Current preclinical and clinical TGF-β-based therapeutic strategies

Class	Drug	Target	Disease	References/Trial ID
Blocking antibodies	CAT-192	TGF-β_1	Systemic sclerosis	NCT00043706
			Myelofibrosis	NTC01291784
	CAT-152	TGF-β_2	Trabeculectomy, renal fibrosis	Hill et al. (2001), Grehn et al. (2007), Khaw et al. (2007)
	LY238770	TGF-β_1	Diabetes	NTC01113801
	GC1008	Pan-TGF-β	Systemic sclerosis	NCT01284322
			Myelofibrosis	NCT01291784
			Idiopathic pulmonary fibrosis	NCT00125385
			Glomerulosclerosis (FSGS)	NCT00464321
Peptide inhibitor	sTβRII	Pan-TGF-β	Liver fibrosis	Sullivan et al. (2010), Yao et al. (2010), Nakamuta et al. (2005), Cui et al. (2003), Yata et al. (2002), Ueno et al. (2000), George et al. (1999)
	P144	TGF-β_1	Skin fibrosis	NCT00781053
			Myocardial and liver fibrosis	Ezquerro et al. (2003), Hermida et al. (2009)
Kinase inhibitors	GW388788	TβRII/ALK5	Infarction, renal and skin fibrosis	Lagares et al. (2010), Petersen et al. (2008), Tan et al. (2010)
	SKI2162	ALK5	Peyronie's disease	Piao et al. (2010)
	GW6604	ALK5	Liver fibrosis	De Gouville et al. (2005)
	LY2109761	TβRII/ALK5	Several cancers, pulmonary fibrosis	Connolly et al. (2011), Fransvea et al. (2008), Flechsig et al. (2012), Ganapathy et al. (2010), Lacher et al. (2006), Zhang et al. (2010, 2011)
	SD208	ALK5	Scleroderma	Chen et al. (2006)
	SM16	ALK5	Vascular fibrosis	Fu et al. (2008)
	IN-1130	ALK5	Renal fibrosis, Peyronie's disease	Moon et al. (2006), Ryu et al. (2009)
Smad inhibitors	HSc025	Smad-dependent transcription	Systemic sclerosis	Hasegawa et al. (2009)
			Liver fibrosis	Higashi et al. (2011)
	SiS3	Smad3	Liver fibrosis	Matsubara et al. (2011)

(continued)

Table 11.1 (continued)

Class	Drug	Target	Disease	References/Trial ID
Indirect inhibitors	STX-100	αvβ6 integrin	Idiopathic pulmonary fibrosis	NCT01371305
	Tranilast	Not known	Rheumatoid arthritis	NCT00882024
			Lupus (LMDF)	Koike et al. (2011)
			Crohn's disease	Oshitani et al. (2007)
			Diabetes	Martin et al. (2005), Kelly et al. (2007)
			Hypertension	Hocher et al. (2002), Kagitani et al. (2004)
			Liver fibrosis	Ikeda et al. (1996), Uno et al. (2008), Said et al. (2012)

soluble type II receptor. This antagonist, consisting of a chimeric IgG containing the extracellular portion of the TGF-β type II receptor, was able to inhibit several fibrosis markers when tested in the model of bile duct ligation in mice (George et al. 1999). Importantly, it also showed to be dose-dependently effective in models of chemical liver fibrosis with carbon tetrachloride or DMN at either short- or long-term evaluation (Yata et al. 2002; Nakamuta et al. 2005). Interestingly, in this last publication authors demonstrated that remote delivery of the compound in the muscle is effective in inhibiting hepatic alterations. Recently, effectiveness of this compound has been improved using a novel strategy consisting of a fusion protein formed by the soluble portion of TβRII and IFN-γ administered intraperitoneally in a model of experimental fibrosis in rats (Yao et al. 2010). Alternatively, several studies have exploited a genetic approach using adenovirus containing the ectodomain of the TβRII alone or fused with other proteins. This approach was tested in vitro by infection of primary rat HSC, showing an inhibitory effect on the autocrine TGF-β production concomitant with transdifferentiation into myofibroblasts (Cui et al. 2003). In addition, the adenovirus strategy has shown to be effective in vivo with no apparent side effects. In this case, adenovirus expressing the entire ectodomain of the human TβRII fused to the Fc portion of human IgG (AdTbeta-ExR) was injected in the skeletal muscle in rats (Ueno et al. 2000).

Additionally, a soluble form of the type III receptor (betaglycan) has also been tested both at the preclinical and at the clinical levels for treatment of different fibrosis-related diseases, such as pulmonary, cardiac, and skin fibrosis (Liu et al. 2002; Hermida et al. 2009). Regarding the liver, a short peptide derived from this receptor, P144, showed in vitro efficacy blocking TGF-β-dependent stimulation of the human α2(I) collagen promoter. Importantly, intraperitoneal administration of P144 was able to diminish histological fibrosis markers and the number of myofibroblats in rats treated with carbon tetrachloride, with no apparent side effects (Ezquerro et al. 2003).

Although inhibiting the kinase activity of TGF-β receptors is clearly the most recurrent strategy used in current preclinical and clinical therapies related to this cytokine, no significant studies referring to liver fibrosis have been performed. However, a great number of inhibitors of the ATP binding site of TβRI kinase (activin receptor-like kinase 5: ALK5) have been designed and preclinically tested in various fibrosis-related diseases and, may be eventually, used for liver fibrosis treatment. Up to date, only one kinase inhibitor, GW6604, has shown beneficial effects preventing liver damage in both an acute model of liver disease and a chronic model of dimethylnitrosamine (DMN)-induced liver fibrosis (de Gouville et al. 2005). In this last chronic model, where DMN was administered for 6 weeks and GW6604 dosed for the last 3 weeks, mortality was prevented, which correlated with reduced matrix deposition and decreased liver function deterioration. Another compound, IN-1130, has also potential applicability for liver fibrosis treatment. Indeed, this compound, which has been positively tested for renal fibrosis and Peyronie's disease (Moon et al. 2006; Ryu et al. 2009), is preferentially accumulated in the liver upon oral administration (Kim et al. 2008).

Alternatively, new compounds have been recently discovered which act by inhibiting Smad-dependent transcriptional inhibition. The indol-derivative SiS3 is a specific inhibitor of Smad3 phosphorylation and activity, and it was first described to be effective in inhibiting the activated phenotype of scleroderma fibroblast (Jinnin et al. 2006). Regarding liver fibrosis, only one in vitro preclinical study using a cholestatic disease model has been published so far. In this work, SiS3 treatment was able to inhibit the expression of several genes related to cholestasis development (Matsubara et al. 2011). In addition, the compound Hsc025 is a Smad-dependent transcriptional inhibitor that has been effectively tested in mice models of skin, pulmonary and hepatic fibrosis by oral administration (Hasegawa et al. 2009; Higashi et al. 2011). Indeed, in vitro treatment with HSc025 significantly suppressed collagen gene expression in cultured HSC, while oral administration of HSc025 improved liver injury and hepatic fibrosis degree in mice treated with carbon tetrachloride (Higashi et al. 2011).

Although its effects at the molecular level are not fully understood, tranilast is a drug mainly described as a collagen expression inhibitor, thus possessing antifibrotic properties. Indeed, while it can also inhibit the production of other cytokines, the major effect described for tranilast is the inhibition of both TGF-β expression and action (Miyazawa et al. 1995; Ikeda et al. 1996; Platten et al. 2001). This compound showed potential for hepatic fibrosis treatment several years ago, when it was described that tranilast treatment inhibited the expression of pro-collagen and TGF--β (Ikeda et al. 1996). Importantly, this compound also was effective in two different in vivo models of liver fibrosis. First, using a dietary model of NASH where obese diabetic and nondiabetic rats were fed with a methionine-deficient and choline-deficient diet, treatment with tranilast was effective at two different levels (Uno et al. 2008): on one side, it was able to inhibit fibrosis development and the activation of stellate cells, downregulating the expression of TGF-β, pro-collagen, and plasminogen activator-1; on the other side, it attenuated hepatic inflammation and Kupffer cell recruitment, downregulating the expression of TNFα. In the second model, tranilast was able to improve hepatic fibrosis due to schistosomal infection, proved by a significant improvement of hepatic functions, reduction of the histopathological changes and collagen content, and decreased TGF-β1 levels in serum (Said et al. 2012).

11.7 Conclusions

Liver fibrosis is one of the main causes of mortality worldwide. Nowadays, a lot of effort is being made in order to increase the knowledge of the molecular mechanisms underlying this complicated disease, in which TGF-β seems to play a determinant role. Indeed, the active implication of TGF-β signaling in the progression of liver fibrosis, regardless of its original etiology, makes this cytokine an attractive therapeutic target for the development of new treatments. In addition to the increasing number of compounds aimed at direct inhibition of the TGF-β pathway, the recent discovery of new downstream molecules with crucial roles in liver fibrosis development, such as NADPH oxidases, is opening the therapeutic perspectives. Indeed, specific targeting of these molecules could be an important step forward in the treatment of the disease, since its inhibition may be effective enough avoiding the possible side effects of TGF-β systemic inhibition.

References

Abhilash PA, Harikrishnan R, Indira M (2012) Ascorbic acid supplementation down-regulates the alcohol induced oxidative stress, hepatic stellate cell activation, cytotoxicity and mRNA levels of selected fibrotic genes in guinea pigs. Free Radic Res 46:204–213

Albright CD, Salganik RI, Craciunescu CN, Mar MH, Zeisel SH (2003) Mitochondrial and microsomal derived reactive oxygen species mediate apoptosis induced by transforming growth factor-β1 in immortalized rat hepatocytes. J Cell Biochem 89:254–261

Bedard K, Krause KH (2007) The NOX family of ROS-generating NADPH oxidases: physiology and pathophysiology. Physiol Rev 87:245–313

Bondi CD, Manickam N, Lee DY, Block K, Gorin Y, Abboud HE, Barnes JL (2010) NAD(P)H oxidase mediates TGF-β1-induced activation of kidney myofibroblasts. J Am Soc Nephrol 21:93–102

Boudreau HE, Emerson SU, Korzeniowska A, Jendrysik MA, Leto TL (2009) Hepatitis C virus (HCV) proteins induce NADPH oxidase 4 expression in a transforming growth factor β-dependent manner: a new contributor to HCV-induced oxidative stress. J Virol 83:12934–12946

Brenner DA (2009) Molecular pathogenesis of liver fibrosis. Trans Am Clin Climatol Assoc 120:361–368

Caja L, Ortiz C, Bertran E, Murillo MM, Miro-Obradors MJ, Palacios E, Fabregat I (2007) Differential intracellular signalling induced by TGF-β in rat adult hepatocytes and hepatoma cells: implications in liver carcinogenesis. Cell Signal 19:683–694

Caja L, Sancho P, Bertran E, Iglesias-Serret D, Gil J, Fabregat I (2009) Overactivation of the MEK/ERK pathway in liver tumor cells confers resistance to TGF-{β}-induced cell death through impairing up-regulation of the NADPH oxidase NOX4. Cancer Res 69:7595–7602

Caja L, Bertran E, Campbell J, Fausto N, Fabregat I (2011) The transforming growth factor-β (TGF-β) mediates acquisition of a mesenchymal stem cell-like phenotype in human liver cells. J Cell Physiol 226:1214–1223

Carmona-Cuenca I, Roncero C, Sancho P, Caja L, Fausto N, Fernandez M, Fabregat I (2008) Upregulation of the NADPH oxidase NOX4 by TGF-β in hepatocytes is required for its pro-apoptotic activity. J Hepatol 49:965–976

Carr BI, Hayashi I, Branum EL, Moses HL (1986) Inhibition of DNA synthesis in rat hepatocytes by platelet-derived type β transforming growth factor. Cancer Res 46:2330–2334

Chen Y, Shi-wen X, Eastwood M, Black CM, Denton CP, Leask A, Abraham DJ (2006) Contribution of activin receptor-like kinase 5 (transforming growth factor β receptor type I) signaling to the fibrotic phenotype of scleroderma fibroblasts. Arthritis Rheum 54:1309–1316

Chu AS, Diaz R, Hui JJ, Yanger K, Zong Y, Alpini G, Stanger BZ, Wells RG (2011) Lineage tracing demonstrates no evidence of cholangiocyte epithelial-to-mesenchymal transition in murine models of hepatic fibrosis. Hepatology 53:1685–1695

Connolly EC, Saunier EF, Quigley D, Luu MT, De Sapio A, Hann B, Yingling JM, Akhurst RJ (2011) Outgrowth of drug-resistant carcinomas expressing markers of tumor aggression after long-term TβRI/II kinase inhibition with LY2109761. Cancer Res 71:2339–2349

Constandinou C, Henderson N, Iredale JP (2005) Modeling liver fibrosis in rodents. Methods Mol Med 117:237–250

Cucoranu I, Clempus R, Dikalova A, Phelan PJ, Ariyan S, Dikalov S, Sorescu D (2005) NAD(P)H oxidase 4 mediates transforming growth factor-β1-induced differentiation of cardiac fibroblasts into myofibroblasts. Circ Res 97:900–907

Cui X, Shimizu I, Lu G, Itonaga M, Inoue H, Shono M, Tamaki K, Fukuno H, Ueno H, Ito S (2003) Inhibitory effect of a soluble transforming growth factor β type II receptor on the activation of rat hepatic stellate cells in primary culture. J Hepatol 39:731–737

Cui W, Matsuno K, Iwata K, Ibi M, Matsumoto M, Zhang J, Zhu K, Katsuyama M, Torok NJ, Yabe-Nishimura C (2011) NOX1/nicotinamide adenine dinucleotide phosphate, reduced form (NADPH) oxidase promotes proliferation of stellate cells and aggravates liver fibrosis induced by bile duct ligation. Hepatology 54:949–958

de Gouville AC, Boullay V, Krysa G, Pilot J, Brusq JM, Loriolle F, Gauthier JM, Papworth SA, Laroze A, Gellibert F, Huet S (2005) Inhibition of TGF-β signaling by an ALK5 inhibitor protects rats from dimethylnitrosamine-induced liver fibrosis. Br J Pharmacol 145:166–177

De Minicis S, Seki E, Paik YH, Osterreicher CH, Kodama Y, Kluwe J, Torozzi L, Miyai K, Benedetti A, Schwabe RF, Brenner DA (2010) Role and cellular source of nicotinamide adenine dinucleotide phosphate oxidase in hepatic fibrosis. Hepatology 52:1420–1430

Diesen DL, Kuo PC (2010) Nitric oxide and redox regulation in the liver: Part I. General considerations and redox biology in hepatitis. J Surg Res 162:95–109

Dionisio N, Garcia-Mediavilla MV, Sanchez-Campos S, Majano PL, Benedicto I, Rosado JA, Salido GM, Gonzalez-Gallego J (2009) Hepatitis C virus NS5A and core proteins induce oxidative stress-mediated calcium signalling alterations in hepatocytes. J Hepatol 50:872–882

Dooley S, ten Dijke P (2012) TGF-β in progression of liver disease. Cell Tissue Res 347:245–256

Dooley S, Delvoux B, Lahme B, Mangasser-Stephan K, Gressner AM (2000) Modulation of transforming growth factor β response and signaling during transdifferentiation of rat hepatic stellate cells to myofibroblasts. Hepatology 31:1094–1106

Dooley S, Delvoux B, Streckert M, Bonzel L, Stopa M, ten Dijke P, Gressner AM (2001) Transforming growth factor β signal transduction in hepatic stellate cells via Smad2/3 phosphorylation, a pathway that is abrogated during in vitro progression to myofibroblasts. TGFβ signal transduction during transdifferentiation of hepatic stellate cells. FEBS Lett 502:4–10

Dooley S, Hamzavi J, Breitkopf K, Wiercinska E, Said HM, Lorenzen J, ten Dijke P, Gressner AM (2003) Smad7 prevents activation of hepatic stellate cells and liver fibrosis in rats. Gastroenterology 125:178–191

Dooley S, Hamzavi J, Ciuclan L, Godoy P, Ilkavets I, Ehnert S, Ueberham E, Gebhardt R, Kanzler S, Geier A, Breitkopf K, Weng H, Mertens PR (2008) Hepatocyte-specific Smad7 expression attenuates TGF-β-mediated fibrogenesis and protects against liver damage. Gastroenterology 135:642–659

Ezquerro IJ, Lasarte JJ, Dotor J, Castilla-Cortazar I, Bustos M, Penuelas I, Blanco G, Rodriguez C, Lechuga Mdel C, Greenwel P, Rojkind M, Prieto J, Borras-Cuesta F (2003) A synthetic peptide from transforming growth factor β type III receptor inhibits liver fibrogenesis in rats with carbon tetrachloride liver injury. Cytokine 22:12–20

Fang Y, Han SI, Mitchell C, Gupta S, Studer E, Grant S, Hylemon PB, Dent P (2004) Bile acids induce mitochondrial ROS, which promote activation of receptor tyrosine kinases and signaling pathways in rat hepatocytes. Hepatology 40:961–971

Fickert P, Fuchsbichler A, Wagner M, Zollner G, Kaser A, Tilg H, Krause R, Lammert F, Langner C, Zatloukal K, Marschall HU, Denk H, Trauner M (2004) Regurgitation of bile acids from leaky bile ducts causes sclerosing cholangitis in Mdr2 (Abcb4) knockout mice. Gastroenterology 127:261–274

Flechsig P, Dadrich M, Bickelhaupt S, Jenne J, Hauser K, Timke C, Peschke P, Hahn EW, Grone HJ, Yingling J, Lahn M, Wirkner U, Huber PE (2012) LY2109761 Attenuates Radiation-Induced Pulmonary Murine Fibrosis via Reversal of TGF-β and BMP-Associated Proinflammatory and Proangiogenic Signals. Clin Cancer Res 18:3616–3627

Foo NP, Lin SH, Lee YH, Wu MJ, Wang YJ (2011) alpha-Lipoic acid inhibits liver fibrosis through the attenuation of ROS-triggered signaling in hepatic stellate cells activated by PDGF and TGF-β. Toxicology 282:39–46

Franco DL, Mainez J, Vega S, Sancho P, Murillo MM, de Frutos CA, Del Castillo G, Lopez-Blau C, Fabregat I, Nieto MA (2010) Snail1 suppresses TGF-β-induced apoptosis and is sufficient to trigger EMT in hepatocytes. J Cell Sci 123:3467–3477

Franklin CC, Rosenfeld-Franklin ME, White C, Kavanagh TJ, Fausto N (2003) TGFβ1-induced suppression of glutathione antioxidant defenses in hepatocytes: caspase-dependent post-translational and caspase-independent transcriptional regulatory mechanisms. FASEB J 17:1535–1537

Fransvea E, Angelotti U, Antonaci S, Giannelli G (2008) Blocking transforming growth factor-β up-regulates E-cadherin and reduces migration and invasion of hepatocellular carcinoma cells. Hepatology 47:1557–1566

Friedman SL (2010) Evolving challenges in hepatic fibrosis. Nat Rev Gastroenterol Hepatol 7:425–436

Fu K, Corbley MJ, Sun L, Friedman JE, Shan F, Papadatos JL, Costa D, Lutterodt F, Sweigard H, Bowes S, Choi M, Boriack-Sjodin PA, Arduini RM, Sun D, Newman MN, Zhang X, Mead JN, Chuaqui CE, Cheung HK, Zhang X, Cornebise M, Carter MB, Josiah S, Singh J, Lee WC, Gill A, Ling LE (2008) SM16, an orally active TGF-β type I receptor inhibitor prevents myofibroblast induction and vascular fibrosis in the rat carotid injury model. Arterioscler Thromb Vasc Biol 28:665–671

Ganapathy V, Ge R, Grazioli A, Xie W, Banach-Petrosky W, Kang Y, Lonning S, McPherson J, Yingling JM, Biswas S, Mundy GR, Reiss M (2010) Targeting the transforming growth factor-β pathway inhibits human basal-like breast cancer metastasis. Mol Cancer 9:122

George J, Roulot D, Koteliansky VE, Bissell DM (1999) In vivo inhibition of rat stellate cell activation by soluble transforming growth factor β type II receptor: a potential new therapy for hepatic fibrosis. Proc Natl Acad Sci USA 96:12719–12724

Gotzmann J, Huber H, Thallinger C, Wolschek M, Jansen B, Schulte-Hermann R, Beug H, Mikulits W (2002) Hepatocytes convert to a fibroblastoid phenotype through the cooperation of TGF-β1 and Ha-Ras: steps towards invasiveness. J Cell Sci 115:1189–1202

Graf D, Kurz AK, Fischer R, Reinehr R, Haussinger D (2002) Taurolithocholic acid-3 sulfate induces CD95 trafficking and apoptosis in a c-Jun N-terminal kinase-dependent manner. Gastroenterology 122:1411–1427

Grehn F, Hollo G, Khaw P, Overton B, Wilson R, Vogel R, Smith Z (2007) Factors affecting the outcome of trabeculectomy: an analysis based on combined data from two phase III studies of an antibody to transforming growth factor β2, CAT-152. Ophthalmology 114:1831–1838

Guo J, Loke J, Zheng F, Hong F, Yea S, Fukata M, Tarocchi M, Abar OT, Huang H, Sninsky JJ, Friedman SL (2009) Functional linkage of cirrhosis-predictive single nucleotide polymorphisms of Toll-like receptor 4 to hepatic stellate cell responses. Hepatology 49:960–968

Hammerich L, Heymann F, Tacke F (2011) Role of IL-17 and Th17 cells in liver diseases. Clin Dev Immunol 2011:345803

Hasegawa M, Matsushita Y, Horikawa M, Higashi K, Tomigahara Y, Kaneko H, Shirasaki F, Fujimoto M, Takehara K, Sato S (2009) A novel inhibitor of Smad-dependent transcriptional activation suppresses tissue fibrosis in mouse models of systemic sclerosis. Arthritis Rheum 60:3465–3475

Hayashi H, Sakai T (2011) Animal models for the study of liver fibrosis: new insights from knockout mouse models. Am J Physiol Gastrointest Liver Physiol 300:G729–G738

Hayashi H, Sakai T (2012) Biological Significance of Local TGF-β Activation in Liver Diseases. Front Physiol 3:12

Hecker L, Vittal R, Jones T, Jagirdar R, Luckhardt TR, Horowitz JC, Pennathur S, Martinez FJ, Thannickal VJ (2009) NADPH oxidase-4 mediates myofibroblast activation and fibrogenic responses to lung injury. Nat Med 15:1077–1081

Hellerbrand C, Stefanovic B, Giordano F, Burchardt ER, Brenner DA (1999) The role of TGFβ1 in initiating hepatic stellate cell activation in vivo. J Hepatol 30:77–87

Hermida N, Lopez B, Gonzalez A, Dotor J, Lasarte JJ, Sarobe P, Borras-Cuesta F, Diez J (2009) A synthetic peptide from transforming growth factor-β1 type III receptor prevents myocardial fibrosis in spontaneously hypertensive rats. Cardiovasc Res 81:601–609

Herrera B, Alvarez AM, Sanchez A, Fernandez M, Roncero C, Benito M, Fabregat I (2001) Reactive oxygen species (ROS) mediates the mitochondrial-dependent apoptosis induced by transforming growth factor (β) in fetal hepatocytes. FASEB J 15:741–751

Herrera B, Murillo MM, Alvarez-Barrientos A, Beltran J, Fernandez M, Fabregat I (2004) Source of early reactive oxygen species in the apoptosis induced by transforming growth factor-β in fetal rat hepatocytes. Free Radic Biol Med 36:16–26

Higashi K, Tomigahara Y, Shiraki H, Miyata K, Mikami T, Kimura T, Moro T, Inagaki Y, Kaneko H (2011) A novel small compound that promotes nuclear translocation of YB-1 ameliorates experimental hepatic fibrosis in mice. J Biol Chem 286:4485–4492

Hill C, Flyvbjerg A, Rasch R, Bak M, Logan A (2001) Transforming growth factor-β2 antibody attenuates fibrosis in the experimental diabetic rat kidney. J Endocrinol 170:647–651

Hocher B, Godes M, Olivier J, Weil J, Eschenhagen T, Slowinski T, Neumayer HH, Bauer C, Paul M, Pinto YM (2002) Inhibition of left ventricular fibrosis by tranilast in rats with renovascular hypertension. J Hypertens 20:745–751

Ikeda H, Inao M, Fujiwara K (1996) Inhibitory effect of tranilast on activation and transforming growth factor β 1 expression in cultured rat stellate cells. Biochem Biophys Res Commun 227:322–327

Ikeda R, Ishii K, Hoshikawa Y, Azumi J, Arakaki Y, Yasui T, Matsuura S, Matsumi Y, Kono Y, Mizuta Y, Kurimasa A, Hisatome I, Friedman SL, Kawasaki H, Shiota G (2011) Reactive oxygen species and NADPH oxidase 4 induced by transforming growth factor β1 are the therapeutic targets of polyenylphosphatidylcholine in the suppression of human hepatic stellate cell activation. Inflamm Res 60:597–604

Inagaki Y, Kushida M, Higashi K, Itoh J, Higashiyama R, Hong YY, Kawada N, Namikawa K, Kiyama H, Bou-Gharios G, Watanabe T, Okazaki I, Ikeda K (2005) Cell type-specific intervention of transforming growth factor β/Smad signaling suppresses collagen gene expression and hepatic fibrosis in mice. Gastroenterology 129:259–268

Jiang JX, Venugopal S, Serizawa N, Chen X, Scott F, Li Y, Adamson R, Devaraj S, Shah V, Gershwin ME, Friedman SL, Torok NJ (2010) Reduced nicotinamide adenine dinucleotide phosphate oxidase 2 plays a key role in stellate cell activation and liver fibrogenesis in vivo. Gastroenterology 139:1375–1384

Jiang JX, Chen X, Serizawa N, Szyndralewicz C, Page P, Schroder K, Brandes RP, Devaraj S, Torok NJ (2012) Liver fibrosis and hepatocyte apoptosis are attenuated by GKT137831, a novel NOX4/NOX1 inhibitor in vivo. Free Radic Biol Med 53:289–296

Jinnin M, Ihn H, Tamaki K (2006) Characterization of SIS3, a novel specific inhibitor of Smad3, and its effect on transforming growth factor-β1-induced extracellular matrix expression. Mol Pharmacol 69:597–607

Jobling MF, Mott JD, Finnegan MT, Jurukovski V, Erickson AC, Walian PJ, Taylor SE, Ledbetter S, Lawrence CM, Rifkin DB, Barcellos-Hoff MH (2006) Isoform-specific activation of latent transforming growth factor β (LTGF-β) by reactive oxygen species. Radiat Res 166:839–848

Kagitani S, Ueno H, Hirade S, Takahashi T, Takata M, Inoue H (2004) Tranilast attenuates myocardial fibrosis in association with suppression of monocyte/macrophage infiltration in DOCA/salt hypertensive rats. J Hypertens 22:1007–1015

Kaimori A, Potter J, Kaimori JY, Wang C, Mezey E, Koteish A (2007) Transforming growth factor-β1 induces an epithelial-to-mesenchymal transition state in mouse hepatocytes in vitro. J Biol Chem 282:22089–22101

Kanzler S, Lohse AW, Keil A, Henninger J, Dienes HP, Schirmacher P, Rose-John S, zum Buschenfelde KH, Blessing M (1999) TGF-β1 in liver fibrosis: an inducible transgenic mouse model to study liver fibrogenesis. Am J Physiol 276:G1059–G1068

Kanzler S, Meyer E, Lohse AW, Schirmacher P, Henninger J, Galle PR, Blessing M (2001) Hepatocellular expression of a dominant-negative mutant TGF-β type II receptor accelerates chemically induced hepatocarcinogenesis. Oncogene 20:5015–5024

Kelly DJ, Zhang Y, Connelly K, Cox AJ, Martin J, Krum H, Gilbert RE (2007) Tranilast attenuates diastolic dysfunction and structural injury in experimental diabetic cardiomyopathy. Am J Physiol Heart Circ Physiol 293:H2860–H2869

Khaw P, Grehn F, Hollo G, Overton B, Wilson R, Vogel R, Smith Z (2007) A phase III study of subconjunctival human anti-transforming growth factor β monoclonal antibody (CAT-152) to prevent scarring after first-time trabeculectomy. Ophthalmology 114:1822–1830

Kim YW, Kim YK, Lee JY, Chang KT, Lee HJ, Kim DK, Sheen YY (2008) Pharmacokinetics and tissue distribution of 3-((5-(6-methylpyridin-2-yl)-4-(quinoxalin-6-yl)-1H-imidazol-2-yl) methyl)benzamid e; a novel ALK5 inhibitor and a potential anti-fibrosis drug. Xenobiotica 38:325–339

Kitamura K, Nakamoto Y, Akiyama M, Fujii C, Kondo T, Kobayashi K, Kaneko S, Mukaida N (2002) Pathogenic roles of tumor necrosis factor receptor p55-mediated signals in dimethylnitrosamine-induced murine liver fibrosis. Lab Invest 82:571–583

Koike Y, Hatamochi A, Koyano S, Namikawa H, Hamasaki Y, Yamazaki S (2011) Lupus miliaris disseminatus faciei successfully treated with tranilast: report of two cases. J Dermatol 38:588–592

Kukielka E, Dicker E, Cederbaum AI (1994) Increased production of reactive oxygen species by rat liver mitochondria after chronic ethanol treatment. Arch Biochem Biophys 309:377–386

Lacher MD, Tiirikainen MI, Saunier EF, Christian C, Anders M, Oft M, Balmain A, Akhurst RJ, Korn WM (2006) Transforming growth factor-β receptor inhibition enhances adenoviral infectability of carcinoma cells via up-regulation of Coxsackie and Adenovirus Receptor in conjunction with reversal of epithelial-mesenchymal transition. Cancer Res 66:1648–1657

Lagares D, Garcia-Fernandez RA, Jimenez CL, Magan-Marchal N, Busnadiego O, Lamas S, Rodriguez-Pascual F (2010) Endothelin 1 contributes to the effect of transforming growth factor β1 on wound repair and skin fibrosis. Arthritis Rheum 62:878–889

Lee CG, Homer RJ, Zhu Z, Lanone S, Wang X, Koteliansky V, Shipley JM, Gotwals P, Noble P, Chen Q, Senior RM, Elias JA (2001) Interleukin-13 induces tissue fibrosis by selectively stimulating and activating transforming growth factor β. J Exp Med 194:809–821

Liu RM, Gaston Pravia KA (2010) Oxidative stress and glutathione in TGF-β-mediated fibrogenesis. Free Radic Biol Med 48:1–15

Liu M, Suga M, Maclean AA, St George JA, Souza DW, Keshavjee S (2002) Soluble transforming growth factor-β type III receptor gene transfection inhibits fibrous airway obliteration in a rat model of Bronchiolitis obliterans. Am J Respir Crit Care Med 165:419–423

Lopez-Novoa JM, Nieto MA (2009) Inflammation and EMT: an alliance towards organ fibrosis and cancer progression. EMBO Mol Med 1:303–314

Louis H, Le Moine A, Quertinmont E, Peny MO, Geerts A, Goldman M, Le Moine O, Deviere J (2000) Repeated concanavalin A challenge in mice induces an interleukin 10-producing phenotype and liver fibrosis. Hepatology 31:381–390

MacDonald GA, Bridle KR, Ward PJ, Walker NI, Houglum K, George DK, Smith JL, Powell LW, Crawford DH, Ramm GA (2001) Lipid peroxidation in hepatic steatosis in humans is associated with hepatic fibrosis and occurs predominately in acinar zone 3. J Gastroenterol Hepatol 16:599–606

Martin J, Kelly DJ, Mifsud SA, Zhang Y, Cox AJ, See F, Krum H, Wilkinson-Berka J, Gilbert RE (2005) Tranilast attenuates cardiac matrix deposition in experimental diabetes: role of transforming growth factor-β. Cardiovasc Res 65:694–701

Masamune A, Watanabe T, Kikuta K, Satoh K, Shimosegawa T (2008) NADPH oxidase plays a crucial role in the activation of pancreatic stellate cells. Am J Physiol Gastrointest Liver Physiol 294:G99–G108

Matsubara T, Tanaka N, Patterson AD, Cho JY, Krausz KW, Gonzalez FJ (2011) Lithocholic acid disrupts phospholipid and sphingolipid homeostasis leading to cholestasis in mice. Hepatology 53:1282–1293

Matsuzaki K (2009) Modulation of TGF-β signaling during progression of chronic liver diseases. Front Biosci 14:2923–2934

Miyazawa K, Kikuchi S, Fukuyama J, Hamano S, Ujiie A (1995) Inhibition of PDGF- and TGF-β 1-induced collagen synthesis, migration and proliferation by tranilast in vascular smooth muscle cells from spontaneously hypertensive rats. Atherosclerosis 118:213–221

Moon JA, Kim HT, Cho IS, Sheen YY, Kim DK (2006) IN-1130, a novel transforming growth factor-β type I receptor kinase (ALK5) inhibitor, suppresses renal fibrosis in obstructive nephropathy. Kidney Int 70:1234–1243

Murillo MM, del Castillo G, Sanchez A, Fernandez M, Fabregat I (2005) Involvement of EGF receptor and c-Src in the survival signals induced by TGF-β1 in hepatocytes. Oncogene 24:4580–4587

Murillo MM, Carmona-Cuenca I, Del Castillo G, Ortiz C, Roncero C, Sanchez A, Fernandez M, Fabregat I (2007) Activation of NADPH oxidase by transforming growth factor-β in hepatocytes mediates up-regulation of epidermal growth factor receptor ligands through a nuclear factor-κB-dependent mechanism. Biochem J 405:251–259

Murphy MP (2009) How mitochondria produce reactive oxygen species. Biochem J 417:1–13

Nakamuta M, Morizono S, Tsuruta S, Kohjima M, Kotoh K, Enjoji M (2005) Remote delivery and expression of soluble type II TGF-β receptor in muscle prevents hepatic fibrosis in rats. Int J Mol Med 16:59–64

Nikolaou K, Tsagaratou A, Eftychi C, Kollias G, Mosialos G, Talianidis I (2012) Inactivation of the deubiquitinase CYLD in hepatocytes causes apoptosis, inflammation, fibrosis, and cancer. Cancer Cell 21:738–750

Oberhammer F, Fritsch G, Pavelka M, Froschl G, Tiefenbacher R, Purchio T, Schulte-Hermann R (1992) Induction of apoptosis in cultured hepatocytes and in the regressing liver by transforming growth factor-β 1 occurs without activation of an endonuclease. Toxicol Lett 64–65 Spec No:701-4

Ortiz C, Caja L, Bertran E, Gonzalez-Rodriguez A, Valverde AM, Fabregat I, Sancho P (2012) Protein-tyrosine phosphatase 1B (PTP1B) deficiency confers resistance to transforming growth factor-β (TGF-β)-induced suppressor effects in hepatocytes. J Biol Chem 287:15263–15274

Oshitani N, Yamagami H, Watanabe K, Higuchi K, Arakawa T (2007) Long-term prospective pilot study with tranilast for the prevention of stricture progression in patients with Crohn's disease. Gut 56:599–600

Paik YH, Iwaisako K, Seki E, Inokuchi S, Schnabl B, Osterreicher CH, Kisseleva T, Brenner DA (2011) The nicotinamide adenine dinucleotide phosphate oxidase (NOX) homologues NOX1 and NOX2/gp91(phox) mediate hepatic fibrosis in mice. Hepatology 53:1730–1741

Pawlak K, Zolbach K, Borawski J, Mysliwiec M, Kovalchuk O, Chyczewski L, Pawlak D (2008) Chronic viral hepatitis C, oxidative stress and the coagulation/fibrinolysis system in haemodialysis patients. Thromb Res 123:166–170

Perlemuter G, Letteron P, Carnot F, Zavala F, Pessayre D, Nalpas B, Brechot C (2003) Alcohol and hepatitis C virus core protein additively increase lipid peroxidation and synergistically trigger hepatic cytokine expression in a transgenic mouse model. J Hepatol 39:1020–1027
Petersen M, Thorikay M, Deckers M, van Dinther M, Grygielko ET, Gellibert F, de Gouville AC, Huet S, ten Dijke P, Laping NJ (2008) Oral administration of GW788388, an inhibitor of TGF-β type I and II receptor kinases, decreases renal fibrosis. Kidney Int 73:705–715
Piao S, Choi MJ, Tumurbaatar M, Kim WJ, Jin HR, Shin SH, Tuvshintur B, Yin GN, Song JS, Kwon MH, Lee SJ, Han JY, Kim SJ, Ryu JK, Suh JK (2010) Transforming growth factor (TGF)-β type I receptor kinase (ALK5) inhibitor alleviates profibrotic TGF-β1 responses in fibroblasts derived from Peyronie's plaque. J Sex Med 7:3385–3395
Platten M, Wild-Bode C, Wick W, Leitlein J, Dichgans J, Weller M (2001) N-[3,4-dimethoxycinnamoyl]-anthranilic acid (tranilast) inhibits transforming growth factor-β relesase and reduces migration and invasiveness of human malignant glioma cells. Int J Cancer 93:53–61
Pociask DA, Sime PJ, Brody AR (2004) Asbestos-derived reactive oxygen species activate TGF-β1. Lab Invest 84:1013–1023
Proell V, Carmona-Cuenca I, Murillo MM, Huber H, Fabregat I, Mikulits W (2007) TGF-β dependent regulation of oxygen radicals during transdifferentiation of activated hepatic stellate cells to myofibroblastoid cells. Comp Hepatol 6:1
Rowe RG, Lin Y, Shimizu-Hirota R, Hanada S, Neilson EG, Greenson JK, Weiss SJ (2011) Hepatocyte-derived Snail1 propagates liver fibrosis progression. Mol Cell Biol 31:2392–2403
Ryu JK, Piao S, Shin HY, Choi MJ, Zhang LW, Jin HR, Kim WJ, Han JY, Hong SS, Park SH, Lee SJ, Kim IH, Lee CR, Kim DK, Mamura M, Kim SJ, Suh JK (2009) IN-1130, a novel transforming growth factor-β type I receptor kinase (activin receptor-like kinase 5) inhibitor, promotes regression of fibrotic plaque and corrects penile curvature in a rat model of Peyronie's disease. J Sex Med 6:1284–1296
Said E, Said SA, Elkashef WF, Gameil NM, Ammar EM (2012) Tranilast ameliorates impaired hepatic functions in Schistosoma mansoni-infected mice. Inflammopharmacology 20:77–87
Sampson N, Koziel R, Zenzmaier C, Bubendorf L, Plas E, Jansen-Durr P, Berger P (2011) ROS signaling by NOX4 drives fibroblast-to-myofibroblast differentiation in the diseased prostatic stroma. Mol Endocrinol 25:503–515
Sanchez A, Alvarez AM, Benito M, Fabregat I (1996) Apoptosis induced by transforming growth factor-β in fetal hepatocyte primary cultures:involvement of reactive oxygen intermediates. J Biol Chem 271:7416–7422
Sancho P, Fabregat I (2011) The NADPH oxidase inhibitor VAS2870 impairs cell growth and enhances TGF-β-induced apoptosis of liver tumor cells. Biochem Pharmacol 81:917–924
Sancho P, Bertran E, Caja L, Carmona-Cuenca I, Murillo MM, Fabregat I (2009) The inhibition of the epidermal growth factor (EGF) pathway enhances TGF-β-induced apoptosis in rat hepatoma cells through inducing oxidative stress coincident with a change in the expression pattern of the NADPH oxidases (NOX) isoforms. Biochim Biophys Acta 1793:253–263
Sancho P, Martin-Sanz P, Fabregat I (2011) Reciprocal regulation of NADPH oxidases and the cyclooxygenase-2 pathway. Free Radic Biol Med 51:1789–1798
Sancho P, Mainez J, Roncero C, Fernandez-Rodriguez CM, Pinedo F, Huber H, Eferl R, Mikulits W, Fabregat I (2012) NADPH oxidase NOX4 mediates stellate cell activation and hepatocyte cell death during liver fibrosis development. PLoS One 7:e45285
Schnabl B, Kweon YO, Frederick JP, Wang XF, Rippe RA, Brenner DA (2001) The role of Smad3 in mediating mouse hepatic stellate cell activation. Hepatology 34:89–100
Sedeek M, Callera G, Montezano A, Gutsol A, Heitz F, Szyndralewiez C, Page P, Kennedy CR, Burns KD, Touyz RM, Hebert RL (2010) Critical role of Nox4-based NADPH oxidase in glucose-induced oxidative stress in the kidney:implications in type 2 diabetic nephropathy. Am J Physiol Renal Physiol 299:F1348–F1358
Seki S, Kitada T, Sakaguchi H (2005) Clinicopathological significance of oxidative cellular damage in non-alcoholic fatty liver diseases. Hepatol Res 33:132–134

Seki E, De Minicis S, Osterreicher CH, Kluwe J, Osawa Y, Brenner DA, Schwabe RF (2007) TLR4 enhances TGF-β signaling and hepatic fibrosis. Nat Med 13:1324–1332

Sies H, Cadenas E (1985) Oxidative stress:damage to intact cells and organs. Philos Trans R Soc Lond B Biol Sci 311:617–631

Sullivan BP, Weinreb PH, Violette SM, Luyendyk JP (2010) The coagulation system contributes to αVβ6 integrin expression and liver fibrosis induced by cholestasis. Am J Pathol 177:2837–2849

Tan SM, Zhang Y, Connelly KA, Gilbert RE, Kelly DJ (2010) Targeted inhibition of activin receptor-like kinase 5 signaling attenuates cardiac dysfunction following myocardial infarction. Am J Physiol Heart Circ Physiol 298:H1415–H1425

Taura K, Miura K, Iwaisako K, Osterreicher CH, Kodama Y, Penz-Osterreicher M, Brenner DA (2010) Hepatocytes do not undergo epithelial-mesenchymal transition in liver fibrosis in mice. Hepatology 51:1027–1036

Teratani T, Tomita K, Suzuki T, Oshikawa T, Yokoyama H, Shimamura K, Tominaga S, Hiroi S, Irie R, Okada Y, Kurihara C, Ebinuma H, Saito H, Hokari R, Sugiyama K, Kanai T, Miura S, Hibi T (2012) A high-cholesterol diet exacerbates liver fibrosis in mice via accumulation of free cholesterol in hepatic stellate cells. Gastroenterology 142:152–164

Thiery JP, Acloque H, Huang RY, Nieto MA (2009) Epithelial-mesenchymal transitions in development and disease. Cell 139:871–890

Ueberham E, Low R, Ueberham U, Schonig K, Bujard H, Gebhardt R (2003) Conditional tetracycline-regulated expression of TGF-β1 in liver of transgenic mice leads to reversible intermediary fibrosis. Hepatology 37:1067–1078

Ueno H, Sakamoto T, Nakamura T, Qi Z, Astuchi N, Takeshita A, Shimizu K, Ohashi H (2000) A soluble transforming growth factor β receptor expressed in muscle prevents liver fibrogenesis and dysfunction in rats. Hum Gene Ther 11:33–42

Uno M, Kurita S, Misu H, Ando H, Ota T, Matsuzawa-Nagata N, Kita Y, Nabemoto S, Akahori H, Zen Y, Nakanuma Y, Kaneko S, Takamura T (2008) Tranilast, an antifibrogenic agent, ameliorates a dietary rat model of nonalcoholic steatohepatitis. Hepatology 48:109–118

Valdes F, Alvarez AM, Locascio A, Vega S, Herrera B, Fernandez M, Benito M, Nieto MA, Fabregat I (2002) The epithelial mesenchymal transition confers resistance to the apoptotic effects of transforming growth factor β in fetal rat hepatocytes. Mol Cancer Res 1:68–78

Valdes F, Murillo MM, Valverde AM, Herrera B, Sanchez A, Benito M, Fernandez M, Fabregat I (2004) Transforming growth factor-β activates both pro-apoptotic and survival signals in fetal rat hepatocytes. Exp Cell Res 292:209–218

Wang L, Clutter S, Benincosa J, Fortney J, Gibson LF (2005) Activation of transforming growth factor-β1/p38/Smad3 signaling in stromal cells requires reactive oxygen species-mediated MMP-2 activity during bone marrow damage. Stem Cells 23:1122–1134

Wells RG (2011) The epithelial-to-mesenchymal transition in liver fibrosis:here today, gone tomorrow? Hepatology 51:737–740

Wynn TA (2008) Cellular and molecular mechanisms of fibrosis. J Pathol 214:199–210

Wynn TA, Barron L (2010) Macrophages:master regulators of inflammation and fibrosis. Semin Liver Dis 30:245–257

Yadav D, Hertan HI, Schweitzer P, Norkus EP, Pitchumoni CS (2002) Serum and liver micronutrient antioxidants and serum oxidative stress in patients with chronic hepatitis C. Am J Gastroenterol 97:2634–2639

Yao H, Pan J, Qian Y, Pei Z, Bader A, Brockmeyer NH, Altmeyer P, Zhang L (2010) Enhanced effect of soluble transforming growth factor-β receptor II and IFN-γ fusion protein in reversing hepatic fibrosis. Eur J Med Res 15:152–161

Yata Y, Gotwals P, Koteliansky V, Rockey DC (2002) Dose-dependent inhibition of hepatic fibrosis in mice by a TGF-β soluble receptor:implications for antifibrotic therapy. Hepatology 35:1022–1030

Yoshida K, Matsuzaki K (2012) Differential regulation of TGF-β/Smad signaling in hepatic stellate cells between acute and chronic liver injuries. Front Physiol 3:53

Yoshida K, Matsuzaki K, Mori S, Tahashi Y, Yamagata H, Furukawa F, Seki T, Nishizawa M, Fujisawa J, Okazaki K (2005) Transforming growth factor-β and platelet-derived growth factor signal via c-Jun N-terminal kinase-dependent Smad2/3 phosphorylation in rat hepatic stellate cells after acute liver injury. Am J Pathol 166:1029–1039

Zeisberg M, Yang C, Martino M, Duncan MB, Rieder F, Tanjore H, Kalluri R (2007) Fibroblasts derive from hepatocytes in liver fibrosis via epithelial to mesenchymal transition. J Biol Chem 282:23337–23347

Zhang B, Halder SK, Kashikar ND, Cho YJ, Datta A, Gorden DL, Datta PK (2010) Antimetastatic role of Smad4 signaling in colorectal cancer. Gastroenterology 138:969–980

Zhang M, Kleber S, Rohrich M, Timke C, Han N, Tuettenberg J, Martin-Villalba A, Debus J, Peschke P, Wirkner U, Lahn M, Huber PE (2011) Blockade of TGF-β signaling by the TGFβR-I kinase inhibitor LY2109761 enhances radiation response and prolongs survival in glioblastoma. Cancer Res 71:7155–7167

Zhu H, Jia Z, Misra H, Li YR (2012) Oxidative stress and redox signaling mechanisms of alcoholic liver disease:updated experimental and clinical evidence. J Dig Dis 13:133–142

Zhuge J, Cederbaum AI (2006) Increased toxicity by transforming growth factor-β 1 in liver cells overexpressing CYP2E1. Free Radic Biol Med 41:1100–1112

Chapter 12
TGF-β and Inhibitory Smads in Inflammation

Seong-Jin Kim and Seok Hee Park

Abstract TGF-β is a multifunctional cytokine involved in diverse cellular functions, including cell growth, apoptosis, and immune responses, although its effect is likely to be dependent on cell context. Among a variety of cellular functions exerted by TGF-β, recent advances emphasize the importance of TGF-β and its signaling pathway in innate immunity and adaptive immunity. The typical examples of immune regulations by TGF-β include suppression of toll-like receptor (TLR) signaling which recognizes the invading pathogens and T cell differentiations such as Th17 and Treg. In particular, much attention has been paid to anti-inflammatory function of TGF-β, which is mediated by the inhibitory Smads, Smad6 and Smad7. In this review, we mainly discuss the anti-inflammatory role of TGF-β suppressing inflammatory responses and the underlying mechanism mediated by the inhibitory Smads.

Keywords Inflammation • Inhibitory Smads • Innate immune responses • Smad6 • Smad7

12.1 Introduction

Although extensive studies have been performed to reveal essential roles of TGF-β and the related molecular mechanism during the past 25 years (Blobe et al. 2000; Derynck and Zhang 2003; Heldin et al. 1997; Letterio and Roberts 1997; Massague 1998; Miyazono et al. 2000), the functions of TGF-β and its signaling pathway are

---

S.-J. Kim (✉)
CHA Cancer Institute, CHA University of Medical Science, Seoul 135-081, South Korea
e-mail: kimsj@cha.ac.kr

S.H. Park
Department of Biological Sciences, Sungkyunkwan University, Suwon, 440-746, South Korea

A. Moustakas and K. Miyazawa (eds.), *TGF-β in Human Disease*,
DOI 10.1007/978-4-431-54409-8_12, © Springer Japan 2013

very complicated and controversial. The major reason is that TGF-β signaling pathways are involved in most cellular functions showing cell context-dependent fashion and forming highly intricate signaling networks through cross talking with other cellular signaling pathways.

There are three TGF-β members (TGF-β1, TGF-β2, and TGF-β3) present in mammals. Among them, TGF-β1 is the major form found in the immune system. TGF-β signaling pathways are mainly divided into canonical and noncanonical pathways (Derynck and Zhang 2003; Mu et al. 2012). TGF-β canonical pathway is mediated by heterodimeric receptors with serine/threonine receptor kinases and cytoplasmic proteins called Smads. Upon binding to TGF-β, type II (TβRII) and type I (TβRI) receptors initiate the signaling process with their intrinsic kinase activities, and then the activated type I receptor transmits intracellular signals through the phosphorylation of receptor-activated Smad proteins (R-Smads), Smad2 and Smad3. The phosphorylated R-Smads form heteromeric complexes with common partner Smad4 (Co-Smad) and translocate into the nucleus to modulate expressions of diverse target genes (Derynck and Zhang 2003; Massague and Wotton 2000; Miyazono et al. 2000). The more interesting finding is that this canonical TGF-β signaling has its own regulatory system, called inhibitory Smads (I-Smads). The I-Smads include Smad7 and Smad6 proteins (Hayashi et al. 1997; Imamura et al. 1997; Nakao et al. 1997). Smad7 is transcriptionally induced by TGF-β and acts as an intracellular antagonist through stable interaction of activated TβRI receptor, resulting in the formation of autoregulatory negative feedback loop of TGF-β signaling (Hayashi et al. 1997; Nakao et al. 1997). Smad6, another inhibitory Smad, is also induced by TGF-β in several cell lines such as mink lung epithelial (Mv1Lu) cells, human keratinocyte (HaCaT) cells, CMT-93 mouse intestinal epithelial cells, and mouse primary peritoneal macrophages (Afrakhte et al. 1998; Choi et al. 2006). It has been initially identified as an inhibitor of Smad2 and Smad1 phosphorylation by TGF-β superfamily (Imamura et al. 1997; Ishida et al. 2000), and it is likely to be an important regulator involved in TGF-β signaling (Afrakhte et al. 1998; Choi et al. 2006; Kleeff et al. 1999).

In contrast, TGF-β noncanonical signaling is independent of the activations of R-Smads, and TGF-β utilizes a variety of cellular signaling pathways such as p38 and JNK mitogen-activated protein kinases (MAPK), PI3 kinase, and TRAF6-TAK1 pathways (Mu et al. 2012; Zhang 2009). Therefore, TGF-β noncanonical pathways complicate the understanding of the precise molecular mechanisms underlying TGF-β-mediated cellular functions.

Inflammation is defined as a protective response to eliminate detrimental stimuli and to repair damaged tissue (Medzhitov 2008). Inflammatory stimuli include a variety of factors such as microbial infections, necrotic tissue injuries, and cytokines (Takeuchi and Akira 2010). In terms of tissue homeostasis, inflammation should be resolved in a timely manner and damaged tissues should restore their normal structural and functional state. If inflammatory responses is not resolved, persistent inflammation may become a key factor causing chronic inflammatory diseases such as cancer, diabetes, sepsis, and atherosclerosis (Nathan and Ding 2010).

The innate immune system mediates inflammatory responses induced by microbial infection or tissue damage (Akira et al. 2006; Beutler et al. 2006; Takeuchi and Akira 2010). The innate immune responses are initiated by pattern recognition receptors (PRRs) which recognize pathogen-associated molecular patterns (PAMP) or damage-associated molecular patterns (DAMPs) (Takeuchi and Akira 2010). Among PRRs, toll-like receptors (TLRs) recognize PAMPs and transmit signals to downstream proteins through adaptor MyD88-dependent or MyD88-independent pathways, resulting in the activation of NF-κB or IRF3 protein to increase expression of pro-inflammatory cytokines and interferon-β1 (O'Neill and Bowie 2007; Takeuchi and Akira 2010).

Therefore, to maintain tissue homeostasis, TLR activation causing the inflammatory responses must be regulated, and aberrant regulation of TLR signaling is known to induce hyper-inflammatory diseases (Cook et al. 2004). Interestingly, anti-inflammatory activities of TGF-β cytokine have been reported in knockout mice studies, but the molecular mechanisms as to how TGF-β signaling is interconnected with innate immune responses and inflammations have not been addressed as much as other mechanisms mediated by TGF-β such as cell growth and apoptosis. Therefore, we review recent progresses on the TGF-β signaling pathway regulating inflammation and innate immune responses.

12.2 TGF-β and Inflammation

12.2.1 Effects of TGF-β Signaling Components on Inflammation

Previous studies on animal models of TGF-β signaling deficiency have clearly substantiated the importance of TGF-β signaling for immune functions and inflammation. The TGF-β1-null mutation in mice triggered an autoimmune inflammatory condition involving nuclear autoantibodies and autoreactive T cells. These mice had excessive inflammatory responses and developed a multifocal inflammatory disease resulting in cardiopulmonary complications that were ultimately lethal (Kulkarni et al. 1995; Shull et al. 1992). Histopathological analysis of these mice revealed massive infiltration of lymphocytes and macrophages in many organs, especially in the heart and lungs. In the spleen and the lymph nodes, inflammatory lesions included proliferation of immunoblasts and lymphoblasts in B and T cell zones. Inflammatory lesions were also seen in the pancreas, salivary glands, colon, and stomach of some of the TGF-β1(−/−) mice. In addition, conditional KO mice of TβRII gene, when it was disrupted by poly I:C induction in hematopoietic cells, showed a lethal inflammatory disease affecting multiple organs (Leveen et al. 2002). Bone marrow from conditional knockout mice of TβRII transferred to normal recipient mice caused a similar lethal inflammation, regardless of whether induction of TGF-β receptor deficiency occurred in donor animals before or in recipient animals after transplantation. These results demonstrate that TGF-β signaling deficiency within cells of hematopoietic origin is sufficient enough to cause a lethal inflammatory disorder in mice. Another kidney-specific conditional KO mice of TβRII gene enhanced NF-κB signaling and renal inflammation including IL-1β and TNF-α in the model of unilateral urethral obstructive (UOO) nephropathy (Meng et al. 2012a). The important role of TGF-β in immune homeostasis has been previously suggested by the observation of multifocal immune-mediated inflammation in CD4-dnTGF-βRII transgenic mice similar to that of TGF-β1 KO mice. The CD4-dnTGF-βRII transgenic mice developed autoimmune disease characterized by inflammatory infiltration in several organs and the presence of circulating autoimmune antibodies, indicating that T cell homeostasis and prevention of inflammatory infiltration require TGF-β signaling in T cells (Gorelik and Flavell 2000). Among Smad proteins, Smad3 KO mice displayed impaired mucosal immunity and reduced T cell responsiveness (Ashcroft et al. 1999; Yang et al. 1999). Smad3-deficient mice were normal during embryonic and early postnatal development. After weaning, Smad3 KO mice invariably developed an illness associated with progressive leukocytosis, periodontitis, gastritis, colitis, and chronic infection with abscess formation adjacent to mucosal surfaces, suggesting that Smad3 has an important role in TGF-β-mediated regulation of T cell activation and mucosal immunity. The important role of Smad4 in inflammation has recently been suggested in a mouse model of unilateral urethral obstruction using conditional Smad4 knockout mice and in isolated Smad4 mutant macrophages and fibroblasts (Meng et al. 2012b). Disruption of Smad4 significantly enhanced renal inflammation as evidenced by a greater CD45(+) leukocyte and F4/80(+) macrophage infiltration and upregulation of IL-1β, TNF-α, MCP-1, and ICAM-1 in the obstructed kidney and in IL-1β-stimulated macrophages. These results suggest that Smad4 may also be a key regulator for the diverse roles of TGF-β1 in inflammation. The studies of animal models described in this review demonstrate the significant role of TGF-β in the regulation of T cell homeostasis and prevention of immune inflammation in many organs. These mice provide good models to study the pathogenic mechanisms in animal deficient of TGF-β signaling and to elucidate the specific cellular and molecular mechanisms of TGF-β that maintain homeostasis within the immune system.

Evidence of the anti-inflammatory role of TGF-β in human was shown in the study using biopsy specimens. Blockade of TGF-β in normal intestinal biopsies grown ex vivo and lamina propria mononuclear cells (LPMCs) downregulated T cell apoptosis and induced a significant increase in pro-inflammatory cytokines, including IFN-γ, TNF-α, IL-12, IL-6, IL-18, and IL-17, supporting an anti-inflammatory role of TGF-β in dampening T cell-mediated tissue-damaging responses in the human gut (Di Sabatino et al. 2008). On the contrary, the opposite results were observed in colonic biopsies of patients with ulcerative colitis (UC) or Crohn's disease (CD) (Babyatsky et al. 1996). The area of active inflammation in UC or CD showed an increased level of TGF-β expression. Surprisingly, the increased level of TGF-β in UC or CD did not exert an anti-inflammatory function (Monteleone et al. 2001). The reason was due to the overexpression of Smad7, an inhibitory Smad, in intestinal mucosa derived from active CD and most UC patients

(Monteleone et al. 2001). The overexpressed Smad7 blocked Smad3 activation through interfering with TGF-β type I receptor in these inflammatory diseases. Even though a number of studies are still performed to understand biological functions of TGF-β, these results basically suggest that in vivo functions of TGF-β signaling pathway are critically connected into immunological responses including innate immunity and adaptive immunity.

12.2.2 Anti-Inflammatory Role of TGF-β Signaling Pathway Suppressing Innate Immune Responses

TLRs have been known to be an important defense system responsible for invading pathogens. However, the molecular mechanism behind how their signaling cascades are regulated by anti-inflammatory cytokine TGF-β has recently been demonstrated in several reports. The initial report about an anti-inflammatory role of TGF-β suppressing TLR4 signaling pathway stated that TGF-β inhibited MyD88-dependent TLR4 signaling through decreasing the MyD88 protein levels in dose- and time-dependent manner (Naiki et al. 2005). However, the molecular mechanism behind the polyubiquitination and subsequent degradation of MyD88 by TGF-β was not clearly demonstrated in this report. Recently, the inhibitory Smad6 protein was shown to be responsible for the polyubiquitination and selective degradation of MyD88 through the recruitment of E3 ubiquitin ligases Smurf proteins (Fig. 12.1a) (Lee et al. 2011). These observations suggest that MyD88-dependent TLR signaling pathways such as TLR4 and TLR2 are negatively regulated by the Smad6-Smurf pathway when treated with TGF-β. The more interesting finding was that another inhibitory Smad, Smad7, is not involved in MyD88 degradation, indicating that the two inhibitory Smads, Smad6 and Smad7, have distinct roles as anti-inflammatory mediators (Lee et al. 2011). However, Smad6 is likely to be targeting another adaptor protein Pellino-1 when suppressing TLR4 signaling (Choi et al. 2006). TGF-β-induced Smad6 protein binds to Pellino-1, an adaptor protein binding to IRAK1 (Grosshans et al. 1999; Jiang et al. 2003; Moynagh 2009) and in turn disrupts IRAK1-mediated signaling complex in TLR4 signaling (Fig. 12.1a). The disruption of IRAK1-mediated signaling complexes through sequestering Pellino-1 by Smad6 resulted in the inhibition of NF-κB-mediated pro-inflammatory gene expressions (Choi et al. 2006). These findings strongly suggest that Smad6 is a pivotal component of TGF-β-mediated anti-inflammatory network and targets MyD88-dependent and Pellino-1-mediated TLR signaling. Interestingly, recent reports showed that Pellino-1 has additional binding partners in TLR signaling as well as IRAK1. Pellino-1 binds to TRIF-associated RIP1 protein (Chang et al. 2009). Another report indicated that Pellino-1 interacted with the IKKε/TBK1 complex which phosphorylates IRF3 (Smith et al. 2011). These findings suggest that TGF-β-induced Smad6 disrupts the RIP1- and IKKε/TBK1-mediated signaling complexes and subsequently inhibits MyD88-independent TLR3 signaling pathway.

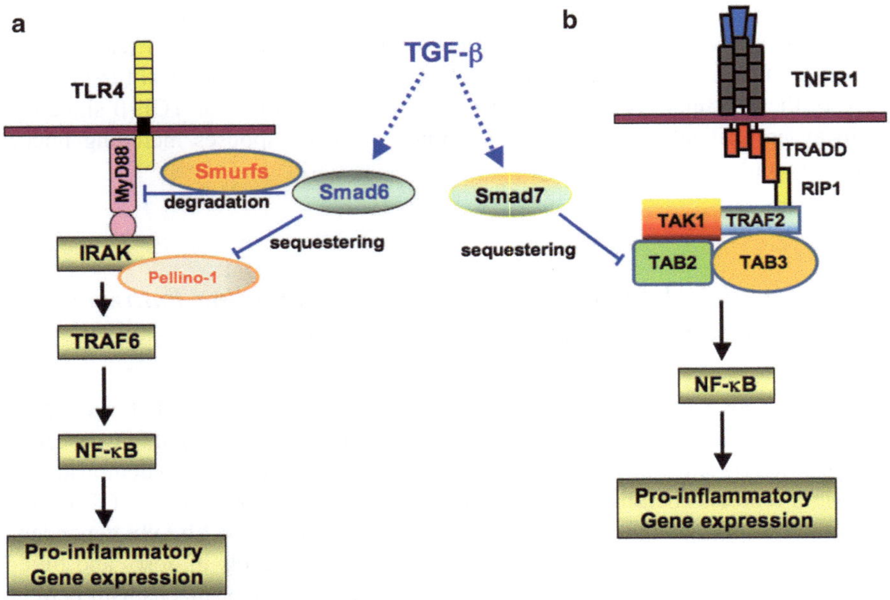

Fig. 12.1 Roles of the inhibitory Smads as mediators of anti-inflammatory TGF-β signaling. (**a**) TGF-β-induced Smad6 selectively degrades MyD88 protein in TLR4 signaling through recruiting E3 ubiquitin ligase Smurf proteins and also disrupts IRAK1-mediated signaling complex through sequestering the Pellino-1 protein. (**b**) TGF-β-induced Smad7 protein binds to TAB2 and TAB3 and subsequently suppresses TNF-α signaling through blocking the recruitment of the kinase TAK1 to the TRAF2 protein

In the case of Smad7, this molecule has been demonstrated to block pro-inflammatory TNF-α signaling (Hong et al. 2007). TGF-β-induced Smad7 interacts with the adaptors TAB2 and TAB3, and this interaction leads to the inhibition of TNF-α-induced NF-κB activity by blocking TAK1-associating proteins, TAB2 and TAB3, from forming a complex with TRAF2 (Fig. 12.1b) (Hong et al. 2007). Furthermore, Smad7 transgene expression in mouse skin under control of the keratin 5 promoter (K5-Smad7 mice) markedly suppressed inflammation and NF-κB nuclear translocation (Hong et al. 2007), suggesting that Smad7 is a critical mediator of the TGF-β pathway that blocks pro-inflammatory TNF-α signaling. Interestingly, Smad6 was not involved in TGF-β-mediated suppression of TNF-α signaling. TGF-β did not inhibit TNF-α-induced IL-6 expression in Smad7 knockdown primary peritoneal macrophages, whereas TGF-β still inhibited TNF-α-induced IL-6 expression in Smad6 knockdown primary peritoneal macrophages (Hong et al. 2007). Smad6 did not interact with TAB2. The Smad7 amino acids that are critical for TAB2-/TAB3-Smad7 interaction are not well conserved in Smad6. These findings strongly imply that Smad6 and Smad7 are differentially responsible for mediating anti-inflammatory TGF-β activity.

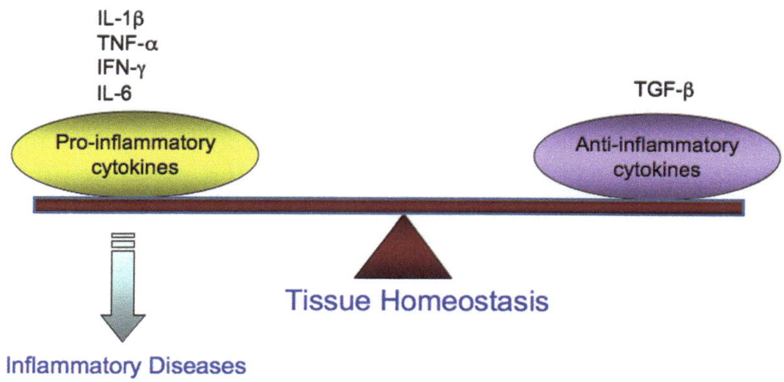

Fig. 12.2 Homeostatic balance between pro-inflammatory signals and anti-inflammatory TGF-β signaling under normal physiological conditions. The disruption of homeostatic balance between pro-inflammatory and anti-inflammatory signals, caused by excessive pro-inflammatory signals or abnormal reduction of anti-inflammatory signals, may be an important reason leading to the inflammatory diseases

Based on these findings, it is obvious that Smad6 and Smad7 play important roles in suppressing TLR4 signaling and pro-inflammatory TNF-α signaling, respectively. Although the TGF-β signaling pathway has been extensively studied, the physiological role of Smad7 and Smad6 requires further studies. The generation of Smad7 conditional knockout mice has been recently reported (Tojo et al. 2012); however, cell type-specific Smad6 conditional knockout mice are still unavailable. In vivo studies of these mice will elucidate how these molecules mediate TGF-β anti-inflammatory signaling.

12.2.3 The Inhibitory Smads as Cross Talkers of Intracellular Signaling Pathways

Recent reports suggest that inhibitory Smads may act as adaptors to facilitate cross talk between TGF-β signaling and pro-inflammatory signaling pathways. In this regard, Smad7 was extensively studied in the field of reciprocal regulation of TGF-β signaling and IFN-γ signaling. The pleiotropic cytokine IFN-γ suppressed TGF-β signaling through the increased expression of Smad7 by the activation of JAK-STAT1 pathway, which prevents the interaction of Smad3 with the TGF-β receptor and downregulates TGF-β signaling (Ulloa et al. 1999). Other groups reported another mechanism of IFN-γ-mediated suppression of TGF-β signaling (Ghosh et al. 2001). IFN-γ inhibited TGF-β signaling pathway through the sequestration of nuclear coactivator p300/CBP, which has been known to be associated with R-Smads (Ghosh et al. 2001). These results imply that the suppression of TGF-β signaling by IFN-γ will amplify the pro-inflammatory activity of IFN-γ under normal physiological conditions (Fig. 12.2).

In contrast, it has been reported that TGF-β inhibited IFN-γ-induced STAT1-dependent gene expression by enhancing STAT1-PIAS (a protein inhibitor of activated STAT1) interactions (Reardon and McKay 2007). However, these effects were shown in epithelial cells but not macrophages. Another group has shown that TGF-β inhibited IFN-γ-induced nitric oxide production in macrophages through the interaction of TGF-β type I receptor with IFN-γ receptor 1(IFNGR1) (Takaki et al. 2006). These results indicate that TGF-β suppresses the pro-inflammatory signal of IFN-γ to maximize its anti-inflammatory activity. Therefore, the anti-inflammatory and pro-inflammatory signals should be critically regulated to maintain tissue homeostasis (Fig. 12.2).

In addition, it was reported that TGF-β signaling is suppressed by NF-κB/RelA-dependent pathways (Bitzer et al. 2000). Activation of NF-κB/RelA by a variety of pathogenic and pro-inflammatory stimuli (TNF-α and IL-1β) increased transcription of the *Smad7* gene. This effect promoted the association of inhibitory Smad7 and type I TGF-β receptor upon TGF-β stimulation, leading to the suppression of TGF-β/Smad signaling. This gives further evidence that pro-inflammatory signals counteract anti-inflammatory signaling thresholds for TGF-β, dependent on physiological requirements to potentiate inflammatory responses. Similarly, a recent study supported the importance of a balance mechanism between TGF-β anti-inflammatory signal and pro-inflammatory signals. In this study, IL-1β or lipopolysaccharide (LPS) suppressed TGF-β-induced anti-inflammatory signaling in a NF-κB-independent manner through the interaction of TRAF6 with TGF-β type III receptor upon TGF-β stimulation, which then leads to pro-inflammatory factor-mediated attenuation of Smad2/3 phosphorylation (Lim et al. 2012). Although the antagonistic effects of IL-1β/LPS on TGF-β1 signaling may occur at multiple molecular levels, a fine modulation of the TGF-β receptors may critically affect cell fates by changing the kinetics of Smad2/3 phosphorylation. By this mechanism, IL-1β may potentiate inflammatory responses.

However, a recent report indicates that TRAF6-mediated polyubiquitination of transforming growth factor-β-associated kinase 1 (TAK1) activates NF-κB in HepG2 cells upon TGF-β1 treatment independent of p38 MAP kinase (Hamidi et al. 2012). This finding makes it difficult to understand the role of TGF-β1 in inflammatory responses. TAK1 is an important kinase in TGF-β noncanonical pathway, and TGF-β-induced Lys-63-linked polyubiquitination of TAK1 by TRAF6 is crucial for its activation and subsequent activation of p38 MAP kinase (Sorrentino et al. 2008). In this report, inflammatory stimuli, including IL-1β, TNF-α, and LPS, are able to induce TAK1 polyubiquitination and subsequent NF-κB activation in prostate cancer cells and RAW264.7 macrophages (Hamidi et al. 2012), suggesting that the increase of pro-inflammatory cytokines such as IL-6 may contribute to the formation of a tumor microenvironment. The interesting thing in this report is that TGF-β 1 also induces the same polyubiquitination of TAK1 and subsequent activation of NF-κB (Hamidi et al. 2012), suggesting that this phenomena may activate NF-κB survival signal in tumorigenesis. However, the authors did not show production of the pro-inflammatory cytokines in HepG2 cells upon TGF-β1 treatment as well as whether the activation of NF-κB upon TGF-β1 treatment is consistently observed in immune cells.

If TGF-β1-induced TAK1 and NF-κB activations are detected in immune cells such as macrophages and dendritic cells (DC), TGF-β1 is likely to have dual activities in inflammatory responses, as inflammatory or anti-inflammatory mediator and its activity might be differentially regulated according to cellular context or the related microenvironment.

On the other hand, the induction of Smad6 expression by pro-inflammatory signals is not reported, although Smad6 is transcriptionally induced by the anti-inflammatory cytokine TGF-β and responsible for TGF-β-mediated anti-inflammatory process (Afrakhte et al. 1998; Choi et al. 2006; Lee et al. 2011). A recent report showed that bacterial endotoxin LPS inhibits TGF-β-stimulated Smad6 expression in RAW264.7 macrophage cells (Kim and Kim 2011). The inhibition of TGF-β-induced Smad6 expression by LPS was mediated by phoshorylation of the Smad3 linker region through a TLR4-IRAK1-ERK1/2 pathway. The induction of Smad6 gene expression by pro-inflammatory signals has not been reported; however, we cannot exclude this possibility. It will be worth investigating whether pro-inflammatory signals potentiate inflammatory responses by stimulating Smad6 expression, which may counteract anti-inflammatory signaling thresholds for TGF-β/BMP.

12.2.4 A Role of TGF-β in T Cell Biology

TGF-β1 is one of the regulatory cytokines with pivotal functions in the control of inflammation. TGF-β1 directly targets T cells to ensure immune tolerance to self- and environmental antigens. Studies using animal models such as transgenic mice-expressing dominant-negative mutants of TβRII gene and T cell-specific TβRII KO mice indicated a principle role of TGF-β in T cell biology (Gorelik and Flavell 2000; Li et al. 2006; Lucas et al. 2000; Marie et al. 2006; Li and Flavell 2008). Generally, $CD4^+$ effector (helper) T cells are divided into two subsets, T helper type 1 (Th1) and T helper type 2 (Th2) cells, depending on the functions and cytokine production patterns. Th1 cells are characterized by their production of IFN-γ, a potent activator of cell-mediated immunity and thus responsible for removal of intracellular pathogens. Th2 cells that enhance eradication of parasitic infection are characterized by production of IL-4, IL-5, and IL-13. These helper T cells are critically regulated by regulatory T cells (Treg). Another subset of T helper cells is called Th17 cells which produce IL-17, a potent pro-inflammatory cytokine.

Recent studies have defined TGF-β as a critical regulator of thymic T cell development as well as a crucial player in peripheral T cell homeostasis, tolerance to self antigens, and T cell differentiation during the immune response (Li and Flavell 2008). Certain reports indicated that TGF-β has been implicated in the development of thymus-derived naturally occurring Treg cells (nTreg) (Liu et al. 2008). In addition, TGF-β promoted the differentiation of induced Treg (iTreg) cells from naïve T cells in peripheral tissues, which is enhanced by retinoic acid and IL-2 (Fig. 12.3) (Benson et al. 2007; Coombes et al. 2007; Davidson et al. 2007; Mucida et al. 2007;

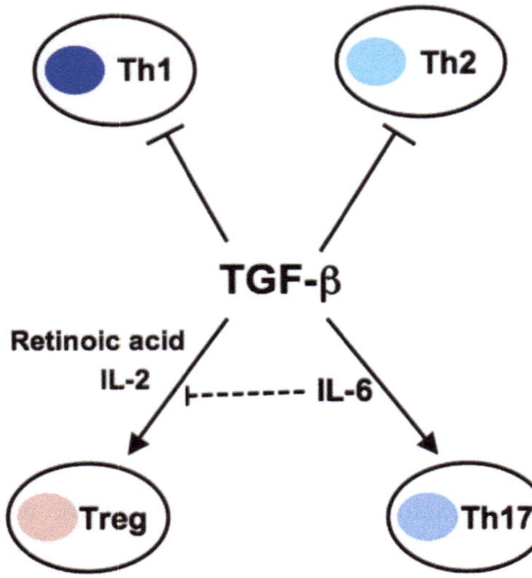

Fig. 12.3 A role of TGF-β in T cell differentiation and tolerance. Active TGF-β directly inhibits the differentiation of Th1 and Th2 cells. TGF-β promotes the generation of the Th17 cells from naïve T cells in the presence of IL-6 and the differentiation of naïve T cells into iTreg cells is simultaneously inhibited

Zheng et al. 2007). It has been known that TGF-β induces the expression of Foxp3, a master regulator of Treg cells (Chen et al. 2003). Foxp3 expression in Treg cells inhibited secretion of pro-inflammatory cytokines, including IL-2, IFN-γ, IL-4, and IL-17. Foxp3 expression also enhanced expression of anti-inflammatory cytokines, IL-10 and TGF-β, and increased an inhibitor for co-stimulation, CTLA4 (Bettelli et al. 2005; Fontenot et al. 2003; Zhou et al. 2008), eventually controlling and dampening inflammation as well as promoting tolerance. In particular, the finding that TGF-β in breast milk induces tolerance through $CD4^+$ Treg cells strongly supports the importance of TGF-β-Treg axis in the immune system (Verhasselt et al. 2008). That is, airborne antigens were efficiently transferred from the mother to the neonate through breast milk and breastfeeding-induced tolerance during lactation was mediated by $CD4^+$ Treg cells depending on the presence of TGF-β (Verhasselt et al. 2008). In addition, it has been reported that TGF-β administrated through the oral route retains sufficient biological activity in the intestinal mucosa and enhances the induction of oral tolerance (Ando et al. 2007). These studies collectively emphasize the crucial role that TGF-β plays to control the immune responses, such as tolerance induction.

While the differentiation of iTreg cells by TGF-β is inhibited in the presence of pro-inflammatory cytokine IL-6, TGF-β promotes the generation of the Th17 cells from naïve T cells in the presence of IL-6 (Fig. 12.3) (Bettelli et al. 2006). When naïve T cells differentiated into Th17 cells upon TGF-β plus IL-6, the differentiation of naïve T cells into iTreg cells was simultaneously inhibited (Fig. 12.3). This differential regulation of naïve T cells by TGF-β implies that the two T cell lineages are closely interconnected showing a mutually exclusive pattern, and has important functions in immune responses. The mechanisms underlying the regulation of Th17 and iTreg cell differentiation remains to be understood.

Alternatively, another study showed that dendritic cells (DCs) are involved in T cell responses regulated by TGF-β. Upon stimulation of naïve T cells by DCs in the presence of TGF-β, antigen-specific Foxp3⁺ Treg cells were generated (Yamazaki et al. 2006). Treg cells secreted the latent form of TGF-β1 associated with the latency-associated protein (LAP) (Nakamura et al. 2001; Oida et al. 2003), although it is still controversial whether Treg cells are the primary source for TGF-β production. In turn, αvβ8 integrin, which was expressed on DCs, degraded LAP protein through interaction with LAP and released active TGF-β (Taylor 2009). Conditional deletion of integrin αvβ8 on DCs showing autoimmune diseases (Travis et al. 2007) strongly supported the importance of interaction between Treg and DC cells. Active TGF-β finally inhibited the differentiation of Th1 and Th2 cells from naïve T cells and promoted the differentiation to iTreg (Fig. 12.3). Furthermore, certain reports indicate that the mRNA of transmembrane protein called glycoprotein A repetitions predominant (GARP) is selectively induced by TCR stimulation in human Foxp3⁺ Treg cells and the GARP protein acts as a receptor for latent TGF-β (Wang et al. 2008; Probst-Kepper et al. 2009; Stockis et al. 2009; Wang et al. 2009). These results imply that GARP is a highly specific marker which could be used for the identification of activated Treg cells, and the binding of TGF-β to GARP is an important clue for explaining the activation of Treg cells. In conclusion, a number of studies about the roles of TGF-β in T cells using in vivo or in vitro systems strongly emphasize that TGF-β's immunological function might be more important to understand the pathogenesis of human diseases.

12.3 Conclusion

Numerous studies about TGF-β's biological functions suggest that TGF-β exerts powerful anti-inflammatory activity and is the master controller of immune responses. Therefore, TGF-β-mediated signaling pathway should be critically controlled together with other essential intracellular signaling pathways to maintain homeostatic balance under normal physiological conditions. If this homeostatic balance is disrupted by unexpected stimuli or genetic mutations of TGF-β signaling-related genes, human diseases such as autoimmune and inflammatory diseases may develop. The typical examples are inflammatory bowel diseases such as Crohn's disease. Recently, many researchers are interested in autoimmune diseases and inflammatory diseases exerted by Th17 cells. In particular, Th17 cells and their effector molecules such as IL-17, IL-21, IL-22, and GM-CSF are known to be associated with the pathogenesis of rheumatoid arthritis, systematic lupus erythemaosus, multiple sclerosis, psoriasis, IBD, allergy, and asthma (Miossec et al. 2009; Wilke et al. 2011). Although the TGF-β/Smad signaling pathway is extensively studied in T cell differentiation, Treg cells, and Th17 cells, it has not yet been addressed as to how the inhibitory Smads, Smad6 and Smad7, act on the homeostasis of T cells. Therefore, it will be interesting to study whether inhibitory Smads mediate the role of TGF-β in T cell biology.

Drugs have been and are being developed to inhibit TGF-β activity, including receptor kinase antagonists, neutralizing antibodies, and antisense oligonucleotides. Since inhibitory Smads mediate some of TGF-β's anti-inflammatory activities, they may have a therapeutic potential for treatment of TLR4- or inflammation-related diseases. Therefore, if the detailed mechanisms of TGF-β's anti-inflammation activity are revealed in the future, it will be quite promising for drug development targeting or mimicking TGF-β or its signaling components.

Acknowledgments The authors acknowledge a number of key contributions which are not cited because of space limitations. S.-J.K. and S.H.P. are supported by a National Research Foundation grant of Korea (2009–0081756) funded by the Korea government.

References

Afrakhte M, Moren A, Jossan S, Itoh S, Sampath K, Westermark B, Heldin CH, Heldin NE, ten Dijke P (1998) Induction of inhibitory Smad6 and Smad7 mRNA by TGF-β family members. Biochem Biophys Res Commun 249:505–511

Akira S, Uematsu S, Takeuchi O (2006) Pathogen recognition and innate immunity. Cell 124:783–801

Ando T, Hatsushika K, Wako M, Ohba T, Koyama K, Ohnuma Y, Katoh R, Ogawa H, Okumura K, Luo J, Wyss-Coray T, Nakao A (2007) Orally administered TGF-β is biologically active in the intestinal mucosa and enhances oral tolerance. J Allergy Clin Immunol 120:916–923

Ashcroft GS, Yang X, Glick AB, Weinstein M, Letterio JL, Mizel DE, Anzano M, Greenwell-Wild T, Wahl SM, Deng C, Roberts AB (1999) Mice lacking Smad3 show accelerated wound healing and an impaired local inflammatory response. Nat Cell Biol 1:260–266

Babyatsky MW, Rossiter G, Podolsky DK (1996) Expression of transforming growth factors α and β in colonic mucosa in inflammatory bowel disease. Gastroenterology 110:975–984

Benson MJ, Pino-Lagos K, Rosemblatt M, Noelle RJ (2007) All-trans retinoic acid mediates enhanced T reg cell growth, differentiation, and gut homing in the face of high levels of co-stimulation. J Exp Med 204:1765–1774

Bettelli E, Carrier Y, Gao W, Korn T, Strom TB, Oukka M, Weiner HL, Kuchroo VK (2006) Reciprocal developmental pathways for the generation of pathogenic effector TH17 and regulatory T cells. Nature 441:235–238

Bettelli E, Dastrange M, Oukka M (2005) Foxp3 interacts with nuclear factor of activated T cells and NF-κB to repress cytokine gene expression and effector functions of T helper cells. Proc Natl Acad Sci USA 102:5138–5143

Beutler B, Jiang Z, Georgel P, Crozat K, Croker B, Rutschmann S, Du X, Hoebe K (2006) Genetic analysis of host resistance: toll-like receptor signaling and immunity at large. Annu Rev Immunol 24:353–389

Bitzer M, von Gersdorff G, Liang D, Dominguez-Rosales A, Beg AA, Rojkind M, Bottinger EP (2000) A mechanism of suppression of TGF-β/SMAD signaling by NF-κB/ RelA. Genes Dev 14:187–197

Blobe GC, Schiemann WP, Lodish HF (2000) Role of transforming growth factor β in human disease. N Engl J Med 342:1350–1358

Chang M, Jin W, Sun SC (2009) Peli1 facilitates TRIF-dependent toll-like receptor signaling and proinflammatory cytokine production. Nat Immunol 10:1089–1095

Chen W, Jin W, Hardegen N, Lei KJ, Li L, Marinos N, McGrady G, Wahl SM (2003) Conversion of peripheral CD4+CD25- naive T cells to CD4+CD25+ regulatory T cells by TGF-β induction of transcription factor Foxp3. J Exp Med 198:1875–1886

Choi KC, Lee YS, Lim S, Choi HK, Lee CH, Lee EK, Hong S, Kim IH, Kim SJ, Park SH (2006) Smad6 negatively regulates interleukin 1-receptor-toll-like receptor signaling through direct interaction with the adaptor Pellino-1. Nat Immunol 7:1057–1065

Cook DN, Pisetsky DS, Schwartz DA (2004) Toll-like receptors in the pathogenesis of human disease. Nat Immunol 5:975–979

Coombes JL, Siddiqui KR, Arancibia-Carcamo CV, Hall J, Sun CM, Belkaid Y, Powrie F (2007) A functionally specialized population of mucosal CD103+ DCs induces Foxp3+ regulatory T cells via a TGF-β and retinoic acid-dependent mechanism. J Exp Med 204:1757–1764

Davidson TS, DiPaolo RJ, Andersson J, Shevach EM (2007) Cutting Edge: IL-2 is essential for TGF-β-mediated induction of Foxp3+ T regulatory cells. J Immunol 178:4022–4026

Derynck R, Zhang YE (2003) Smad-dependent and Smad-independent pathways in TGF-β family signalling. Nature 425:577–584

Di Sabatino A, Pickard KM, Rampton D, Kruidenier L, Rovedatti L, Leakey NA, Corazza GR, Monteleone G, MacDonald TT (2008) Blockade of transforming growth factor β upregulates T-box transcription factor T-bet, and increases T helper cell type 1 cytokine and matrix metalloproteinase-3 production in the human gut mucosa. Gut 57:605–612

Fontenot JD, Gavin MA, Rudensky AY (2003) Foxp3 programs the development and function of CD4+CD25+ regulatory T cells. Nat Immunol 4:330–336

Ghosh AK, Yuan W, Mori Y, Chen S, Varga J (2001) Antagonistic regulation of type I collagen gene expression by interferon-gamma and transforming growth factor-β. Integration at the level of p300/CBP transcriptional coactivators. J Biol Chem 276:11041–11048

Gorelik L, Flavell RA (2000) Abrogation of TGFβ signaling in T cells leads to spontaneous T cell differentiation and autoimmune disease. Immunity 12:171–181

Grosshans J, Schnorrer F, Nusslein-Volhard C (1999) Oligomerisation of tube and pelle leads to nuclear localisation of dorsal. Mech Dev 81:127–138

Hamidi A, von Bulow V, Hamidi R, Winssinger N, Barluenga S, Heldin CH, Landström M (2012) Polyubiquitination of transforming growth factor β (TGFβ)-associated kinase 1 mediates nuclear factor-κB activation in response to different inflammatory stimuli. J Biol Chem 287:123–133

Hayashi H, Abdollah S, Qiu Y, Cai J, Xu YY, Grinnell BW, Richardson MA, Topper JN, Gimbrone MA Jr, Wrana JL, Falb D (1997) The MAD-related protein Smad7 associates with the TGFβ receptor and functions as an antagonist of TGFβ signaling. Cell 89:1165–1173

Heldin CH, Miyazono K, ten Dijke P (1997) TGF-β signalling from cell membrane to nucleus through SMAD proteins. Nature 390:465–471

Hong S, Lim S, Li AG, Lee C, Lee YS, Lee EK, Park SH, Wang XJ, Kim SJ (2007) Smad7 binds to the adaptors TAB2 and TAB3 to block recruitment of the kinase TAK1 to the adaptor TRAF2. Nat Immunol 8:504–513

Imamura T, Takase M, Nishihara A, Oeda E, Hanai J, Kawabata M, Miyazono K (1997) Smad6 inhibits signalling by the TGF-β superfamily. Nature 389:622–626

Ishida W, Hamamoto T, Kusanagi K, Yagi K, Kawabata M, Takehara K, Sampath TK, Kato M, Miyazono K (2000) Smad6 is a Smad1/5-induced smad inhibitor. Characterization of bone morphogenetic protein-responsive element in the mouse Smad6 promoter. J Biol Chem 275:6075–6079

Jiang Z, Johnson HJ, Nie H, Qin J, Bird TA, Li X (2003) Pellino 1 is required for interleukin-1 (IL-1)-mediated signaling through its interaction with the IL-1 receptor-associated kinase 4 (IRAK4)-IRAK-tumor necrosis factor receptor-associated factor 6 (TRAF6) complex. J Biol Chem 278:10952–10956

Kim EY, Kim BC (2011) Lipopolysaccharide inhibits transforming growth factor-β1-stimulated Smad6 expression by inducing phosphorylation of the linker region of Smad3 through a TLR4-IRAK1-ERK1/2 pathway. FEBS Lett 585:779–785

Kleeff J, Maruyama H, Friess H, Buchler MW, Falb D, Korc M (1999) Smad6 suppresses TGF-β-induced growth inhibition in COLO-357 pancreatic cancer cells and is overexpressed in pancreatic cancer. Biochem Biophys Res Commun 255:268–273

Kulkarni AB, Ward JM, Yaswen L, Mackall CL, Bauer SR, Huh CG, Gress RE, Karlsson S (1995) Transforming growth factor-β1 null mice. An animal model for inflammatory disorders. Am J Pathol 146:264–275

Lee YS, Park JS, Kim JH, Jung SM, Lee JY, Kim SJ, Park SH (2011) Smad6-specific recruitment of Smurf E3 ligases mediates TGF-β1-induced degradation of MyD88 in TLR4 signalling. Nat Commun 2:460. doi:10.1038/ncomms1469

Letterio JJ, Roberts AB (1997) TGF-β: a critical modulator of immune cell function. Clin Immunol Immunopathol 84:244–250

Leveen P, Larsson J, Ehinger M, Cilio CM, Sundler M, Sjostrand LJ, Holmdahl R, Karlsson S (2002) Induced disruption of the transforming growth factor β type II receptor gene in mice causes a lethal inflammatory disorder that is transplantable. Blood 100:560–568

Li MO, Flavell RA (2008) Contextual regulation of inflammation: a duet by transforming growth factor-β and interleukin-10. Immunity 28:468–476

Li MO, Sanjabi S, Flavell RA (2006) Transforming growth factor-β controls development, homeostasis, and tolerance of T cells by regulatory T cell-dependent and -independent mechanisms. Immunity 25:455–471

Lim S, Bae E, Kim HS, Kim TA, Byun K, Kim B, Hong S, Im JP, Yun C, Lee B, Park SH, Letterio J, Kim SJ (2012) TRAF6 mediates IL-1β/LPS-induced suppression of TGF-β signaling through its interaction with the type III TGF-β receptor. PLoS One 7:e32705

Liu Y, Zhang P, Li J, Kulkarni AB, Perruche S, Chen W (2008) A critical function for TGF-β signaling in the development of natural CD4+CD25+Foxp3+ regulatory T cells. Nat Immunol 9:632–640

Lucas PJ, Kim SJ, Melby SJ, Gress RE (2000) Disruption of T cell homeostasis in mice expressing a T cell-specific dominant negative transforming growth factor β II receptor. J Exp Med 191:1187–1196

Marie JC, Liggitt D, Rudensky AY (2006) Cellular mechanisms of fatal early-onset autoimmunity in mice with the T cell-specific targeting of transforming growth factor-β receptor. Immunity 25:441–454

Massague J (1998) TGF-β signal transduction. Annu Rev Biochem 67:753–791

Massague J, Wotton D (2000) Transcriptional control by the TGF-β/Smad signaling system. EMBO J 19:1745–1754

Medzhitov R (2008) Origin and physiological roles of inflammation. Nature 454:428–435

Meng XM, Huang XR, Xiao J, Chen HY, Zhong X, Chung AC, Lan HY (2012a) Diverse roles of TGF-β receptor II in renal fibrosis and inflammation in vivo and in vitro. J Pathol 227:175–188

Meng XM, Huang XR, Xiao J, Chung AC, Qin W, Chen HY, Lan HY (2012b) Disruption of Smad4 impairs TGF-β/Smad3 and Smad7 transcriptional regulation during renal inflammation and fibrosis in vivo and in vitro. Kidney Int 81:266–279

Miossec P, Korn T, Kuchroo VK (2009) Interleukin-17 and type 17 helper T cells. N Engl J Med 361:888–898

Miyazono K, ten Dijke P, Heldin CH (2000) TGF-β signaling by Smad proteins. Adv Immunol 75:115–157

Monteleone G, Kumberova A, Croft NM, McKenzie C, Steer HW, MacDonald TT (2001) Blocking Smad7 restores TGF-β1 signaling in chronic inflammatory bowel disease. J Clin Invest 108:601–609

Moynagh PN (2009) The Pellino family: IRAK E3 ligases with emerging roles in innate immune signalling. Trends Immunol 30:33–42

Mu Y, Gudey SK, Landstrom M (2012) Non-Smad signaling pathways. Cell Tissue Res 347:11–20

Mucida D, Park Y, Kim G, Turovskaya O, Scott I, Kronenberg M, Cheroutre H (2007) Reciprocal TH17 and regulatory T cell differentiation mediated by retinoic acid. Science 317:256–260

Naiki Y, Michelsen KS, Zhang W, Chen S, Doherty TM, Arditi M (2005) Transforming growth factor-β differentially inhibits MyD88-dependent, but not TRAM- and TRIF-dependent, lipopolysaccharide-induced TLR4 signaling. J Biol Chem 280:5491–5495

Nakamura K, Kitani A, Strober W (2001) Cell contact-dependent immunosuppression by CD4(+) CD25(+) regulatory T cells is mediated by cell surface-bound transforming growth factor β. J Exp Med 194:629–644
Nakao A, Afrakhte M, Moren A, Nakayama T, Christian JL, Heuchel R, Itoh S, Kawabata M, Heldin NE, Heldin CH, ten Dijke P (1997) Identification of Smad7, a TGFβ-inducible antagonist of TGF-β signalling. Nature 389:631–635
Nathan C, Ding A (2010) Nonresolving inflammation. Cell 140:871–882
O'Neill LA, Bowie AG (2007) The family of five: TIR-domain-containing adaptors in toll-like receptor signalling. Nat Rev Immunol 7:353–364
Oida T, Zhang X, Goto M, Hachimura S, Totsuka M, Kaminogawa S, Weiner HL (2003) CD4+CD25-T cells that express latency-associated peptide on the surface suppress CD4+CD45RBhigh-induced colitis by a TGF-β-dependent mechanism. J Immunol 170:2516–2522
Probst-Kepper M, Geffers R, Kröger A, Viegas N, Eric C, Hecht HJ, Lünsdorf H, Roubin R, Moharregh-Khiabani D, Wagner K, Ocklenburg F, Jeron A, Garritsen H, Arstila TP, Kekäläinen E, Balling R, Hauser H, Buer J, Weiss S (2009) GARP: a key receptor controlling FOXP3 in human regulatory T cells. J Cell Mol Med 13:3343–3357
Reardon C, McKay DM (2007) TGF-β suppresses IFN-gamma-STAT1-dependent gene transcription by enhancing STAT1-PIAS1 interactions in epithelia but not monocytes/macrophages. J Immunol 178:4284–4295
Shull MM, Ormsby I, Kier AB, Pawlowski S, Diebold RJ, Yin M, Allen R, Sidman C, Proetzel G, Calvin D et al (1992) Targeted disruption of the mouse transforming growth factor-β1 gene results in multifocal inflammatory disease. Nature 359:693–699
Smith H, Liu XY, Dai L, Goh ET, Chan AT, Xi J, Seh CC, Qureshi IA, Lescar J, Ruedl C, Gourlay R, Morton S, Hough J, McIver EG, Cohen P, Cheung PC (2011) The role of TBK1 and IKKε in the expression and activation of Pellino 1. Biochem J 434:537–548
Sorrentino A, Thakur N, Grimsby S, Marcusson A, von Bulow V, Schuster N, Zhang S, Heldin CH, Landström M (2008) The type I TGF-β receptor engages TRAF6 to activate TAK1 in a receptor kinase-independent manner. Nat Cell Biol 10:1199–1207
Stockis J, Colau D, Coulie PG, Lucas S (2009) Membrane protein GARP is a receptor for latent TGF-β on the surface of activated human Treg. Eur J Immunol 39:3315–3322
Takaki H, Minoda Y, Koga K, Takaesu G, Yoshimura A, Kobayashi T (2006) TGF-β1 suppresses IFN-γ-induced NO production in macrophages by suppressing STAT1 activation and accelerating iNOS protein degradation. Genes Cells 11:871–882
Takeuchi O, Akira S (2010) Pattern recognition receptors and inflammation. Cell 140:805–820
Taylor AW (2009) Review of the activation of TGF-β in immunity. J Leukoc Biol 85:29–33
Tojo M, Takebe A, Takahashi S, Tanaka K, Imamura T, Miyazono K, Chiba T (2012) Smad7-deficient mice show growth retardation with reduced viability. J Biochem 151(6):621–631, Feb 29 Epub ahead of print
Travis MA, Reizis B, Melton AC, Masteller E, Tang Q, Proctor JM, Wang Y, Bernstein X, Huang X, Reichardt LF, Bluestone JA, Sheppard D (2007) Loss of integrin α(v)β8 on dendritic cells causes autoimmunity and colitis in mice. Nature 449:361–365
Ulloa L, Doody J, Massague J (1999) Inhibition of transforming growth factor-β/SMAD signalling by the interferon-γ/STAT pathway. Nature 397:710–713
Verhasselt V, Milcent V, Cazareth J, Kanda A, Fleury S, Dombrowicz D, Glaichenhaus N, Julia V (2008) Breast milk-mediated transfer of an antigen induces tolerance and protection from allergic asthma. Nat Med 14:170–175
Wang R, Wan Q, Kozhaya L, Fujii H, Unutmaz D (2008) Identification of a regulatory T cell specific cell surface molecule that mediates suppressive signals and induces Foxp3 expression. PLoS One 3:e2705
Wang R, Kozhaya L, Mercer F, Khaitan A, Fujii H, Unutmaz D (2009) Expression of GARP selectively identifies activated human FOXP3+ regulatory T cells. Proc Natl Acad Sci USA 106:13439–13444

Wilke CM, Bishop K, Fox D, Zou W (2011) Deciphering the role of Th17 cells in human disease. Trends Immunol 32:603–611

Yamazaki S, Patel M, Harper A, Bonito A, Fukuyama H, Pack M, Tarbell KV, Talmor M, Ravetch JV, Inaba K, Steinman RM (2006) Effective expansion of alloantigen-specific Foxp3+ CD25+ CD4+ regulatory T cells by dendritic cells during the mixed leukocyte reaction. Proc Natl Acad Sci USA 103:2758–2763

Yang X, Letterio JJ, Lechleider RJ, Chen L, Hayman R, Gu H, Roberts AB, Deng C (1999) Targeted disruption of SMAD3 results in impaired mucosal immunity and diminished T cell responsiveness to TGF-β. EMBO J 18:1280–1291

Zhang YE (2009) Non-Smad pathways in TGF-β signaling. Cell Res 19:128–139

Zheng SG, Wang J, Wang P, Gray JD, Horwitz DA (2007) IL-2 is essential for TGF-β to convert naive CD4+ CD25- cells to CD25+ Foxp3+ regulatory T cells and for expansion of these cells. J Immunol 178:2018–2027

Zhou L, Lopes JE, Chong MM, Ivanov II, Min R, Victora GD, Shen Y, Du J, Rubtsov YP, Rudensky AY, Ziegler SF, Littman DR (2008) TGF-β-induced Foxp3 inhibits Th17 cell differentiation by antagonizing RORγt function. Nature 453:236–240

Part III
Cardiovascular Diseases

Chapter 13
TGF-β and Cardiovascular Disorders

Laurens A. van Meeteren, Marie-José Goumans, and Peter ten Dijke

Abstract Transforming growth factor β (TGF-β) is a secreted pleiotropic cytokine that is involved in a wide range of biological processes and has essential roles in development and tissue homeostasis. TGF-β elicits cellular effects by activating serine/threonine kinase receptors located at the plasma membrane and intracellular Smad effector proteins. TGF-β is an important (cardio)vascular regulator as shown by many in vitro studies on cultured endothelial and mural cells, in vivo studies in animal models, and genetic studies in which mutations in human TGF-β signaling components have been causally linked to cardiovascular diseases. Here we review recent progress in our understanding of the (dys)function of TGF-β in the cardiovascular system.

Keywords Angiogenesis • Cardiovascular • Endothelium • TGF-β

13.1 Introduction

Transforming growth factor β (TGF-β) is one of the most studied members of a large family of structurally related secreted pleiotropic cell to cell signaling molecules. Different members of the family include the activins and bone morphogenetic proteins (BMPs). Many of these cytokines have essential roles in numerous processes during development, but also in maintenance of tissue homeostasis and tissue repair in the adult (Massagué 1998). Not surprisingly,

L.A. van Meeteren • M.-J. Goumans • P. ten Dijke (✉)
Department of Molecular Cell Biology and Centre for Biomedical Genetics, Leiden University Medical Center, Postzone S-1-P, PO Box 9600, Leiden 2300, The Netherlands
e-mail: p.ten_dijke@lumc.nl

A. Moustakas and K. Miyazawa (eds.), *TGF-β in Human Disease*,
DOI 10.1007/978-4-431-54409-8_13,

when their action is perturbed, they have an important role in pathological conditions, like vascular diseases, cancer, and fibrosis (Blobe et al. 2000; ten Dijke and Arthur 2007).

The ligands of the TGF-β family mediate their effects by binding specific single transmembrane receptors at the cell membrane (Feng and Derynck 2005). Upon ligand-induced heteromeric complex formation, the type I receptor is phosphorylated by the type II receptor. There are seven different type I receptors (ALK1 to ALK7) (also known as activin receptor-like kinases (ALK)) and five type II receptors (activin receptor type IIA (ActRIIA), activin receptor type IIB (ActRIIB), BMP type II receptor (BMPRII), TGF-β type II receptor (TGF-βRII), and AMH type II receptor (AMHRII)). Different ligands can bind different combinations of type I and type II receptor thereby creating specificity of signaling. TGF-β signals mostly via ALK5 and TGF-βRII, activins via ALK4 with ActRIIA or ActRIIB, and BMPs signal via ALK1, 2, 3, and 6 together with BMPRII, ActRIIA, or ActRIIB. For regulation of endothelial function by TGF-β, ALK1 and ALK5 signaling are most important (van Meeteren and ten Dijke 2012).

After binding of the ligand the type I receptor phosphorylates specific transcription factors, receptor regulated (R)-Smads. Upon activation by type I receptors, R-Smads form heteromeric complexes with the common mediator (Co)-Smad (Smad4) and these heteromeric complexes accumulate into the nucleus, where they regulate the transcription of specific target genes (Moustakas and Heldin 2009).

Inhibitory Smads (I-Smads) inhibit the activation of R-Smads by competing for type I receptor interaction and by recruiting phosphatases and ubiquitin ligases to the activated receptor complex leading to dephosphorylation or proteosomal degradation of the receptor complex (Itoh and ten Dijke 2007). R-Smads can be divided into two groups based on the type I receptor that is activating them. The first group consists of Smad1, 5, and 8 and these are activated by ALK1, 2, 3, and 6. The second group contains Smad2 and 3 and is activated by ALK4, 5, and 7. In addition to Smad signaling, TGF-β and BMP signaling can result in activation of pathways where Smads are not directly involved. Non-Smad pathways include various branches of MAP kinase pathways, Rho-like GTPase signaling pathways, and PI3K/AKT pathways (Moustakas and Heldin 2005; Zhang 2009). Non-Smad signaling pathway have been found to modulate Smad signaling and vice versa (Moustakas and Heldin 2005) (Figs. 13.1).

Co-receptors are receptors that do not signal by themselves since they lack intracellular enzymatic motifs such as kinase domains (Kirkbride et al. 2005). For TGF-β signaling co-receptors endoglin and betaglycan (also called TGF-β receptor III) have been identified (Cheifetz et al. 1992; Wang et al. 1991). Both receptors are structurally related and have a small intracellular tail, a single transmembrane domain, and a large extracellular domain. Endoglin is highly expressed in proliferating endothelial cells, hence its name endoglin.

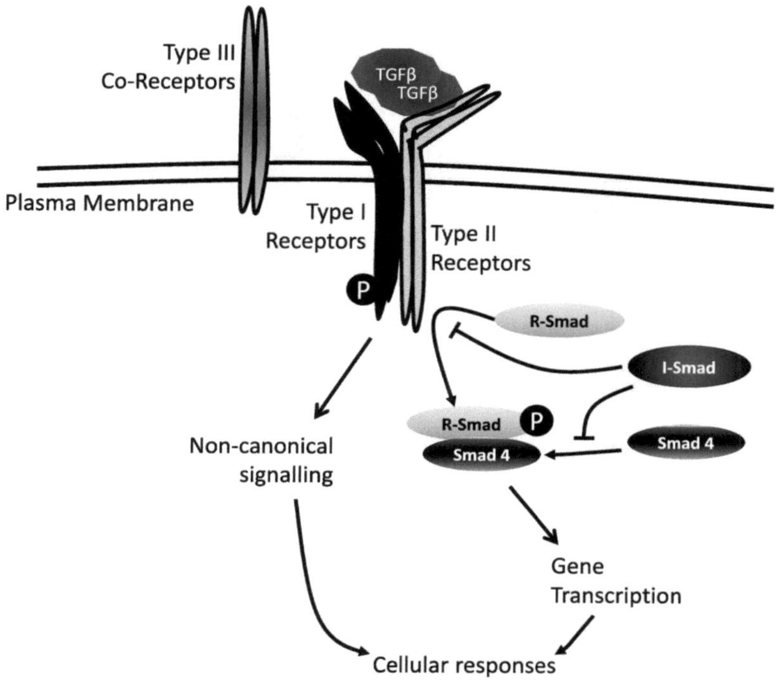

Fig. 13.1 Basic signaling unit of the TGF-β system. TGF-β family ligands such as TGF-β and BMPs bind a complex of type I receptors, type II receptors, and possibly a type III co-receptor leading to activation of R-Smads. R-Smads then form a complex with Smad4 leading to specific gene-transcription. I-Smads inhibit the activation of R-Smads and complex formation of Smad4 with R-Smads. In addition, the receptor complex can also activate non-Smad pathways

13.2 TGF-β Signaling in Vascular Biology

13.2.1 Regulating Endothelial Function by TGF-β

Genetic studies in mice and humans revealed the importance of components of the TGF-β signaling pathway in vascular morphogenesis and angiogenesis. Deletion of components of the TGF-β pathway (ligands, receptors, and intracellular Smad mediators) in mice leads to embryonic lethality due to vascular abnormalities (see below) (van Meeteren et al. 2011).

To regulate the activation state of endothelial cells TGF-β can differentially activate two type I receptors, ALK1 and ALK5 (Goumans et al. 2002; Oh et al. 2000). ALK5 is expressed in most tissues, but ALK1 expression is restricted to endothelial cells. TGF-β-induced ALK5 signaling activates Smad2 and Smad3 phosphorylation resulting in inhibition of endothelial cell proliferation, migration, and organization

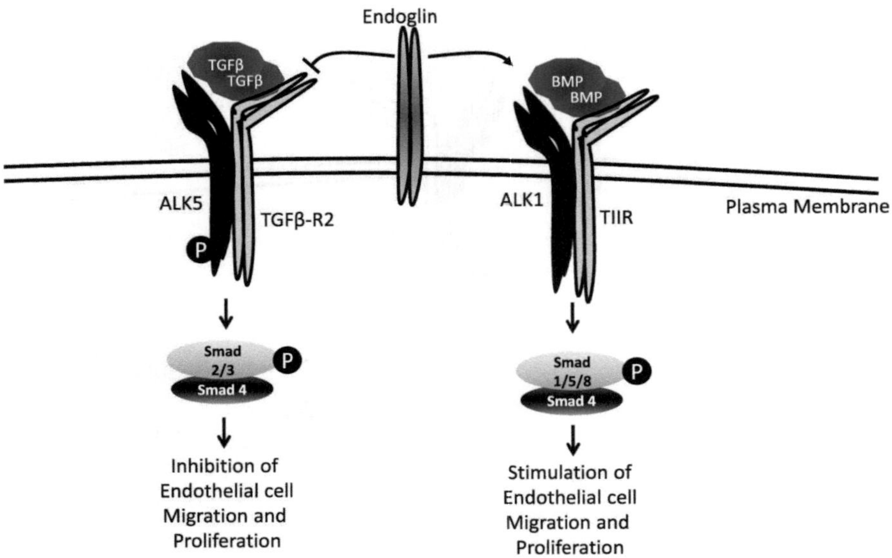

Fig. 13.2 TGF-β signaling in endothelial cells. BMP binding to ALK1 (in complex with a type II receptor and endoglin) activates Smad1, 5, and 8 leading to stimulation of endothelial cell migration and proliferation. ALK1 is antagonized by ALK5 that is activated by TGF-β leading to Smad2 and 3 activation and endothelial cell quiescence

(Goumans et al. 2003) (Fig. 13.2). The ALK5 kinase inhibitor SB-431542 enabled proliferation and sheet formation of embryonic stem cell-derived endothelial cells. SB-431542 upregulated the expression of claudin-5, an endothelial-specific component of tight junctions, suggesting a role of ALK5 signaling in regulating vascular permeability (Watabe et al. 2003). In addition, VEGF and inhibitors of TGF-β type I receptor kinase synergistically promoted blood vessel formation by inducing β 5-integrin expression (Liu et al. 2009). In contradiction to these findings, in vivo, SB-431542 has been reported to induce enhanced delivery of nanoparticles to the tumor tissue extravascular space due to increased vessel leakiness (Kano et al. 2007). Indeed, ALK5 has been reported before to be important for TGF-β-induced permeability and cytoskeletal remodeling of endothelial cells (Birukova et al. 2005). Interestingly a major component of endothelial cell–cell junctions, VE-cadherin, was found to interact with ALK5, ALK1, endoglin, and TGF-βRII, which possibly reveals the link between permeability and TGF-β signaling (Rudini et al. 2008). In conclusion, ALK5 signaling results in keeping endothelial cells in a quiescent state.

TGF-β-induced ALK1 signaling activates Smad1 and Smad5 leading to endothelial cell proliferation, migration, and organization (Goumans et al. 2003). A central intracellular effector of ALK1 is Id1; its upregulation was shown to be required for TGF-β/ALK1-induced endothelial cell migration and tube formation (Valdimarsdottir et al. 2002) and inhibition of ALK1 was shown to inhibit endothelial cell sprouting (van Meeteren et al. 2012). However, inhibitory effects of ALK1 signaling on endothelial cells have also been reported (David et al. 2007; Lamouille et al. 2002; Mallet et al. 2006). The effect of ALK1 is likely dependent on cellular context.

Although ALK1 and ALK5 have opposing effects on endothelial cells, they do interact with each other physically. ALK5-deficient endothelial cells are not only defective in ALK5 signaling but also show impaired ALK1 responses; ALK5 was found to be essential for recruitment of ALK1 into a TGF-β receptor complex, and the kinase activity of ALK5 is essential for full ALK1 activation (Goumans et al. 2003). In the presence of fibronectin ALK1 activity is not dependent on ALK5 activity since in the presence of fibronectin an ALK5 inhibitor has no effect on ALK1-mediated Smad1/5/8 phosphorylation (Tian et al. 2012). On the other hand, ALK1 can directly antagonize ALK5 signaling at the level of Smads (Goumans et al. 2002; Oh et al. 2000). Taken together, the cross talk between ALK1 and ALK5 signaling provides endothelial cells with a sophisticated mechanism to fine-tune endothelial function.

Endothelial cells can express both the TGF-β co-receptors endoglin and betaglycan. Endoglin positive but betaglycan negative endothelial cells are responsive to TGF-β-1 and -3, but not -2 (Cheifetz et al. 1990). In endothelial cells that express both co-receptors endoglin can form a complex with betaglycan and the TGF-β receptor complex simultaneously (Wong et al. 2000).

Endoglin is predominantly expressed in highly proliferating endothelial cells, but is also expressed in hematopoietic cells, syncytiotrophoblasts of term placenta, stromal cells, smooth muscle cells, and mesenchymal cells (Bot et al. 2009; Dağdeviren et al. 1998). Ectopic expression of endoglin inhibits TGF-β-induced growth inhibition of endothelial cells, monocytes and myoblasts (Lastres 1996; Lebrin et al. 2004), and extracellular matrix deposition (Obreo et al. 2004). Endoglin plays an important role in balancing ALK1 and ALK5 signaling pathways. Endoglin overexpression potentiates TGF-β/ALK1 signaling and inhibits TGF-β/ALK5 signaling, whereas endoglin knockdown resulted in impaired TGF-β/ALK1 signaling responses (Blanco et al. 2005; Lebrin et al. 2004). Endothelial cells isolated from endoglin heterozygote or homozygote embryos displayed significantly reduced proliferation and migration, increased collagen production, and decreased nitric oxide (NO) synthase expression and VEGF secretion (Jerkic et al. 2006). Hypoxia upregulates endoglin levels and this upregulation was shown to protect endothelial cells from hypoxia-induced apoptosis (Li et al. 2003).

A general feature of co-receptors is that they can be cleaved from the cell membrane with the consequence that they exist as soluble forms. Indeed, also betaglycan and endoglin can be shedded by membrane type matrix metalloprotease-1 (MT-MMP) from the plasma membrane giving rise to a soluble betaglycan and soluble endoglin (sol-endoglin) (Hawinkels et al. 2010; Velasco-Loyden et al. 2004). Interestingly, sol-endoglin levels in plasma are a prognostic marker for several pathological conditions, including pregnant women suffering from preeclampsia (Levine et al. 2006; Liu et al. 2012). Furthermore, several labs have reported increased levels of sol-endoglin in serum from cancer patients as a marker of poor prognosis (Li et al. 2000a).

13.2.2 Regulating Mural Cell Function by TGF-β

Both endothelial cells and their supporting cells, such as vascular smooth muscle cells (VSMCs) and pericytes, are needed for proper endothelial function. Tight regulation and close coordination between endothelial cells and VSMCs and pericytes are needed to form a mature vascular network (Armulik et al. 2005, 2011). VSMCs express type I and type II receptors for TGF-β family members (Bobik 2006), and TGF-β is a potent stimulator of VSMC differentiation (Owens 1995). VSMCs form gap junctions with endothelial cells and upon receiving signals from VSMCs, endothelial cells line up and recruit more VSMCs. TGF-β is an important regulator of the interaction between endothelial cells and the supporting cells. Endothelial cells produce latent TGF-β that upon endothelial–VSMC interaction can be activated leading to VSMC differentiation and function (Ding et al. 2004). Knockout mice of several TGF-β signaling players reveal the significance of TGF-β signaling in VSMC–endothelial cell interaction. ALK1- and endoglin knockout mice show defective remodeling of the primary capillary plexus of the yolk sac and failure in SMC development (Arthur et al. 2000; Goumans and Mummery 2000; Oh et al. 2000). TGF-βRII, Smad5, Smad1, and TGF-β1 knockout mice show defects in vasculature structure or blood vessel organization indicative of a defect in endothelial cell lining and impaired VSMC development (ten Dijke and Arthur 2007).

13.2.3 Lessons from Knockout Mice

TGF-β ligands receptors and downstream signaling components are involved in various aspects of vascular biology, in physiology as well as pathophysiology and in development as well as the adult stage (ten Dijke et al. 2008).

Knockout studies in mice offered the first indication that TGF-β signaling plays an essential role in vascular function and development. We will focus mainly on the animal models with disrupted TGF-β signaling components that exhibit defects in angiogenesis and vasculogenesis. For a summary of these knockouts, see Table 13.1.

The first reports on TGF-β knockout mice showed that TGF-β was only necessary for postnatal survival and not for embryonic development (Kulkarni et al. 1993; Shull et al. 1992). However, subsequently it was reported that on the genetic background used for the first studies placental and lactational transfer of maternal TGF-β1 from heterozygous mothers to knockout embryos and pups could have rescued their embryonic lethality (Kallapur et al. 1999; Letterio et al. 1994). In a mixed 129×NIH/Ola×C57BL/6 background, 50 % of the mutant embryos die around embryonic day 10.5 (E10.5) due to inadequate yolk sac development (Bonyadi et al. 1997; Dickson et al. 1995) as a result of failure in both vasculogenesis and hematopoiesis. A defect in terminal differentiation of the endothelial cells in the yolk sac affecting endothelial tube formation and/or its integrity resulting in insufficient capillary tube formation and weak vessels was observed. However, there was no

Table 13.1 List of TGF-β signaling component knockout mice that show vascular defects

Gene	Phenotype knockout mice	References
Ligands		
TGF-β1	Embryonic lethal (E10.5) with vascular defects such as inadequate yolk sac development, defective vasculogenesis, and hematopoiesis	Dickson et al. (1995), Kulkarni et al. (1993), Letterio et al. (1994), Shull et al. (1992)
Receptors		
ALK1	Embryonic lethal (E11.5) due to severe vascular abnormalities such as large shunts between arteries and veins. Also defects in differentiation and recruitment of VSMCs	Oh et al. (2000), Urness et al. (2000)
ALK5	Embryonic lethal (E11.5), severe defects in vascular development of the yolk sac and placenta, and absence of circulating red blood cells	Larsson et al. (2001)
Co-receptors		
Endoglin	Embryonic lethal (E10.5) due to angiogenesis defects and defects in VSMC differentiation	Arthur et al. (2000), Bourdeau et al. (1999), Carvalho et al. (2004), Li (1999)
Betaglycan	Embryonic lethal (E13.5) with defects in coronary vessels and increased apoptosis in the liver	Compton et al. (2007), Stenvers et al. (2003)
Smads		
Smad4	Endothelial-specific knockout is embryonic lethal (E10.5) due to cardiovascular defects and defects in VSMC differentiation and recruitment	Lan et al. (2007)
Smad5	Embryonic lethal (E10.5) due to angiogenesis defects in the yolk sac and enlarged blood vessels that failed to recruit VSMCs	Yang et al. (1999)

obvious defect in vascular development within the TGF-β1-deficient embryos at E9.5. TGF-β2 and TGF-β3 knockout mice die perinatal with no obvious vascular defects (Kaartinen et al. 1995; Sanford et al. 1997).

Acvrl1 (encoding for mouse ALK1) homozygous knockout embryos die at midgestation (around E11.5) due to severe vascular abnormalities such as large shunts between arteries and veins and hyperdilation of large vessels. Furthermore, defects in differentiation and recruitment of supporting cells such as VSMCs were observed (Oh et al. 2000; Urness et al. 2000). Interestingly, a zebrafish with a mutation in ALK1 suffers from an abnormal circulation pattern in which blood flows through a limited number of dilated cranial vessels and failure to perfuse the trunk and tail (Roman et al. 2002).

Mice deficient for ALK5 die also at midgestation, exhibiting severe defects in vascular development of the yolk sac and placenta, and an absence of circulating red blood cells. Endothelial cells from ALK5-deficient embryos show enhanced cell proliferation, improper migratory behavior, and impaired fibronectin production in vitro, defects that are associated with the vascular defects seen in vivo (Larsson et al. 2001).

When *Alk5* was knocked out in only ALK1 expressing endothelial cells, no vascular phenotype was observed. This would suggest that ALK5 plays no important roles in endothelial cells (Park et al. 2008) in line with earlier reports from the same lab that found no detectable expression of ALK5 in endothelial cells (Seki et al. 2006). However, it is in conflict with earlier studies that showed that conditional *Alk5* crossed with the endothelial-specific *tie-1*-cre mice did show a lethal vascular phenotype (Carvalho et al. 2004). This discrepancy is most likely due to different expression pattern between the *tie-1*-cre and the *alk1*-cre. *Tie-1*-cre is supposedly expressed earlier in endothelial cells, and therefore it could mean that ALK5 plays an essential role in the earliest stages of mouse vascular development. Indeed, in the *alk1*-cre mice cre is only expressed at a stage when the *tie-1*-cre ALK5 embryos have already died.

Knockout mice for many other ALKs are also embryonic lethal but at stages before vascular development. For example knockout mice for *Alk2*, *Alk3*, and *Alk4* die around E7.5 (Goumans and Mummery 2000). Therefore it cannot be excluded that also these have roles in the development of the vascular system. By crossing conditional knockout mice for these receptors with an endothelial-specific inducer these roles could be revealed in the future.

Mice deficient for the ALK1 co-receptor endoglin were described by three independent groups. All endoglin knockout mice showed early embryonic lethality at E10.5–11.5 days caused by angiogenesis defects of the yolk sac and abnormal heart development (Arthur et al. 2000; Bourdeau et al. 1999; Li 1999). The formation of the primitive vascular plexus in the yolk sac occurred normally but the successive branching and remodeling associated with angiogenesis fails to proceed. Subsequent analysis of the endoglin knockout mice showed that endoglin is required for downstream TGF-β signaling from the endothelial cell to the adjacent smooth muscle cell to promote smooth muscle cell differentiation (Carvalho et al. 2004). Mice deficient for the co-receptor betaglycan die around E13.5 due to defects in the formation of the coronary vessels and increased apoptosis in the liver (Compton et al. 2007; Stenvers et al. 2003).

Smad1, *2*, and *4* deficient mice all die before vascular development starts due to defects and abnormalities in germ layer formation. In addition, the allantois in Smad1 knockouts fails to connect with the placenta (Huang et al. 2002).

Endothelial-specific *Smad4* knockouts die at E10.5 due to cardiovascular defects, including attenuated vessels sprouting and remodeling, collapsed dorsal aortas, enlarged hearts with reduced trabeculae, and failed endocardial cushion formation. Furthermore, vessels that formed showed impaired development of VSMCs and association between endothelial and VSMCs. Noticeably, *Smad4*-deficient endothelial cells from these mice demonstrated an intrinsic defect in tube formation in vitro (Lan et al. 2007).

Smad5 knockout mouse embryos die between days 10.5 and 11.5 due to defects in angiogenesis. The mutant yolk sacs lacked normal vasculature and had irregularly distributed blood cells, although they contained hematopoietic precursors capable of erythroid differentiation. In addition, *Smad5* mutant embryos had enlarged blood vessels surrounded by decreased numbers of VSMCs, suffered massive apoptosis of mesenchymal cells (Yang et al. 1999).

13.2.4 Endothelial-to-Mesenchymal Transition

Recent evidence has demonstrated that endothelial cells can have a remarkable plasticity. By a process called endothelial-to-mesenchymal transition (EndMT) endothelial cells convert to a more mesenchymal cell type that can give rise to cells such as fibroblasts, but also bone cells. EndMT is essential during embryonic development and tissue regeneration (van Meeteren and ten Dijke 2012). Interestingly, it also plays a role in pathological conditions like fibrosis of organs such as the heart and kidney (Goumans et al. 2008). In addition, EndMT contributes to the generation of cancer-associated fibroblasts that are known to influence the tumor-microenvironment favorable for the tumor cells. EndMT is a form of the more widely known and studied epithelial-to-mesenchymal transition (EMT) (Thiery et al. 2009).

Many endothelial cells can be induced to undergo EndMT when stimulated with TGF-β or Notch ligands (Frid et al. 2002; Ishisaki et al. 2003; Noseda et al. 2004; Timmerman et al. 2004; Zeisberg et al. 2007a, b), albeit that embryonic cells are more plastic than adult cells. The molecular mechanism behind TGF-β-induced EndMT has been found to involve the Snail family of transcription repressors. In mouse embryonic stem cell derived endothelial cells TGF-β stimulation induced EndMT and increased the expression of Snail. This upregulation of Snail by TGF-β was shown to be dependent on the activation of Smad, MEK, PI3K, and p38 MAPK by TGF-β (Medici et al. 2010). Subsequent knockdown of Snail blocked the TGF-β-induced EndMT (Kokudo et al. 2008). Although overexpression of Snail was sufficient to induce EMT (Cano et al. 2000), for EndMT Snail expression alone is insufficient. The inhibitor of Snail, GSK-3β, needs to be inhibited by phosphorylation by kinases such as AKT to induce EndMT (Medici et al. 2010).

Notch can, as TGF-β, induce EndMT in endothelial cells in vitro. In this Notch-induced EndMT the Snail family member Slug has been shown to be essential (Leong et al. 2007). Snail and Slug are known to repress the expression of VE-cadherin (Lopez et al. 2009). Since VE-cadherin is essential for endothelial cell–cell junctions, this obviously could provide a link to a mechanism by which EndMT occurs. A different factor involved in TGF-β-induced EndMT was shown to be plasminogen activator inhibitor-1 (PAI-1). Although elevated levels of PAI-1 are implicated in tissue fibrosis (Ghosh and Vaughan 2012), lack of PAI-1 in the heart is associated with the development of cardiac fibrosis in aged mice (Ghosh et al. 2010). It was shown that in the PAI-deficient endothelial cells of these mice both Smad and non-Smad TGF-β signaling is spontaneously activated. This spontaneous activation leads to EndMT and subsequently the fibrosis observed in these animals. In addition, it was recently shown that c-Abelson tyrosine kinase (c-Abl) and Protein Kinase C (PKC)-δ are crucial for TGF-β-induced EndMT, and therefore that imatinib mesylate and rottlerin (inhibitors of c-Abl and PKC-δ, respectively) might be effective therapeutic agents for fibroproliferative pathologies in which EndMT plays a role (Li and Jimenez 2011).

It has also been reported that pathways other than TGF-β can lead to EndMT. In myocardial infarction (MI) for example, canonical (β-catenin-dependent) Wnt signaling is induced 4 days after experimental MI in the subepicardial endothelial cells and perivascular cells. Coincidently with canonical Wnt activation EndMT was also triggered after the infarction. In addition, canonical Wnt signaling induced mesenchymal characteristics in cultured endothelial cells, suggesting a direct role of canonical Wnt signaling in EndMT (Aisagbonhi et al. 2011).

13.3 TGF-β in Vascular Pathologies

Given the importance of TGF-β signaling for blood vessels, it is not surprising that TGF-β signaling is involved in several pathological conditions concerning the endothelium (Fig. 13.3).

13.3.1 Hereditary Hemorrhagic Telangiectasia

Hereditary hemorrhagic telangiectasia (HHT), or Osler–Rendu–Weber syndrome, is an autosomal dominant vascular dysplasia characterized by the development of mucocutaneous telangiectasias and arteriovenous malformations in the lungs, brain, liver, and gastrointestinal tract (Abdalla and Letarte 2006). Many HHT patients suffer from recurrent nosebleeds, which severely affect their quality of life and are clinically difficult to treat. Mutations in three different genes have been shown to be responsible for HHT. Interestingly these are *ENG* (encoding endoglin), *ACVRL1* (encoding ALK1), and *SMAD4*, and the proteins encoded by these genes are all important players in TGF-β signaling in endothelial cells. Mutations in *ENG* cause

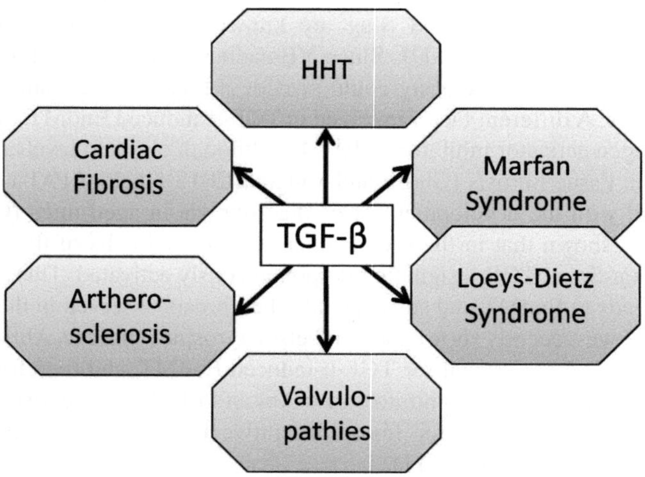

Fig. 13.3 TGF-β plays a central role in vascular pathologies like Marfan and Loeys–Dietz syndrome, hereditary hemorrhagic telangiectasia (HHT), atherosclerosis, cardiac fibrosis, and congenital heart valve defects

HHT type 1 (HHT-1) (McAllister et al. 1994); *ACVRL1* mutations cause HHT-2 (Johnson et al. 1996); and mutations in *SMAD4* cause a syndrome consisting of both juvenile polyposis and a HHT phenotype (Gallione et al. 2006). Endoglin heterozygous knockout mice have dilated and fragile blood vessels which resemble clinical manifestations of HHT-1 patients (Arthur et al. 2000; Bourdeau et al. 2001). Endothelial-specific ALK1 knockout mice suffer from vascular malformations mimicking all pathologic features of HHT-2 (Park et al. 2009). Several clinical trials in HHT patients are ongoing with anti-angiogenesis agents, including bevacizumab, a neutralizing antibody against VEGF and thalidomide to inhibit bleedings and other vascular malformations associated with HHT. Recently, it was reported that thalidomide treatment of a small group of HHT patients reduced the severity and frequency of nosebleeds (epistaxis) in the majority of the patients. In addition, in heterozygous endoglin knockout mice thalidomide treatment stimulated mural cell coverage and thus rescued vessel wall defects. Thalidomide treatment increased platelet-derived growth factor-B (PDGF-B) expression in endothelial cells and stimulated mural cell activation. The effects of thalidomide treatment were partially reversed by pharmacological or genetic interference with PDGF signaling from endothelial cells to pericytes (Lebrin et al. 2010).

13.3.2 Marfan Syndrome and Loeys Dietz Syndrome

Marfan syndrome (MFS) is a genetic disorder that has been linked to mutations in the *FBN1* gene, which encodes for the extracellular matrix glycoprotein Fibrillin-1 (Dietz et al. 1991). This protein forms an important component of elastic fibers in the aorta, ligaments, eye, and other tissues. Inactivating *FBN1* mutations in MFS affect cardiovascular, pulmonary, ocular, and skeletal tissues, among other tissues (Ramirez and Dietz 2007). Most dangerous are the aortic aneurysms, which carry the risk of sudden rupture and death. Initially it was thought that MFS was caused by structural defects in elastic fibers. However, fibrillin-1 not only serves an important role in providing structural integrity of connective tissues, but also acts as reservoir for growth factors such as TGF-β (ten Dijke and Arthur 2007). The latter function appears to be prominently associated with the vascular pathologies of MFS (Bolar et al. 2012). Fibrillin-1 contains multiple latent TGF-β binding motifs, and can associate with LTBP-1 and control the availability of active extracellular TGF-β (Chaudhry et al. 2007). Insufficient or nonfunctional fibrillin-1 as result of a *FBN1* mutation may lead to the release of bioactive TGF-β, and thereby trigger elevated TGF-β/Smad2 signaling (Habashi et al. 2006; Neptune et al. 2003).

A disorder closely related to MFS is Loeys Dietz syndrome (LDS). The latter has been linked to mutations in *ALK5*, *TGFBR2*, and *TGF-β2* (Lindsay et al. 2012; Loeys et al. 2005). LSD overlaps clinically to a large extent with MFS, but LSD has also unique properties, including widely spaced eyes. Some of the TGF-β receptor mutations in LSD were found to be inactivating mutations (Horbelt et al. 2010; Loeys et al. 2006), which paradoxically leads to overactive TGF-β/Smad2 signaling.

Both MFS and LDS demonstrate an upregulation of TGF-β/Smad2 signaling. Consistent with the notion of overactive TGF-β signaling, neutralizing antibodies for TGF-β in animal models for MFS prevented the formation of fragmented elastin fibers and aortic aneurysms (Neptune et al. 2003). Importantly, losartan, an angiotensin inhibitor that inhibits TGF-β signaling through not well-understood mechanisms, also demonstrated a protective effect against aortic aneurysms (Habashi et al. 2006). Losartan has recently moved into phase III clinical trial in patients with MFS. Recently, the non-Smad signaling pathways, i.e., extracellular regulated kinase (ERK) and Jun N-terminal kinase (JNK) MAP kinases, were found to be elevated in MFS and to contribute to aortic aneurysms development (Holm et al. 2011). Selective inhibition of ERK or JNK in a MFS mouse model was found to inhibit aortic growth. Thus, inhibition of ERK and JNK may be beneficial for treatment of MFS.

While aortic aneurysms are one of the complications in MFS and LDS, familial thoracic aortic aneurysms and dissections can also result from mutations in these genes, resulting in loss of signaling through the TGF-β type I and II receptor, in the absence of all features of MFS and LDS (Milewicz et al. 1996; Pannu et al. 2005; Tran-Fadulu et al. 2009). Furthermore, mutations in *SMAD3* have been linked to the aneurysms osteoarthritis syndrome, a form of TAAD with tortuosity of the arterial tree and early onset of osteoarthritis (van de Laar et al. 2012; Regalado et al. 2011). In addition, the Shprintzen–Goldberg syndrome (SGS) that has considerable phenotypic overlap with MFS and LDS, including aortic aneurysm, was found to be linked with mutations in the proto-oncogene SKI, a known repressor of TGF-β activity (Doyle et al. 2012). Whether alterations in these TGF-β signaling components are inhibiting or enhancing TGF-β signaling is still a matter of debate.

13.3.3 TGF-β Signaling and Valvulopathies

TGF-β is crucial for valve development and homeostasis. The development of the valve starts with the formation of the cardiac cushion by EndMT of the endocardial cells. This occurs in the atrioventricular canal, which separates the atria and ventricles, and the outflow tract, which connects the ventricles with the aortic sac of the developing heart (Fig. 13.4) (Kruithof et al. 2012). Endocardial cushions are the primordia of the septa and valves and become mature structures by mesenchymal cell proliferation, remodeling, and valve elongation, resulting in thin leaflets with highly organized extracellular matrix. Valve maturation continues after birth before reaching the adult configuration with the mechanical properties to withstand the continuous changing of the hemodynamic environment during the heart cycle (Kruithof et al. 2012).

TGF-β is one of the key regulators of EndMT and an increased expression of TGF-β1 can be seen in the endocardial cells overlying the cushions, while TGF-β2 is expressed in the myocardium and endocardium flanking the cushions during initiation of cushion development (Akhurst et al. 1990; Molin et al. 2003). The

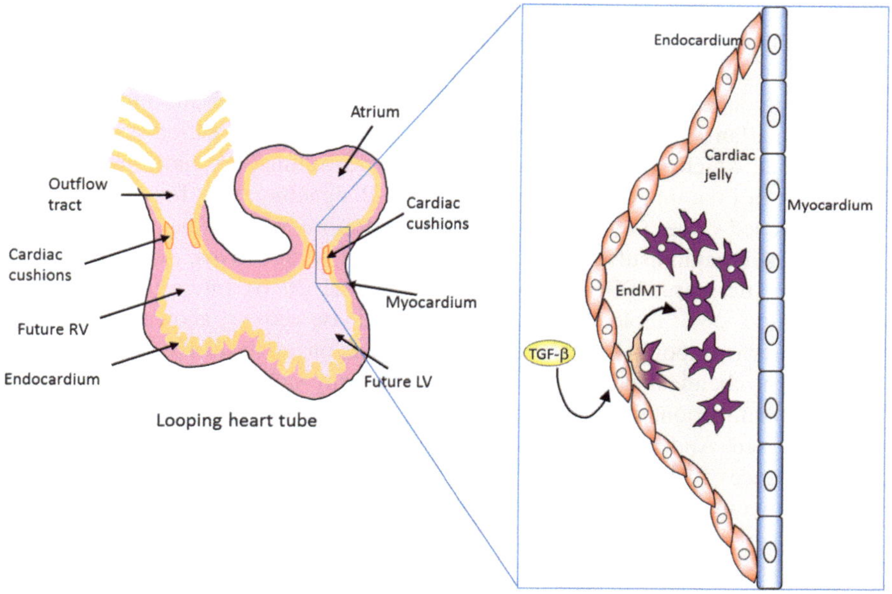

Fig. 13.4 Cardiac valve development. During embryogenesis the cardiac valves develop by the TGF-β-dependent mechanism of EndMT of the endocardial cells

requirement for TGF-β in cushion formation was initially shown using chicken explant cultures. When cultured in the presence of antisense oligo's or TGF-β neutralizing antibodies, EndMT, necessary for development of the cushions, was blocked (Boyer et al. 1999; Potts et al. 1991). Interestingly TGF-β1-deficient embryos, born to TGF-β1-null mothers, demonstrate severe cardiac abnormalities, including disorganized valves (Dickson et al. 1995). Furthermore, the valves in TGF-β2-deficient mice at E18.5 displayed defective remodeling and differentiation resulting in thickened valves (Azhar et al. 2011; Bartram et al. 2001; Sanford et al. 1997). The expression of TGF-β3 increases as valve development proceeds into advanced stages of heart development suggesting that TGF-β3 may play an important role in valve structure and function postnatal (Camenisch et al. 2002). Endocardium-specific deletion of ALK5 resulted in severe hypoplastic AV cushions, consistent with a crucial role for TGF-β signaling in EndMT (Sridurongrit et al. 2008). Endoglin is expressed in endocardial cells just before EndMT and endoglin deficiency results in embryonic lethality between E10.5 and E11.5 due to vascular abnormalities and disturbed cardiac development including hypocellular cushions (Arthur et al. 2000; Bourdeau et al. 1999; Li 1999). The TGF-β type III receptor betaglycan is also expressed in the endocardium during cushion formation (Stenvers et al. 2003), and betaglycan mutant mice die at the end of gestation harboring a thin ventricular wall and poorly developed septa (Compton et al. 2007).

Cardiac valve formation is crucial for proper heart function and septal/valve malformations comprise the largest part of congenital heart defects. Perturbation of TGF-β signaling pathway has been associated with both syndromic and non-syndromic congenital heart disease. Mutations in fibrillin-1, found in the genetic disorder marfan (see above), may cause thickening and elongation of the mitral valve leaflet, which can result in prolapse of the valve function (Judge et al. 2011). Mutations in Filamin-A, able to regulate TGF-β signaling, was found in patients with thickened valves (Kyndt et al. 2007). Furthermore, Williams syndrome, which has an elastin haploinsufficiency, shows elongated aortic valves, increased proliferation, and decreased TGF-β signaling (Hinton et al. 2010). Interestingly, a recent study, using a small molecule to inhibit ALK5 kinase activity, showed that indeed this receptor is important for the effect TGF-β has on valve formation. Using two independent chemical inhibitors lesions in the heart valves consisting of hemorrhages, inflammation, degeneration and interstitial cell activation, and proliferation were observed (Anderton et al. 2011). Most likely normal tissue repair is blocked as a consequence of ALK5 inhibition, and the valves may be particularly affected as a result of ongoing mechanical stress at opening and closure during each heart cycle.

13.3.4 TGF-β Signaling in Atherosclerosis

Atherosclerosis is a chronic inflammatory response in the arterial wall and characterized by vascular plaque formation (Libby et al. 2011). Stable plaques contain more collagen and smooth muscle cells, whereas unstable plaques have more macrophages and contain a large lipid core covered with a thin fibrous cap. Unstable plaques are more likely to rupture and cause clinical symptoms, e.g., cerebrovascular accidents or myocardial infarction. In atherosclerotic lesions, the TGF-β signaling components are detectable in endothelial cells, SMCs, myofibroblast, dendritic cells, T cells, and monocyte/macrophages and are rapidly upregulated during vascular injury (Bobik et al. 1999; Bot et al. 2009; Frostegård et al. 1999; Kalinina et al. 2004). There is some controversial information regarding the role of TGF-β in atherogenesis (Grainger 2004; Toma and McCaffrey 2012). Some reports show that increased TGF-β1 expression in human atherosclerotic plaques correlated with advanced atherosclerosis (Herder et al. 2012; Panutsopulos et al. 2005; Wang 1997), and other studies showed an inverse relationship between TGF-β1 levels in serum and the development of atherosclerosis, suggesting an anti-atherogenic effect (Grainger et al. 1995; Hering et al. 2002; Mallat et al. 2001). Furthermore, inhibition of TGF-β activity by various approaches resulted in pro-atherogenic changes in the vessel wall in animal models of atherosclerosis. Also an important role of TGF-β as an immune modulating cytokine in atherosclerosis was reported. Inhibiting TGF-β signaling in Apoe$^{-/-}$ mice by using a recombinant soluble TGF-RII (Lutgens 2002) or a blocking TGF-β1 antibody (Mallat et al. 2001) resulted in accelerated atherosclerosis. Lesions exhibited an unstable phenotype that contained low amounts of fibrosis, an increased amount of inflammatory cells, and even intraplaque hemorrhages.

Cardiac overexpression of TGF-β1, resulting in increased plasma levels of TGF-β, limited plaque growth and induced plaque stabilization (Frutkin et al. 2009). Mice with abrogated TGF-β signaling in T cells (Apoe$^{-/-}$CD4-dnTGF-βRII) (Gojova et al. 2003; Robertson et al. 2003) or DC cells (Apoe$^{-/-}$CD11cDNR) (Lievens et al. 2012) both showed accelerated lesion progression, with plaques containing abundant inflammatory cells paralleled by a decrease in plaque fibrosis.

It might be proposed that TGF-β signaling participates in the development of atherosclerosis, but may more interestingly promote a stable lesion phenotype, suggesting its role in the protection of acute ischemic situations like myocardial infarction. Interestingly, human aortic plaque vascular smooth muscles cells were shown to have enhanced levels of endoglin, supporting a role for endoglin in vascular wall integrity (Bot et al. 2009; Conley et al. 2000). In addition, soluble endoglin levels were found elevated in patients with coronary artery disease and atherosclerosis (Li et al. 2000b), and changes in soluble endoglin plasma levels post-myocardial infarction (MI) have predictive value for acute mortality in this patient group (Cruz-Gonzalez et al. 2008).

13.3.5 TGF-β Signaling in Cardiac Fibrosis

The development of heart failure starts with an acute MI or chronic injury, resulting in reduced ventricular performance and increased wall stress. Pathological deposition of extracellular matrix and myocardial hypertrophy are compensatory changes to reduce wall stress as the ventricle dilates. This will eventually lead to heart failure (Cohn et al. 2000). The process of cardiac fibrosis can be divided into two types: reactive fibrosis in response to inflammatory processes and reparative or replacement fibrosis in cell response to cell death and loss of myocardial tissue (Beltrami et al. 1994; Park et al. 2009; Silver et al. 1990). Cardiac fibroblasts are the interconnected cells that lie within the myocytes and extracellular space, and are the primary source of collagen in the heart (Souders et al. 2009). Activation of these fibroblasts or change of the fibroblast phenotype to myofibroblasts is what drives ECM accumulation and the development of pathological fibrosis. The local increase of TGF-β by various stimuli including inflammation, ischemia, mechanical stress, and vasoactive hormones like angiopoietin II is a key driver of this profibrotic process (Creemers and Pinto 2011; Dobaczewski et al. 2011; Goumans et al. 2008). Expression levels of TGF-β are increased in the left ventricle of both hypertrophic and idiopathic-dilated cardiomyopathy (Li et al. 1997; Martin et al. 2005; Pauschinger et al. 1999; Villarreal and Dillmann 1992), and in the borderzone of the infarcted region following myocardial infarction (Chuva de Sousa Lopes et al. 2004; Frantz et al. 2008) and due to diabetes (Connelly et al. 2009; Westermann et al. 2007).

That indeed TGF-β promoted myocardial hypertrophy and fibrosis was shown by overexpressing TGF-β1 in mice (Rosenkranz et al. 2002). These animals developed significant cardiac hypertrophy accompanied by interstitial fibrosis. In addition,

TGF-β1 heterozygous animals were protected from age-associated cardiac fibrosis and diastolic dysfunction (Brooks and Conrad 2000). Furthermore, treating animals with an anti-TGF-β neutralizing antibody prevented collagen accumulation following pressure overload and attenuated diastolic dysfunction without affecting cardiac hypertrophy (Ellmers et al. 2008; Kuwahara et al. 2002; Okada et al. 2005). Interestingly, mice overexpressing a dominant negative TGF-βRII showed markedly reduced collagen deposition following pressure overload, resulting in increased left ventricular dilatation and diastolic dysfunction (Lucas et al. 2010; Okada et al. 2005). Endoglin is also expressed in cardiac fibroblast and found to be increased during Ang-2-induced cardiac fibrosis. Recently, Kapur and coworkers showed that when treating animals with soluble endoglin 1 day before aortic bending, cardiac fibrosis and the development of heart failure were reduced (Kapur et al. 2012).

The functional effects of TGF-β signaling are likely to be biphasic. TGF-β signaling is necessary to preserve cardiac structure to protect the heart from the increased pressure and uncontrolled matrix degradation, while excessive TGF-β signaling results in enhanced collagen deposition, increased myocardial stiffness, and diastolic dysfunction. That TGF-β may function as a master switch is nicely demonstrated by inhibiting TGF-β signaling post MI. Early TGF-β inhibition within 24 h after occlusion of the coronary artery enhanced the inflammatory response and increased mortality (Frantz et al. 2008; Ikeuchi et al. 2004), while late TGF-β inhibition after the inflammatory phase of infarct healing reduced the number of myofibroblasts and improved cardiac output (Okada et al. 2005).

Both the intracellular Smad signaling pathways as well as non-Smad pathways are thought to play a role in the pathological remodeling of the heart. Smad3 was shown to be involved in myofibroblast transformation and mediated the TGF-β-induced ECM production and tissue inhibitor of matrix metalloprotease (TIMP) upregulation (Dobaczewski et al. 2010). In addition, TGF-β-activated kinase (TAK1) is activated in the pressure overloaded myocardium, and activated TAK expressed in the mouse myocardium induces cardiac hypertrophy and severe systolic dysfunction (Zhang et al. 2000). Finally, an indirect effect of TGF-β on cardiac fibrosis, by inducing, e.g., connective tissue growth factor (CTGF) and endothelin (Leask 2010), cannot be excluded.

13.4 Concluding Remarks

TGF-β plays a pivotal role in controlling cardiovascular homeostasis. Mutations in human TGF-β signaling components such as TGF-β receptors and Smads have been directly linked to cardiovascular disorders. Moreover, genetic studies in mouse models revealed that too little or overactive TGF-β signaling in ECs and SMCs leads to vascular dysfunction. Several small molecular weight compounds and antibodies that modulate TGF-β or BMP signaling are being evaluated in (pre)clinical trials for cardiovascular disorders. Examples are: losartan to inhibit overactive TGF-β signaling in Marfan syndrome (Möberg et al. 2012), Bevacizumab (a VEGF-A neutralizing antibody) and thalidomide to normalize and stabilize the weak vessel phenotype of HHT patients (Dupuis-Girod et al. 2012; Lebrin et al. 2010), and TGF-β receptor kinase inhibitors to correct the BMP/TGF-β imbalance seen in pulmonary hypertension (Long et al. 2009). However, with TGF-β being such a multifunctional model with effects on many different cell types that are highly context dependent, TGF-β signaling components remain very challenging therapeutic targets. Current treatments that target TGF-β or its receptors are not selective for only the pathological signaling pathways triggered by TGF-β. Careful selection and monitoring of patients are needed to guard for side effects.

There remains to be urgent need to more precisely dissect the mechanisms of TGF-β in vascular cells and identify cell type-specific regulators of TGF-β signaling, thereby enabling strategies to selectively target "bad" responses, while leaving "good" effects of TGF-β intact. The use of conditional knockouts and knockins for TGF-β signaling components in different cells, different tissues, and different times will be very instrumental to achieve this. Moreover, induced pluripotent stem cells (iPSCs) technology may be used to generate cardiovascular patient-derived endothelial and smooth muscle cells (Reed et al. 2012). These cells can be used not only to study pathology, but can also be valuable as tools to identify small molecule drugs that rescue the disease cell phenotype. Advances in all the above research fields will provide new opportunities for treatment of increasing number of vascular pathologies that are associated with dysregulation in TGF-β signaling.

Acknowledgments Research on TGF-β in vascular disorders in our laboratory is supported by grants from Dutch Cancer Society, Netherlands organization for scientific research, KNAW, and the LeDucq foundation.

References

Abdalla SA, Letarte M (2006) Hereditary haemorrhagic telangiectasia: current views on genetics and mechanisms of disease. J Med Genet 43:97–110. doi:10.1136/jmg.2005.030833

Aisagbonhi O, Rai M, Ryzhov S et al (2011) Experimental myocardial infarction triggers canonical Wnt signaling and endothelial-to-mesenchymal transition. Dis Model Mech 4:469–483. doi:10.1242/dmm.006510

Akhurst RJ, Lehnert SA, Faissner A, Duffie E (1990) TGF-β in murine morphogenetic processes: the early embryo and cardiogenesis. Development 108:645–656

Anderton MJ, Mellor HR, Bell A et al (2011) Induction of heart valve lesions by small-molecule ALK5 inhibitors. Toxicol Pathol 39:916–924. doi:10.1177/0192623311416259

Armulik A, Abramsson A, Betsholtz C (2005) Endothelial/pericyte interactions. Circ Res 97:512–523. doi:10.1161/01.RES.0000182903.16652.d7

Armulik A, Genové G, Betsholtz C (2011) Pericytes: developmental, physiological, and pathological perspectives, problems, and promises. Dev Cell 21:193–215. doi:10.1016/j.devcel.2011.07.001

Arthur HM, Ure J, Smith AJ et al (2000) Endoglin, an ancillary TGF-β receptor, is required for extraembryonic angiogenesis and plays a key role in heart development. Dev Biol 217:42–53. doi:10.1006/dbio.1999.9534

Azhar M, Brown K, Gard C et al (2011) Transforming growth factor β2 is required for valve remodeling during heart development. Dev Dyn 240:2127–2141. doi:10.1002/dvdy.22702

Bartram U, Molin DG, Wisse LJ et al (2001) Double-outlet right ventricle and overriding tricuspid valve reflect disturbances of looping, myocardialization, endocardial cushion differentiation, and apoptosis in TGF-β(2)-knockout mice. Circulation 103:2745–2752. doi:10.1161/01.CIR.103.22.2745

Beltrami CA, Finato N, Rocco M et al (1994) Structural basis of end-stage failure in ischemic cardiomyopathy in humans. Circulation 89:151–163. doi:10.1161/01.CIR.89.1.151

Birukova AA, Adyshev D, Gorshkov B et al (2005) ALK5 and Smad4 are involved in TGF-β1-induced pulmonary endothelial permeability. FEBS Lett 579:4031–4037. doi:10.1016/j.febslet.2005.06.018

Blanco FJ, Santibanez JF, Guerrero-Esteo M et al (2005) Interaction and functional interplay between endoglin and ALK-1, two components of the endothelial transforming growth factor-β receptor complex. J Cell Physiol 204:574–584. doi:10.1002/jcp.20311

Blobe GC, Schiemann WP, Lodish HF (2000) Role of transforming growth factor β in human disease. N Engl J Med 342:1350–1358. doi:10.1056/NEJM200005043421807

Bobik A (2006) Transforming growth factor-βs and vascular disorders. Arterioscler Thromb Vasc Biol 26:1712–1720. doi:10.1161/01.ATV.0000225287.20034.2c

Bobik A, Agrotis A, Kanellakis P et al (1999) Distinct patterns of transforming growth factor-β isoform and receptor expression in human atherosclerotic lesions. Colocalization implicates TGF-β in fibrofatty lesion development. Circulation 99:2883–2891. doi:10.1161/01.CIR.99.22.2883

Bolar N, Van Laer L, Loeys BL (2012) Marfan syndrome: from gene to therapy. Curr Opin Pediatr 24:498–504. doi:10.1097/MOP.0b013e3283557d4c

Bonyadi M, Rusholme SA, Cousins FM et al (1997) Mapping of a major genetic modifier of embryonic lethality in TGF-β1 knockout mice. Nat Genet 15:207–211. doi:10.1038/ng0297-207

Bot PTG, Hoefer IE, Sluijter JPG et al (2009) Increased expression of the transforming growth factor-β signaling pathway, endoglin, and early growth response-1 in stable plaques. Stroke 40:439–447. doi:10.1161/STROKEAHA.108.522284

Bourdeau A, Dumont DJ, Letarte M (1999) A murine model of hereditary hemorrhagic telangiectasia. J Clin Invest 104:1343–1351. doi:10.1172/JCI8088

Bourdeau A, Faughnan ME, McDonald ML et al (2001) Potential role of modifier genes influencing transforming growth factor-β1 levels in the development of vascular defects in endoglin heterozygous mice with hereditary hemorrhagic telangiectasia. Am J Pathol 158:2011–2020. doi:10.1016/S0002-9440(10)64673-1

Boyer AS, Erickson CP, Runyan RB (1999) Epithelial-mesenchymal transformation in the embryonic heart is mediated through distinct pertussis toxin-sensitive and TGF-β signal transduction mechanisms. Dev Dyn 214:81–91. doi:10.1002/(SICI)1097-0177(199901)214:1<81::AID-DVDY8>3.0.CO;2-3

Brooks WW, Conrad CH (2000) Myocardial fibrosis in transforming growth factor β(1)heterozygous mice. J Mol Cell Cardiol 32:187–195. doi:10.1006/jmcc.1999.1065

Camenisch TD, Molin DGM, Person A et al (2002) Temporal and distinct TGF-β ligand requirements during mouse and avian endocardial cushion morphogenesis. Dev Biol 248:170–181. doi:10.1006/dbio.2002.0731

Cano A, Pérez-Moreno MA, Rodrigo I et al (2000) The transcription factor snail controls epithelial-mesenchymal transitions by repressing E-cadherin expression. Nat Cell Biol 2:76–83. doi:10.1038/35000025

Carvalho RLC, Jonker L, Goumans M-J et al (2004) Defective paracrine signalling by TGF-β in yolk sac vasculature of endoglin mutant mice: a paradigm for hereditary haemorrhagic telangiectasia. Development 131:6237–6247. doi:10.1242/dev.01529

Chaudhry SS, Cain SA, Morgan A et al (2007) Fibrillin-1 regulates the bioavailability of TGF-β1. J Cell Biol 176:355–367. doi:10.1083/jcb.200608167

Cheifetz S, Hernandez H, Laiho M et al (1990) Distinct transforming growth factor-β (TGF-β) receptor subsets as determinants of cellular responsiveness to three TGF-β isoforms. J Biol Chem 265:20533–20538

Cheifetz S, Bellón T, Calés C et al (1992) Endoglin is a component of the transforming growth factor-β receptor system in human endothelial cells. J Biol Chem 267:19027–19030

Chuva de Sousa Lopes SM, Feijen A, Korving J et al (2004) Connective tissue growth factor expression and Smad signaling during mouse heart development and myocardial infarction. Dev Dyn 231:542–550. doi:10.1002/dvdy.20162

Cohn JN, Ferrari R, Sharpe N (2000) Cardiac remodeling—concepts and clinical implications: a consensus paper from an international forum on cardiac remodeling. J Am Coll Cardiol 35:569–582. doi:10.1016/S0735-1097(99)00630-0

Compton LA, Potash DA, Brown CB, Barnett JV (2007) Coronary vessel development is dependent on the type III transforming growth factor β receptor. Circ Res 101:784–791. doi:10.1161/CIRCRESAHA.107.152082

Conley BA, Smith JD, Guerrero-Esteo M et al (2000) Endoglin, a TGF-β receptor-associated protein, is expressed by smooth muscle cells in human atherosclerotic plaques. Atherosclerosis 153:323–335. doi:10.1016/S0021-9150(00)00422-6

Connelly KA, Kelly DJ, Zhang Y et al (2009) Inhibition of protein kinase C-β by ruboxistaurin preserves cardiac function and reduces extracellular matrix production in diabetic cardiomyopathy. Circ Heart Fail 2:129–137. doi:10.1161/CIRCHEARTFAILURE.108.765750

Creemers EE, Pinto YM (2011) Molecular mechanisms that control interstitial fibrosis in the pressure-overloaded heart. Cardiovasc Res 89:265–272. doi:10.1093/cvr/cvq308

Cruz-Gonzalez I, Pabón P, Rodríguez-Barbero A et al (2008) Identification of serum endoglin as a novel prognostic marker after acute myocardial infarction. J Cell Mol Med 12:955–961. doi:10.1111/j.1582-4934.2008.00156.x

Dağdeviren A, Müftüoğlu SF, Nur Çakar A, Örs Ü (1998) Endoglin (CD 105) expression in human lymphoid organs and placenta. Ann Anat 180:461–469. doi:10.1016/S0940-9602(98)80109-X

David L, Mallet C, Mazerbourg S et al (2007) Identification of BMP9 and BMP10 as functional activators of the orphan activin receptor-like kinase 1 (ALK1) in endothelial cells. Blood 109:1953–1961. doi:10.1182/blood-2006-07-034124

Dickson MC, Martin JS, Cousins FM et al (1995) Defective haematopoiesis and vasculogenesis in transforming growth factor-β 1 knock out mice. Development 121:1845–1854

Dietz HC, Cutting GR, Pyeritz RE et al (1991) Marfan syndrome caused by a recurrent de novo missense mutation in the fibrillin gene. Nature 352:337–339. doi:10.1038/352337a0

Ding R, Darland DC, Parmacek MS, D'Amore PA (2004) Endothelial–mesenchymal interactions in vitro reveal molecular mechanisms of smooth muscle/pericyte differentiation. Stem Cells Dev 13:509–520. doi:10.1089/1547328042417336

Dobaczewski M, Bujak M, Li N et al (2010) Smad3 signaling critically regulates fibroblast phenotype and function in healing myocardial infarction. Circ Res 107:418–428. doi:10.1161/CIRCRESAHA.109.216101

Dobaczewski M, Chen W, Frangogiannis NG (2011) Transforming growth factor (TGF)-β signaling in cardiac remodeling. J Mol Cell Cardiol 51:600–606. doi:10.1016/j.yjmcc.2010.10.033

Doyle AJ, Doyle JJ, Bessling SL et al (2012) Mutations in the TGF-β repressor SKI cause Shprintzen-Goldberg syndrome with aortic aneurysm. Nat Genet. doi:10.1038/ng.2421

Dupuis-Girod S, Ginon I, Saurin J-C et al (2012) Bevacizumab in patients with hereditary hemorrhagic telangiectasia and severe hepatic vascular malformations and high cardiac output. JAMA 307:948–955. doi:10.1001/jama.2012.250

Ellmers LJ, Scott NJA, Medicherla S et al (2008) Transforming growth factor-β blockade down-regulates the renin-angiotensin system and modifies cardiac remodeling after myocardial infarction. Endocrinology 149:5828–5834. doi:10.1210/en.2008-0165

Feng X-H, Derynck R (2005) Specificity and versatility in TGF-β signaling through Smads. Annu Rev Cell Dev Biol 21:659–693. doi:10.1146/annurev.cellbio.21.022404.142018

Frantz S, Hu K, Adamek A et al (2008) Transforming growth factor β inhibition increases mortality and left ventricular dilatation after myocardial infarction. Basic Res Cardiol 103:485–492. doi:10.1007/s00395-008-0739-7

Frid MG, Kale VA, Stenmark KR (2002) Mature vascular endothelium can give rise to smooth muscle cells via endothelial-mesenchymal transdifferentiation: in vitro analysis. Circ Res 90:1189–1196. doi:10.1161/01.RES.0000021432.70309.28

Frostegård J, Ulfgren A-K, Nyberg P et al (1999) Cytokine expression in advanced human atherosclerotic plaques: dominance of pro-inflammatory (Th1) and macrophage-stimulating cytokines. Atherosclerosis 145:33–43. doi:10.1016/S0021-9150(99)00011-8

Frutkin AD, Otsuka G, Stempien-Otero A et al (2009) TGF-β1 limits plaque growth, stabilizes plaque structure, and prevents aortic dilation in apolipoprotein E-null mice. Arterioscler Thromb Vasc Biol 29:1251–1257. doi:10.1161/ATVBAHA.109.186593

Gallione CJ, Richards JA, Letteboer TGW et al (2006) SMAD4 mutations found in unselected HHT patients. J Med Genet 43:793–797. doi:10.1136/jmg.2006.041517

Ghosh AK, Vaughan DE (2012) PAI-1 in tissue fibrosis. J Cell Physiol 227:493–507. doi:10.1002/jcp.22783

Ghosh AK, Bradham WS, Gleaves LA et al (2010) Genetic deficiency of plasminogen activator inhibitor-1 promotes cardiac fibrosis in aged mice: involvement of constitutive transforming growth factor-β signaling and endothelial-to-mesenchymal transition. Circulation 122:1200–1209. doi:10.1161/CIRCULATIONAHA.110.955245

Gojova A, Brun V, Esposito B et al (2003) Specific abrogation of transforming growth factor-β signaling in T cells alters atherosclerotic lesion size and composition in mice. Blood 102:4052–4058. doi:10.1182/blood-2003-05-1729

Goumans MJ, Mummery C (2000) Functional analysis of the TGF-β receptor/Smad pathway through gene ablation in mice. Int J Dev Biol 44:253–265

Goumans M-J, Valdimarsdottir G, Itoh S et al (2002) Balancing the activation state of the endothelium via two distinct TGF-β type I receptors. EMBO J 21:1743–1753. doi:10.1093/emboj/21.7.1743

Goumans M-J, Valdimarsdottir G, Itoh S et al (2003) Activin receptor-like kinase (ALK)1 Is an antagonistic mediator of lateral TGF-β/ALK5 signaling. Mol Cell 12:817–828. doi:10.1016/S1097-2765(03)00386-1

Goumans M-J, van Zonneveld AJ, ten Dijke P (2008) Transforming growth factor β-induced endothelial-to-mesenchymal transition: a switch to cardiac fibrosis? Trends Cardiovasc Med 18:293–298. doi:10.1016/j.tcm.2009.01.001

Grainger DJ (2004) Transforming growth factor β and atherosclerosis: so far, so good for the protective cytokine hypothesis. Arterioscler Thromb Vasc Biol 24:399–404. doi:10.1161/01.ATV.0000114567.76772.33

Grainger DJ, Kemp PR, Metcalfe JC et al (1995) The serum concentration of active transforming growth factor-β is severely depressed in advanced atherosclerosis. Nat Med 1:74–79. doi:10.1038/nm0195-74

Habashi JP, Judge DP, Holm TM et al (2006) Losartan, an AT1 antagonist, prevents aortic aneurysm in a mouse model of Marfan syndrome. Science 312:117–121. doi:10.1126/science.1124287

Hawinkels LJAC, Kuiper P, Wiercinska E et al (2010) Matrix metalloproteinase-14 (MT1-MMP)-mediated endoglin shedding inhibits tumor angiogenesis. Cancer Res 70:4141–4150. doi:10.1158/0008-5472.CAN-09-4466

Herder C, Peeters W, Zierer A et al (2012) TGF-β1 content in atherosclerotic plaques, TGF-β1 serum concentrations and incident coronary events. Eur J Clin Invest 42:329–337. doi:10.1111/j.1365-2362.2011.02587.x

Hering S, Jost C, Schulz H et al (2002) Circulating transforming growth factor β1 (TGF-β1) is elevated by extensive exercise. Eur J Appl Physiol 86:406–410. doi:10.1007/s00421-001-0537-5

Hinton RB, Adelman-Brown J, Witt S et al (2010) Elastin haploinsufficiency results in progressive aortic valve malformation and latent valve disease in a mouse model. Circ Res 107:549–557. doi:10.1161/CIRCRESAHA.110.221358

Holm TM, Habashi JP, Doyle JJ et al (2011) Noncanonical TGF-β signaling contributes to aortic aneurysm progression in Marfan syndrome mice. Science 332:358–361. doi:10.1126/science.1192149

Horbelt D, Guo G, Robinson PN, Knaus P (2010) Quantitative analysis of TGFBR2 mutations in Marfan-syndrome-related disorders suggests a correlation between phenotypic severity and Smad signaling activity. J Cell Sci 123:4340–4350. doi:10.1242/jcs.074773

Huang S, Tang B, Usoskin D et al (2002) Conditional knockout of the Smad1 gene. Genesis 32:76–79. doi:10.1002/gene.10059

Ikeuchi M, Tsutsui H, Shiomi T et al (2004) Inhibition of TGF-β signaling exacerbates early cardiac dysfunction but prevents late remodeling after infarction. Cardiovasc Res 64:526–535. doi:10.1016/j.cardiores.2004.07.017

Ishisaki A, Hayashi H, Li A-J, Imamura T (2003) Human umbilical vein endothelium-derived cells retain potential to differentiate into smooth muscle-like cells. J Biol Chem 278:1303–1309. doi:10.1074/jbc.M207329200

Itoh S, ten Dijke P (2007) Negative regulation of TGF-β receptor/Smad signal transduction. Curr Opin Cell Biol 19:176–184. doi:10.1016/j.ceb.2007.02.015

Jerkic M, Rodríguez-Barbero A, Prieto M et al (2006) Reduced angiogenic responses in adult Endoglin heterozygous mice. Cardiovasc Res 69:845–854. doi:10.1016/j.cardiores.2005.11.020

Johnson DW, Berg JN, Baldwin MA et al (1996) Mutations in the activin receptor-like kinase 1 gene in hereditary haemorrhagic telangiectasia type 2. Nat Genet 13:189–195. doi:10.1038/ng0696-189

Judge DP, Rouf R, Habashi J, Dietz HC (2011) Mitral valve disease in Marfan syndrome and related disorders. J Cardiovasc Transl Res 4:741–747. doi:10.1007/s12265-011-9314-y

Kaartinen V, Voncken JW, Shuler C et al (1995) Abnormal lung development and cleft palate in mice lacking TGF-β 3 indicates defects of epithelial-mesenchymal interaction. Nat Genet 11:415–421. doi:10.1038/ng1295-415

Kalinina N, Agrotis A, Antropova Y et al (2004) Increased expression of the DNA-binding cytokine HMGB1 in human atherosclerotic lesions: role of activated macrophages and cytokines. Arterioscler Thromb Vasc Biol 24:2320–2325. doi:10.1161/01.ATV.0000145573.36113.8a

Kallapur S, Ormsby I, Doetschman T (1999) Strain dependency of TGF-β1 function during embryogenesis. Mol Reprod Dev 52:341–349. doi:10.1002/(SICI)1098-2795(199904)52:4<341::AID-MRD2>3.0.CO;2-N

Kano MR, Bae Y, Iwata C et al (2007) Improvement of cancer-targeting therapy, using nanocarriers for intractable solid tumors by inhibition of TGF-β signaling. Proc Natl Acad Sci USA 104:3460–3465. doi:10.1073/pnas.0611660104

Kapur NK, Wilson S, Yunis AA et al (2012) Reduced endoglin activity limits cardiac fibrosis and improves survival in heart failure. Circulation 125:2728–2738. doi:10.1161/CIRCULATIONAHA.111.080002

Kirkbride KC, Ray BN, Blobe GC (2005) Cell-surface co-receptors: emerging roles in signaling and human disease. Trends Biochem Sci 30:611–621. doi:10.1016/j.tibs.2005.09.003

Kokudo T, Suzuki Y, Yoshimatsu Y et al (2008) Snail is required for TGF-β-induced endothelial-mesenchymal transition of embryonic stem cell-derived endothelial cells. J Cell Sci 121:3317–3324. doi:10.1242/jcs.028282

Kruithof BPT, Duim SN, Moerkamp AT, Goumans M-J (2012) TGF-β and BMP signaling in cardiac cushion formation: lessons from mice and chicken. Differentiation 84:89–102. doi:10.1016/j.diff.2012.04.003

Kulkarni AB, Huh CG, Becker D et al (1993) Transforming growth factor β 1 null mutation in mice causes excessive inflammatory response and early death. Proc Natl Acad Sci USA 90:770–774

Kuwahara F, Kai H, Tokuda K et al (2002) Transforming growth factor-β function blocking prevents myocardial fibrosis and diastolic dysfunction in pressure-overloaded rats. Circulation 106:130–135. doi:10.1161/01.CIR.0000020689.12472.E0

Kyndt F, Gueffet J-P, Probst V et al (2007) Mutations in the gene encoding filamin A as a cause for familial cardiac valvular dystrophy. Circulation 115:40–49. doi:10.1161/CIRCULATIONAHA.106.622621

Lamouille S, Mallet C, Feige J-J, Bailly S (2002) Activin receptor-like kinase 1 is implicated in the maturation phase of angiogenesis. Blood 100:4495–4501. doi:10.1182/blood.V100.13.4495

Lan Y, Liu B, Yao H et al (2007) Essential role of endothelial Smad4 in vascular remodeling and integrity. Mol Cell Biol 27:7683–7692. doi:10.1128/MCB.00577-07

Larsson J, Goumans MJ, Sjöstrand LJ et al (2001) Abnormal angiogenesis but intact hematopoietic potential in TGF-β type I receptor-deficient mice. EMBO J 20:1663–1673. doi:10.1093/emboj/20.7.1663

Lastres P (1996) Endoglin modulates cellular responses to TGF-β 1. J Cell Biol 133:1109–1121. doi:10.1083/jcb.133.5.1109

Leask A (2010) Potential therapeutic targets for cardiac fibrosis: TGF-β, angiotensin, endothelin, CCN2, and PDGF, partners in fibroblast activation. Circ Res 106:1675–1680. doi:10.1161/CIRCRESAHA.110.217737

Lebrin F, Goumans M-J, Jonker L et al (2004) Endoglin promotes endothelial cell proliferation and TGF-β/ALK1 signal transduction. EMBO J 23:4018–4028. doi:10.1038/sj.emboj.7600386

Lebrin F, Srun S, Raymond K et al (2010) Thalidomide stimulates vessel maturation and reduces epistaxis in individuals with hereditary hemorrhagic telangiectasia. Nat Med 16:420–428. doi:10.1038/nm.2131

Leong KG, Niessen K, Kulic I et al (2007) Jagged1-mediated notch activation induces epithelial-to-mesenchymal transition through slug-induced repression of E-cadherin. J Exp Med 204:2935–2948. doi:10.1084/jem.20071082

Letterio J, Geiser A, Kulkarni A et al (1994) Maternal rescue of transforming growth factor-β 1 null mice. Science 264:1936–1938. doi:10.1126/science.8009224

Levine RJ, Lam C, Qian C et al (2006) Soluble endoglin and other circulating antiangiogenic factors in preeclampsia. N Engl J Med 355:992–1005. doi:10.1056/NEJMoa055352

Li DY (1999) Defective angiogenesis in mice lacking endoglin. Science 284:1534–1537. doi:10.1126/science.284.5419.1534

Li Z, Jimenez SA (2011) Protein kinase Cδ and c-Abl kinase are required for transforming growth factor β induction of endothelial-mesenchymal transition in vitro. Arthritis Rheum 63:2473–2483. doi:10.1002/art.30317

Li RK, Li G, Mickle DA et al (1997) Overexpression of transforming growth factor-β1 and insulin-like growth factor-I in patients with idiopathic hypertrophic cardiomyopathy. Circulation 96:874–881. doi:10.1161/01.CIR.96.3.874

Li C, Guo B, Wilson PB et al (2000a) Plasma levels of soluble CD105 correlate with metastasis in patients with breast cancer. Int J Cancer 89:122–126. doi:10.1002/(SICI)1097-0215(20000320)89:2<122::AID-IJC4>3.0.CO;2-M

Li C, Bethell H, Wilson P et al (2000b) The significance of CD105, TGF-β and CD105/TGF-β complexes in coronary artery disease. Atherosclerosis 152:249–256. doi:10.1016/S0021-9150(99)00476-1

Li C, Issa R, Kumar P et al (2003) CD105 prevents apoptosis in hypoxic endothelial cells. J Cell Sci 116:2677–2685. doi:10.1242/jcs.00470

Libby P, Ridker PM, Hansson GK (2011) Progress and challenges in translating the biology of atherosclerosis. Nature 473:317–325. doi:10.1038/nature10146

Lievens D, Habets KL, Robertson A-K et al (2012) Abrogated transforming growth factor β receptor II (TGFβRII) signalling in dendritic cells promotes immune reactivity of T cells resulting in enhanced atherosclerosis. Eur Heart J. doi:10.1093/eurheartj/ehs106

Lindsay ME, Schepers D, Bolar NA et al (2012) Loss-of-function mutations in TGF-β2 cause a syndromic presentation of thoracic aortic aneurysm. Nat Genet 44:922–927. doi:10.1038/ng.2349

Liu Z, Kobayashi K, van Dinther M et al (2009) VEGF and inhibitors of TGF-β type-I receptor kinase synergistically promote blood-vessel formation by inducing α5-integrin expression. J Cell Sci 122:3294–3302. doi:10.1242/jcs.048942

Liu Z, Afink GB, ten Dijke P (2012) Soluble fms-like tyrosine kinase 1 and soluble endoglin are elevated circulating anti-angiogenic factors in pre-eclampsia. Pregnancy Hypertens Int J Wom Cardiovasc Health 2:358–367. doi:10.1016/j.preghy.2012.06.003

Loeys BL, Chen J, Neptune ER et al (2005) A syndrome of altered cardiovascular, craniofacial, neurocognitive and skeletal development caused by mutations in TGFBR1 or TGFBR2. Nat Genet 37:275–281. doi:10.1038/ng1511

Loeys BL, Schwarze U, Holm T et al (2006) Aneurysm syndromes caused by mutations in the TGF-β receptor. N Engl J Med 355:788–798. doi:10.1056/NEJMoa055695
Long L, Crosby A, Yang X et al (2009) Altered bone morphogenetic protein and transforming growth factor-β signaling in rat models of pulmonary hypertension: potential for activin receptor-like kinase-5 inhibition in prevention and progression of disease. Circulation 119:566–576. doi:10.1161/CIRCULATIONAHA.108.821504
Lopez D, Niu G, Huber P, Carter WB (2009) Tumor-induced upregulation of twist, snail, and slug represses the activity of the human VE-cadherin promoter. Arch Biochem Biophys 482:77–82. doi:10.1016/j.abb.2008.11.016
Lucas JA, Zhang Y, Li P et al (2010) Inhibition of transforming growth factor-β signaling induces left ventricular dilation and dysfunction in the pressure-overloaded heart. Am J Physiol Heart Circ Physiol 298:H424–H432. doi:10.1152/ajpheart.00529.2009
Lutgens E (2002) Transforming growth factor-β mediates balance between inflammation and fibrosis during plaque progression. Arterioscler Thromb Vasc Biol 22:975–982. doi:10.1161/01.ATV.0000019729.39500.2F
Mallat Z, Gojova A, Marchiol-Fournigault C et al (2001) Inhibition of transforming growth factor-β signaling accelerates atherosclerosis and induces an unstable plaque phenotype in mice. Circ Res 89:930–934. doi:10.1161/hh2201.099415
Mallet C, Vittet D, Feige J-J, Bailly S (2006) TGF-β1 induces vasculogenesis and inhibits angiogenic sprouting in an embryonic stem cell differentiation model: respective contribution of ALK1 and ALK5. Stem Cells 24:2420–2427. doi:10.1634/stemcells.2005-0494
Martin J, Kelly DJ, Mifsud SA et al (2005) Tranilast attenuates cardiac matrix deposition in experimental diabetes: role of transforming growth factor-β. Cardiovasc Res 65:694–701. doi:10.1016/j.cardiores.2004.10.041
Massagué J (1998) TGF-β signal transduction. Annu Rev Biochem 67:753–791. doi:10.1146/annurev.biochem.67.1.753
McAllister KA, Grogg KM, Johnson DW et al (1994) Endoglin, a TGF-β binding protein of endothelial cells, is the gene for hereditary haemorrhagic telangiectasia type 1. Nat Genet 8:345–351. doi:10.1038/ng1294-345
Medici D, Shore EM, Lounev VY et al (2010) Conversion of vascular endothelial cells into multipotent stem-like cells. Nat Med 16:1400–1406. doi:10.1038/nm.2252
Milewicz DM, Michael K, Fisher N et al (1996) Fibrillin-1 (FBN1) mutations in patients with thoracic aortic aneurysms. Circulation 94:2708–2711. doi:10.1161/01.CIR.94.11.2708
Möberg K, De Nobele S, Devos D et al (2012) The Ghent Marfan Trial—a randomized, double-blind placebo controlled trial with losartan in Marfan patients treated with β-blockers. Int J Cardiol 157:354–358. doi:10.1016/j.ijcard.2010.12.070
Molin DGM, Bartram U, Van der Heiden K et al (2003) Expression patterns of TGF-β1-3 associate with myocardialisation of the outflow tract and the development of the epicardium and the fibrous heart skeleton. Dev Dyn 227:431–444. doi:10.1002/dvdy.10314
Moustakas A, Heldin C-H (2005) Non-Smad TGF-β signals. J Cell Sci 118:3573–3584. doi:10.1242/jcs.02554
Moustakas A, Heldin C-H (2009) The regulation of TGF-β signal transduction. Development 136:3699–3714. doi:10.1242/dev.030338
Neptune ER, Frischmeyer PA, Arking DE et al (2003) Dysregulation of TGF-β activation contributes to pathogenesis in Marfan syndrome. Nat Genet 33:407–411. doi:10.1038/ng1116
Noseda M, McLean G, Niessen K et al (2004) Notch activation results in phenotypic and functional changes consistent with endothelial-to-mesenchymal transformation. Circ Res 94:910–917. doi:10.1161/01.RES.0000124300.76171.C9
Obreo J, Díez-Marques L, Lamas S et al (2004) Endoglin expression regulates basal and TGF-β1-induced extracellular matrix synthesis in cultured L6E9 myoblasts. Cell Physiol Biochem 14:301–310. doi:10.1159/000080340
Oh SP, Seki T, Goss KA et al (2000) Activin receptor-like kinase 1 modulates transforming growth factor-β 1 signaling in the regulation of angiogenesis. Proc Natl Acad Sci 97:2626–2631. doi:10.1073/pnas.97.6.2626

Okada H, Takemura G, Kosai K et al (2005) Postinfarction gene therapy against transforming growth factor-β signal modulates infarct tissue dynamics and attenuates left ventricular remodeling and heart failure. Circulation 111:2430–2437. doi:10.1161/01.CIR.0000165066.71481.8E

Owens GK (1995) Regulation of differentiation of vascular smooth muscle cells. Physiol Rev 75:487–517

Pannu H, Fadulu VT, Chang J et al (2005) Mutations in transforming growth factor-β receptor type II cause familial thoracic aortic aneurysms and dissections. Circulation 112:513–520. doi:10.1161/CIRCULATIONAHA.105.537340

Panutsopulos D, Papalambros E, Sigala F et al (2005) Protein and mRNA expression levels of VEGF-A and TGF-β1 in different types of human coronary atherosclerotic lesions. Int J Mol Med 15:603–610

Park SO, Lee YJ, Seki T et al (2008) ALK5- and TGFBR2-independent role of ALK1 in the pathogenesis of hereditary hemorrhagic telangiectasia type 2. Blood 111:633–642. doi:10.1182/blood-2007-08-107359

Park SO, Wankhede M, Lee YJ et al (2009) Real-time imaging of de novo arteriovenous malformation in a mouse model of hereditary hemorrhagic telangiectasia. J Clin Invest 119:3487–3496. doi:10.1172/JCI39482

Pauschinger M, Knopf D, Petschauer S et al (1999) Dilated cardiomyopathy is associated with significant changes in collagen type I/III ratio. Circulation 99:2750–2756. doi:10.1161/01.CIR.99.21.2750

Potts JD, Dagle JM, Walder JA et al (1991) Epithelial-mesenchymal transformation of embryonic cardiac endothelial cells is inhibited by a modified antisense oligodeoxynucleotide to transforming growth factor β 3. Proc Natl Acad Sci USA 88:1516–1520

Ramirez F, Dietz HC (2007) Marfan syndrome: from molecular pathogenesis to clinical treatment. Curr Opin Genet Dev 17:252–258. doi:10.1016/j.gde.2007.04.006

Reed DM, Foldes G, Harding SE, Mitchell JA (2012) Stem cell derived endothelial cells for cardiovascular disease; a therapeutic perspective. Br J Clin Pharmacol. doi:10.1111/j.1365-2125.2012.04361.x

Regalado ES, Guo D-C, Villamizar C et al (2011) Exome sequencing identifies SMAD3 mutations as a cause of familial thoracic aortic aneurysm and dissection with intracranial and other arterial aneurysms. Circ Res 109:680–686. doi:10.1161/CIRCRESAHA.111.248161

Robertson A-KL, Rudling M, Zhou X et al (2003) Disruption of TGF-β signaling in T cells accelerates atherosclerosis. J Clin Invest 112:1342–1350. doi:10.1172/JCI18607

Roman BL, Pham VN, Lawson ND et al (2002) Disruption of acvrl1 increases endothelial cell number in zebrafish cranial vessels. Development 129:3009–3019

Rosenkranz S, Flesch M, Amann K et al (2002) Alterations of β-adrenergic signaling and cardiac hypertrophy in transgenic mice overexpressing TGF-β(1). Am J Physiol Heart Circ Physiol 283:H1253–H1262. doi:10.1152/ajpheart.00578.2001

Rudini N, Felici A, Giampietro C et al (2008) VE-cadherin is a critical endothelial regulator of TGF-β signalling. EMBO J 27:993–1004. doi:10.1038/emboj.2008.46

Sanford LP, Ormsby I, Gittenberger-de Groot AC et al (1997) TGF-β2 knockout mice have multiple developmental defects that are non-overlapping with other TGF-β knockout phenotypes. Development 124:2659–2670

Seki T, Hong K-H, Oh SP (2006) Nonoverlapping expression patterns of ALK1 and ALK5 reveal distinct roles of each receptor in vascular development. Lab Invest 86:116–129. doi:10.1038/labinvest.3700376

Shull MM, Ormsby I, Kier AB et al (1992) Targeted disruption of the mouse transforming growth factor-β 1 gene results in multifocal inflammatory disease. Nature 359:693–699. doi:10.1038/359693a0

Silver MA, Pick R, Brilla CG et al (1990) Reactive and reparative fibrillar collagen remodelling in the hypertrophied rat left ventricle: two experimental models of myocardial fibrosis. Cardiovasc Res 24:741–747. doi:10.1093/cvr/24.9.741

Souders CA, Bowers SLK, Baudino TA (2009) Cardiac fibroblast: the renaissance cell. Circ Res 105:1164–1176. doi:10.1161/CIRCRESAHA.109.209809

Sridurongrit S, Larsson J, Schwartz R et al (2008) Signaling via the TGF-β type I receptor Alk5 in heart development. Dev Biol 322:208–218. doi:10.1016/j.ydbio.2008.07.038

Stenvers KL, Tursky ML, Harder KW et al (2003) Heart and liver defects and reduced transforming growth factor 2 sensitivity in transforming growth factor type III receptor-deficient embryos. Mol Cell Biol 23:4371–4385. doi:10.1128/MCB.23.12.4371-4385.2003

ten Dijke P, Arthur HM (2007) Extracellular control of TGF-β signalling in vascular development and disease. Nat Rev Mol Cell Biol 8:857–869. doi:10.1038/nrm2262

ten Dijke P, Goumans M-J, Pardali E (2008) Endoglin in angiogenesis and vascular diseases. Angiogenesis 11:79–89. doi:10.1007/s10456-008-9101-9

Thiery JP, Acloque H, Huang RYJ, Nieto MA (2009) Epithelial-mesenchymal transitions in development and disease. Cell 139:871–890. doi:10.1016/j.cell.2009.11.007

Tian H, Mythreye K, Golzio C et al (2012) Endoglin mediates fibronectin/α5β1 integrin and TGF-β pathway crosstalk in endothelial cells. EMBO J 31:3885–3900. doi:10.1038/emboj.2012.246

Timmerman LA, Grego-Bessa J, Raya A et al (2004) Notch promotes epithelial-mesenchymal transition during cardiac development and oncogenic transformation. Genes Dev 18:99–115. doi:10.1101/gad.276304

Toma I, McCaffrey TA (2012) Transforming growth factor-β and atherosclerosis: interwoven atherogenic and atheroprotective aspects. Cell Tissue Res 347:155–175. doi:10.1007/s00441-011-1189-3

Tran-Fadulu V, Pannu H, Kim DH et al (2009) Analysis of multigenerational families with thoracic aortic aneurysms and dissections due to TGFBR1 or TGFBR2 mutations. J Med Genet 46:607–613. doi:10.1136/jmg.2008.062844

Urness LD, Sorensen LK, Li DY (2000) Arteriovenous malformations in mice lacking activin receptor-like kinase-1. Nat Genet 26:328–331. doi:10.1038/81634

Valdimarsdottir G, Goumans M-J, Rosendahl A et al (2002) Stimulation of Id1 expression by bone morphogenetic protein is sufficient and necessary for bone morphogenetic protein-induced activation of endothelial cells. Circulation 106:2263–2270. doi:10.1161/01.CIR.0000033830.36431.46

van de Laar IMBH, van der Linde D, Oei EHG et al (2012) Phenotypic spectrum of the SMAD3-related aneurysms-osteoarthritis syndrome. J Med Genet 49:47–57. doi:10.1136/jmedgenet-2011-100382

van Meeteren LA, ten Dijke P (2012) Regulation of endothelial cell plasticity by TGF-β. Cell Tissue Res 347:177–186. doi:10.1007/s00441-011-1222-6

van Meeteren LA, Goumans M-J, ten Dijke P (2011) TGF-β receptor signaling pathways in angiogenesis; emerging targets for anti-angiogenesis therapy. Curr Pharm Biotechnol 12:2108–2120. doi:BSP/CPB/E-Pub/000240-12-16 [pii]

van Meeteren LA, Thorikay M, Bergqvist S et al (2012) Anti-human activin receptor-like kinase 1 (ALK1) antibody attenuates bone morphogenetic protein 9 (BMP9)-induced ALK1 signaling and interferes with endothelial cell sprouting. J Biol Chem 287:18551–18561. doi:10.1074/jbc.M111.338103

Velasco-Loyden G, Arribas J, López-Casillas F (2004) The shedding of betaglycan is regulated by pervanadate and mediated by membrane type matrix metalloprotease-1. J Biol Chem 279:7721–7733. doi:10.1074/jbc.M306499200

Villarreal FJ, Dillmann WH (1992) Cardiac hypertrophy-induced changes in mRNA levels for TGF-β 1, fibronectin, and collagen. Am J Physiol 262:H1861–H1866

Wang X (1997) Circulating transforming growth factor β1 and coronary artery disease. Cardiovasc Res 34:404–410. doi:10.1016/S0008-6363(97)00033-3

Wang X-F, Lin HY, Ng-Eaton E et al (1991) Expression cloning and characterization of the TGF-β type III receptor. Cell 67:797–805. doi:10.1016/0092-8674(91)90074-9

Watabe T, Nishihara A, Mishima K et al (2003) TGF-β receptor kinase inhibitor enhances growth and integrity of embryonic stem cell-derived endothelial cells. J Cell Biol 163:1303–1311. doi:10.1083/jcb.200305147

Westermann D, Van Linthout S, Dhayat S et al (2007) Cardioprotective and anti-inflammatory effects of interleukin converting enzyme inhibition in experimental diabetic cardiomyopathy. Diabetes 56:1834–1841. doi:10.2337/db06-1662

Wong SH, Hamel L, Chevalier S, Philip A (2000) Endoglin expression on human microvascular endothelial cells. Eur J Biochem 267:5550–5560. doi:10.1046/j.1432-1327.2000.01621.x

Yang X, Castilla LH, Xu X et al (1999) Angiogenesis defects and mesenchymal apoptosis in mice lacking SMAD5. Development 126:1571–1580

Zeisberg EM, Potenta S, Xie L et al (2007a) Discovery of endothelial to mesenchymal transition as a source for carcinoma-associated fibroblasts. Cancer Res 67:10123–10128. doi:10.1158/0008-5472.CAN-07-3127

Zeisberg EM, Tarnavski O, Zeisberg M et al (2007b) Endothelial-to-mesenchymal transition contributes to cardiac fibrosis. Nat Med 13:952–961. doi:10.1038/nm1613

Zhang YE (2009) Non-Smad pathways in TGF-β signaling. Cell Res 19:128–139. doi:10.1038/cr.2008.328

Zhang D, Gaussin V, Taffet GE et al (2000) TAK1 is activated in the myocardium after pressure overload and is sufficient to provoke heart failure in transgenic mice. Nat Med 6:556–563. doi:10.1038/75037

Chapter 14
TGF-β Signaling in Physiological and Pathological Angiogenesis

Sara I. Cunha and Kristian Pietras

Abstract Members of the transforming growth factor-β (TGF-β) family exert their effect virtually on all cell types in the body, producing diverse and complex cellular responses. TGF-β signaling is deregulated and hyperactive in many malignant conditions, making it an appealing target in the combat of cancer disease. The predominantly endothelial TGF-β receptors, ALK1 and endoglin, which are activated during neoangiogenesis both during development and pathological conditions, pose attractive modulating opportunities to impair tumor vessel formation and cancer progression. However, the precise function of TGF-β family signaling in ECs is difficult to predict, as it appears highly context dependent due to the many ligands and receptors influencing the final outcome. Furthermore, TGF-β is involved in autocrine and intricate dynamic paracrine signaling events in the context of the tumor microenvironment. Pharmacological inhibitors for ALK1, endoglin, and TGF-β or its receptors have been developed and will facilitate more comprehensive studies on the exact function of the TGF-β family in the endothelium, and more specifically in tumor angiogenesis. Here, we will summarize the current knowledge on TGF-β signaling in the regulation of the formation and function of the vascular network.

Keywords Angiogenesis • BMP • Cancer • Targeted therapy • TGF-β

S.I. Cunha
Division of Vascular Biology, Department of Medical Biochemistry and Biophysics, Karolinska Institutet, Stockholm, Sweden

K. Pietras (✉)
Division of Vascular Biology, Department of Medical Biochemistry and Biophysics, Karolinska Institutet, Stockholm, Sweden

Department of Laboratory Medicine Malmö, Lund University Cancer Center, Lund University, Lund, Sweden
e-mail: Kristian.Pietras@med.lu.se

A. Moustakas and K. Miyazawa (eds.), *TGF-β in Human Disease*,
DOI 10.1007/978-4-431-54409-8_14, © Springer Japan 2013

Abbreviations

ALK	Activin receptor-like kinase
AVM	Arteriovenous malformations
BMP	Bone morphogenetic proteins
EC	Endothelial cell
EMT	Epithelial-to-mesenchymal transition
EndMT	Endothelial-to-mesenchymal transition
FGF	Fibroblast growth factor
GDF	Growth and differentiation factor
HHT	Hereditary hemorrhagic telangiectasia
MMP	Matrix metalloproteinase
TGF-β	Transforming growth factor-β
VEGF	Vascular endothelial growth factor
VSMC	Vascular smooth muscle cell

14.1 Introduction

14.1.1 The TGF-β-Rich Tumor Microenvironment

The tumor microenvironment, or stroma, influences the growth of the tumor and its ability to progress and metastasize. The stroma thus constitutes an important aspect to consider when developing therapeutic approaches, as it can alter interstitial fluid pressure, limit the access of therapeutics to the tumor, change drug metabolism, or even contribute to the development of drug resistance (Egeblad et al. 2010).

Despite the importance of tumor–host stroma interactions, there is limited understanding of the stromal milieu composition, and the complexity and dynamics in the relationship between the tumor malignant cells and the surrounding host cells. It is, however, acknowledged that tumor cells and their stroma coevolve during tumorigenesis and progression (Pietras and Östman 2010). Nonetheless, the precise nature of the cells that comprise the normal stroma, how these cells or newly recruited cells are altered during tumor progression, and how they reciprocally influence tumor initiation and progression, is poorly understood.

The large family of TGF-β extracellular pleiotropic cytokines exerts influence essentially on all cellular strata in the body, namely in epithelial cells, fibroblasts, immune, endothelial, lymphatic and perivascular cells (Fig. 14.1) (Elliott and Blobe 2005). TGF-β is the prototypical element of an extensive ligand group that also includes bone morphogenetic proteins (BMPs), activins, inhibins, nodal, and growth and differentiation factors (GDF)s that elicit signaling activity through a collection of five type II receptors and seven type I receptors (Padua and Massague 2009). Different ligand-receptor II–receptor I combinations can be assembled, hence delineating one of the TGF-β family hallmarks of complexity.

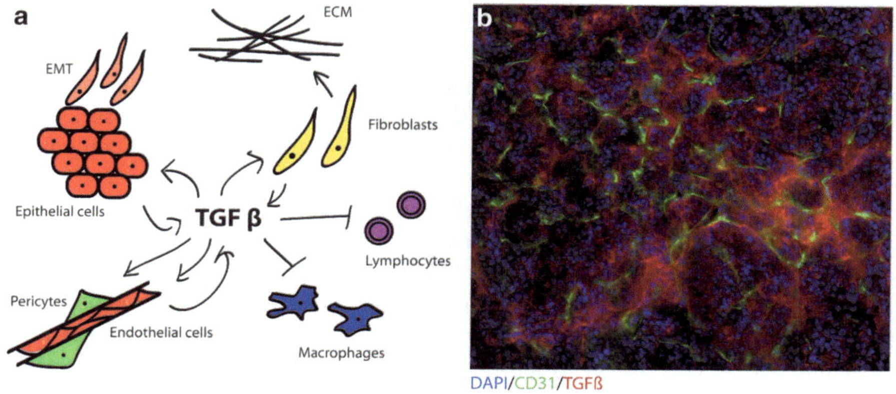

Fig. 14.1 (**a**) TGF-β secreted from various sources is acting on a wide range of cell types in the tumor microenvironment. TGF-β promotes tumor epithelial cell invasiveness and metastasis in an autocrine fashion. TGF-β also affects tumor matrix remodeling by cancer-associated fibroblasts and regulates angiogenesis by acting on ECs and pericytes. Tumor infiltration by leukocytes, macrophages, and bone marrow-derived endothelial, mesenchymal, and myeloid precursor cells is also mediated by TGF-β. TGF-β suppresses proliferation and differentiation of lymphocytes including T cells, natural killer cells, and macrophages, preventing immune surveillance control over the developing tumor. (**b**) Immunostaining of an experimental pancreatic neuroendocrine tumor from RIP1-Tag2 mice for TGF-β (*red*) and the endothelial cell marker CD31 (*green*) illustrates the abundance of TGF-β throughout the tumor microenvironment

In normal, unstressed tissue, sustained basal release of TGF-β by local sources regulates homeostasis. In pathological conditions, TGF-β is abundantly released in the tumor microenvironment (Fig. 14.1), initially as a signal to avoid premalignant progression, but eventually as a cue that cancer cells utilize to their own advantage in later stages of malignancy (Massague 2008; Wakefield and Roberts 2002). Most human tumors overproduce TGF-β whose autocrine and paracrine actions promote tumor epithelial cell invasiveness and metastasis, functioning as a differentiation switch required for transient but reversible invasiveness of carcinoma cells through epithelial-to-mesenchymal mechanisms (Pardali and Moustakas 2007). In addition to eliciting mitogenic signals toward the carcinoma cells, TGF-β also affects cancer-associated fibroblast induction of tumor matrix remodeling and regulates angiogenesis by acting on endothelial cells (ECs) and pericytes. Finally, TGF-β suppresses proliferation and differentiation of lymphocytes including T cells, natural killer cells, and macrophages, thus preventing immune surveillance control over the developing tumor.

In a nutshell, TGF-β signaling is intimately implicated in tumor development and contributes to most hallmarks of cancer described by Hanahan and Weinberg (2000, 2011). It is, thus, of vital importance to carefully analyze the role of the TGF-β family members in the tumor microenvironment and how signaling circuits arising from tumor and stromal interactions can be efficiently modulated in cancer therapy.

14.1.2 Involvement of TGF-β in Vascular Syndromes

In vascular biology, TGF-β has traditionally been seen as a differentiation regulator for vascular smooth muscle cells (VSMCs), ultimately contributing to vessel stabilization and maturation, by inducing ECM deposition and inhibiting EC migration and proliferation.

The critical relevance of TGF-β signaling in vascular development was, however, recognized by identification of mutations in TGF-β receptor genes in familial vascular syndromes, in the type I endothelial specific receptor, activin receptor-like kinase (ALK)1, and the type III or co-receptor, endoglin (ten Dijke and Arthur 2007). Germline loss-of-function mutations in ALK1 or endoglin are causal in the development of the human syndrome of hereditary hemorrhagic telangiectasia (HHT). HHT is characterized by cutaneous telangiectases and gastrointestinal hemorrhage (Berg et al. 1997; McAllister et al. 1994). In addition, major arteriovenous malformations occur in lung, liver, or brain, which may ultimately cause severe morbidity and mortality. While *Alk1* and *endoglin* null mice die at midgestation as a result of severe arteriovenous malformations (AVMs) (Arthur et al. 2000; Oh et al. 2000), mice lacking one copy of the gene for either *Alk1* (Alk1+/−; HHT2) or *endoglin* (Eng+/−; HHT1) recapitulate with age the HHT phenotype (Srinivasan et al. 2003; Torsney et al. 2003).

There is additional strong evidence gathered from in vivo studies on genetically manipulated mouse models, for a prominent role of the TGF-β pathway in vasculo- and angiogenesis mechanisms (ten Dijke and Arthur 2007).

Nearly half of TGF-β1$^{-/-}$ and 25 % of TGF-β1+/− mouse embryos die in utero due to defective hematopoiesis and vasculogenesis (Dickson et al. 1995). TGF-β1-deficient mice that do survive until birth die at about 3 weeks of age of a multifocal inflammatory disorder, primarily affecting heart and lungs, although mild liver inflammation has been also observed (Shull et al. 1992). Supporting these reports, the classical TGF-βRII and TGF-βRI/ALK5 receptors also exhibit defective vasculogenesis and lead to mid gestation lethality (Larsson et al. 2001; Oshima et al. 1996).

In fact, deletion of TGF-β signaling mediators results (Table 14.1), for most cases, in serious implications on the developing embryonic vasculature leading ultimately to lethality in mice (Bertolino et al. 2005; ten Dijke and Arthur 2007).

A lot of attention has been focused on the endothelial cell-specific TGF-β receptors ALK1 and endoglin not only because they already have documented involvement in the human syndrome HHT, but more importantly they hold therapeutic promise in pathological conditions, such as cancer, where their upregulated expression in the vascular component plays a critical role in cancer development.

Table 14.1 TGF-β signaling mediators implicated in vascular abnormalities in mice and human pathological conditions

Gene	Mouse phenotype	Human pathological condition	References
TGF-β1	Null: embryonic lethality with angiogenesis defects; some are born and die after a few weeks due to inflammatory disorders		Shull et al. (1992), Dickson et al. (1995)
TGF-β2	Null: aortic arch defects, cardiac septal defects, abnormal cardiac cushions, persistent truncus arteriosus, perinatal lethality		Shull et al. (1992), Sanford et al. (1997)
BMP 2	Null: embryonic lethal with defects in amnion/chorion and cardiac development.		Zhang and Bradley (1996)
	Het: viable; susceptible to hypoxic pulmonary hypertension		Anderson et al. 2010)
BMP 4	Het: viable; less severe hypoxic pulmonary hypertension; impaired vascular remodeling		Anderson et al. (2010)
TGF-βRI/ALK5	Null: embryonic lethal, angiogenesis defects	Aortic aneurysm; LDS	Larsson et al. (2001)
TGF-βRII	Null: embryonic lethal with defects in vasculogenesis and hematopoiesis	MFS2, LDS	Oshima et al. (1996)
BMPRII	Null: pre-angiogenesis lethality	PAH	Beppu et al. (2004), Song et al. (2005)
	Het: Susceptible to pulmonary hypertension, defective vascular remodeling, abnormal vascular tone		Rudarakanchana et al. (2002)
TβRIII	Embryonic lethal at E13.5 with proliferative defects in the heart and apoptosis in the liver; defects in coronary vasculogenesis		Stenvers et al. (2003), Compton et al. (2007)
ALK1	Null: Embryonic lethal at E11.5 with angiogenesis defects; dilated vessels, AVMs, impaired VSMC recruitment	HHT	Oh et al. (2000) Urness et al. (2000)
	Het: model for HHT		Park et al. (2006)

(continued)

Table 14.1 (continued)

Gene	Mouse phenotype	Human pathological condition	References
	EC-specific deletion: vascular dysplasia, susceptibility to arteriovenous malformations		
ALK2	Null: lethality, defects in cardiovascular embryogenesis, nervous system, skeleton, growth and size; impaired bicuspid aortic valve	FOP	Shore et al. (2006) Thomas et al. (2012)
Endoglin	Null: Embryonic lethal with angiogenesis and cardiac defects	HHT	Arthur et al. (2000) Bourdeau et al. (2000)
	Het: model for HHT, spontaneous pulmonary hypertension, abnormal vascular tone		Mahmoud et al. (2010)
	Conditional deletion in EC: arteriovenous malformations		
ALK3/BMPRIa	Conditional deletion in the mesoderm: embryonic lethal at E10.5 with hemorrhage, cardiac defects, and impaired smooth muscle cell recruitment		Park et al. (2006)
	Conditional deletion in embryonic SMCs: embryonic lethality at E11.5 with cardiac defects, hemorrhage, impaired vascular remodeling		El-Bizri et al. (2008a), El-Bizri et al. (2008b)
	Conditional deletion in adult SMCs: viable with impaired vascular remodeling		
Smad 1	Null: embryonic lethal at E9.5 with defects in fusion of the chorioallantoic membrane and yolk sac angiogenesis		Lechleider et al. (2001)
Smad 2	Null: pre-angiogenesis lethality		Goumans and Mummery (2000)
Smad 4	Null: pre-angiogenesis lethality	Juvenile polyposis; HHT	Gallione et al. (2004)

(continued)

Table 14.1 (continued)

Gene	Mouse phenotype	Human pathological condition	References
	Conditional deletion in EC: embryonic lethal at E10.5 with cardiac and angiogenesis defects		Lan et al. (2007)
Smad 5	Null: embryonic lethal at E9.5 with cardiac/ angiogenesis defects		Chang et al. (1999), Yang et al. (1999)
Smad 6	Null: some viability; cardiac defects, vascular calcification and systemic hypertension		Galvin et al. (2000)
Smad 8	Null: viable with spontaneous pulmonary vascular remodeling		Huang et al. (2009)
EMILIN 1	Null: reduced arterial diameter, increased vascular resistance, emphysema, increased TGF-β signaling in the vascular wall	Hypertension	Zacchigna et al. (2006)
Fibrilin 1	Null: die perinatality with ruptured aortic aneurysm, impaired pulmonary function	MFS1	Carta et al. (2006), Judge et al. (2004)
Fibronectin 1	Null: deformed heart and embryonic vessels, defective extraembryonic vasculature	Ehlers–Danlos syndrome	George et al. (1993)
LTBP1	Null: persistent truncus arteriosus and cardiac valve development		Todorovic et al. (2007)
Thrombospondin 1	Null: pneumonia, alveolar hemorrhage, VSMC hyperplasia		Lawler et al. (1998), Crawford et al. (1998)

HHT hereditary hemorrhagic telangiectasia, *LDS* Loeys–Dietz syndrome, *MFS2* Marfan syndrome type 2, *PAH* pulmonary arterial hypertension, *vSMC* vascular smooth muscle cell, *FOP* fibrodysplasia ossificans

14.1.3 TGF-β Signaling in ECs: ALK1/ALK5 Interconnecting Crosstalk

Following synthesis, secretion, and activation, the mature TGF-β dimeric ligand is released from the ECM to trigger specific serine/threonine type I and type II kinase receptor heterotetrameric complexes (Wrana et al. 1994). In ECs, TGF-β signaling has been described to signal via the globally expressed type I receptor, ALK5 or alternatively through the predominantly endothelial type I receptor, ALK1.

The prevailing recruited type I receptor dictates the activation of a particular Smad transducing cascade. ALK1 activation generates phosphorylation of Smad1, 5, and 8, whereas ALK5 leads to Smad 2 and 3 signaling activation (Goumans et al. 2002). The selected activated Smad subset independently forms a heteromeric complex with a related molecule, Smad4, which translocates the complexes into the nucleus to launch transcription of specific target genes (Massague 2000; ten Dijke et al. 2000), involved in distinct angiogenic responses.

TGF-β ligands also interact with co-receptors or type III receptors, represented by betaglycan (TGF-βRIII) and the typically endothelial endoglin. However, because these type III receptors lack the kinase domain, they essentially hold an accessory role in ligand binding and signaling activation, adding yet another level of regulation to the TGF-β complex signaling web.

Far from consensual, the TGF-β contribution to vascular biology knowledge has been constantly under debate due to numerous paradoxical reports (Goumans et al. 2002; Lamouille et al. 2002; Valdimarsdottir et al. 2002).

A crucial role in angiogenesis for ALK1 was first described in a study reporting ALK1 as pivotal for SMC recruitment, implying a vital role for TGF-β/ALK1 signaling axis in the maturation phase of angiogenesis (Oh et al. 2000). On the other hand, signaling derived from ALK5 was more pronounced during the activation phase of angiogenesis, when ECs degrade perivascular basement membrane, invade and migrate into the newly available space, through active proliferation and lumen formation. The balance theory was then hypothesized for the first time, speculating that different levels of TGF-β ligand availability would determine the sequential angiogenic fate and control the properties of the endothelium during angiogenesis. Also, in parallel studies another laboratory presented evidence for development of shunts between arteries and veins and severe arteriovenous malformations due to fusion of major arteries and veins in mice lacking ALK1 (Urness et al. 2000).

The balance working model was quickly challenged when another laboratory proposed that TGF-β engages in the activation of ALK1 signaling via Smad1/5, which concomitantly inhibits ALK5 signaling through Smad2/3 (Goumans et al. 2002). ALK5, while critical for ALK1 signaling, demonstrated by studies on ALK5-deficient mouse embryonic ECs, commits to an anti-angiogenic cascade of events, while ALK1 mediates pro-angiogenic activation (Goumans et al. 2003). These studies indicated that TGF-β stimulatory effects on either ALK5 or ALK1 are mutually exclusive inducing differential transcriptional activation of PAI-1 and Id1, respectively, which ultimately elicit a different set of physiological responses.

Motivated by the fact that both ALK1 and ALK5 null mice render an embryonic lethal phenotype due to extensive vascular abnormalities (Larsson et al. 2001; Oh et al. 2000; Urness et al. 2000), ALK1 dependency on ALK5, by means of signaling or by mere anchoring, has been questioned and addressed with reservation by multiple laboratories.

Transcriptional profiling of human umbilical vein ECs expressing constitutively active adenoviral constructs of ALK1 or ALK5 demonstrated substantial differences in the transcriptional output from either signaling pathway (Ota et al. 2002), validating previously described downstream gene regulation. Interestingly, the non-overlapping expression patterns of ALK1 and ALK5 in vivo (Seki et al. 2006) by thorough analysis of a knockin mouse line carrying a lacZ reporter in the *Alk5* gene locus ($Alk5^{lacZ}$), also lends support to divergent roles in vascular development for each of the two type I receptors expressed by ECs.

ALK5 suppression, by genetic silencing or small molecule inhibition, was shown not to interfere with BMP9/ALK1-induced phosphorylation of Smad1/5/8 in bovine aortic ECs (BAECs) (Shao et al. 2009). Instead, silencing of Alk1 or any of its downstream molecular effectors, by means of siRNA transfection, rather induce a potent ALK5 signaling upregulation. In agreement with the ALK5-independent action of ALK1 is the notion that ALK5 is present in ECs in vivo either at low levels, or only expressed by the neighboring VSMCs, suggesting that ALK5 may only participate in ALK1-dependent angiogenesis in a paracrine fashion (Park et al. 2008; Shao et al. 2009). Congruent with these results, EC-specific ablation of Alk5 does not inflict vascular abnormalities in mice or zebrafish (Park et al. 2008). However, embryos from knockin mice carrying a mutation on L45 loop in Alk5 rescued to some extent the earliest vascular defects observed in ALK5 mouse knockouts (Itoh et al. 2009), probably because this mutation, yet interfering with ALK5 kinase ability to phosphorylate Smad2, inherently preserves ALK5 competence to mediate non-Smad signaling and lateral signaling to ALK1. In agreement with these findings, ALK1 signal inhibition proved to interfere, in pathological conditions, not only with its own target genes but ALK5 signal transduction also exhibited a suppressive modulation in a model of pancreatic neuroendocrine cancer (Cunha et al. 2010).

More recently, it has been demonstrated that selective deletion of ALK5 in ECs using an $Alk5^{GFPCre}$ mouse line resulted in embryonic lethality due to brain vessel pathological morphology and intracerebral hemorrhage (Nguyen et al. 2011). Independent observations of EC-specific deletion of Smad2/3 using Tie2-Cre transgenic mice revealed critical hemorrhaging and embryonic lethality around E12.5. In this study, vascular maturation was incomplete owing to inadequate assembly of mural cells in the vasculature, most likely because of impaired expression of PDGF-B by the Smad2/3 ablated endothelium (Itoh et al. 2012). These observations substantiate the vital relevance of ALK5 in the endothelium either as a signaling anchor or by actively participating in the vasculogenic process.

Collectively, these reports demonstrate that ALK5 signaling is indeed relevant for endothelial homeostasis. Further studies aiming at dichotomizing ALK1 versus ALK5 signaling in endothelial and in perivascular cells during development and in the tumor microenvironment are thus required to clarify and reconciliate previous paradoxical observations and infer about the benefit or risk of clinically targeting such pathways without proper amendments. It is plausible that the relative stoichiometry of ALK1 and ALK5 signaling may be crucial for proper regulation of gene expression (Cunha and Pietras 2011).

14.1.4 Bone Morphogenetic Proteins

The BMPs, a subcategory of the TGF-β superfamily, were first identified in extracts from bone matrix and characterized by their ability to induce ectopic bone formation when implanted subcutaneously in rats (Wozney et al. 1988). It soon became clear that BMPs play a key role in vertebrate organogenesis, as well as in embryonic vascularization (Kishigami and Mishina 2005; Zhao 2003).

The BMP family, including the GDFs, comprises a group of 20 ligands that activate a classical BMP pathway in vertebrates (Lowery and de Caestecker 2010; Wagner et al. 2010). In the canonical BMP signaling pathway, three type II receptors (BMPRII, ActRIIa, and ActRII2b) and four type I receptors (ALK1, ALK2, ALK3, and ALK6) can be activated (Miyazono et al. 2010; Moustakas and Heldin 2009). In addition to primarily triggering Smad1, 5, and 8, BMP cues may also activate Smad2 (Upton et al. 2009) and Smad-independent signaling (Sieber et al. 2009).

BMP9 has been implicated in hematopoiesis, hepato-, osteo-, chondro-, and adipogenesis (Canalis et al. 2003; Li et al. 2003; Lord et al. 2010; Ploemacher et al. 1999; Sieber et al. 2009). It has also been described as a regulator of glucose metabolism (Chen et al. 2003) and as a differentiation factor for cholinergic neurons in the central nervous system (Lopez-Coviella et al. 2000). More recently, BMP9 was pinpointed as the physiologically functional high affinity ligand for the predominantly endothelial receptor, ALK1. This fact highlighted BMP9 as a critical modulator of angiogenesis (David et al. 2007, 2008; Scharpfenecker et al. 2007).

BMP9 was originally cloned from a rodent cDNA library obtained from mouse liver, where it was shown to be highly expressed (Miller et al. 2000). Accordingly, the liver was characterized as the major source of human and mouse BMP9, expressed by hepatocytes and intrahepatic biliary epithelial cells, while brain and lung only express it at much lower levels (Bidart et al. 2012). In line with these observations, the Human Protein Atlas profile for BMP9 in normal tissues indicates that it is highly expressed in liver, pancreas, placenta, lung, epididimus, gastrointestinal tract, gall bladder, and thyroid, but also in hematopoietic cells (www.proteinatlas.org).

Interestingly, it is now clearly described that pulmonary and cerebral arterial malformations occur more often in HHT1, while hepatic arterial malformations are more frequent in HHT2 (Letteboer et al. 2006). In fact, in HHT2, the frequency of hepatic arterial malformations is between 38 % and 41 %, while in HHT1, it ranges between 2.5 % and 8 % (Gallione et al. 2006). The specific expression of the ALK1 ligand, BMP9, predominantly in the liver reflects a seemingly tissue-specific manifestation in HHT2.

The *modus operandus* of BMP ligand interaction with their receptors differs from that of TGF-β: while TGF-β exhibits higher affinity for type II receptors and do not stably interact with type I receptors alone, BMPs bind independently to both type I and type II receptors (Groppe et al. 2008; Lin et al. 2006). The BMP ligands can also display affinity to the co-receptors endoglin and betaglycan (Lowery and de Caestecker 2010). In fact, BMP9 can directly bind endoglin (Scharpfenecker et al. 2007).

BMP9 is synthesized as a precursor protein, which is then cleaved by furin, a serine-endoprotease, forming a short dimeric mature form to which the prodomain can remain non-covalently associated (Bidart et al. 2012). Until recently, neither BMP9 nor its closely related family member BMP10 was found to be negatively regulated by common BMP pathway antagonists (David et al. 2008; Seemann et al. 2009). However, recent studies show that ALK1 activation by BMP9 induces expression of matrix Gla protein and crossveinless 2 (CV2), both known as antagonists of BMP4-induced angiogenesis (Yao et al. 2006, 2011, 2012). CV2, a member of the Chordin family, preferentially binds and inhibits BMP9 thereby providing strong feedback inhibition on ALK1 (Yao et al. 2012), suggesting a critical mutual regulation by BMP9 and CV2 in vasculature regulation.

Analogous to TGF-β signaling mediated by ALK1, BMP9 has also been reported to have incongruent effects on ECs. For example, BMP9 exhibits anti-angiogenic effects counteracting fibroblast growth factor (FGF)-induced angiogenesis in ex vivo metatarsal models (David et al. 2007; Scharpfenecker et al. 2007) and acts as a circulating vascular quiescent factor (David et al. 2008). Nevertheless, multiple types of ECs activate their proliferative status in vitro in response to BMP9, which pro-angiogenic properties also activate matrigel plug vascularization and tumor angiogenesis in a pancreatic cancer xenograft model (Suzuki et al. 2010).

In order to unmistakably clarify the effects on the endothelium by BMP9 stimulation and its specific downstream mediators, an extensive analysis of BMP9 downstream activation in comparison to other ligands on ECs is mandatory in the field. According to the present knowledge, BMP-induced responses have as common denominator the Smad1/5/8 pathway and Id1, 2, and 3 as target genes, suggesting other differentially responsive genes may exist, more specifically induced by each BMP. In a recent study such efforts were initiated where EC-specific Smad1/5 target genes were characterized and upregulation of Notch signaling-related genes were identified upon BMP9 stimulation (Morikawa et al. 2011).

Despite the paucity of detailed studies, the Human Protein Atlas profile for BMP9 in cancer disease indicates that BMP9 expression is increased in colorectal cancer, head and neck squamous carcinoma, and pancreatic and liver cancers (www.proteinatalas.com). Interestingly, BMP9 is primarily expressed in the islets of Langerhans by the tumor cellular compartment in mouse neuroendocrine pancreatic tumors (Cunha et al. 2010).

Similar to BMP9, BMP10 has also been identified as a functional activator of ALK1 in ECs, inducing comparable cellular effects. In agreement with a 65 % amino acid sequence homology between both ligands, BMP10, much like BMP9, exhibits angiostatic properties on dermal HMVEC (David et al. 2007). Nevertheless, BMP10 binds to ALK1 with lower affinity than BMP9 (David et al. 2007) and is mainly expressed in the murine developing and postnatal heart. The impaired lethal cardiac growth and physiology in the BMP10 knockout mouse coupled to normal vascular development of embryo and yolk sac propose a critical role for BMP10 in cardiogenesis (Chen et al. 2004). Interestingly, it has also been demonstrated that BMP10 can additionally bind to ALK3 (Mazerbourg et al. 2005). Of note, ALK3 targeted deletion in neural crest cells generates embryonic heart failure (Stottmann et al. 2004).

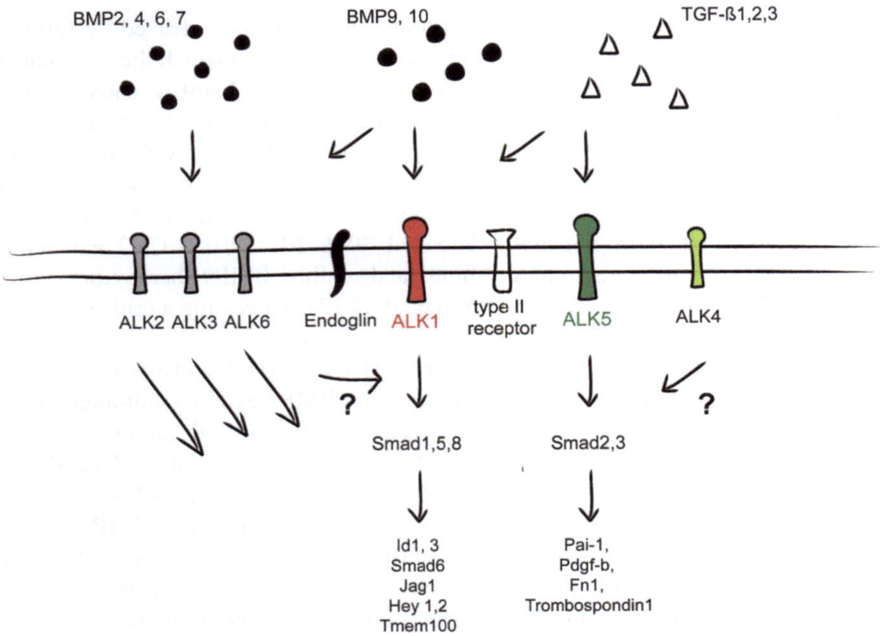

Fig. 14.2 Multiple TGF-β family ligands and receptors are acting on endothelial cells to shape the angiogenic response

All in all, these observations suggest the cardiac-specific nature of BMP10 signaling most likely through ALK3, rather than ALK1. However, the direct effect of BMP10 on the vasculature should not be overlooked, as it has potential to cooperate or even compensate for BMP9 signaling through ALK1. In fact, very recently Ricard et al. unveiled that *Bmp9*-KO mice do not exhibit defective vascularization in the retina (Ricard et al. 2012). However, injection of the extracellular domain of ALK1 or a neutralizing anti-BMP10 antibody impaired retinal vascularization in Bmp9-KO neonates, reducing retinal vascular expansion and exacerbating vascular density (Ricard et al. 2012). These data thus sustain a cooperative or compensatory role for BMP9 and BMP10 in postnatal vascular remodeling of the retina. Whether this cooperative role occurs in the context of cancer remains to be determined.

14.1.5 Interplay Between Type I Receptors

As mentioned earlier, ALK1 shares similar properties in terms of BMP-dependent activation of Smad1/5/8 signaling with the related BMP type I receptors ALK2, ALK3, and ALK6 (Fig. 14.2). Ligand specificity has not been carefully elucidated, and many ligands, including BMP2, BMP4, BMP6, BMP7, BMP9, and BMP10, exhibit a multitude of effects on ECs, ranging from metabolism, endothelial-to-mesenchymal transition (EndMT), and tumor angiogenesis (Fig. 14.2) (Bostrom et al. 2011; Heinke et al. 2008; Langenfeld and Langenfeld 2004; Medici et al. 2010; Ramoshebi and Ripamonti 2000; Yao et al. 2008; Zeisberg et al. 2007). While described as the physiological ligand for ALK1, BMP9 has also documented binding ability toward ALK2 in non-EC, such as myoblasts and breast tumor cells (Scharpfenecker et al. 2007), with the BMP9/ALK2 signaling axis being also linked to promotion of proliferation of ovarian cancer cells (Herrera et al. 2009). Lending support to the need of substantial analysis of the signaling arising from these receptors in the endothelium, vascular ECs have been shown to transform into multipotent stem-like cells in an ALK2-dependent fashion, in lesions from individuals with *fibrodysplasia ossificans progressiva* (FOP) (Medici et al. 2010). This disabling disorder occurs as a result of gain-of-function mutations in ALK2 in humans or mirrored in mice by constitutive activation of ALK2 signaling on chondrocytes and osteoblasts. Lineage tracing of heterotopic ossification in mice using a Tie2-Cre construct disclosed the endothelial origin of these cell types (Medici et al. 2010). In agreement with this finding, ECs conditionally deficient for ALK2 do not succeed to undergo EndMT during endocardial cushion formation in embryogenesis (Wang et al. 2005). Of note, ALK2 has been demonstrated to upregulate ALK1 in ECs in response to high-density lipoproteins, after which ALK1 in turn promotes survival by inducing expression of vascular endothelial growth factor (VEGF)-A (Yao et al. 2008). Glucose level augmentation co-regulates ALK1 and ALK2 expression in human aortic ECs (Bostrom et al. 2011). Also, BMP/TGF-β receptors appear to be activated and function sequentially: ALK3, ALK2, ALK1, and ALK5, where each receptor can possibly entail a distinct function and correlate to a specific stage in vascular growth and development (Bostrom et al. 2011; Shao et al. 2009; Yao et al. 2008).

Thus, the interplay and/or compensatory crosstalk primarily between ALK1 and ALK2, but also with ALK3 and ALK6, which is of critical importance in a therapeutic context, promptly begs for more detailed tumor studies.

14.1.6 TGF-β and BMP Signaling Pathways: Competitive or Synergistic?

Closely connected to receptor interplay is the role specifically played by the ligands. Classically, BMPs and TGF-βs have long been described to exert parallel antagonistic effect on the other pathway in a variety of biological contexts (Gronroos et al. 2012).

Moreover, in physiological conditions, cells in the body are exposed to multiple ligands simultaneously, which may render alternative responses than what is customarily studied when analyzing the effects of each ligand in isolation. In an attempt to clarify the role of ALK1 signaling in EC, we recently described an unanticipated synergistic effect of TGF-β with BMP9 on tumor angiogenesis. We demonstrated in vitro and in vivo in various systems, that while either cytokine on its own exerted suppressive action on the endothelium, both ligands in combination boosted the EC response toward other pro-angiogenic stimuli (Cunha et al. 2010). On a molecular level, simultaneous ECs induction with TGF-β and BMP9 induces a synergistic response on ALK5 target gene expression (e.g., PAI-1 and PDGF-B).

Another publication demonstrated that BMP2 synergistically enhances TGF-β3-induced initial phenotypic changes associated with EndMT, taking place during endocardial cushion formation (Yamagishi et al. 1999). Of note, BMPs and their receptors are expressed at many sites in which epithelial or endothelial-to-mesenchymal transition occurs during developmental organogenesis (Dewulf et al. 1995; Jones et al. 1991; Lyons et al. 1989).

TGF-β and BMP7 also coadjuvantly stimulate angiogenesis in the chick chorio-allantoic membrane assay (Ramoshebi and Ripamonti 2000), and collaborately activate prostate cancer cells (Buijs et al. 2007) and osteoblast differentiation. In contrast, BMP7 counteracts TGF-β-induced EndMT in a model of cardiac fibrosis, rendering the EC capable of preserving their endothelial identity (Zeisberg et al. 2007), suggesting a context dependency of the synergy between TGF-βs and BMPs.

The numerous BMP ligands and type I receptors exert a variety of effects on ECs, yet the fact that different ligands utilize common pathway components raises important questions, which may have been neglected until recently: how cells respond specifically to individual ligands, and how cells integrate and interpret signals received from multiple ligands? Concerning this context, worthy of note are studies suggesting that preformed BMP receptor complexes or BMP-induced oligomerization of type I and type II receptors predominantly activate Smad-dependent and -independent signaling, respectively (Nohe et al. 2002; 2004). Also, the choice of type II receptor can persuade the signaling outcome of BMP stimulation as downstream-specific binding of Limk1 to the BMP type II receptor, but not to TGF-β or activin type II receptors (Foletta et al. 2003; Lee-Hoeflich et al. 2004). More recently, different R-Smad complex formation, Smad1/5–Smad2 versus Smad1/5–Smad3, was described (Gronroos et al. 2012), opening the possibility that the novel complexes may be the source of antagonistic versus synergistic responses in different studies.

Evidently, signaling through non-Smad effectors, the recruitment of distinct type II receptors and perhaps more importantly the variability created by alternative Smad complex formation should be further examined as the explanations for the diverse effects.

14.2 Physiological Role of the TGF-β Family in the Vasculature

14.2.1 Physiological Role of ALK1 in the Vasculature

The importance of this receptor became obvious, when Alk1 loss of function studies revealed that its complete loss causes embryonic lethality at midgestation, due to severe vascular abnormalities, which included vessels hyperdilation, AVMs resulting from fusion of major arteries and veins, and impaired recruitment of VSMC (Oh et al. 2000; Urness et al. 2000). Mutations in the Alk1 gene have been identified as an underlying cause for development of HHT, a rare, human autosomal dominant disease characterized by the presence of recurrent epistaxis and small characteristic malformations of the peripheral blood vessels near the surface of the skin or mucosal linings (Geirdal et al. 2012). AVMs of the lung, liver, and central nervous system are also known clinical findings. Interestingly, EC-specific deletion of the Alk1 gene in the mouse results in neonatal lethality at P5, with the pups exhibiting hemorrhaging brain, lung, and gastrointestinal tract (Park et al. 2008). In attempts to evaluate the contribution of ALK1 to vascular homeostasis in adult mice, Park et al. (2009) deleted the Alk1 gene by tamoxifen administration in 2 months old $R26^{+/CreER}$ $Acvrl1^{2loxP/loxP}$ mice. Tamoxifen-induced Alk1 deletion resulted in severe internal hemorrhage in lung, small intestine, and uterine vessels, and ultimately fatality.

Strong expression of ALK1 has been reported during developmental and neonatal stages, while suppressed during adulthood, except in certain organs, e.g., the lungs (Park et al. 2008). Supportive of that is the observation that ALK1 is fundamental for umbilical and placental blood vessel formation (Hong et al. 2007). However, ALK1 expression is induced in feeding arteries and newly formed blood vessels during wound healing, in adult subdermal blood vessels (Park et al. 2009). AVMs appearing only in subdermal blood vessels where a wound was inflicted provides in vivo experimental evidence that genetic predisposition by endoglin or ALK1 mutations is not enough for development of de novo AVMs in HHT (Park et al. 2008). Interestingly, only selected vascular beds in HHT patients develop telangiectasias or AVM lesions, while other areas (>99.9 %) remain normal (Sadick et al. 2006).

Inhibition of ALK1 by systemic injection of an ALK1 soluble extracellular domain efficiently impaired retinal neonatal angiogenesis, described and validated in three independent studies (Larrivee et al. 2012; Niessen et al. 2010; Ricard et al. 2012). These data implied that ALK1 signaling blockade induced retinal hypervascularization and appearance of AVMs in neonatal mice. Incidentally, the most recent studies also report a cooperative effect of ALK1 and Notch signaling pathways (Larrivee et al. 2012; Morikawa et al. 2011; Ricard et al. 2012). The reported synergy between ALK1 and Notch pathways generated exacerbation of the hypervascularization phenotype, inducing potentiated expression of Notch target genes in the stalk cells, which concomitantly suppress VEGF signaling to the endothelial tip cell (Larrivee et al. 2012). In parallel studies, endothelial-specific inactivation of Smad1/5 in mouse embryonic development yields impaired Dll4/Notch signaling and augmented tip cell in detriment to stalk cell number (Moya et al. 2012). These studies put forward a regulatory crosstalk loop among BMP9/ALK1/Smad1/5 and Notch signaling coordinating tip versus stalk cell specification.

Additionally to being expressed by blood ECs, ALK1 is also expressed by lymphatic ECs (Niessen et al. 2010). In vitro stimulation of lymphatic ECs by BMP9 generates downstream target gene transactivation. Furthermore, inhibition of ALK1 signaling by means of an ALK1-Fc soluble fusion protein diminishes neonatal retinal lymphangiogenesis, while the use of ALK1 targeting monoclonal antibody also impairs Lyve-1 positive lymphangiogenesis in mammary fat pad–implanted MDA-MB-231 breast carcinoma xenografts (Hu-Lowe et al. 2011). Of note, lymph vessel development comprises coordinate and synergistic ALK1 and VEGFR3 signaling regulation, evocative of the crosstalk observed between ALK1/ALK5, ALK1/ALK2, and VEGF-receptor signaling in blood vessel angiogenesis (Cunha et al. 2010; Mitchell et al. 2010).

14.2.2 Physiological Role of Endoglin in the Vasculature

Endoglin, an auxiliary receptor for TGF-β, is required for angiogenesis during development (Li et al. 1999). It is expressed primarily in ECs and its expression is substantially incremented during EC activation, inflammation, ischemia, and tumor angiogenesis (Docherty et al. 2006; Jonker and Arthur 2002; Torsney et al. 2002). The mechanisms involved in endoglin upregulation are presumably multifactorial, but hypoxia is a probable inducer as it prevails in most pathophysiological environments where endoglin is enhanced (Bernabeu et al. 2009).

Endoglin associates with type II receptors of the TGF-β family in the presence of ligand and with the type I signaling receptors, ALK1 and ALK5, even in the absence of exogenous ligand (Barbara et al. 1999). Despite possessing no enzymatic activity, endoglin has been reported to be necessary to modulate ligand-receptor interaction in ALK1, but not in ALK5 signaling (Cheifetz et al. 1992; Lebrin et al. 2004; Pece-Barbara et al. 2005; Scharpfenecker et al. 2007). More recently, ALK5 was shown to phosphorylate the cytoplasmic domain of endoglin in ECs (Ray et al. 2010). Depending on the serine phosphorylation status, only on 646 or on both 646 and 649 serine residues, results in loss of endoglin-mediated inhibition or activation of Smad1/5/8 signaling, respectively, in response to TGF-β/BMP9 signaling (Ray et al. 2010). Taken together, these results indicate that endoglin phosphorylation by ALK5 is an important mechanism for regulating TGF-β and BMP signaling in ECs.

Even though endoglin has an undeniably well-documented connection to ALK1 and its signaling enhancement (Blanco et al. 2005; Lebrin et al. 2004), it is imperative to mention that it interacts not only with other ligands than TGF-β and BMP9 (TGF-β3, activin A, BMP2, and BMP7) but also with several different type I and type II receptors involved in BMP and TGF-β signaling (Barbara et al. 1999). Alternatively, endoglin has also been implicated in interactions with cytoplasmic proteins such as Zyxin, ZRP-1, β-Arrestin, and Tctex2β, which may further generate additional cellular outcomes (Bernabeu et al. 2009). The signaling unrestrictiveness of endoglin consequently adds an extra degree of intricacy to the elaborated signaling networks deriving from the TGF-β family in the angiogenesis field. Moreover, the fact that endoglin is positively associated with ECs proliferation, while weakly expressed in quiescent endothelium, has focused the interest on endoglin as a potential target for cancer in vivo.

14.2.3 Physiological Role of TGF-β in the Vasculature

TGF-β effect in vascular biology is traditionally regarded as an accessory pathway that primarily participates in VMCs differentiation and maturation. However, apart from the genetic evidence for TGF-β relevant direct role in the ECs, there is additional in vitro evidence revealing that several key angiogenic mediators such as VEGF and connective tissue growth factor are direct targets of the TGF-β signaling pathway (Padua and Massague 2009; Sanchez-Elsner et al. 2001). Interestingly, hypoxic conditions present at the core of a tumor juxtaposed to TGF-β signaling can induce robust levels of VEGF mRNA through the activation of hypoxia-inducible factor (HIF)-1 and Smad proteins (Sanchez-Elsner et al. 2001). TGF-β can also regulate the expression, secretion, and activity of matrix metalloproteinase (MMP)-2 and MMP-9, and downregulate the expression of the protease inhibitor TIMP in the tumor and ECs (Derynck et al. 2001; Hagedorn et al. 2001). This TGF-β-mediated metalloprotease activity can subsequently enhance the migratory and invasive properties of ECs required for angiogenesis.

Multiple genetic mouse models suggestive of a direct role for TGF-β signaling pathway in angiogenesis further corroborate the need to more thoroughly analyze this pathway in the tumor context. Increased expression of TGF-β in either prostate carcinoma cells or Chinese hamster ovary cells resulted in robust angiogenic responses, which could be blocked by TGF-β neutralizing antibodies (Tuxhorn et al. 2002; Ueki et al. 1992). In sharp contrast, LNCaP human prostate xenografts treated with TGF-β neutralizing antibodies exhibited a 2- to 3.5-fold reduction in microvessel density vessels relative to control animals, whereas blood lakes were found in some areas of the tumors (Tuxhorn et al. 2002). Additionally, a TGF-β small molecule inhibitor LY364947 impaired tumor angiogenesis and growth in a transgenic multistep pancreatic islet tumor model (Cunha, SI, unpublished observation; Fig. 14.3). Another study using the orthotopic xenograft model of breast cancer MDA-MB-231 reported that ALK5-mediated TGF-β signaling is critical for metastasis dissemination and tumor angiogenesis in MMP9-regulated fashion (Safina et al. 2007). In parallel, Smad2 and Smad3 were reported to have opposing roles in breast cancer bone metastasis by differentially affecting VEGF-mediated tumor angiogenesis in an experimental mouse model of bone metastasis, where osteotropic MDA-MB-231-luc cells were intracardially inoculated (Petersen et al. 2010).

To sum up, both direct and indirect effects of TGF-β on the tumor microenvironment stimulate tumor angiogenesis.

TGF-β affects tumor angiogenesis per se but it also regulates vascular permeability. Neutralizing antibodies to TGF-β1 or an ALK5 inhibitor significantly augment vessel permeability. In two distinct transgenic mouse tumor models, MMTV-PyMT and K14-HPV16, breast and epidermal squamous cell cancers, respectively, inhibition of ALK5 further enhanced vascular leakage into the interstitium and facilitated increased delivery of high molecular weight compounds into premalignant and malignant tissue (Lammerts et al. 2002; Sounni et al. 2010). These data strongly suggest that ALK5 antagonists can be therapeutically exploited to improve the

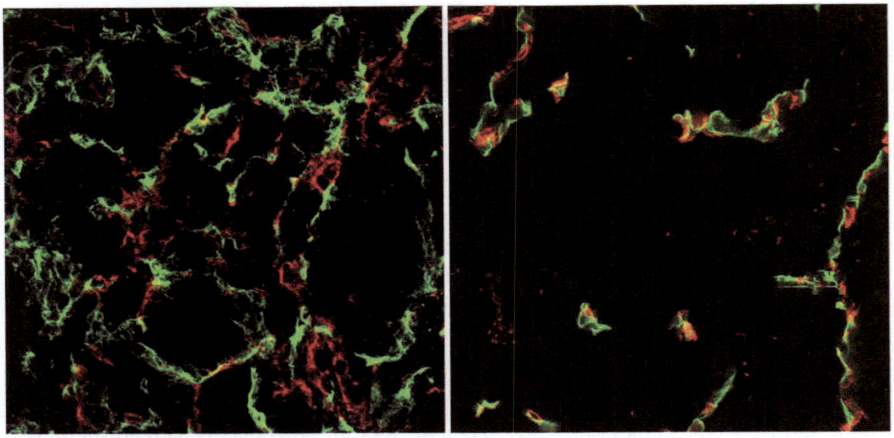

Fig. 14.3 Inhibition of ALK5 signaling by the small molecule LY364947 inhibits tumor angiogenesis in the RIP1-Tag2 mouse model of pancreatic neuroendocrine cancer (immunostaining for the endothelial cell marker podocalyxin, *green* and the pericyte marker NG2, *red*)

delivery of drugs or molecular contrast agents into tissues where chronic damage or neoplastic disease limits their efficient delivery. Along with these observations, another study compared the effects of two different tyrosine kinase inhibitors, imatinib and sorafenib, with an ALK5 inhibitor (LY364947) on extravasation of a modeled nanoparticle, 2 MDa dextran in two tumor models: the CT26 colon cancer model and the BxPC3 pancreatic cancer model (Kano et al. 2009). In fact, sorafenib most potently enhanced the accumulation of nanoparticles in the CT26 colon cancer model, whereas TGF-β inhibitor exhibited a stronger effect on the BxPC3 pancreatic cancer model, suggesting that while ALK5 inhibitors are an appropriate strategy to enhance delivery of nanoparticle-delivered drugs in pancreatic tumors, this may not be longitudinally optimal for all tumor types.

A small molecule ALK5 inhibitor at a low dose was used for treating several experimental intractable solid tumors, including pancreatic adenocarcinoma and diffuse-type gastric cancer, characterized by hypovascularity and thick fibrotic tumor stroma. Low doses of ALK5 inhibitor altered TGF-β signaling neither in the malignant cells nor in the cancer-associated fibroblast components. However, the ALK5 inhibitor decreased pericyte coverage on the endothelium without reducing endothelial area specifically in the tumor vasculature. As a result, ALK5 inhibition promoted accumulation of anticancer nanocarriers in the tumors. In the absence of ALK5 inhibitor, the anticancer nanocarriers exhibited poor growth-inhibitory effects (Kano et al. 2007).

The vasculature of solid tumors is abnormal, both in terms of vessel architecture and blood flow dynamics. Permeable heterogeneous vessel walls permit the leakage of proteins and fluid that, coupled with inefficient lymphatic drainage, hinder drug delivery. As acute blockade of TGF-β signaling transiently alters vessel stability, permeabilization of tumor vascular beds improves intravenous delivery of high molecular weight compounds to the tumor such as antibodies or nanoparticles that are therapeutically selective, collectively rendering an improved clinical outcome. This effect has particular relevance when treating stroma-rich intractable solid tumors. Furthermore, by promotion of a more pervious tumor vessel phenotype, not only the delivery of standard therapeutic agents can be improved but also diagnostic molecular imaging agents can be more easily monitored in tumor tissue where TGF-β signaling has been transiently impaired.

Most studies on the role of TGF-β in cancer primarily rely on the effect of this cytokine on the cancerous epithelial cells and on how it affects cellular plasticity, more specifically epithelial-to-mesenchymal transition (EMT) and metastatic dissemination. Here, we focus essentially on the effects of TGF-β signaling on the tumor vasculature; however some of the studies on metastasis may have affected TGF-β-mediated vasculature leniency to tumor cells intra- and extravasation. In fact, some laboratories have shown that the metastatic spread of the 4 T1 breast cancer transplantable model of metastatic breast cancer can be efficiently suppressed by administering an antibody that targets all three isoforms of the TGF-β ligand. This work went on to show that TGF-β neutralizing antibodies have multiple cooperative effects on angiogenesis, immune cell function, and tumor cell viability, eventually leading to effective tumor growth control and reduction of metastatic foci (Nam et al. 2008).

14.3 Perspective

The formation of new blood vessels in pathological processes, such as cancer, leads to unstable, proliferative suboptimally functional vessels in a dynamic state of remodeling. Tumor vascular networks support tumor growth, in spite of a range of abnormal features that compromise their physiology. Tumor vessels are commonly tortuous and leaky, causing hemorrhage and increased interstitial fluid pressure (Carmeliet and Jain 2000). Inefficient blood flow caused by poor hierarchical anatomy and organization of the tumor vasculature leads to ischemia and necrosis, which are common characteristics of rapidly growing tumors.

Its inherent instability should make tumor vascular beds distinct from normal vessels and therefore more prone to be exquisitely affected by selective targeting anti-angiogenic drugs (Baluk et al. 2005). During the past few years we have witnessed a suboptimal outcome from anti-angiogenic treatment of human cancer implying the need for improvement and suggesting that alternative therapeutic avenues should be explored. Better knowledge of molecular and cellular mechanisms taking place in the tumor microenvironment and in the tumor vasculature is essential to achieve such goals.

The complexity of the TGF-β and BMP pathways by means of redundancy, cooperation, or by simply having different levels of regulation reflects the uniqueness and tight tissue specificity of this intriguing pathway. However, despite being complex, the TGF-β signaling pathway may hold central roles in tumor angiogenesis, as suggested by studies targeting its endothelial-specific receptors ALK1 and endoglin. Inhibitors for these receptors have already been generated and they hold great promise in impairment of de novo formed tumor vessels, while leaving normal idle vasculature undisturbed. The preclinical development and clinical use of inhibitors of pro-angiogenic TGF-β signaling will be discussed in detail in Chap. 19.

References

Anderson L, Lowery JW, Frank DB et al (2010) Bmp2 and Bmp4 exert opposing effects in hypoxic pulmonary hypertension. Am J Physiol Regul Integr Comp Physiol 298:R833–R842

Arthur HM, Ure J, Smith AJ et al (2000) Endoglin, an ancillary TGFβ receptor, is required for extraembryonic angiogenesis and plays a key role in heart development. Dev Biol 217:42–53

Baluk P, Hashizume H, McDonald DM (2005) Cellular abnormalities of blood vessels as targets in cancer. Curr Opin Genet Dev 15:102–111

Barbara NP, Wrana JL, Letarte M (1999) Endoglin is an accessory protein that interacts with the signaling receptor complex of multiple members of the transforming growth factor-β superfamily. J Biol Chem 274:584–594

Beppu H, Ichinose F, Kawai N et al (2004) BMPR-II heterozygous mice have mild pulmonary hypertension and an impaired pulmonary vascular remodeling response to prolonged hypoxia. Am J Physiol Lung Cell Mol Physiol 287:L1241–L1247

Berg JN, Gallione CJ, Stenzel TT et al (1997) The activin receptor-like kinase 1 gene: genomic structure and mutations in hereditary hemorrhagic telangiectasia type 2. Am J Hum Genet 61:60–67

Bernabeu C, Lopez-Novoa JM, Quintanilla M (2009) The emerging role of TGF-β superfamily coreceptors in cancer. Biochim Biophys Acta 1792:954–973

Bertolino P, Deckers M, Lebrin F, ten Dijke P (2005) Transforming growth factor-β signal transduction in angiogenesis and vascular disorders. Chest 128:585S–590S

Bidart M, Ricard N, Levet S et al (2012) BMP9 is produced by hepatocytes and circulates mainly in an active mature form complexed to its prodomain. Cell Mol Life Sci 69:313–324

Blanco FJ, Santibanez JF, Guerrero-Esteo M et al (2005) Interaction and functional interplay between endoglin and ALK-1, two components of the endothelial transforming growth factor-β receptor complex. J Cell Physiol 204:574–584

Bostrom KI, Jumabay M, Matveyenko A et al (2011) Activation of vascular bone morphogenetic protein signaling in diabetes mellitus. Circ Res 108:446–457

Bourdeau A, Faughnan ME, Letarte M (2000) Endoglin-deficient mice, a unique model to study hereditary hemorrhagic telangiectasia. Trends Cardiovasc Med 10:279–285

Buijs JT, Rentsch CA, van der Horst G et al (2007) BMP7, a putative regulator of epithelial homeostasis in the human prostate, is a potent inhibitor of prostate cancer bone metastasis in vivo. Am J Pathol 171:1047–1057

Canalis E, Economides AN, Gazzerro E (2003) Bone morphogenetic proteins, their antagonists, and the skeleton. Endocr Rev 24:218–235

Carmeliet P, Jain RK (2000) Angiogenesis in cancer and other diseases. Nature 407:249–257

Carta L, Pereira L, Arteaga-Solis E et al (2006) Fibrillins 1 and 2 perform partially overlapping functions during aortic development. J Biol Chem 281:8016–8023

Chang H, Huylebroeck D, Verschueren K et al (1999) Smad5 knockout mice die at mid-gestation due to multiple embryonic and extraembryonic defects. Development 126:1631–1642

Cheifetz S, Bellon T, Cales C et al (1992) Endoglin is a component of the transforming growth factor-β receptor system in human endothelial cells. J Biol Chem 267:19027–19030

Chen C, Grzegorzewski KJ, Barash S et al (2003) An integrated functional genomics screening program reveals a role for BMP-9 in glucose homeostasis. Nat Biotechnol 21:294–301
Chen H, Shi S, Acosta L et al (2004) BMP10 is essential for maintaining cardiac growth during murine cardiogenesis. Development 131:2219–2231
Compton LA, Potash DA, Brown CB, Barnett JV (2007) Coronary vessel development is dependent on the type III transforming growth factor β receptor. Circ Res 101:784–791
Crawford SE, Stellmach V, Murphy-Ullrich JE et al (1998) Thrombospondin-1 is a major activator of TGF-β1 in vivo. Cell 93:1159–1170
Cunha SI, Pietras K (2011) ALK1 as an emerging target for antiangiogenic therapy of cancer. Blood 117:6999–7006
Cunha SI, Pardali E, Thorikay M et al (2010) Genetic and pharmacological targeting of activin receptor-like kinase 1 impairs tumor growth and angiogenesis. J Exp Med 207:85–100
David L, Mallet C, Mazerbourg S et al (2007) Identification of BMP9 and BMP10 as functional activators of the orphan activin receptor-like kinase 1 (ALK1) in endothelial cells. Blood 109:1953–1961
David L, Mallet C, Keramidas M et al (2008) Bone morphogenetic protein-9 is a circulating vascular quiescence factor. Circ Res 102:914–922
Derynck R, Akhurst RJ, Balmain A (2001) TGF-β signaling in tumor suppression and cancer progression. Nat Genet 29:117–129
Dewulf N, Verschueren K, Lonnoy O et al (1995) Distinct spatial and temporal expression patterns of two type I receptors for bone morphogenetic proteins during mouse embryogenesis. Endocrinology 136:2652–2663
Dickson MC, Martin JS, Cousins FM et al (1995) Defective haematopoiesis and vasculogenesis in transforming growth factor-β 1 knock out mice. Development 121:1845–1854
Docherty NG, Lopez-Novoa JM, Arevalo M et al (2006) Endoglin regulates renal ischaemia-reperfusion injury. Nephrol Dial Transplant 21:2106–2119
Egeblad M, Nakasone ES, Werb Z (2010) Tumors as organs: complex tissues that interface with the entire organism. Dev Cell 18:884–901
El-Bizri N, Guignabert C, Wang L et al (2008a) SM22α-targeted deletion of bone morphogenetic protein receptor 1A in mice impairs cardiac and vascular development, and influences organogenesis. Development 135:2981–2991
El-Bizri N, Wang L, Merklinger SL et al (2008b) Smooth muscle protein 22α-mediated patchy deletion of Bmpr1a impairs cardiac contractility but protects against pulmonary vascular remodeling. Circ Res 102:380–388
Elliott RL, Blobe GC (2005) Role of transforming growth factor β in human cancer. J Clin Oncol 23:2078–2093
Foletta VC, Lim MA, Soosairajah J et al (2003) Direct signaling by the BMP type II receptor via the cytoskeletal regulator LIMK1. J Cell Biol 162:1089–1098
Gallione CJ, Repetto GM, Legius E et al (2004) A combined syndrome of juvenile polyposis and hereditary haemorrhagic telangiectasia associated with mutations in MADH4 (SMAD4). Lancet 363:852–859
Gallione CJ, Richards JA, Letteboer TG et al (2006) SMAD4 mutations found in unselected HHT patients. J Med Genet 43:793–797
Galvin KM, Donovan MJ, Lynch CA et al (2000) A role for smad6 in development and homeostasis of the cardiovascular system. Nat Genet 24:171–174
Geirdal AO, Dheyauldeen S, Bachmann-Harildstad G, Heimdal K (2012) Quality of life in patients with hereditary hemorrhagic telangiectasia in Norway: A population based study. Am J Med Genet A 158A:1269–1278
George EL, Georges-Labouesse EN, Patel-King RS et al (1993) Defects in mesoderm, neural tube and vascular development in mouse embryos lacking fibronectin. Development 119:1079–1091
Goumans MJ, Mummery C (2000) Functional analysis of the TGFβ receptor/Smad pathway through gene ablation in mice. Int J Dev Biol 44:253–265

Goumans MJ, Valdimarsdottir G, Itoh S et al (2002) Balancing the activation state of the endothelium via two distinct TGF-β type I receptors. EMBO J 21:1743–1753
Goumans MJ, Valdimarsdottir G, Itoh S et al (2003) Activin receptor-like kinase (ALK)1 is an antagonistic mediator of lateral TGFβ/ALK5 signaling. Mol Cell 12:817–828
Gronroos E, Kingston IJ, Ramachandran A et al (2012) TGF-β inhibits BMP-induced transcription through novel phosphorylated Smad1/5-Smad3 complexes. Mol Cell Biol 32:2904–2916
Groppe J, Hinck CS, Samavarchi-Tehrani P et al (2008) Cooperative assembly of TGF-β superfamily signaling complexes is mediated by two disparate mechanisms and distinct modes of receptor binding. Mol Cell 29:157–168
Hagedorn HG, Bachmeier BE, Nerlich AG (2001) Synthesis and degradation of basement membranes and extracellular matrix and their regulation by TGF-β in invasive carcinomas (Review). Int J Oncol 18:669–681
Hanahan D, Weinberg RA (2000) The hallmarks of cancer. Cell 100:57–70
Hanahan D, Weinberg RA (2011) Hallmarks of cancer: the next generation. Cell 144:646–674
Heinke J, Wehofsits L, Zhou Q et al (2008) BMPER is an endothelial cell regulator and controls bone morphogenetic protein-4-dependent angiogenesis. Circ Res 103:804–812
Herrera B, van Dinther M, Ten Dijke P, Inman GJ (2009) Autocrine bone morphogenetic protein-9 signals through activin receptor-like kinase-2/Smad1/Smad4 to promote ovarian cancer cell proliferation. Cancer Res 69:9254–9262
Hong KH, Seki T, Oh SP (2007) Activin receptor-like kinase 1 is essential for placental vascular development in mice. Lab Invest 87:670–679
Huang Z, Wang D, Ihida-Stansbury K et al (2009) Defective pulmonary vascular remodeling in Smad8 mutant mice. Hum Mol Genet 18:2791–2801
Hu-Lowe DD, Chen E, Zhang L et al (2011) Targeting activin receptor-like kinase 1 inhibits angiogenesis and tumorigenesis through a mechanism of action complementary to anti-VEGF therapies. Cancer Res 71:1362–1373
Itoh F, Itoh S, Carvalho RL et al (2009) Poor vessel formation in embryos from knock-in mice expressing ALK5 with L45 loop mutation defective in Smad activation. Lab Invest 89:800–810
Itoh F, Itoh S, Adachi T et al (2012) Smad2/Smad3 in endothelium is indispensable for vascular stability via S1PR1 and N-cadherin expressions. Blood 119:5320–5328
Jones CM, Lyons KM, Hogan BL (1991) Involvement of Bone Morphogenetic Protein-4 (BMP-4) and Vgr-1 in morphogenesis and neurogenesis in the mouse. Development 111:531–542
Jonker L, Arthur HM (2002) Endoglin expression in early development is associated with vasculogenesis and angiogenesis. Mech Dev 110:193–196
Judge DP, Biery NJ, Keene DR et al (2004) Evidence for a critical contribution of haploinsufficiency in the complex pathogenesis of Marfan syndrome. J Clin Invest 114:172–181
Kano MR, Bae Y, Iwata C et al (2007) Improvement of cancer-targeting therapy, using nanocarriers for intractable solid tumors by inhibition of TGF-β signaling. Proc Natl Acad Sci USA 104:3460–3465
Kano MR, Komuta Y, Iwata C et al (2009) Comparison of the effects of the kinase inhibitors imatinib, sorafenib, and transforming growth factor-β receptor inhibitor on extravasation of nanoparticles from neovasculature. Cancer Sci 100:173–180
Kishigami S, Mishina Y (2005) BMP signaling and early embryonic patterning. Cytokine Growth Factor Rev 16:265–278
Lammerts E, Roswall P, Sundberg C et al (2002) Interference with TGF-β1 and -β3 in tumor stroma lowers tumor interstitial fluid pressure independently of growth in experimental carcinoma. Int J Cancer 102:453–462
Lamouille S, Mallet C, Feige JJ, Bailly S (2002) Activin receptor-like kinase 1 is implicated in the maturation phase of angiogenesis. Blood 100:4495–4501
Lan Y, Liu B, Yao H et al (2007) Essential role of endothelial Smad4 in vascular remodeling and integrity. Mol Cell Biol 27:7683–7692
Langenfeld EM, Langenfeld J (2004) Bone morphogenetic protein-2 stimulates angiogenesis in developing tumors. Mol Cancer Res 2:141–149

Larrivee B, Prahst C, Gordon E et al (2012) ALK1 signaling inhibits angiogenesis by cooperating with the Notch pathway. Dev Cell 22:489–500

Larsson J, Goumans MJ, Sjostrand LJ et al (2001) Abnormal angiogenesis but intact hematopoietic potential in TGF-β type I receptor-deficient mice. EMBO J 20:1663–1673

Lawler J, Sunday M, Thibert V et al (1998) Thrombospondin-1 is required for normal murine pulmonary homeostasis and its absence causes pneumonia. J Clin Invest 101:982–992

Lebrin F, Goumans MJ, Jonker L et al (2004) Endoglin promotes endothelial cell proliferation and TGF-β/ALK1 signal transduction. EMBO J 23:4018–4028

Lechleider RJ, Ryan JL, Garrett L et al (2001) Targeted mutagenesis of Smad1 reveals an essential role in chorioallantoic fusion. Dev Biol 240:157–167

Lee-Hoeflich ST, Causing CG, Podkowa M et al (2004) Activation of LIMK1 by binding to the BMP receptor, BMPRII, regulates BMP-dependent dendritogenesis. EMBO J 23:4792–4801

Letteboer TG, Mager JJ, Snijder RJ et al (2006) Genotype-phenotype relationship in hereditary haemorrhagic telangiectasia. J Med Genet 43:371–377

Li DY, Sorensen LK, Brooke BS et al (1999) Defective angiogenesis in mice lacking endoglin. Science 284:1534–1537

Li JZ, Hankins GR, Kao C et al (2003) Osteogenesis in rats induced by a novel recombinant helper-dependent bone morphogenetic protein-9 (BMP-9) adenovirus. J Gene Med 5:748–756

Lin SJ, Lerch TF, Cook RW et al (2006) The structural basis of TGF-β, bone morphogenetic protein, and activin ligand binding. Reproduction 132:179–190

Lopez-Coviella I, Berse B, Krauss R et al (2000) Induction and maintenance of the neuronal cholinergic phenotype in the central nervous system by BMP-9. Science 289:313–316

Lord E, Bergeron E, Senta H et al (2010) Effect of BMP-9 and its derived peptide on the differentiation of human white preadipocytes. Growth Factors 28:149–156

Lowery JW, de Caestecker MP (2010) BMP signaling in vascular development and disease. Cytokine Growth Factor Rev 21:287–298

Lyons KM, Pelton RW, Hogan BL (1989) Patterns of expression of murine Vgr-1 and BMP-2a RNA suggest that transforming growth factor-β-like genes coordinately regulate aspects of embryonic development. Genes Dev 3:1657–1668

Mahmoud M, Allinson KR, Zhai Z et al (2010) Pathogenesis of arteriovenous malformations in the absence of endoglin. Circ Res 106:1425–1433

Massague J (2000) How cells read TGF-β signals. Nat Rev Mol Cell Biol 1:169–178

Massague J (2008) TGFβ in Cancer. Cell 134:215–230

Mazerbourg S, Sangkuhl K, Luo CW et al (2005) Identification of receptors and signaling pathways for orphan bone morphogenetic protein/growth differentiation factor ligands based on genomic analyses. J Biol Chem 280:32122–32132

McAllister KA, Grogg KM, Johnson DW et al (1994) Endoglin, a TGF-β binding protein of endothelial cells, is the gene for hereditary haemorrhagic telangiectasia type 1. Nat Genet 8:345–351

Medici D, Shore EM, Lounev VY et al (2010) Conversion of vascular endothelial cells into multipotent stem-like cells. Nat Med 16:1400–1406

Miller AF, Harvey SA, Thies RS, Olson MS (2000) Bone morphogenetic protein-9. An autocrine/paracrine cytokine in the liver. J Biol Chem 275(24):17937–17945.

Mitchell D, Pobre EG, Mulivor AW et al (2010) ALK1-Fc inhibits multiple mediators of angiogenesis and suppresses tumor growth. Mol Cancer Ther 9:379–388

Miyazono K, Kamiya Y, Morikawa M (2010) Bone morphogenetic protein receptors and signal transduction. J Biochem 147:35–51

Morikawa M, Koinuma D, Tsutsumi S et al (2011) ChIP-seq reveals cell type-specific binding patterns of BMP-specific Smads and a novel binding motif. Nucleic Acids Res 39:8712–8727

Moustakas A, Heldin CH (2009) The regulation of TGFβ signal transduction. Development 136:3699–3714

Moya IM, Umans L, Maas E et al (2012) Stalk cell phenotype depends on integration of Notch and Smad1/5 signaling cascades. Dev Cell 22:501–514

Nam JS, Terabe M, Mamura M et al (2008) An anti-transforming growth factor β antibody suppresses metastasis via cooperative effects on multiple cell compartments. Cancer Res 68:3835–3843

Nguyen HL, Lee YJ, Shin J et al (2011) TGF-β signaling in endothelial cells, but not neuroepithelial cells, is essential for cerebral vascular development. Lab Invest 91:1554–1563

Niessen K, Zhang G, Ridgway JB et al (2010) ALK1 signaling regulates early postnatal lymphatic vessel development. Blood 115:1654–1661

Nohe A, Hassel S, Ehrlich M et al (2002) The mode of bone morphogenetic protein (BMP) receptor oligomerization determines different BMP-2 signaling pathways. J Biol Chem 277:5330–5338

Nohe A, Keating E, Knaus P, Petersen NO (2004) Signal transduction of bone morphogenetic protein receptors. Cell Signal 16:291–299

Oh SP, Seki T, Goss KA et al (2000) Activin receptor-like kinase 1 modulates transforming growth factor-β 1 signaling in the regulation of angiogenesis. Proc Natl Acad Sci USA 97:2626–2631

Oshima M, Oshima H, Taketo MM (1996) TGF-β receptor type II deficiency results in defects of yolk sac hematopoiesis and vasculogenesis. Dev Biol 179:297–302

Ota T, Fujii M, Sugizaki T et al (2002) Targets of transcriptional regulation by two distinct type I receptors for transforming growth factor-β in human umbilical vein endothelial cells. J Cell Physiol 193:299–318

Padua D, Massague J (2009) Roles of TGFβ in metastasis. Cell Res 19:89–102

Pardali K, Moustakas A (2007) Actions of TGF-β as tumor suppressor and pro-metastatic factor in human cancer. Biochim Biophys Acta 1775:21–62

Park C, Lavine K, Mishina Y et al (2006) Bone morphogenetic protein receptor 1A signaling is dispensable for hematopoietic development but essential for vessel and atrioventricular endocardial cushion formation. Development 133:3473–3484

Park SO, Lee YJ, Seki T et al (2008) ALK5- and TGFBR2-independent role of ALK1 in the pathogenesis of hereditary hemorrhagic telangiectasia type 2. Blood 111:633–642

Park SO, Wankhede M, Lee YJ et al (2009) Real-time imaging of de novo arteriovenous malformation in a mouse model of hereditary hemorrhagic telangiectasia. J Clin Invest 119:3487–3496

Pece-Barbara N, Vera S, Kathirkamathamby K et al (2005) Endoglin null endothelial cells proliferate faster and are more responsive to transforming growth factor β1 with higher affinity receptors and an activated Alk1 pathway. J Biol Chem 280:27800–27808

Petersen M, Pardali E, van der Horst G et al (2010) Smad2 and Smad3 have opposing roles in breast cancer bone metastasis by differentially affecting tumor angiogenesis. Oncogene 29:1351–1361

Pietras K, Östman A (2010) Hallmarks of cancer: interactions with the tumor stroma. Exp Cell Res 316:1324–1331

Ploemacher RE, Engels LJ, Mayer AE et al (1999) Bone morphogenetic protein 9 is a potent synergistic factor for murine hemopoietic progenitor cell generation and colony formation in serum-free cultures. Leukemia 13:428–437

Ramoshebi LN, Ripamonti U (2000) Osteogenic protein-1, a bone morphogenetic protein, induces angiogenesis in the chick chorioallantoic membrane and synergizes with basic fibroblast growth factor and transforming growth factor-β1. Anat Rec 259:97–107

Ray BN, Lee NY, How T, Blobe GC (2010) ALK5 phosphorylation of the endoglin cytoplasmic domain regulates Smad1/5/8 signaling and endothelial cell migration. Carcinogenesis 31:435–441

Ricard N, Ciais D, Levet S et al (2012) BMP9 and BMP10 are critical for postnatal retinal vascular remodeling. Blood 119(25):6162–6171

Rudarakanchana N, Flanagan JA, Chen H et al (2002) Functional analysis of bone morphogenetic protein type II receptor mutations underlying primary pulmonary hypertension. Hum Mol Genet 11:1517–1525

Sadick H, Sadick M, Gotte K et al (2006) Hereditary hemorrhagic telangiectasia: an update on clinical manifestations and diagnostic measures. Wien Klin Wochenschr 118:72–80

Safina A, Vandette E, Bakin AV (2007) ALK5 promotes tumor angiogenesis by upregulating matrix metalloproteinase-9 in tumor cells. Oncogene 26:2407–2422
Sanchez-Elsner T, Botella LM, Velasco B et al (2001) Synergistic cooperation between hypoxia and transforming growth factor-β pathways on human vascular endothelial growth factor gene expression. J Biol Chem 276:38527–38535
Sanford LP, Ormsby I, Gittenberger-de Groot AC et al (1997) TGFβ2 knockout mice have multiple developmental defects that are non-overlapping with other TGFβ knockout phenotypes. Development 124:2659–2670
Scharpfenecker M, van Dinther M, Liu Z et al (2007) BMP-9 signals via ALK1 and inhibits bFGF-induced endothelial cell proliferation and VEGF-stimulated angiogenesis. J Cell Sci 120:964–972
Seemann P, Brehm A, Konig J et al (2009) Mutations in GDF5 reveal a key residue mediating BMP inhibition by NOGGIN. PLoS Genet 5:e1000747
Seki T, Hong KH, Oh SP (2006) Nonoverlapping expression patterns of ALK1 and ALK5 reveal distinct roles of each receptor in vascular development. Lab Invest 86:116–129
Shao ES, Lin L, Yao Y, Bostrom KI (2009) Expression of vascular endothelial growth factor is coordinately regulated by the activin-like kinase receptors 1 and 5 in endothelial cells. Blood 114:2197–2206
Shore EM, Xu M, Feldman GJ et al (2006) A recurrent mutation in the BMP type I receptor ACVR1 causes inherited and sporadic fibrodysplasia ossificans progressiva. Nat Genet 38:525–527
Shull MM, Ormsby I, Kier AB et al (1992) Targeted disruption of the mouse transforming growth factor-β 1 gene results in multifocal inflammatory disease. Nature 359:693–699
Sieber C, Kopf J, Hiepen C, Knaus P (2009) Recent advances in BMP receptor signaling. Cytokine Growth Factor Rev 20:343–355
Song Y, Jones JE, Beppu H et al (2005) Increased susceptibility to pulmonary hypertension in heterozygous BMPR2-mutant mice. Circulation 112:553–562
Sounni NE, Dehne K, van Kempen L et al (2010) Stromal regulation of vessel stability by MMP14 and TGFβ. Dis Model Mech 3:317–332
Srinivasan S, Hanes MA, Dickens T et al (2003) A mouse model for hereditary hemorrhagic telangiectasia (HHT) type 2. Hum Mol Genet 12:473–482
Stenvers KL, Tursky ML, Harder KW et al (2003) Heart and liver defects and reduced transforming growth factor β2 sensitivity in transforming growth factor β type III receptor-deficient embryos. Mol Cell Biol 23:4371–4385
Stottmann RW, Choi M, Mishina Y et al (2004) BMP receptor IA is required in mammalian neural crest cells for development of the cardiac outflow tract and ventricular myocardium. Development 131:2205–2218
Suzuki Y, Ohga N, Morishita Y et al (2010) BMP-9 induces proliferation of multiple types of endothelial cells in vitro and in vivo. J Cell Sci 123:1684–1692
ten Dijke P, Arthur HM (2007) Extracellular control of TGFβ signalling in vascular development and disease. Nat Rev Mol Cell Biol 8:857–869
ten Dijke P, Miyazono K, Heldin CH (2000) Signaling inputs converge on nuclear effectors in TGF-β signaling. Trends Biochem Sci 25:64–70
Thomas PS, Sridurongrit S, Ruiz-Lozano P, Kaartinen V (2012) Deficient signaling via Alk2 (Acvr1) leads to bicuspid aortic valve development. PLoS One 7:e35539
Todorovic V, Frendewey D, Gutstein DE et al (2007) Long form of latent TGF-β binding protein 1 (Ltbp1L) is essential for cardiac outflow tract septation and remodeling. Development 134:3723–3732
Torsney E, Charlton R, Parums D et al (2002) Inducible expression of human endoglin during inflammation and wound healing in vivo. Inflamm Res 51:464–470
Torsney E, Charlton R, Diamond AG et al (2003) Mouse model for hereditary hemorrhagic telangiectasia has a generalized vascular abnormality. Circulation 107:1653–1657
Tuxhorn JA, McAlhany SJ, Yang F et al (2002) Inhibition of transforming growth factor-β activity decreases angiogenesis in a human prostate cancer-reactive stroma xenograft model. Cancer Res 62:6021–6025

Ueki N, Nakazato M, Ohkawa T et al (1992) Excessive production of transforming growth-factor β 1 can play an important role in the development of tumorigenesis by its action for angiogenesis: validity of neutralizing antibodies to block tumor growth. Biochim Biophys Acta 1137:189–196
Upton PD, Davies RJ, Trembath RC, Morrell NW (2009) Bone morphogenetic protein (BMP) and activin type II receptors balance BMP9 signals mediated by activin receptor-like kinase-1 in human pulmonary artery endothelial cells. J Biol Chem 284:15794–15804
Urness LD, Sorensen LK, Li DY (2000) Arteriovenous malformations in mice lacking activin receptor-like kinase-1. Nat Genet 26:328–331
Valdimarsdottir G, Goumans MJ, Rosendahl A et al (2002) Stimulation of Id1 expression by bone morphogenetic protein is sufficient and necessary for bone morphogenetic protein-induced activation of endothelial cells. Circulation 106:2263–2270
Wagner DO, Sieber C, Bhushan R et al (2010) BMPs: from bone to body morphogenetic proteins. Sci Signal 3:mr1
Wakefield LM, Roberts AB (2002) TGF-β signaling: positive and negative effects on tumorigenesis. Curr Opin Genet Dev 12:22–29
Wang J, Sridurongrit S, Dudas M et al (2005) Atrioventricular cushion transformation is mediated by ALK2 in the developing mouse heart. Dev Biol 286:299–310
Wozney JM, Rosen V, Celeste AJ et al (1988) Novel regulators of bone formation: molecular clones and activities. Science 242:1528–1534
Wrana JL, Attisano L, Wieser R et al (1994) Mechanism of activation of the TGF-β receptor. Nature 370:341–347
Yamagishi T, Nakajima Y, Miyazono K, Nakamura H (1999) Bone morphogenetic protein-2 acts synergistically with transforming growth factor-β3 during endothelial-mesenchymal transformation in the developing chick heart. J Cell Physiol 180:35–45
Yang X, Castilla LH, Xu X et al (1999) Angiogenesis defects and mesenchymal apoptosis in mice lacking SMAD5. Development 126:1571–1580
Yao Y, Zebboudj AF, Shao E et al (2006) Regulation of bone morphogenetic protein-4 by matrix GLA protein in vascular endothelial cells involves activin-like kinase receptor 1. J Biol Chem 281:33921–33930
Yao Y, Shao ES, Jumabay M et al (2008) High-density lipoproteins affect endothelial BMP-signaling by modulating expression of the activin-like kinase receptor 1 and 2. Arterioscler Thromb Vasc Biol 28:2266–2274
Yao Y, Jumabay M, Wang A, Bostrom KI (2011) Matrix Gla protein deficiency causes arteriovenous malformations in mice. J Clin Invest 121:2993–3004
Yao Y, Jumabay M, Ly A et al (2012) Crossveinless 2 regulates bone morphogenetic protein 9 in human and mouse vascular endothelium. Blood 119:5037–5047
Zacchigna L, Vecchione C, Notte A et al (2006) Emilin1 links TGF-β maturation to blood pressure homeostasis. Cell 124:929–942
Zeisberg EM, Potenta S, Xie L et al (2007) Discovery of endothelial to mesenchymal transition as a source for carcinoma-associated fibroblasts. Cancer Res 67:10123–10128
Zhang H, Bradley A (1996) Mice deficient for BMP2 are nonviable and have defects in amnion/chorion and cardiac development. Development 122:2977–2986
Zhao GQ (2003) Consequences of knocking out BMP signaling in the mouse. Genesis 35:43–56

Chapter 15
TGF-β Signaling Pathway and MicroRNAs in Cardiovascular Disease

Nisha Marathe and Akiko Hata

Abstract Transforming growth factor-β (TGF-β) signaling has long been known to influence development and function of different organs, including the cardiovascular system. In addition to the canonical pathways of Smad-dependent transcriptional regulation of protein-coding genes, recent studies have shown TGF-β influence on microRNA (miRNA) expression to play a critical role in TGF-β-dependent biological effects. A number of miRNAs, such as miR-143, miR-145, miR-21, miR-155, and miR-26a, are identified as critical downstream mediators of the TGF-β-family signaling. Deregulation of miRNA expression can result in various developmental defects and diseases. In this review, we summarize the latest view of the effects of miRNAs associated with TGF-β-family signaling on the cardiovascular system.

Keywords Bone morphogenetic protein • MicroRNA • Precursor microRNA • Smad-primary microRNA • Transforming growth factor-β • Vascular smooth muscle cell

Abbreviations

BMP	Bone morphogenetic protein
miRNA	MicroRNA
pre-miRNA	Precursor microRNA
pri-miRNA	Primary microRNA
TGF-β	Transforming growth factor-β
VSMC	Vascular smooth muscle cell

N. Marathe • A. Hata (✉)
Cardiovascular Research Institute, University of California, San Francisco, CA 94158, USA
e-mail: akiko.hata@ucsf.edu

A. Moustakas and K. Miyazawa (eds.), *TGF-β in Human Disease*,
DOI 10.1007/978-4-431-54409-8_15, © Springer Japan 2013

15.1 Introduction

The TGF-β superfamily of growth factors, including TGF-βs and bone morphogenetic proteins (BMPs), is involved in multiple cellular processes such as cell differentiation, proliferation, and apoptosis (Bobik 2006). Aberrant gene expression in vascular smooth muscle cells (VSMC), endothelial cells, fibroblasts, and cardiac myocytes—key cellular components of the cardiovascular system—contributes to heart attack, stroke, peripheral vascular disease, hypertension as well as congenital heart defects (Toma and McCaffrey 2012). Both TGF-β and BMP signaling are required for healthy cardiac development and function (Bobik 2006). TGF-β acts upon ligand binding to the TGF-β receptor type-II (TGF-βRII) which then recruits and phosphorylates the TGF-β receptor type-I (TGF-βRI) to form a type I/type II receptor complex. Activation of the BMP receptor heterodimer is accomplished in a similar manner (Shi and Massague 2003). The receptor complex then phosphorylates a receptor-specific signal transducer known as R-Smad—the TGF-β receptor complex transmits a signal through Smad 2 and 3, while Smad 1, 5, and 8 are responsible for the BMP signaling. Upon phosphorylation by the type I receptor kinase, R-Smads form a complex with the common-Smad (co-Smad), Smad4, and translocate into the nucleus where the R-Smad/co-Smad complex binds to the DNA sequence known as the Smad binding element (SBE) together with cofactors which associate with the sequence adjacent to the SBE, and modulates transcription positively or negatively (Butz et al. 2012) (Fig. 15.1). More recently, R-Smads are shown to regulate miRNA biogenesis at the processing step in the nucleus and modulate their expression (Davis et al. 2008) (Fig. 15.3).

MiRNAs are small (20–23 nucleotides, nt) noncoding RNAs involved in post-translational regulation of gene expression (Blahna and Hata 2012). MiRNAs typically bind to the partially complementary sequence in the 3′-untranslated region (3′-UTR) of messenger RNAs (mRNAs) and negatively regulate their expression through either translational inhibition or promotion of mRNA degradation (Filipowicz et al. 2008). Genome-wide analyses of 3′-UTRs suggest more than 60 % of mRNAs can be targeted by miRNAs (Friedman et al. 2009). Thus, miRNAs are important modulators of protein expression. Of the 20–23 nt of the mature miRNA sequence, a 6–8 nt sequence at the 5′ end, known as the "seed sequence," is usually sufficient to confer target specificity, although under certain conditions other nucleotides of the mature miRNA sequence are also important (Grimson et al. 2007). Regions in the 3′-UTR complementary to the miRNA seed region are known as "miRNA recognition element (MRE)" and result in downregulation of the gene in the presence of corresponding miRNAs. Due to the relatively short miRNA recognition sequence, a large number of genes can be targeted by a single miRNA; conversely, a single mRNA can be targeted by multiple miRNAs.

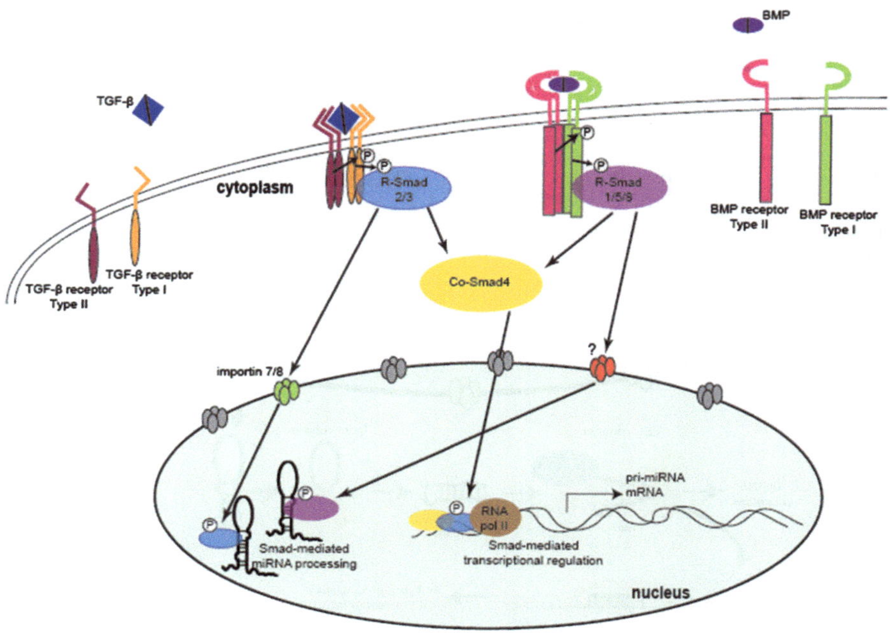

Fig. 15.1 Schematic representation of TGF-β/ BMP signaling. TGF-β/ BMP bind to the Type II receptor which recruits and transphosphorylates the Type I receptor. Phosphorylation of R-Smads by the activated Type I receptor kinase induces association with Co-Smad4 and translocation into the nucleus to regulate transcription. Alternately, R-Smads can regulate miRNA biogenesis in the nucleus. Smad2/3 are imported to the nucleus via importins 7 and 8. The mechanism of nuclear import of R-Smad1/5/8 is yet to be characterized

15.1.1 MiRNA Biogenesis

MiRNAs are synthesized as a long primary transcript (pri-miRNA) by RNA polymerase II. Pri-miRNAs contain a 5′-7-methyl-guanosine cap and 3′-polyadenylation similar to mRNAs (Fig. 15.2). A majority of miRNA genes are encoded in intergenic regions in the DNA and are transcribed independently; however, 30–50 % of miRNAs lie within the introns and are transcribed and regulated in conjunction with the host gene (Rodriguez et al. 2004). The pri-miRNA is subsequently processed in the nucleus by the Drosha complex whose core components include an RNase III enzyme, Drosha, and a cofactor, DiGeorge critical region 8 (DGCR8) (Lee et al. 2003; Gregory et al. 2004) (Fig. 15.2). Other proteins associated with the Drosha complex include RNA helicases, such as p68 and p72 (Fukuda et al. 2007). Drosha cleaves the pri-miRNA into a shorter precursor-miRNA (pre-miRNA). DGCR8 is a double-stranded RNA binding protein and stabilizes the interaction between Drosha and the pri-miRNA (Yeom et al. 2006). p68 (also known as DDX5) and p72

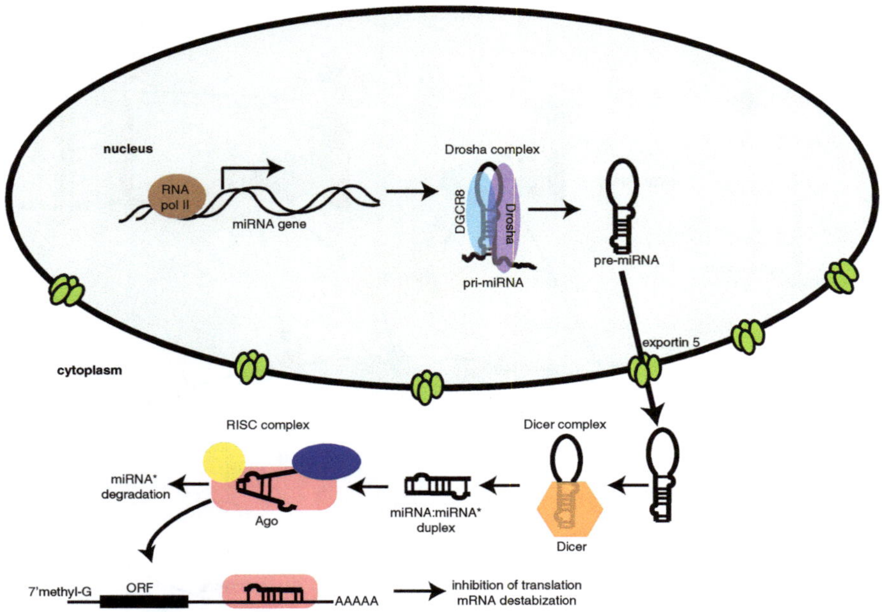

Fig. 15.2 MiRNA biogenesis pathway. MiRNA is transcribed into a long primary transcript known as pri-miRNA by RNA pol II. Pri-miRNA is then cleaved by the Drosha complex, which is composed of Drosha, DGCR8, and other cofactors to generate precursor miRNA (pre-miRNA). Pre-miRNA is then exported from the nucleus through exportin 5. In the cytoplasm, the pre-miRNA is cleaved by the Dicer complex to generate a mature miRNA duplex. The miRNA duplex is then loaded onto the RISC in which one of the two strands (known as the guide strand) is selected for target gene silencing. The other strand (known as the passenger strand) is rapidly degraded

(also known as DDX17) are both DEAD-box RNA helicases and the helicase activity is essential for miRNA processing (Salzman et al. 2007). p68 and p72 each is known to control processing of distinct, but overlapping, subsets of miRNAs (Fukuda et al. 2007).

The pre-miRNA is composed of a stem-loop structure with the mature miRNA sequence encoded in the stem region (Davis and Hata 2009) (Fig. 15.3 shown in red line). Following cleavage by the Drosha complex, the pre-miRNA is exported to the cytoplasm through exportin-5 and further processed into the mature miRNA by the cytoplasmic RNase III enzyme called Dicer (Fig. 15.2). Dicer recognizes the stem-loop structure of the pre-miRNA and cleaves the pre-miRNA into a ~22 bp RNA duplex. Argonaut proteins (Ago1-4 in human) associate with the miRNA duplex, separate the two strands, and choose one of the strands, which is called the "guide strand," and present it to the RNA-induced silencing complex (RISC), which then targets and silences the mRNA (Davis and Hata 2009). The strand that is not selected by Argonaut is known as the "passenger strand" and is rapidly degraded. In some cases, both the guide and passenger strands actively participate in mRNA silencing (Fig. 15.2).

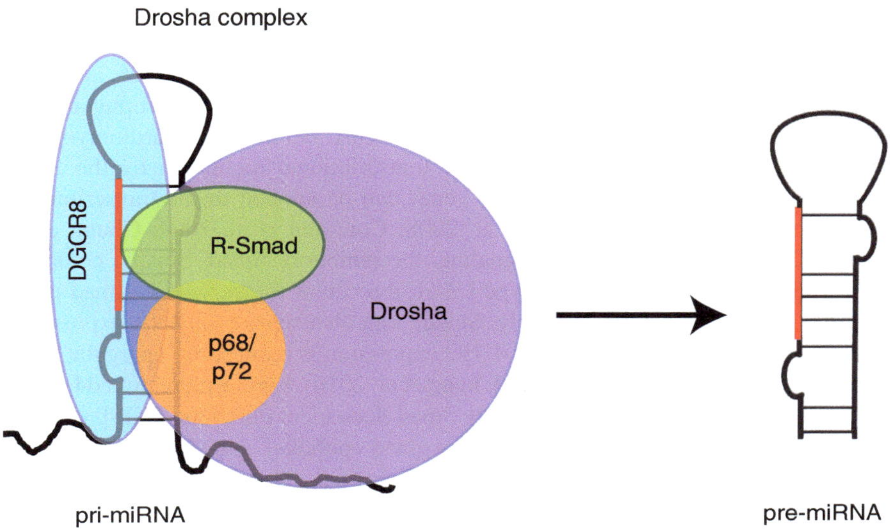

Fig. 15.3 MiRNA processing by Smads. Processing of a subset of pri-miRNAs is mediated by the Drosha complex in a p68 and R-Smad-dependent manner. R-Smads facilitate the recruitment of Drosha and promote the cleavage of pri-miRNAs by Drosha. The red line in the stem region of the pri-miRNA indicates a sequence encoding mature miRNA

In contrast with transcriptional regulation of genes, which can mediate dramatic changes, miRNAs often effect subtle changes in gene expression. MiRNA-mediated gene silencing typically reduces protein and/or mRNA expression by ~30–50 % (Selbach et al. 2008). Despite the relatively small effect, miRNAs are critical for development and the maintenance of homeostasis (Selbach et al. 2008; Baek et al. 2008; Hata and Davis 2010). Indeed, mice deficient in DGCR8, p68, p72, Dicer, or Ago2 die early in gestation due to gross developmental defects (Wang et al. 2007; Fukuda et al. 2007; Bernstein et al. 2003; Morita et al. 2007). MiRNA expression can be controlled at the transcriptional level, as well as at either cropping step by Drosha or Dicer (Bauersachs and Thum 2011; Davis et al. 2008; Adams et al. 2009). This allows for fine-tuning of miRNA expression and, subsequently, silencing of the miRNA target genes. In this review, we summarize the functional relationship between the TGF-β signaling pathway and miRNAs.

15.1.2 Transcriptional Regulation of miRNAs by TGF-β

TGF-β is known to regulate transcription via the R-Smad/co-Smad complex (Shi and Massague 2003) (Fig. 15.1). TGF-β2 controls endothelial-to-mesenchymal transition (EndMT) of mouse endothelial cells isolated from heart microvasculature.

In response to TGF-β2 treatment, cells become more fibroblast-like and miRNAs associated with fibroblasts are induced (Ghosh et al. 2012). Interestingly, deregulation of these miRNAs, including miR-125b, miR-21, miR-30b, miR-195, and miR-375, is often associated with cardiovascular diseases, such as hypertrophy, cardiac fibrosis, arrhythmia, myocardial infarction, heart failure, and cardiomyopathy (Ghosh et al. 2012). The same rules of transcriptional regulation by the Smad complex that apply to protein coding genes can be applied to the transcriptional regulation of miRNAs (Ozsolak et al. 2008; Corcoran et al. 2009; Chung et al. 2010). During canonical TGF-β signaling, the inhibitory Smad, Smad7, competes with R-Smads for binding at the type I TGF-β receptor to decrease R-Smad phosphorylation and signaling; similarly, Smad7 can also repress miRNA expression. For example, Smad3-dependent miR-192 expression is reduced by upregulation of Smad7 in rat tubular epithelial cells (Chung et al. 2010). Furthermore, Smad4 seems to be a key regulator of transcriptionally Smad-dependent miRNAs. TGF-β stimulation of Smad4-deficient murine mammary gland epithelial cells resulted in deregulation of 28 miRNAs (Kong et al. 2008). Since Smad4 is required upon ligand stimulation for the nuclear translocation of R-Smads and binding to the SBE and is involved in both TGF-β and BMP signaling, it is not surprising that Smad4 is critical for transcriptional control of miRNAs by TGF-β and BMP signaling.

15.1.3 Posttranscriptional Regulation of miRNAs by TGF-β

In addition to transcriptional regulation, miRNAs are also subject to regulation during maturation—namely, at the cleavage of pri-miRNAs to generate pre-miRNAs by the Drosha complex in the nucleus (Fig. 15.3). In VSMC, TGF-β, and BMP have a critical role in regulating VSMC phenotype by modulating miRNA expression. In response to TGF-β and BMP signaling, R-Smads bind to the stem region in the pri-miR-21 encoding mature miR-21 and facilitate its processing to pre-miR-21 (Davis et al. 2008). Upon BMP or TGF-β stimulation, no change of pri-miR-21 levels is observed; however, an increase in pre-miR-21 levels can be detected, suggesting a posttranscriptional regulation of miR-21 by Smads (Davis et al. 2008). RNA-immunoprecipitation assays place R-Smads in a complex with p68 and Drosha upon TGF-β or BMP stimulation (Davis et al. 2010). Interestingly, TGF-β or BMP stimulation also results in increased Drosha recruitment to pri-miR-21, suggesting that R-Smads may facilitate the recruitment and association of Drosha with specific pri-miRNAs, which in turn leads to more efficient processing of the pri-miR-21 (Davis et al. 2010). This mechanism is proposed to extend to ~20 miRNAs in VSMC (Davis et al. 2010).

Unlike transcriptional regulation, Smad4 does not seem to be required for TGF-β-mediated regulation of miRNA processing, as downregulation of Smad4 by small inhibitory RNA (siRNA) in VSMC does not affect the processing of pri-miR-21 (Davis et al. 2008). Smad4 was previously shown to be required for the nuclear translocation of R-Smads; however, recent studies show that R-Smads and Smad4 can be translocated to the nucleus via an alternate nuclear import machinery known as importin (Imp) 7/8 (Xu et al. 2007; Yao et al. 2008). It can be speculated that the R-Smads that translocate to the nucleus independently from Smad4 are responsible for miRNA processing while those translocating to the nucleus as a complex with Smad4 are involved in transcriptional regulation. The *Arabidopsis* homologue of Smad nuclear interacting protein 1 (SNIP1), DAWDLE (DDL), plays a role in the expression of both pri-miRNAs and pre-miRNAs through transcriptional control. It is proposed that DDL enables access or recognition of pri-miRNAs by Dicer-like 1 (DCL1), the *Arabidopsis* homologue of Dicer (Yu et al. 2008). Since SNIP1 was originally identified as a Smad-interacting protein, it is plausible that SNIP1 promotes miRNA biogenesis through association with Smads.

Interestingly, the mechanism of regulation of miRNA processing in the nucleus is not limited to R-Smads. Other transcription factors, such as estrogen receptor-α (ER-α) and p53, have been reported to regulate the catalytic activity of Drosha (Suzuki et al. 2009; Yamagata et al. 2009). Upon activation of ER-α or p53, these proteins interact with the Drosha complex through association with p68 and modulate processing of a distinct set of miRNAs (Yamagata et al. 2009; Suzuki et al. 2009).

15.1.4 Regulation of TGF-β Signaling by miRNAs

Smads themselves are subject to regulation by miRNAs. The transcripts encoding Smad proteins have been identified as targets of miRNA(s) (Blahna and Hata 2012). This introduces an additional level of complexity to the regulation of miRNAs by enabling positive- or negative-feedback loops. For example, Smad5 is a target of miR-135 (Li et al. 2008). Upon BMP2 stimulation in murine pre-myoblast cells, miR-135 is downregulated, enabling elevated expression of Smad5 and increased BMP signaling (Li et al. 2008). Conversely, miR-26a forms a negative-feedback loop with Smad1. In human adipose-derived stem cells, miR-26a expression is increased during differentiation thereby limiting the duration of Smad signaling by targeting Smad1 (Luzi et al. 2008). In addition to Smads, other proteins involved in TGF-β or BMP signal transduction are also known to be targeted by miRNAs. These include ZEB2 (also known as Smad interacting protein 1, SIP1) (Kato et al. 2007; Hu et al. 2010), the TGF-β receptor type II (TGF-βRII) (Kim et al. 2009), the BMP receptor type II (BMPRII) (Qin et al. 2009), and Smad7 (Liu et al. 2010). The TGF-β2 ligand is also targeted by miR-200a (Wang et al. 2011).

15.1.5 TGF-β and Cardiovascular Disease

Pathways activated by TGF-β family ligands have been implicated in a number of cardiovascular diseases. Elevated TGF-β expression leads to transcriptional activation of connective tissue growth factor (CTGF) implicated in the development of atherosclerotic plaques (Ohyama et al. 2012). CTGF is a key component of the extracellular matrix (ECM) and promotes a pathological phenotype by inducing ECM deposit and monocyte infiltration (Ohyama et al. 2012). BMP also leads to vascular inflammation and calcification resulting in the formation of atherosclerotic plaques (Derwall et al. 2012). BMPs can promote differentiation of multipotent cells to the osteoblastic lineage, thereby increasing vascular calcification (Cai et al. 2012). Therefore, negative control of TGF-β family signaling is crucial for preventing the formation of atherosclerotic lesions.

TGF-β is also a key effector of myocardial fibrosis. In a pressure-overload model, activation of fibroblasts leads to myocardial fibrosis, which can be alleviated by treatment with a neutralizing antibody against TGF-β (Kuwahara et al. 2002). TGF-β also induces cells to undergo EndMT. Aberrant TGF-β signaling can lead to cardiac fibrosis due to the influx of mesenchymal cells in the heart (Kovacic et al. 2012). It was also recently reported that TGF-β-induced activation of Rho-associated kinase (ROCK) leads to suppression of BMP signaling and increases fibrosis (Wang et al. 2012d). In addition, angiotensin II-induced IL-6 increases TGF-β expression and signaling, leading to cardiac fibrosis (Ma et al. 2012).

15.2 TGF-β Regulates VSMC Phenotype

Arterial hypertension is another pathological condition associated with abnormal TGF-β signaling. In patients, increased TGF-β expression correlates with increased arterial pressure. Increased TGF-β signaling in Emilin1 knockout animals causes hypertension, which can be controlled with genetic ablation of TGF-β1 (Gordon and Blobe 2008). TGF-β signaling is suppressed by hypoxia-inducible factor 1 (HIF-1) to prevent myocardial hypertrophy in a transaortic constriction mouse model (Wei et al. 2012). Additionally, pulmonary artery hypertension (PAH) is highly associated with mutations in the gene encoding BMPRII (Machado et al. 2006; Santibanez et al. 2011). About 70–80 % of heritable PAH and 15–40 % of idiopathic PAH are associated with a BMPRII mutation (Machado et al. 2006; Thomson et al. 2000). It has been suggested that a number of the nonsense or missense mutations identified in *BMPRII* produce mRNA transcripts which would be susceptible to nonsense-mediated mRNA decay, or inactivation of BMP receptor function as a result of (a) accumulation in the endoplasmic reticulum and reduced expression on the cell surface, (b) lack of ligand binding, or (c) inactivation of serine/threonine kinase activity (Morrell 2010), which in turn cause downregulation of the BMP signal transduction.

VSMC compose the medial layer of the blood vessel wall and are critical for contraction and relaxation of the vessel to control blood pressure. Although differentiated, VSMC are plastic and undergo a phenotype switch in response to cellular stimuli or injury—the differentiated "contractile" state is characterized by a low proliferation rate, expression of smooth muscle cell (SMC)-specific contractile genes, and the capacity to contract upon stimulation, while the dedifferentiated "synthetic" state exhibits greater proliferation, migration, and reduced expression of contractile genes (Owens et al. 2004). This phenotype switching allows VSMC to repair injuries by facilitating replacement of damaged cells; however, prolonged time in the synthetic state can lead to progression of vascular proliferative diseases, such as atherosclerosis, PAH, or restenosis after angioplasty (Owens et al. 2004). TGF-β and BMP promote VSMC towards the contractile phenotype exhibited by an elongated, spindle-like morphology, inhibition of proliferation and migration, and the expression of the SMC-specific contractile genes α-actin (α-SMA), SM22α, and calponin (Rensen et al. 2007).

Recent studies have implicated the involvement of multiple miRNAs in the control of the VSMC phenotype (Davis et al. 2009; Liu et al. 2009; Cheng et al. 2009; Kang and Hata 2012). Mice deficient in Dicer in SMC result in embryonic lethality with severe defects in the vessel wall structure and in the medial layer of the vessel, indicating the critical role of miRNA in SMC during embryogenesis (Albinsson et al. 2010). Additionally, mice carrying an inducible SMC-specific Dicer deletion display a loss of contractile function in their umbilical arteries (Albinsson et al. 2010). MiRNAs play a key role in the development and function of SMC in association with TGF-β signaling. In the following section, we review individual miRNAs downstream of TGF-β or BMP signaling which play important roles in cardiovascular biology.

15.3 MiRNAs Regulated by TGF-β in Cardiovascular Tissues

15.3.1 MiR-143/145

MiR-143 and miR-145 are exclusively expressed in VSMCs and are directly and indirectly regulated transcriptionally by TGF-β and BMP (Long and Miano 2011; Davis-Dusenbery et al. 2011). Activation of Smads by TGF-β and BMP leads to upregulation of myocardin and myocardin-related transcription factors (MRTFs), respectively. These factors form a complex with the serum response factor (SRF) and induce transcription of miR-143 and miR-145 through the DNA sequence known as the CArG box (Cordes et al. 2009; Xin et al. 2009). Both miR-143 and miR-145 are originally transcribed as a single pri-miRNA which is processed into two distinct miRNAs (Iio et al. 2010). Both miR-143 and miR-145 promote the contractile phenotype in VSMCs by allowing CArG box-dependent contractile gene expression and inhibiting VSMC proliferation. Among the targets of miR-143 and miR-145 are Kruppel-like factor 4 (KLF4), Kruppel-like factor 5 (KLF5), and E twenty-six (ETS)-like transcription factor 1 (ELK1), which promote the synthetic phenotype (Davis-Dusenbery et al. 2011; Rangrez et al. 2011). Knockdown of miR-143/145 in SMC inhibits differentiation of VSMC and prevents contraction in the presence of the appropriate cellular stimuli (Davis-Dusenbery et al. 2011; Long and

Miano 2011). MiR-143/145 knockout mice exhibit a reduced medial layer, a general decrease in all contractile gene expression and disorganized actin stress fibers in VSMC (Xin et al. 2009; Boettger et al. 2009; Elia et al. 2009). Additionally, mice exposed to chronic hypoxia develop PAH and express high levels of miR-145 in the smooth muscle layer of vessels, whereas miR-145 knockout mice are protected from PAH (Caruso et al. 2012). PAH patients with *BMPRII* mutations also exhibit high levels of miR-145 in SMC (Caruso et al. 2012).

15.3.2 MiR-21

MiR-21 expression showcases the posttranscriptional mode of regulation of miRNAs by TGF-β and BMP. Upon TGF-β and BMP stimulation, Smads associate with Drosha and facilitate the processing from pri-miR-21 to pre-miR-21, while no observable increase is detected in the pri-miR-21 (Davis et al. 2008). One of the functional targets of miR-21in VSMC is programmed cell death protein 4 (PDCD4), which promotes the contractile gene expression in VSMC (Davis et al. 2008). MiR-21 is also controlled by Fos-related antigen 1 (Fra-1), which is a target of miR-143 (Horita et al. 2011). Therefore, TGF-β signaling induces SRF, which increases miR-143 expression and decreases Fra-1 expression and controls miR-21. Additionally, miR-21 is known to inhibit the downstream target of Fra-1, phosphatase and tensin homolog (PTEN), creating a unique crosstalk between miR-21-PTEN and miR-143-Fra1 and ultimately reducing cell proliferation. Other targets of miR-21 include Smad7 and Dedicator of cytokinesis (DOCK) 4, 5, and 7, which are known to enhance proliferation and migration of VSMC (Kang et al. 2012; Liu et al. 2010). Hence, induction of miR-21 by TGF-β downregulates VSMC proliferation and migration and promotes the contractile phenotype (Kang et al. 2012).

15.3.3 MiR-29

The miR-29 family of miRNAs (miR-29a/b/c) targets the 3′ UTRs of various collagen proteins including collagen I and IV. The loss of miR-29 has been associated with an increase in pulmonary fibrosis (Xiao et al. 2012). MiR-29 is transcriptionally downregulated by TGF-β (Xiao et al. 2012). Upon TGF-β treatment, inhibition of miR-29 results in elevated expression of the ECM genes (Xiao et al. 2012). Smad3-deficient mice are protected from pulmonary fibrosis due to downregulation of miR-29 (Xiao et al. 2012). Overexpression of miR-29 also suppresses expression of TGF-β1 and CTGF, thereby protecting against bleomycin-induced lung fibrosis (Xiao et al. 2012). Decreased levels of miR-29 are also found in renal fibrosis (Wang et al. 2012a).

15.3.4 *MiR-125b*

MiR-125b expression is induced transcriptionally by TGF-β in mouse endothelial cells isolated from heart microvasculature which results in increased expression of ECM genes (Ghosh et al. 2012). p53 is a target of miR-125b (Le et al. 2009). p53 functions to regulate the TGF-β-mediated fibrosis by blocking type I collagen synthesis and Smad3-dependent collagen gene transcription by disrupting the association of Smad3 with a transcriptional cofactor, p300 (Ghosh and Vaughan 2012). Thus, increased expression of miR-125b is essential for the progression of fibrosis.

15.3.5 *MiR-27b*

TGF-β signaling transcriptionally inhibits miR-27b expression (Wang et al. 2012c). Increased levels of miR-27b are found in cardiac myocytes in which Smad4 is deleted leading to cardiac hypertrophy (Wang et al. 2012c). *In vitro* results also confirm a reduction in miR-27b expression upon TGF-β stimulation (Wang et al. 2012c). Conversely, inhibition of miR-27b in vivo using antisense RNA oligonucleotides against miR-27b (antagomir-27b) resulted in less pronounced cardiac dysfunction using a pressure-overload model in mice, indicating a role for miR-27b in cardiovascular disease. Among the validated downstream targets of miR-27b is the peroxisome proliferator-activated receptor γ (PPARγ), a critical modulator of cardiac hypertrophy (Wang et al. 2012c). Activation of PPARγ protects cardiac myoctyes from hypertrophy as measured by cell size and atrial natriuretic factor (ANF) expression (Amin et al. 2010). TGF-β-induced downregulation of miR-27b derepresses the level of PPARγ and protects against cardiac hypertrophy (Wang et al. 2012c). It is of note that TGF-β can also induce cardiac hypertrophy by propagating through Smad-independent signaling involving TGF-β-activated kinase 1 (TAK1) (Xiao and Zhang 2008). Smad signaling is protective against cardiac hypertrophy, revealing the complex nature of TGF-β signaling in cardiomyocytes (Xiao and Zhang 2008; Wang et al. 2005).

15.4 MiRNA Regulation of TGF-β Signaling in Cardiovascular Tissues

15.4.1 *MiR-155*

MiR-155 targets Smad2, which leads to an inhibition of TGF-β signaling (Kong et al. 2008). Targets of miR-155 include the G-protein coupled receptor, angiotensin receptor (AT1R), involved in the signal transduction of the vasoconstricting peptide, angiotensin II (Kong et al. 2008). Expression of miR-155 in aortic adventitial fibroblasts leads to the decrease of contractile gene expression and function (Zheng et al. 2010). Interestingly, TGF-β induces expression of miR-155 in epithelial cells through transcriptional activation by Smad4 (Kong et al. 2008). Thus, miR-155 contributes to the ablation of TGF-β signaling through a negative feedback mechanism (Kong et al. 2008).

15.4.2 MiR-26a

It has been demonstrated that miR-26a targets both Smad1 and Smad4, modulating both the BMP and the TGF-β signaling cascades (Leeper et al. 2011). PTEN, which is a validated target of miR-21, is another target of miR-26a (Huse et al. 2009). MiR-26a could be an additional player in the interesting crosstalk between miR-26a/miR-21-PTEN and miR-143-Fra-1 axis. MiR-26a has been shown to be a mechanosensitive miRNA and is upregulated by mechanical stretch in human airway SMC (Mohamed et al. 2010); thus, vascular constriction in hypertension could be attributed to the induction of miR-26a, which in turn downregulates the BMP and/or the TGF-β signaling.

15.4.3 MiR-200

Expression of the miR-200 family of miRNAs, miR-200a/b/c, in alveolar epithelial cells (AEC) prevents TGF-β-induced epithelial–mesenchymal transition (EMT) and induction of pulmonary fibroblasts (Yang et al. 2012). MiR-200 acts by targeting GATA3 and the zinc finger E-box-binding homeobox proteins ZEB1 and ZEB2, critical transcription factors in the TGF-β-mediated EMT. GATA3, ZEB1, and ZEB2 increase transcription of mesenchymal markers, such as Vimentin or N-cadherin while repressing the transcription of epithelial markers, such as E-cadherin (Gemmill et al. 2011). Additionally, miR-200 expression reduces lung fibrosis in both in vitro and in vivo models (Yang et al. 2012). Interestingly, ZEB1 and ZEB2 also suppress expression of the miR-200 family of miRNAs (Gregory et al. 2011), indicating a positive-regulatory loop between ZEB1/ZEB2 and the miR-200 family of miRNAs. When TGF-β signal is activated, AEC cells undergo EMT as a result of TGF-β-dependent repression of the miR-200 family of miRNAs (Gregory et al. 2011).

15.4.4 MiR-24

TGF-β induces miR-24 expression which, in turn, reduces TGF-β secretion and Smad2/3 phosphorylation and activation (Dogar et al. 2011). MiR-24 targets Furin, a protease that converts the latent TGF-β to mature, bioactive TGF-β, in cardiac fibroblasts; thus, TGF-β signaling is reduced (Wang et al. 2012b). Additionally, when exogenous miR-24 is delivered in vivo, it can reduce cardiac fibrosis. This feedback mechanism provides a novel scheme for TGF-β regulation.

MiR-24 is also induced by platelet-derived growth factor-BB (PDGF-BB), a potent inducer of the synthetic phenotype, and regulates BMP signaling by targeting Tribbles3 (Trb3), a protein which interacts with BMPRII and facilitates BMP signaling via downregulation of Smad ubiquitin-regulatory factor-1 (Smurf1) in VSMC (Chan et al. 2010). Smurf1 negatively regulates BMP signaling by targeting R-Smads (Derynck and Zhang 2003). Therefore, induction of miR-24 by PDGF-BB ultimately downregulates BMP signaling and stimulates proliferation and migration of VSMC (Chan et al. 2010).

15.5 Conclusion

In this review we summarized the current understanding of the role of various miRNAs in TGF-β and BMP-regulated signaling pathways (Tables 15.1 and 15.2). MiRNAs control nearly all aspects of cell biology, including cell death, migration, proliferation, and differentiation. MiRNAs upstream or downstream of TGF-β and BMP signaling pathways are keys to the differentiation and maintenance of the contractile phenotype of VSCM exhibited by a decrease in proliferation and an increase in contractile markers. Aberrant activities of TGF-β or BMP are associated with an array of cardiovascular diseases, and there is a critical need to modulate the activities of TGF-β and BMP signaling to prevent or ameliorate cardiovascular diseases. Since TGF-β and BMP have a variety of essential functions in different tissues, targeting a core molecule in the TGF-β or BMP signaling pathway is not ideal for the therapeutic approach. We speculate that the modulation of the levels of TGF-β or BMP-regulated miRNAs or miRNAs that target Smads could be a better therapeutic option because many miRNAs are expressed in a tissue-specific manner and, therefore, may exhibit minimal side effects in other tissues or organs. Preliminary studies in animal models illustrate that tail vein or subcutaneous injection of chemically modified stable short RNA oligonucleotides with sense or antisense miRNA sequence can be delivered to various organs and modulate gene expression in the recipient tissues (Montgomery et al. 2011). In one study, subcutaneous injection of an antagomir to miR-208a, whose expression is known to be induced upon heart failure, improved cardiac function, overall health, and survival after ingesting a high salt diet previously shown to cause hypertension (Montgomery et al. 2011; Rodenbaugh et al. 2007). Thus, miRNAs might be a viable therapeutic option for cardiovascular diseases. Further studies on miRNAs related to the TGF-β signaling pathways will grant a better understanding of the role of these pathways in cardiac development and homeostasis and, hence, offer potential for miRNA-based therapies.

Table 15.1 TGF-β/BMP regulated miRNAs

miRNA	Regulation	Biological activities of miRNA
miR-143/145	TGF-β ↑ BMP ↑	TGF-β/BMP induces myocardin/ MRTFs which leads to transcriptional activation of miR-143/145 (Cordes et al. 2009; Xin et al. 2009). MiR-143/145 downregulates KLF4, KLF5, and ELK1, promoting the contractile phenotype of VSMC (Davis-Dusenbery et al. 2011; Rangrez et al. 2011)
miR-21	TGF-β ↑ BMP ↑	TGF-β/ BMP stimulation increases processing of pri-miR-21 to pre-miR-21 through association with p68 and Drosha (Davis et al. 2008) MiR-21 targets PDCD4 and DOCK4/5/7 to promote the contractile phenotype of VSMC (Davis et al. 2008; Kang et al. 2012)
miR-29	TGF-β ↓	TGF-β transcriptionally downregulates miR-29 expression through Smad3 (Xiao et al. 2012). MiR-29 targets many collagen genes and, therefore, decreases expression of miR-29 by TGF-β increases expression of ECM genes (Xiao et al. 2012)
miR-125b	TGF-β ↑	TGF-β transcriptionally induces miR-125b expression (Ghosh et al. 2012). MiR-125b targets p53. p53 blocks type I collagen synthesis and Smad3-dependent collagen gene transcription (Ghosh and Vaughan 2012)
miR-27b	TGF-β ↓	TGF-β transcriptional downregulated miR-27b expression (Wang et al. 2012c). MiR-27b targets PPARγ, which protects against cardiac hypertrophy (Wang et al. 2012c)
miR-155	TGF-β ↑	MiR-155 is transcriptionally induced by TGF-β (Kong et al. 2008)
miR-24	TGF-β ↑	TGF-β transcriptionally induces miR-24 expression (Dogar et al. 2011). However, miR-24 target Furing and reduces maturation of TGF-β ligand, creating a feedback mechanism (Wang et al. 2012b)

Table 15.2 MiRNAs regulating TFG-β/ BMP signaling

miRNA	Regulation	Activity by TGF-β/BMP
miR-155	↓ TGF-β	MiR-155 targets Smad2 and AT1R (Kong et al. 2008) MiR-155 is also transcriptionally induced my TGF-β (Kong et al. 2008)
miR-26a	↓ TGF-β ↓ BMP	MiR-26a targets Smad1 and Smad4 (Leeper et al. 2011). MiR-26a also targets PTEN, thereby promoting the synthetic phenotype of VSMC (Huse et al. 2009)
miR-200	↓ TGF-β	MiR-200a/b/c prevents TGF-β-induced EMT (Yang et al. 2012). The miR-200 family of miRNAs targets GATA3, ZEB1, and ZEB2 to prevent increased transcription of mesenchymal markers (Gemmill et al. 2011). ZEB1 and ZEB2 suppress expression of miR-200.
miR-24	↓ TGF-β ↓ BMP	MiR-24 targets Furin and reduces maturation of TGF-β ligand (Wang et al. 2012b). However, TGF-β induces miR-24 expression, creating a feedback mechanism (Dogar et al. 2011) MiR-24 targets Trb3, preventing signaling through the BMPRII and promoting the synthetic phenotype in VSMC (Chan et al. 2010)

Acknowledgments Because of space restrictions and the focus of the article, we deeply apologize to those colleagues whose references we have not had the opportunity to discuss. We thank all members of the Hata laboratory for helpful suggestions and critical discussion. We thank Matthew Blahna for critical reading of the manuscript. This work was supported by grants from the National Institute of Health: HL093154 and HL108317, the American Heart Association: 0940095N and the LeDucq foundation Transatlantic network grant to A.H.

References

Adams BD, Claffey KP, White BA (2009) Argonaute-2 expression is regulated by epidermal growth factor receptor and mitogen-activated protein kinase signaling and correlates with a transformed phenotype in breast cancer cells. Endocrinology 150(1):14–23. doi:10.1210/en.2008-0984

Albinsson S, Suarez Y, Skoura A, Offermanns S, Miano JM, Sessa WC (2010) MicroRNAs are necessary for vascular smooth muscle growth, differentiation, and function. Arterioscler Thromb Vasc Biol 30(6):1118–1126. doi:10.1161/ATVBAHA.109.200873

Amin RH, Mathews ST, Alli A, Leff T (2010) Endogenously produced adiponectin protects cardiomyocytes from hypertrophy by a PPARγ-dependent autocrine mechanism. Am J Physiol Heart Circ Physiol 299(3):H690–H698. doi:10.1152/ajpheart.01032.2009

Baek D, Villen J, Shin C, Camargo FD, Gygi SP, Bartel DP (2008) The impact of microRNAs on protein output. Nature 455(7209):64–71. doi:10.1038/nature07242

Bauersachs J, Thum T (2011) Biogenesis and regulation of cardiovascular microRNAs. Circ Res 109(3):334–347. doi:10.1161/CIRCRESAHA.110.228676

Bernstein E, Kim SY, Carmell MA, Murchison EP, Alcorn H, Li MZ, Mills AA, Elledge SJ, Anderson KV, Hannon GJ (2003) Dicer is essential for mouse development. Nat Genet 35(3):215–217. doi:10.1038/ng1253

Blahna MT, Hata A (2012) Smad-mediated regulation of microRNA biosynthesis. FEBS Lett. doi:10.1016/j.febslet.2012.01.041

Bobik A (2006) Transforming growth factor-βs and vascular disorders. Arterioscler Thromb Vasc Biol 26(8):1712–1720. doi:10.1161/01.ATV.0000225287.20034.2c

Boettger T, Beetz N, Kostin S, Schneider J, Kruger M, Hein L, Braun T (2009) Acquisition of the contractile phenotype by murine arterial smooth muscle cells depends on the Mir143/145 gene cluster. J Clin Invest 119(9):2634–2647. doi:10.1172/JCI38864

Butz H, Racz K, Hunyady L, Patocs A (2012) Crosstalk between TGF-β signaling and the microRNA machinery. Trends Pharmacol Sci 33(7):382–393. doi:10.1016/j.tips.2012.04.003

Cai J, Pardali E, Sanchez-Duffhues G, Ten Dijke P (2012) BMP signaling in vascular diseases. FEBS Lett 586(14):1993–2002. doi:10.1016/j.febslet.2012.04.030

Caruso P, Dempsie Y, Stevens H, McDonald RA, Long L, Lu R, White K, Mair K, McClure JD, Southwood M, Upton P, Xin M, van Rooij E, Olson E, Morrell NW, Maclean MR, Baker AH (2012) A role for miR-145 in pulmonary arterial hypertension. Circ Res. doi:10.1161/CIRCRESAHA.112.267591

Chan MC, Hilyard AC, Wu C, Davis BN, Hill NS, Lal A, Lieberman J, Lagna G, Hata A (2010) Molecular basis for antagonism between PDGF and the TGFβ family of signalling pathways by control of miR-24 expression. EMBO J 29(3):559–573. doi:10.1038/emboj.2009.370

Cheng Y, Liu X, Yang J, Lin Y, Xu DZ, Lu Q, Deitch EA, Huo Y, Delphin ES, Zhang C (2009) MicroRNA-145, a novel smooth muscle cell phenotypic marker and modulator, controls vascular neointimal lesion formation. Circ Res 105(2):158–166. doi:10.1161/CIRCRESAHA.109.197517

Chung AC, Huang XR, Meng X, Lan HY (2010) miR-192 mediates TGF-β/Smad3-driven renal fibrosis. J Am Soc Nephrol 21(8):1317–1325. doi:10.1681/ASN.2010020134

Corcoran DL, Pandit KV, Gordon B, Bhattacharjee A, Kaminski N, Benos PV (2009) Features of mammalian microRNA promoters emerge from polymerase II chromatin immunoprecipitation data. PLoS One 4(4):e5279. doi:10.1371/journal.pone.0005279

Cordes KR, Sheehy NT, White MP, Berry EC, Morton SU, Muth AN, Lee TH, Miano JM, Ivey KN, Srivastava D (2009) miR-145 and miR-143 regulate smooth muscle cell fate and plasticity. Nature 460(7256):705–710. doi:10.1038/nature08195

Davis BN, Hata A (2009) Regulation of microRNA biogenesis: a miRiad of mechanisms. Cell Commun Signal 7:18, 10.1186/1478-811X-7-18

Davis BN, Hilyard AC, Lagna G, Hata A (2008) SMAD proteins control DROSHA-mediated microRNA maturation. Nature 454(7200):56–61. doi:10.1038/nature07086

Davis BN, Hilyard AC, Nguyen PH, Lagna G, Hata A (2009) Induction of microRNA-221 by platelet-derived growth factor signaling is critical for modulation of vascular smooth muscle phenotype. J Biol Chem 284(6):3728–3738. doi:10.1074/jbc.M808788200

Davis BN, Hilyard AC, Nguyen PH, Lagna G, Hata A (2010) Smad proteins bind a conserved RNA sequence to promote microRNA maturation by Drosha. Mol Cell 39(3):373–384. doi:10.1016/j.molcel.2010.07.011

Davis-Dusenbery BN, Chan MC, Reno KE, Weisman AS, Layne MD, Lagna G, Hata A (2011) Down-regulation of Kruppel-like factor-4 (KLF4) by microRNA-143/145 is critical for modulation of vascular smooth muscle cell phenotype by transforming growth factor-β and bone morphogenetic protein 4. J Biol Chem 286(32):28097–28110. doi:10.1074/jbc.M111.236950

Derwall M, Malhotra R, Lai CS, Beppu Y, Aikawa E, Seehra JS, Zapol WM, Bloch KD, Yu PB (2012) Inhibition of bone morphogenetic protein signaling reduces vascular calcification and atherosclerosis. Arterioscler Thromb Vasc Biol 32(3):613–622. doi:10.1161/ATVBAHA.111.242594

Derynck R, Zhang YE (2003) Smad-dependent and Smad-independent pathways in TGF-β family signalling. Nature 425(6958):577–584. doi:10.1038/nature02006

Dogar AM, Towbin H, Hall J (2011) Suppression of latent transforming growth factor (TGF)-β1 restores growth inhibitory TGF-β signaling through microRNAs. J Biol Chem 286(18):16447–16458. doi:10.1074/jbc.M110.208652

Elia L, Quintavalle M, Zhang J, Contu R, Cossu L, Latronico MV, Peterson KL, Indolfi C, Catalucci D, Chen J, Courtneidge SA, Condorelli G (2009) The knockout of miR-143 and -145 alters smooth muscle cell maintenance and vascular homeostasis in mice: correlates with human disease. Cell Death Differ 16(12):1590–1598. doi:10.1038/cdd.2009.153

Filipowicz W, Bhattacharyya SN, Sonenberg N (2008) Mechanisms of post-transcriptional regulation by microRNAs: are the answers in sight? Nat Rev Genet 9(2):102–114. doi:10.1038/nrg2290

Friedman RC, Farh KK, Burge CB, Bartel DP (2009) Most mammalian mRNAs are conserved targets of microRNAs. Genome Res 19(1):92–105. doi:10.1101/gr.082701.108

Fukuda T, Yamagata K, Fujiyama S, Matsumoto T, Koshida I, Yoshimura K, Mihara M, Naitou M, Endoh H, Nakamura T, Akimoto C, Yamamoto Y, Katagiri T, Foulds C, Takezawa S, Kitagawa H, Takeyama K, O'Malley BW, Kato S (2007) DEAD-box RNA helicase subunits of the Drosha complex are required for processing of rRNA and a subset of microRNAs. Nat Cell Biol 9(5):604–611. doi:10.1038/ncb1577

Gemmill RM, Roche J, Potiron VA, Nasarre P, Mitas M, Coldren CD, Helfrich BA, Garrett-Mayer E, Bunn PA, Drabkin HA (2011) ZEB1-responsive genes in non-small cell lung cancer. Cancer Lett 300(1):66–78. doi:10.1016/j.canlet.2010.09.007

Ghosh AK, Vaughan DE (2012) Fibrosis: is it a coactivator disease? Front Biosci (Elite Ed) 4:1556–1570

Ghosh AK, Nagpal V, Covington JW, Michaels MA, Vaughan DE (2012) Molecular basis of cardiac endothelial-to-mesenchymal transition (EndMT): differential expression of microRNAs during EndMT. Cell Signal 24(5):1031–1036. doi:10.1016/j.cellsig.2011.12.024

Gordon KJ, Blobe GC (2008) Role of transforming growth factor-β superfamily signaling pathways in human disease. Biochim Biophys Acta 1782(4):197–228. doi:10.1016/j.bbadis.2008.01.006

Gregory RI, Yan KP, Amuthan G, Chendrimada T, Doratotaj B, Cooch N, Shiekhattar R (2004) The Microprocessor complex mediates the genesis of microRNAs. Nature 432(7014):235–240. doi:10.1038/nature03120

Gregory PA, Bracken CP, Smith E, Bert AG, Wright JA, Roslan S, Morris M, Wyatt L, Farshid G, Lim YY, Lindeman GJ, Shannon MF, Drew PA, Khew-Goodall Y, Goodall GJ (2011) An autocrine TGF-β/ZEB/miR-200 signaling network regulates establishment and maintenance of epithelial-mesenchymal transition. Mol Biol Cell 22(10):1686–1698. doi:10.1091/mbc.E11-02-0103

Grimson A, Farh KK, Johnston WK, Garrett-Engele P, Lim LP, Bartel DP (2007) MicroRNA targeting specificity in mammals: determinants beyond seed pairing. Mol Cell 27(1):91–105. doi:10.1016/j.molcel.2007.06.017

Hata A, Davis BN (2010) Regulation of pri-miRNA processing through Smads. Adv Exp Med Biol 700:15–27

Horita HN, Simpson PA, Ostriker A, Furgeson S, Van Putten V, Weiser-Evans MC, Nemenoff RA (2011) Serum response factor regulates expression of phosphatase and tensin homolog through a microRNA network in vascular smooth muscle cells. Arterioscler Thromb Vasc Biol 31(12):2909–2919. doi:10.1161/ATVBAHA.111.233585

Hu M, Xia M, Chen X, Lin Z, Xu Y, Ma Y, Su L (2010) MicroRNA-141 regulates Smad interacting protein 1 (SIP1) and inhibits migration and invasion of colorectal cancer cells. Dig Dis Sci 55(8):2365–2372. doi:10.1007/s10620-009-1008-9

Huse JT, Brennan C, Hambardzumyan D, Wee B, Pena J, Rouhanifard SH, Sohn-Lee C, le Sage C, Agami R, Tuschl T, Holland EC (2009) The PTEN-regulating microRNA miR-26a is amplified in high-grade glioma and facilitates gliomagenesis in vivo. Genes Dev 23(11):1327–1337. doi:10.1101/gad.1777409

Iio A, Nakagawa Y, Hirata I, Naoe T, Akao Y (2010) Identification of non-coding RNAs embracing microRNA-143/145 cluster. Mol Cancer 9:136. doi:10.1186/1476-4598-9-136

Kang H, Hata A (2012) MicroRNA regulation of smooth muscle gene expression and phenotype. Curr Opin Hematol 19(3):224–231. doi:10.1097/MOH.0b013e3283523e57

Kang H, Davis-Dusenbery BN, Nguyen PH, Lal A, Lieberman J, Van Aelst L, Lagna G, Hata A (2012) Bone morphogenetic protein 4 promotes vascular smooth muscle contractility by activating microRNA-21 (miR-21), which down-regulates expression of family of dedicator of cytokinesis (DOCK) proteins. J Biol Chem 287(6):3976–3986. doi:10.1074/jbc.M111.303156

Kato M, Zhang J, Wang M, Lanting L, Yuan H, Rossi JJ, Natarajan R (2007) MicroRNA-192 in diabetic kidney glomeruli and its function in TGF-β-induced collagen expression via inhibition of E-box repressors. Proc Natl Acad Sci USA 104(9):3432–3437. doi:10.1073/pnas.0611192104

Kim YJ, Hwang SJ, Bae YC, Jung JS (2009) MiR-21 regulates adipogenic differentiation through the modulation of TGF-β signaling in mesenchymal stem cells derived from human adipose tissue. Stem Cells 27(12):3093–3102. doi:10.1002/stem.235

Kong W, Yang H, He L, Zhao JJ, Coppola D, Dalton WS, Cheng JQ (2008) MicroRNA-155 is regulated by the transforming growth factor β/Smad pathway and contributes to epithelial cell plasticity by targeting RhoA. Mol Cell Biol 28(22):6773–6784. doi:10.1128/MCB.00941-08

Kovacic JC, Mercader N, Torres M, Boehm M, Fuster V (2012) Epithelial-to-mesenchymal and endothelial-to-mesenchymal transition: from cardiovascular development to disease. Circulation 125(14):1795–1808. doi:10.1161/CIRCULATIONAHA.111.040352

Kuwahara F, Kai H, Tokuda K, Kai M, Takeshita A, Egashira K, Imaizumi T (2002) Transforming growth factor-β function blocking prevents myocardial fibrosis and diastolic dysfunction in pressure-overloaded rats. Circulation 106(1):130–135

Le MT, Teh C, Shyh-Chang N, Xie H, Zhou B, Korzh V, Lodish HF, Lim B (2009) MicroRNA-125b is a novel negative regulator of p53. Genes Dev 23(7):862–876. doi:10.1101/gad.1767609

Lee Y, Ahn C, Han J, Choi H, Kim J, Yim J, Lee J, Provost P, Radmark O, Kim S, Kim VN (2003) The nuclear RNase III Drosha initiates microRNA processing. Nature 425(6956):415–419. doi:10.1038/nature01957

Leeper NJ, Raiesdana A, Kojima Y, Chun HJ, Azuma J, Maegdefessel L, Kundu RK, Quertermous T, Tsao PS, Spin JM (2011) MicroRNA-26a is a novel regulator of vascular smooth muscle cell function. J Cell Physiol 226(4):1035–1043. doi:10.1002/jcp.22422

Li Z, Hassan MQ, Volinia S, van Wijnen AJ, Stein JL, Croce CM, Lian JB, Stein GS (2008) A microRNA signature for a BMP2-induced osteoblast lineage commitment program. Proc Natl Acad Sci USA 105(37):13906–13911. doi:10.1073/pnas.0804438105

Liu X, Cheng Y, Zhang S, Lin Y, Yang J, Zhang C (2009) A necessary role of miR-221 and miR-222 in vascular smooth muscle cell proliferation and neointimal hyperplasia. Circ Res 104(4):476–487. doi:10.1161/CIRCRESAHA.108.185363

Liu G, Friggeri A, Yang Y, Milosevic J, Ding Q, Thannickal VJ, Kaminski N, Abraham E (2010) miR-21 mediates fibrogenic activation of pulmonary fibroblasts and lung fibrosis. J Exp Med 207(8):1589–1597. doi:10.1084/jem.20100035

Long X, Miano JM (2011) Transforming growth factor-β1 (TGF-β1) utilizes distinct pathways for the transcriptional activation of microRNA 143/145 in human coronary artery smooth muscle cells. J Biol Chem 286(34):30119–30129. doi:10.1074/jbc.M111.258814

Luzi E, Marini F, Sala SC, Tognarini I, Galli G, Brandi ML (2008) Osteogenic differentiation of human adipose tissue-derived stem cells is modulated by the miR-26a targeting of the SMAD1 transcription factor. J Bone Miner Res 23(2):287–295. doi:10.1359/jbmr.071011

Ma F, Li Y, Jia L, Han Y, Cheng J, Li H, Qi Y, Du J (2012) Macrophage-stimulated cardiac fibroblast production of IL-6 is essential for TGF β/Smad activation and cardiac fibrosis induced by angiotensin II. PLoS One 7(5):e35144. doi:10.1371/journal.pone.0035144

Machado RD, Aldred MA, James V, Harrison RE, Patel B, Schwalbe EC, Gruenig E, Janssen B, Koehler R, Seeger W, Eickelberg O, Olschewski H, Elliott CG, Glissmeyer E, Carlquist J, Kim M, Torbicki A, Fijalkowska A, Szewczyk G, Parma J, Abramowicz MJ, Galie N, Morisaki H, Kyotani S, Nakanishi N, Morisaki T, Humbert M, Simonneau G, Sitbon O, Soubrier F, Coulet F, Morrell NW, Trembath RC (2006) Mutations of the TGF-β type II receptor BMPR2 in pulmonary arterial hypertension. Hum Mutat 27(2):121–132. doi:10.1002/humu.20285

Mohamed JS, Lopez MA, Boriek AM (2010) Mechanical stretch up-regulates microRNA-26a and induces human airway smooth muscle hypertrophy by suppressing glycogen synthase kinase-3β. J Biol Chem 285(38):29336–29347. doi:10.1074/jbc.M110.101147

Montgomery RL, Hullinger TG, Semus HM, Dickinson BA, Seto AG, Lynch JM, Stack C, Latimer PA, Olson EN, van Rooij E (2011) Therapeutic inhibition of miR-208a improves cardiac function and survival during heart failure. Circulation 124(14):1537–1547. doi:10.1161/CIRCULATIONAHA.111.030932

Morita S, Horii T, Kimura M, Goto Y, Ochiya T, Hatada I (2007) One Argonaute family member, Eif2c2 (Ago2), is essential for development and appears not to be involved in DNA methylation. Genomics 89(6):687–696. doi:10.1016/j.ygeno.2007.01.004

Morrell NW (2010) Role of bone morphogenetic protein receptors in the development of pulmonary arterial hypertension. Adv Exp Med Biol 661:251–264. doi:10.1007/978-1-60761-500-2_16

Ohyama Y, Tanaka T, Shimizu T, Matsui H, Sato H, Koitabashi N, Doi H, Iso T, Arai M, Kurabayashi M (2012) Runx2/Smad3 complex negatively regulates TGF-β-induced connective tissue growth factor gene expression in vascular smooth muscle cells. J Atheroscler Thromb 19(1):23–35

Owens GK, Kumar MS, Wamhoff BR (2004) Molecular regulation of vascular smooth muscle cell differentiation in development and disease. Physiol Rev 84(3):767–801. doi:10.1152/physrev.00041.2003

Ozsolak F, Poling LL, Wang Z, Liu H, Liu XS, Roeder RG, Zhang X, Song JS, Fisher DE (2008) Chromatin structure analyses identify miRNA promoters. Genes Dev 22(22):3172–3183. doi:10.1101/gad.1706508

Qin W, Zhao B, Shi Y, Yao C, Jin L, Jin Y (2009) BMPRII is a direct target of miR-21. Acta Biochim Biophys Sin 41(7):618–623

Rangrez AY, Massy ZA, Metzinger-Le Meuth V, Metzinger L (2011) miR-143 and miR-145: molecular keys to switch the phenotype of vascular smooth muscle cells. Circ Cardiovasc Genet 4(2):197–205. doi:10.1161/CIRCGENETICS.110.958702

Rensen SS, Doevendans PA, van Eys GJ (2007) Regulation and characteristics of vascular smooth muscle cell phenotypic diversity. Neth Heart J 15(3):100–108

Rodenbaugh DW, Wang W, Davis J, Edwards T, Potter JD, Metzger JM (2007) Parvalbumin isoforms differentially accelerate cardiac myocyte relaxation kinetics in an animal model of diastolic dysfunction. Am J Physiol Heart Circ Physiol 293(3):H1705–H1713. doi:10.1152/ajpheart.00232.2007

Rodriguez A, Griffiths-Jones S, Ashurst JL, Bradley A (2004) Identification of mammalian microRNA host genes and transcription units. Genome Res 14(10A):1902–1910. doi:10.1101/gr.2722704

Salzman DW, Shubert-Coleman J, Furneaux H (2007) P68 RNA helicase unwinds the human let-7 microRNA precursor duplex and is required for let-7-directed silencing of gene expression. J Biol Chem 282(45):32773–32779. doi:10.1074/jbc.M705054200

Santibanez JF, Quintanilla M, Bernabeu C (2011) TGF-β/TGF-β receptor system and its role in physiological and pathological conditions. Clin Sci (Lond) 121(6):233–251. doi:10.1042/CS20110086

Selbach M, Schwanhausser B, Thierfelder N, Fang Z, Khanin R, Rajewsky N (2008) Widespread changes in protein synthesis induced by microRNAs. Nature 455(7209):58–63. doi:10.1038/nature07228

Shi Y, Massague J (2003) Mechanisms of TGF-β signaling from cell membrane to the nucleus. Cell 113(6):685–700

Suzuki HI, Yamagata K, Sugimoto K, Iwamoto T, Kato S, Miyazono K (2009) Modulation of microRNA processing by p53. Nature 460(7254):529–533. doi:10.1038/nature08199

Thomson JR, Machado RD, Pauciulo MW, Morgan NV, Humbert M, Elliott GC, Ward K, Yacoub M, Mikhail G, Rogers P, Newman J, Wheeler L, Higenbottam T, Gibbs JS, Egan J, Crozier A, Peacock A, Allcock R, Corris P, Loyd JE, Trembath RC, Nichols WC (2000) Sporadic primary pulmonary hypertension is associated with germline mutations of the gene encoding BMPR-II, a receptor member of the TGF-β family. J Med Genet 37(10):741–745

Toma I, McCaffrey TA (2012) Transforming growth factor-β and atherosclerosis: interwoven atherogenic and atheroprotective aspects. Cell Tissue Res 347(1):155–175. doi:10.1007/s00441-011-1189-3

Wang J, Xu N, Feng X, Hou N, Zhang J, Cheng X, Chen Y, Zhang Y, Yang X (2005) Targeted disruption of Smad4 in cardiomyocytes results in cardiac hypertrophy and heart failure. Circ Res 97(8):821–828. doi:10.1161/01.RES.0000185833.42544.06

Wang Y, Medvid R, Melton C, Jaenisch R, Blelloch R (2007) DGCR8 is essential for microRNA biogenesis and silencing of embryonic stem cell self-renewal. Nat Genet 39(3):380–385. doi:10.1038/ng1969

Wang B, Koh P, Winbanks C, Coughlan MT, McClelland A, Watson A, Jandeleit-Dahm K, Burns WC, Thomas MC, Cooper ME, Kantharidis P (2011) miR-200a Prevents renal fibrogenesis through repression of TGF-β2 expression. Diabetes 60(1):280–287. doi:10.2337/db10-0892

Wang B, Komers R, Carew R, Winbanks CE, Xu B, Herman-Edelstein M, Koh P, Thomas M, Jandeleit-Dahm K, Gregorevic P, Cooper ME, Kantharidis P (2012a) Suppression of microRNA-29 expression by TGF-β1 promotes collagen expression and renal fibrosis. J Am Soc Nephrol 23(2):252–265. doi:10.1681/ASN.2011010055

Wang J, Huang W, Xu R, Nie Y, Cao X, Meng J, Xu X, Hu S, Zheng Z (2012b) MicroRNA-24 regulates cardiac fibrosis after myocardial infarction. J Cell Mol Med. doi:10.1111/j.1582-4934.2012.01523.x

Wang J, Song Y, Zhang Y, Xiao H, Sun Q, Hou N, Guo S, Wang Y, Fan K, Zhan D, Zha L, Cao Y, Li Z, Cheng X, Zhang Y, Yang X (2012c) Cardiomyocyte overexpression of miR-27b induces cardiac hypertrophy and dysfunction in mice. Cell Res 22(3):516–527. doi:10.1038/cr.2011.132

Wang S, Sun A, Li L, Zhao G, Jia J, Wang K, Ge J, Zou Y (2012d) Upregualtion of BMP-2 antagonizes TGF-β1/ROCK-enhanced cardiac fibrotic signaling through activating of Smurf1/Smad6 complex. J Cell Mol Med. doi:10.1111/j.1582-4934.2012.01538.x

Wei H, Bedja D, Koitabashi N, Xing D, Chen J, Fox-Talbot K, Rouf R, Chen S, Steenbergen C, Harmon JW, Dietz HC, Gabrielson KL, Kass DA, Semenza GL (2012) Endothelial expression of hypoxia-inducible factor 1 protects the murine heart and aorta from pressure overload by suppression of TGF-β signaling. Proc Natl Acad Sci USA 109(14):E841–E850. doi:10.1073/pnas.1202081109

Xiao H, Zhang YY (2008) Understanding the role of transforming growth factor-β signalling in the heart: overview of studies using genetic mouse models. Clin Exp Pharmacol Physiol 35(3):335–341. doi:10.1111/j.1440-1681.2007.04876.x

Xiao J, Meng XM, Huang XR, Chung AC, Feng YL, Hui DS, Yu CM, Sung JJ, Lan HY (2012) miR-29 inhibits bleomycin-induced pulmonary fibrosis in mice. Mol Ther 20(6):1251–1260. doi:10.1038/mt.2012.36

Xin M, Small EM, Sutherland LB, Qi X, McAnally J, Plato CF, Richardson JA, Bassel-Duby R, Olson EN (2009) MicroRNAs miR-143 and miR-145 modulate cytoskeletal dynamics and responsiveness of smooth muscle cells to injury. Genes Dev 23(18):2166–2178. doi:10.1101/gad.1842409

Xu L, Yao X, Chen X, Lu P, Zhang B, Ip YT (2007) Msk is required for nuclear import of TGF-{β}/BMP-activated Smads. J Cell Biol 178(6):981–994. doi:10.1083/jcb.200703106

Yamagata K, Fujiyama S, Ito S, Ueda T, Murata T, Naitou M, Takeyama K, Minami Y, O'Malley BW, Kato S (2009) Maturation of microRNA is hormonally regulated by a nuclear receptor. Mol Cell 36(2):340–347. doi:10.1016/j.molcel.2009.08.017

Yang S, Banerjee S, de Freitas A, Sanders YY, Ding Q, Matalon S, Thannickal VJ, Abraham E, Liu G (2012) Participation of miR-200 in pulmonary fibrosis. Am J Pathol 180(2):484–493. doi:10.1016/j.ajpath.2011.10.005

Yao X, Chen X, Cottonham C, Xu L (2008) Preferential utilization of Imp7/8 in nuclear import of Smads. J Biol Chem 283(33):22867–22874. doi:10.1074/jbc.M801320200

Yeom KH, Lee Y, Han J, Suh MR, Kim VN (2006) Characterization of DGCR8/Pasha, the essential cofactor for Drosha in primary miRNA processing. Nucleic Acids Res 34(16):4622–4629. doi:10.1093/nar/gkl458

Yu B, Bi L, Zheng B, Ji L, Chevalier D, Agarwal M, Ramachandran V, Li W, Lagrange T, Walker JC, Chen X (2008) The FHA domain proteins DAWDLE in Arabidopsis and SNIP1 in humans act in small RNA biogenesis. Proc Natl Acad Sci USA 105(29):10073–10078. doi:10.1073/pnas.0804218105

Zheng L, Xu CC, Chen WD, Shen WL, Ruan CC, Zhu LM, Zhu DL, Gao PJ (2010) MicroRNA-155 regulates angiotensin II type 1 receptor expression and phenotypic differentiation in vascular adventitial fibroblasts. Biochem Biophys Res Commun 400(4):483–488. doi:10.1016/j.bbrc.2010.08.067

Part IV
Other Diseases

Chapter 16
TGF-β and Genetic Skeletal Diseases

Shiro Ikegawa, Mitsuko Nakashima, and Naomichi Matsumoto

Abstract Transforming growth factor-β (TGF-β) is a multi-functional growth factor that controls proliferation, differentiation, and other cellular functions in many types of cells. TGF-β and its family molecules play critical roles in development and maintenance of the skeleton, which are underlined by various skeletal phenotypes caused by mutations in genes encoding ligands, receptors, and signaling molecules of the TGF-β family (TGF-β-pathy). Genetic diseases are categorized as monogenic and polygenic diseases. They give us clues to understand mechanisms for development and maintenance of tissues and organs in humans, which is also true to skeleton. In this paper, we review representative monogenic and polygenic TGF-β-pathy of the skeleton with focus on their significance for understanding mechanisms regulating the skeletal system.

Keywords Asporin • Bone • Cartilage • Cartilage intermediate layer protein (CILP) • Extracellular matrix • Growth factor • Monogenic disease • Polygenic disease • Susceptibility gene • TGF-β-pathy

Abbreviations

AKT	v-akt murine thymoma viral oncogene homolog
BMD	Bone mineral density
BMSC	Bone marrow stromal cell

S. Ikegawa (✉)
Laboratory for Bone and Joint Diseases, Center for Integrative Medical Science, RIKEN, Tokyo 108-8639, Japan
e-mail: sikegawa@ims.u-tokyo.ac.jp

M. Nakashima • N. Matsumoto
Department of Human Genetics, Graduate School of Medicine, Yokohama City University, Yokohama, 236-0004, Japan

A. Moustakas and K. Miyazawa (eds.), *TGF-β in Human Disease*,
DOI 10.1007/978-4-431-54409-8_16, © Springer Japan 2013

CED	Camurati–Engelmann disease
CILP	Cartilage intermediate layer protein
D-repeat	Aspartic acid repeat
ECM	Extracellular matrix
LAP	Latency-associated peptide
LDD	Lumbar disc disease
LDS	Loeys–Dietz syndrome
LRR	Leucine-rich repeat
MAPK	Mitogen activated kinase-like protein
OA	Osteoarthritis
OPLL	Ossification of the posterior longitudinal ligament of the spine
PI3K	Phosphatidylinositol-4,5-bisphosphate 3-kinase
ROCK	Rho-associated protein kinase
SLRP	Small leucine-rich proteoglycan

16.1 Role of Transforming Growth Factor-β in Skeletal Tissue

Transforming growth factor-β (TGF-β) is a multifunctional growth factor that controls proliferation, differentiation, and other cellular functions in many types of cells, playing critical roles in development and maintenance of many types of tissues and organs. It is true in the skeleton (skeletal tissue). TGF-β stimulates proliferation and differentiation of osteoblast (bone forming cell) and chondrocyte (cartilage forming cell) in vitro. High levels of *TGFB1* (gene for TGF β1) mRNA and its protein are found in developing cartilage (growth cartilage) and endochondral and membrane bones, suggesting its role in growth, differentiation, and maintenance of the skeleton (Dickinson et al. 1990). The critical roles of *TGFB* and its family genes to the human skeletal system are underlined by various skeletal phenotypes of patients with diseases caused by mutations in genes encoding ligands, receptors, and signaling molecules of the TGF-β family (Table 16.1). These TGF-β related diseases (TGF-β-pathy) form an important subgroup of skeletal dysplasia (Warman et al. 2011).

16.2 TGF-β-Pathy

Genetic diseases are categorized as monogenic and polygenic diseases from their involvement of the gene. The monogenic (Mendelian) disease is caused by a mutation of a single gene (disease gene). The effect of the disease gene to the disease phenotype is decisive in monogenic diseases. Its inheritance is relatively simple and predictable, following Mendel's law. In contrast, the polygenetic disease is multifactorial; its phenotype is determined by combined and concerted effects of a group of genes (susceptibility genes) and the environment. Its inheritance is complex.

Table 16.1 Monogenic diseases of the skeleton caused by mutations of TGF-β and TGF-β related genes

Disease[a]	Inheritance	OMIM ID	Disease gene
Diaphyseal dysplasia Camurati–Engelmann	AD	131300	*TGFB1*
Marfan syndrome (familial thoracic aortic aneurysms and dissection)	AD		*TGFB2*
Brachydactyly type A2	AD	112600	*BMP2*
Grebe dysplasia	AR	200700	*GDF5*
Fibular hypoplasia and complex brachydactyly (Du Pan)	AR	228900	*GDF5*
Brachydactyly type A2	AD	112600	*GDF5*
Brachydactyly type C	AD, AR	113100	*GDF5*
Multiple synostoses syndrome type 2	AD	186500	*GDF5*
Proximal symphalangism type 2	AD	185800	*GDF5*
Klippel–Feil anomaly with laryngeal malformation	AD	148900	*GDF6*
Loeys–Dietz syndrome types 1A, 2A	AD	609192, 610168	*TGFBR1*
Loeys–Dietz syndrome types 1B, 2B	AD	608967, 610380	*TGFBR2*
Acromesomelic dysplasia with genital anomalies	AR	609441	*BMPR1B*
Brachydactyly type A2	AD	112600	*BMPR1B*
Fibrodysplasia ossificans progressiva	AD	135100	*ACVR1*
Loeys–Dietz syndrome type C (aortic aneurysms and dissections with early-onset OA)	AD	613795	*SMAD3*
Myhre syndrome	AD	139210	*SMAD4*
Geleophysic dysplasia	AR	231050	*ADAMTSL2*
Osteopoikilosis/Buschke–Ollendorff syndrome	AD	166700	*LEMD3*
Melorheostosis with osteopoikilosis	AD	155950	*LEMD3*
Sclerosteosis	AR	269500	*SOST*
Endosteal hyperostosis, van Buchem type	AR	239100	*SOST*
Diaphanospondylodysostosis	AR	608022	*BMPER*
Multiple synostoses syndrome type 1	AD	186500	*NOG*
Proximal symphalangism type 1	AD	185800	*NOG*
Cenani–Lenz like syndactyly	Sporadic	212780	*GREM1, FMN1*

[a]Disease names are based on "Nosology and classification of genetic skeletal disorders: 2010 revision" (Warman et al. Am J Med Genet 2011)

OA osteoarthritis, *AD* autosomal dominant, *AR* autosomal recessive, *ADAMTSL2* ADAMTS (a disintegrin and metalloproteinase with thrombospondin motifs)-like 2, *LEMD3* LEM domain-containing 3, *SOST* sclerostin, *BMPER* bone morphogenetic protein-binding endothelial cell precursor-derived regulator, *NOG* noggin, *GREM1* gremlin 1, *FMN1* formin 1

TGF-β-pathy covers both categories of genetic diseases involving the skeletal system. These diseases give us a clue to clarify the roles of TGF-β in skeleton in vivo in human as well as physiological mechanisms controlling the skeletal system.

As their numbers are so many, this review focuses on a limited number of TGF-β-pathy affecting the skeleton. The list of monogenic diseases includes Camurati–Engelmann disease (CED; MIM #131300) and Loeys–Dietz syndrome (LDS; MIM #609192). Polygenic diseases related to TGFB are osteoporosis (MIM #166710), otosclerosis, ossification of the posterior longitudinal ligament of the spine (OPLL; MIM #602475), osteoarthritis (OA, MIM 165720), and lumbar disc disease (LDD). They all are common diseases.

16.3 Monogenic Diseases

16.3.1 Camurati–Engelmann Disease

16.3.1.1 Clinical Features

CED (alias: Engelmann disease, progressive diaphyseal dysplasia) is a rare skeletal dysplasia of autosomal dominant inheritance. CED is characterized by the cortical thickening of the diaphyses of the long bones. Endo- and exo-osteal bone formations are increased (Fig. 16.1a). Increased exo-osteal bone formation (hyperostosis) is bilateral and symmetrical and usually starts at diaphyses of femora or tibiae, expanding to other long tubular bones (Fig. 16.1b, c). As the disease progresses, metaphyses of the bones may be affected as well, but their epiphyses are spared. Sclerotic changes at the skull base are frequently present. Interestingly, the onset of the disease is usually during childhood and almost always before age 30, indicating that the disease process has nothing to do with prenatal and early postnatal skeletal development. Most patients present with limb pain and muscular weakness of the affected limbs, a waddling gait and easy fatigability. The patients present with apparent undernutrition (Marfanoid habitus) (Fig. 16.1d). The muscular weakness is not necessarily progressive and typical bone changes may be found in asymptomatic individuals. Systemic manifestations including anemia, leukopenia, and hepatosplenomegaly occur occasionally (Janssens et al. 2006).

16.3.1.2 Molecular Genetics of CED

CED results from domain-specific heterozygous mutations in *TGFB1*. The group led by Norio Niikawa mapped the CED locus to 19q13.1–q13.3 by a parametric linkage analysis (Ghadami et al. 2000) and subsequently detected heterozygous missense mutations in exon 4 of *TGFB1*, near the carboxy terminus of the latency-associated peptide (LAP) (Kinoshita et al. 2000). Janssens et al. (2000) confirmed the discovery by finding the same and similar mutations in the LAP region as well as other types of mutations. Different mutations affecting the LAP region have been reported (Kinoshita et al. 2004). Genetic heterogeneity is known in CED; i.e., patients with similar bone phenotypes but with no *TGFB1* mutation are considered as CED type II (MIM 606631).

These LAP mutations cause increased TGF-β function in CED. The mutations are predicted to cause LAP insufficiency due to conformational changes of the LAP proteins by affecting their cysteine residues necessary for di-sulfide bond formation. In vitro studies show these mutations result in disrupted association of LAP and

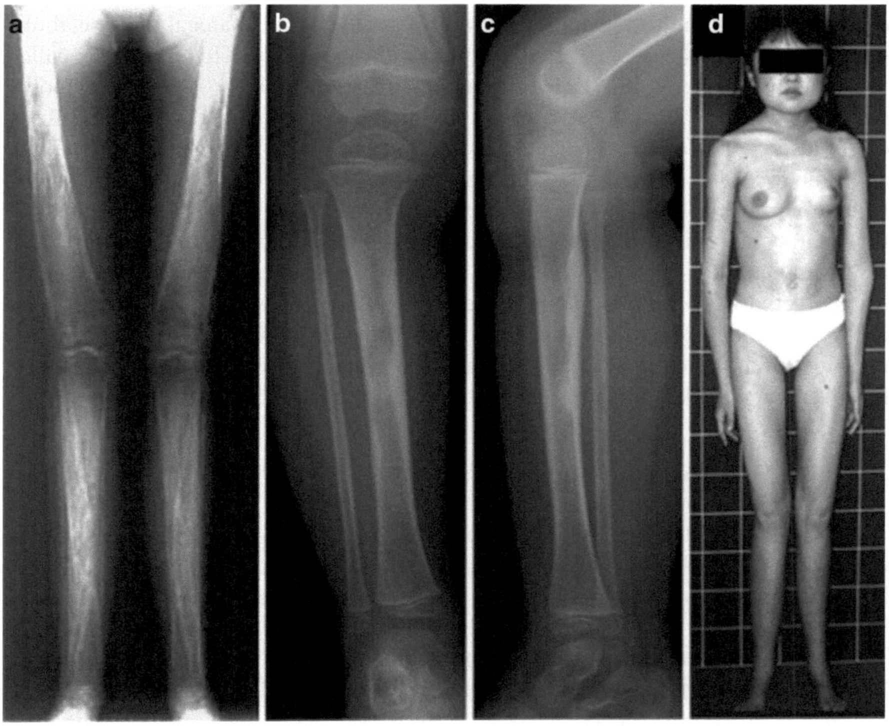

Fig. 16.1 Clinical features of Engelmann disease. (**a**) radiographs of the lower extremities in the late stage bone changes of the Engelmann disease. Symmetric hyperossification of the long bones. Bone widths and bone densities are increased, showing that both exo- and endo-osteal bone formation are increased. (**b**, **c**) the *right* leg A-P (**b**) and lateral (**c**) radiographs showing the early stage bone changes of disease. Irregular thickening of the cortex of tibial diaphysis. (**d**) marfanoid habitus (slender and relatively long limbs) of a patient

TGF-β1 and the subsequent release of the mature (active) TGF-β1 followed by increased TGF-β signaling (Saito et al. 2001). The cell growth of fibroblasts from a CED patient and growth of the mutant gene-transfected fibroblasts was suppressed by blocking TGF-β signaling. Janssens et al. (2003) investigated the effects of the CED mutation using a luciferase reporter assay specific for TGFB-induced transcriptional response. They showed that the mutations increase TGF-β1 activity by increasing active TGF-β1 in the medium of the transfected cells. Thus, the pathomechanism of CED is hyperactivation of TGF-β1 due to LAP insufficiency.

16.3.1.3 Significance of CED—Identification of the "Coupling Factor"

Skeletal phenotypes and mutations of CED gave a hint to solve a long-sought mystery of bone metabolism, "coupling"—a synchronized transition from bone resorption to bone formation in adult bone. Bone is a highly dynamic and specialized organ, constantly adjusting its structure in response to mechanical and metabolic stimulation such as exercise, gravity, and calcium depletion. This process is called "remodeling," where old useless bone is first removed by osteoclasts, bone-resorbing cells of hematopoetic origin, and then new necessary bone is rebuilt by osteoblasts, bone-forming cells of mesenchymal origin. To maintain the structural and functional integrity of bone through this process, the steps of destruction and formation must be coordinated spatiotemporally and quantitatively. The coordination that synchronized the opposing activities is called "coupling" and its failure results in pathological states, including osteoporosis (decreased bone) and osteopetrosis (increased, but functionally useless bone), both of which produce mechanically weak bone liable to fracture. Therefore, its molecular basis has been one of the main research topics in skeletal biology and research of bone and joint diseases; however, even the key molecule(s) controlling it remains enigmatic.

16.3.1.4 TGF-β is the "Coupling" Factor

The TGF-β1 effect in bone metabolism in vivo has long been paradoxical (Iqbal et al. 2009). TGF-β1 stimulates osteoblast proliferation and differentiation in vitro; however, over-expression of TGF-β1 stimulates bone turnover, resulting in bone loss (high-turnover osteoporosis) (Erlebacher et al. 1998). Conversely, blocking of TGF-β signaling causes decreased bone degradation, resulting in excessive bone formation with normal bone formation rates (Filvaroff et al. 1999).

Tang et al. (2009) generated double knockout (KO) mice of *Tgfb1* and the recombination-activating gene-2 (Rag2) because *Tgfb1*–KO mice (single KO mice) develop autoimmune disease and are unavailable for postnatal bone metabolism studies. They showed that TGF-β1 is necessary for the localization of bone marrow stromal cells (BMSCs) and osteoblast precursor cells to sites of bone resorption. TGF-β1 is stored in bone as an inactivated form by binding to LAP. Osteoclasts release and activate TGF-β1 by secreting bone catabolizing enzymes and acid pH during bone resorption. Activated TGF-β1 guides BMSCs correctly to sites of resorption through a Smad signaling pathway. Thus, the coupling signal is valid only at sites of bone resorption. The switch turns on at the right place at the right time. In CED, the mutations deliver active TGF-β1 regardless of the demand of mechanical stress and excessive bones are formed for no purpose. Thus, CED provides an in vivo model to validate the role of TGF-β1 as the "coupling" factor that coordinates bone resorption and formation.

16.3.2 Loeys–Dietz Syndrome

16.3.2.1 Clinical Features

LDS, previously known as Marfan syndrome type 2, is an autosomal dominant syndrome characterized by aortic aneurysm with multiple systemic involvements. Two types are known; LDS type 1 shows a triad of arterial tortuosity and aneurysm,

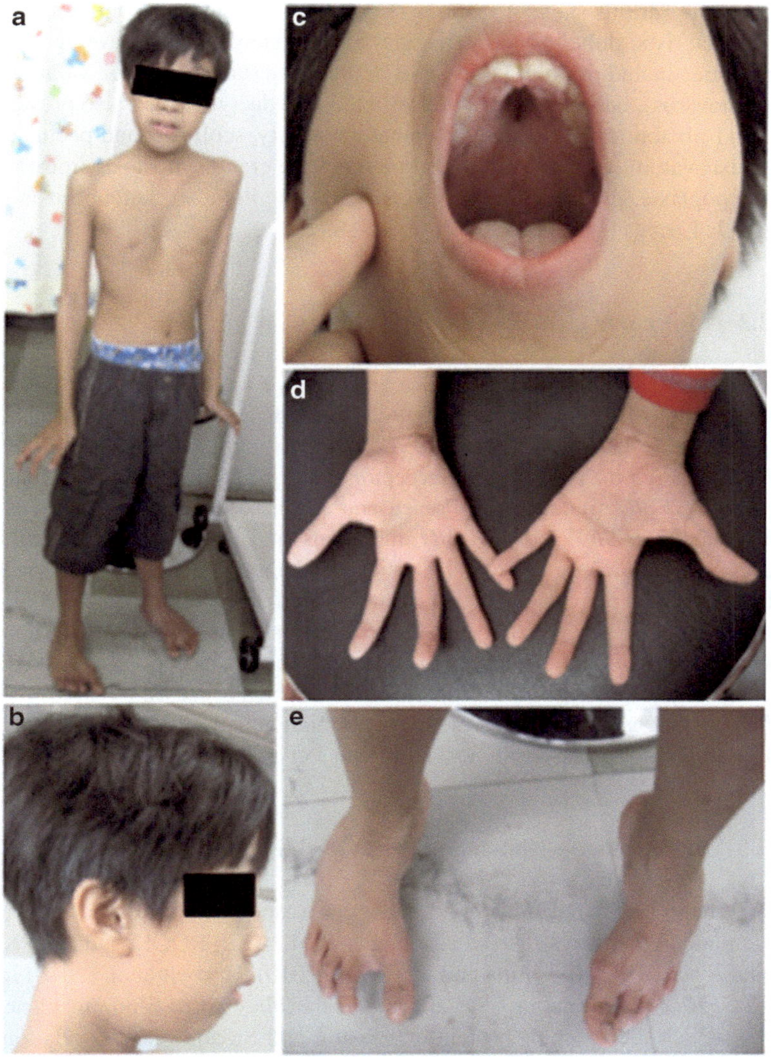

Fig. 16.2 Clinical pictures of Loeys–Dietz syndrome. A patient harboring a *TGFBR2* (for type 2 TGF-β receptor) mutation. (**a**) Slender constitution, pectus excavatum, and dolichostenomelia. (**b**) Dolichocephaly with retrognathia. (**c**) cleft palate (after surgery). (**d**) long and slender fingers. (**e**) talipes equinovarus and sandal gap

hypertelorism, and bifid uvula or cleft palate (Fig. 16.2), while LDS type 2 has only bifid uvula. LDS may lead to intellectual disability. LDS type 1 exhibits intracranial abnormalities, including Chiari malformation type I and hydrocephalus. The natural history of LDS is defined by aggressive arterial aneurysm and dissection.

LDS show various skeletal features (Fig. 16.2). Patients are commonly slender with characteristic dolichostenomelia and arachnodactyly. They have various limb abnormalities including joint laxity, foot deformities (talipes planus, valgus, varus, and equinovarus), hammer toes, and post-axial polydactyly. They also present with spinal and rib abnormalities, cervicospinal instability, scoliosis, and pectus carinatum or excavatum. Other skeletal features include craniosynostosis, retrognathia, malar hypoplasia, protrusio acetabulae, skeletal fragility, and osteoporosis (Kalra et al. 2011; Loeys et al. 2005, 2006).

16.3.2.2 Molecular Genetics

Collod et al. (1994) performed a linkage analysis in a large LDS family without the fibrillin 1 gene (*FBN1*) mutations and found a candidate locus LDS at 3p24.2-p25. Mizuguchi et al. (2004) identified the disruption by chromosomal abnormality of *TGFBR2* encoding the type 2 TGF-β receptor in a Japanese individual and found heterozygous *TGFBR2* mutations in French and Japanese LDS families. Loeys et al. (2005) found *TGFBR2* and *TGFBR1* mutations in LDS. Germline mutations of *TGFBR1* and *TGFBR2* are also found in patients with familial thoracic aortic aneurysms and their dissection demonstrated lack of any skeletal deformities (Matyas et al. 2006; Pannu et al. 2005). As *TGFBR1* and *TGFBR2* are also known as tumor suppressor genes, correlation of these hereditary disorders with carcinogenesis is suspected; however, multiple self-healing squamous epithelioma, which is caused by disease-specific *TGFBR1* mutations, is the only evidence of correlation between the *TGFBR* germline mutation and tumor predisposition so far (Goudie et al. 2011).

16.3.2.3 Significance—Non-canonical TGF-β Signaling Pathway

TGF-β signaling is mediated by Smad-dependent (canonical) and Smad-independent (non-canonical) pathways. The latter includes PI3K-AKT, ROCK, and MAPK (p38, JNK and RAS-ERK) (Ikushima and Miyazono 2010). Horbelt et al. (2010) investigated *TGFBR2* mutations in LDS and found that they show a dominant-negative effect in which the receptor protein expression and Smad2 phosphorylation were reduced in vitro. Although the pathomechanisms of skeletal abnormalities in LDS are far from being understood, studies of aortic aneurysms shed light on possible roles of TGF-β signaling in LDS. Most LDS mutations are missense changes at the serine-threonine kinase domain of *TGFBR1* and *TGFBR2*, which were supposed to result in impaired TGF-β signaling. However, histopathological studies of aortic tissues from LDS patients revealed paradoxical increase in TGF-β signaling with enhanced expression of connective tissue growth factor and collagen deposition. Intra-nuclear phosphorylated Smad2 expression was also increased (Holm et al. 2011; Loeys et al. 2005, 2006). Systemic administration of a TGF-β neutralizing antibody in *Fbn1*-deficient MFS mice showing similar manifestations to LDS can prevent the aberrant enlargement of the aortic vessels (Habashi et al. 2006; Ng et al. 2004). *ERK1/2* were also activated and Losartan, the angiotensin II type 1 receptor

(AT1) blocker, can prevent aortic aneurysms by inhibiting *ERK* pathway in MFS mice (Habashi et al. 2006, 2011). Furthermore, expression of *TGFB2* and *TGFBR3* proteins was elevated and the Smad-independent p38 MAPK pathway was activated in *Tgfbr2* mutant mice (Iwata et al. 2012). These observations implied that not only the canonical signaling but also the non-canonical signaling has an important role in the pathogenesis of LDS.

16.4 Polygenic Diseases

Case–control association studies indicate that *TGFB* is involved in several polygenic diseases affecting the skeleton.

16.4.1 Osteoporosis

Osteoporosis is the pathological decrease of the bone tissue that leads to an increased risk of fracture, mainly in spine (vertebral body), distal radius, and femoral neck. It is represented as reduction of bone mineral density (BMD), but the basic abnormality is a decrease in bone matrix proteins, not in mineralization. Osteoporosis has several clinical types and is classified as primary and secondary osteoporosis based on its etiology. Primary osteoporosis consists of postmenopausal and senile osteoporosis. Postmenopausal osteoporosis is the most common bone and joint disease in women after menopause. Secondary osteoporosis results from various diseases relating to bone metabolism, or prolonged use of medications such as glucocorticoids.

Genetic association of postmenopausal osteoporosis and *TGFB1* has been known. A group reported a significant association of a missense variant in *TGFB1*, c.29T>C (p.Leu10Pro) with BMD and frequency of vertebral fractures in postmenopausal women (Yamada et al. 1998). (Note. c.29T>C: The change of C (cytosine) nucleotide at the position 788 in the cDNA sequence of the gene to T (thymine). p.Leu10Pro: The missense change of Thr (threonine) at position 263 of the protein to Ile (isoleucine)). The group also reported a correlation of the serum concentration of TGF-β1 with the c.29T>C genotype. However, statistical significance of the association is not definite because of the small number of subjects used for the study and of multiple hypothesis testing. Subsequent large-scale association studies in Japanese (Kou et al. 2011) and other ethnic populations have not replicated the association.

16.4.2 Otosclerosis

Otosclerosis is a common bone dysplasia that is unique to the endochondral layer of the otic (labyrinthine) capsule. It is a major cause of acquired hearing loss in adult life, leading to a progressive hearing loss due to stapes fixation. The mean age of its onset is in the third decade (under 50 years of age for 90 % of patients). Genetic heterogeneity of otosclerosis is present; both monogenic and polygenic forms of this disorder are known.

Association of a missense variation in *TGFB1*, c.788C>T (p.Thr 263Ile) with otosclerosis susceptibility has been reported in western European populations (Thys et al. 2007). The association could be replicated in two independent populations (Thys & Camp 2009). A reporter assay showed that the susceptibility variant (Thr 263) has lower Smad signaling activity, although the functional impact of this variant to the pathogenesis of otosclerosis remains unclear. Notably, association of *BMP2* and *BMP4* with otosclerosis is also reported (Thys & Camp 2009).

16.4.3 Ossification of the Posterior Longitudinal Ligament of the Spine

The posterior longitudinal ligament of the spine is a ligament situated at the posterior side of the spinal cord in the spinal canal, a space formed by vertebral bones. OPLL is a disease caused by abnormal (ectopic) ossification of the ligament. OPLL is a localized hyperostosis of this specific ligament, but systemic tendency of abnormal ossification has been reported in a certain proportion of the patients. The incidence of OPLL is 1.9–4.3 % in Japan (Stapleton et al. 2011; Matsunaga and Sakou 2012), and comparable incidence has been reported in other countries, especially in East Asia (Saetia et al. 2011). OPLL causes serious neurological disturbances due to compression of the spinal cord and the nerve roots by the ossified ligament. Its average age of onset is over 50 years with male predominance, affecting motility and quality of life in the patients.

By a case–control study for a group of candidate genes in Japanese OPLL, Horikoshi et al. (2006) found significant association between an intronic SNP in the *TGFB3* gene (rs2268624) and OPLL susceptibility. The statistical significance of the association is not definite to the present standard of the association study, and the functionality of the sequence variation remains unclear. The association data should be checked for their replication in different populations, preferably of different ethnic populations.

16.4.4 Regulatory Molecules of TGF-β in the Pathogenesis of Common Bone and Joint Diseases

16.4.4.1 Extracellular Matrix

Skeletal tissues have large amounts of extracellular matrix (ECM). The primary function of ECM in the skeleton is its contribution to physical and biomechanical strength; however, the ECM also plays important biological roles in skeletal metabolism through regulation of growth factors, including TGF-β (Grimaud et al. 2002; Chen et al. 2004). Many ECM proteins bind to TGF-β and modulate its function. The ECM proteins are mostly tissue-specific, playing a critical role in the tissue-specific, context-dependent function of TGF-β. Not a few of them are associated with common bone and joint diseases.

16.4.4.2 Asporin and Osteoarthritis

Osteoarthritis

Joint is an organ for locomotion and is composed by various tissues including cartilage, bone, synovium, and ligament. OA is the most common disease of the joint, affecting more than 10 % of the population in developed countries all over the world. OA can affect any synovial joint in the human body, mainly in weight-bearing joints of the lower extremities. Histopathologically, OA is characterized by degeneration of articular cartilage that covers the surface of the joint. OA can be resulted from definite causes such as trauma and hereditary and metabolic diseases (secondary OA), but most of them are idiopathic (primary OA), in which genetic factors have been implicated (Dai and Ikegawa 2010). Patients with OA suffer from pain and limitations to the movement of their affected joints. OA is a primary cause of decreased activity in daily living and quality of life after middle age.

Asporin

Asporin is an ECM protein first identified as a specific protein extracted from articular cartilage (Lorenzo et al. 2001). Asporin is a member of the small leucine-rich proteoglycan (SLRP) family (Hocking et al. 1998), which belongs to the leucine-rich repeat (LRR) super-family of proteins. The SLRP family is classified into three classes based on similarities in protein sequence, conserved spacing of the four cysteine cluster in their N-terminal domain, and genomic structure (Iozzo & Murdo 1996; Ameye and Young 2002). Asporin belongs to class I SLRPs which contains ten repeats of the LRR motif, L-X-X-L-X-L/I-X-X-N-X-L/I, in their core proteins (Scholzen et al. 1994; Lorenzo et al. 2001). The N-terminal regions of other class I SLRPs, including decorin and biglycan carry a few chondroitin or dermatan sulfate chains, respectively (Krusius and Ruoslahti 1986; Fisher et al. 1989), whereas asporin contains no consensus sequences for glycosaminoglycan attachment. Therefore, asporin is not a proteoglycan unlike other class I SLRPs. Asporin has a unique aspartic acid repeat (D-repeat) in the N-terminal instead (Henry et al. 2001; Lorenzo et al. 2001).

Association of Asporin with OA

SLRP family genes have been implicated in the etiology of OA. KO mice of fibromodulin (Gill et al. 2002) and double KO mice of fibromodulin and biglycan

(Ameye et al. 2002) develop mild and severe premature OA, respectively. Asporin mRNA is expressed abundantly in OA articular cartilage (Lorenzo et al. 2001; Kizawa et al. 2005). These observations prompted our group to examine the asporin gene (*ASPN*) as a candidate gene for OA.

We made a list of common polymorphisms in *ASP* and carried out a large-scale case–control association study in Japanese OA populations. We identified a significant association between *ASPN* and OA (Kizawa et al. 2005). The D-repeat of asporin is encoded by a polymorphic triplet (GAT) repeat. In the Japanese variants of asporin, there are 10–19 D-repeats. Among them, the allele that encodes 14 repeats (D14 allele) was significantly overrepresented in OA patients. The association was replicated in Han Chinese (Jiang et al. 2006) and European Caucasian populations (Nakamura et al. 2007).

Asporin and TGF-β

SLRPs such as decorin and biglycan have been shown to bind TGF-β and control its accessibility to the receptors (Border et al. 1992; Isaka et al. 1996). A series of experiments using the ATDC5 cell, an in vitro model of chondrocyte differentiation in mouse (Shukunami et al. 1996) and human cartilage cells revealed that asporin co-localizes with TGF-β1 on the cell surface, binds to TGF-β1, and blocks the binding of TGF-β1 to the type II TGF-β receptor outside of the cell. Consequently, phosphorylation of Smads, mainly Smad3 is inhibited and expression of the cartilage marker genes downstream of the TGF-β-Smad signal, including those for type II collagen and aggrecan are decreased (Kizawa et al. 2005; Kou et al. 2007). Knockdown of asporin by siRNA in chondrocyte increases the cartilage marker genes (Nakajima et al. 2007). Thus, asporin works as a negative regulator of chondrocyte differentiation and cartilage ECM formation through negative regulation of *TGFB1*. On the other hand, the group also showed that TGF-β1 could stimulate asporin gene expression, implicating asporin and TGF-β1 into a functional feedback loop that controls homeostasis of the articular cartilage.

Mechanism Causing Susceptibility and Role of Asporin in Articular Cartilage Metabolism

In vitro transfection experiments examining inhibitory effects of different alleles of asporin showed that the D14 asporin variant had the strongest inhibitory effect on TGF-β signaling. Individuals who have the susceptibility allele are suspected to have decreased potential of cartilage recovery and are hence susceptible to OA.

Joints are subjected to daily noxious stimuli. Wear and tear by daily weight bearing and minor or major injuries during daily activity usually cause joint damage. To recover from the damage, cell proliferation and matrix production are necessary. Many growth factors and cytokines support the process. In cartilage, TGF-β is the cardinal player leading to regeneration. Many molecules would enhance TGF-β and its function. However, too much function of TGF-β would result in

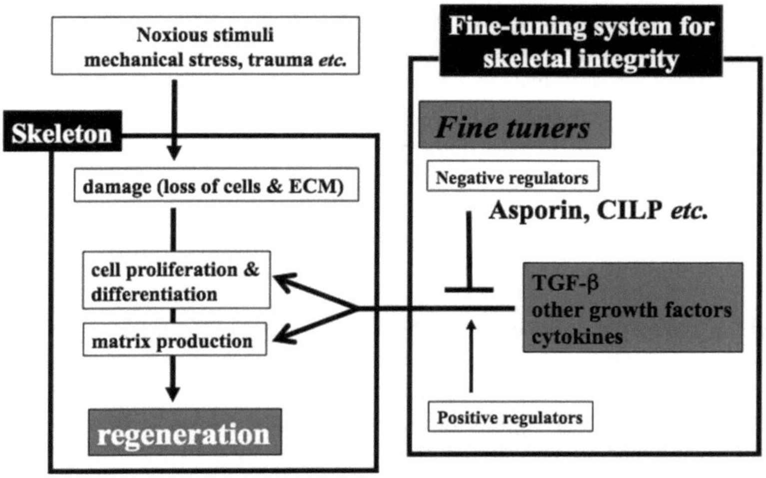

Fig. 16.3 A fine-tuning system for maintaining skeleton. The fine-tuners listed here (asporin and CILP) primarily function through sequestration of TGF-β ligands. They operate at distinct locations, leading to distinct diseases

unwanted results such as non-smooth joint surface and cartilage tumor formation. Therefore, restriction of TGF-β function in appropriate level is necessary. We suppose that asporin is central to the control and fine-tuning of TGF-β function in articular cartilage (Fig. 16.3). The aberration of the fine-tuning system would lead to degeneration.

16.4.4.3 Cartilage Intermediate Layer Protein and LDD

Lumbar Disc Disease

LDD results from degeneration of intervertebral discs of the lumbar spine. LDD is one of the most common musculoskeletal disorders. LDD is a significant health care issue all over the world. Disc degeneration and concomitant disc herniation are a major cause of low back pain and leg pain. Low back pain affects 70–85 % of all people during their lifetime, and lumbar disc herniation is the most common cause of activity limitation in individuals younger than 45 years of age (Andersson 1999). The etiology and pathogenesis of LDD is largely unknown; however, significant roles for genetic factors in the etiology of LDD are indicated by epidemiological studies (Battie et al. 1995; Sambrook et al. 1999).

Cartilage Intermediate Layer Protein (CILP)

CILP is an ECM protein first identified in the intermediate (middle) zone of human articular cartilage (Lorenzo et al. 1998). CILP also localizes to meniscus, tendon, ligament, synovial membrane, and intervertebral disc. Its expression increases substantially in association with aging in human articular cartilage. CILP is one of the few cartilage ECM proteins whose expression is up-regulated in early and late stages of OA (Lorenzo et al. 1998, 2004). *CILP* expression levels were high in human intervertebral discs and increased as disc degeneration progressed (Seki et al. 2005).

Association of CILP with LDD

Several lines of evidence show that ECM proteins play crucial roles in intervertebral disc homeostasis. The association with LDD has been reported for the genes encoding collagen IX (*COL9A2* (Annunen et al. 1999) and *COL9A3* (Paassilta et al. 2001)) and aggrecan (Kawaguchi et al. 1999). These observations underscore the importance of ECM protein metabolism in the etiology of LDD. Therefore, we examined candidate genes encoding ECM proteins of intervertebral discs and found a significant association between a missense SNP, c.1184T>C (p.Ile395Thr) in the *CILP* gene and LDD.

TGF-β in Intervertebral Disc Metabolism and Etiology of LDD

The TGF-β signal is critical for metabolism of the intervertebral disc in vivo. Cells immunopositive for TGF-β1 and the TGF-β type II receptor are present in increased numbers in herniated discs (Tolonen et al. 2001; Specchia et al. 2002). Transgenic mice harboring a dominant-negative TGF-β type II receptor develop skeletal abnormalities and OA (Serra et al. 1997), while *Smad3* KO mice develop joint degeneration and spinal deformity, caused by loss of articular cartilage and decreased production of proteoglycans (Yang et al. 2001). Further, TGF-β1 treatment decreases the expression of matrix metalloproteinases that are involved in tissue remodeling events in disc degeneration (Pattison et al. 2001). Interestingly, association of asporin with LDD is also reported (Song et al. 2008).

Similar to asporin in articular cartilage and OA, CILP functions as a modulator for TGF-β activity in intervertebral disc tissue. CILP bound to TGF-β1 and inhibited TGF-β1-mediated induction of ECM genes in the intervertebral disc (Seki et al. 2005). p.Thr395 CILP produced from the disease-susceptible allele of CILP suppressed TGF-β1-induced gene expression more efficiently than p.Ile395 CILP produced from the non-susceptible allele. The aberrantly increased inhibitory effects of CILP attributed to the susceptibility allele would perturb the balance of TGF-β control over chondrocyte metabolism and intervertebral disc tissue maintenance, leading to a susceptibility to LDD caused by inadequate response of intervertebral disc cells to injury and mechanical stress. Just as in articular cartilage, the fine-tuning system (Fig. 16.3) is necessary in vertebral discs.

16.4.4.4 General Concept: TGF-β-ECM System

ECM proteins expressed abundantly and specifically in skeletal tissues and cells. They bind to TGF-β and positively or negatively regulate TGF-β activity, controlling skeletal tissue metabolism. Then, their abnormalities could cause susceptibility to common bone and joint diseases. The association study underscores the importance of TGF-β in development and maintenance of human skeletal tissue and further highlights the functional link between TGF-β, ECM proteins, and the bone and joint diseases. Therefore, agents controlling and modifying the TGF-β-ECM system would be promising targets for treatment, presenting a novel therapeutic strategy for common bone and joint diseases.

16.5 Further Questions

Among the three isoforms of TGF-β in human, our micro-array data show *TGFB1* is a predominant form in articular cartilage (Ikegawa et al. unpublished data). High levels of *TGFB1* mRNA and protein are found in developing cartilage, endochondral and membrane bones, suggesting its role in the growth and differentiation of bone and cartilage (Dickinson et al. 1990). Although the amino-acid sequence homologies of *TGFB2* and *TGFB3* to *TGFB1* are only 74 % and 77 %, respectively, they have been considered to work in similar ways to *TGFB1*. Their in vivo functions in the human skeleton are unclear.

Acknowledgments Photographs for Fig. 16.2 were kindly provided by Dr. Seiji Mizuno at the Department of Pediatrics, Central Hospital, Aichi Human Service Center, Japan after obtaining written informed consent by the patient's family.

References

Ameye L, Young MF (2002) Mice deficient in small leucine-rich proteoglycans: novel in vivo models for osteoporosis, osteoarthritis, Ehlers–Danlos syndrome, muscular dystrophy, and corneal diseases. Glycobiology 12:107R–116R. doi:10.1093/glycob/cwf065

Ameye L, Aria D, Jepsen K, Oldberg A, Xu T, Young MF (2002) Abnormal collagen fibrils in tendons of biglycan/fibromodulin-deficient mice lead to gait impairment, ectopic ossification, and osteoarthritis. FASEB J 16:673–680. doi:10.1096/fj.01-0848com

Andersson GB (1999) Epidemiological features of chronic low-back pain. Lancet 354:581–585. doi:10.1016/S0140-6736(99)01312-4

Annunen S, Paassilta P, Lohiniva J, Perälä M, Pihlajamaa T, Karppinen J, Tervonen O, Kröqur H, Lähde S, Vanharanta H, Ryhänen L, Göring HH, Ott J, Prockop DJ, Ala-Kokko L (1999) An allele of COL9A2 associated with intervertebral disc disease. Science 285:409–412. doi:10.1126/science.285.5426.409

Battié MC, Haynor DR, Fisher LD, Gill K, Gibbons LE, Videman T (1995) Similarities in degenerative findings on magnetic resonance images of the lumbar spines of identical twins. J Bone Joint Surg Am 77:1662–1670. doi:10.2106/JBJS.K.01342

Border WA, Noble NA, Yamamoto T, Harper JR, Yamauchi Y, Pierschbacher MD, Ruoslahti E (1992) Natural inhibitor of transforming growth factor-β protects against scarring in experimental kidney disease. Nature 360:361–364. doi:10.1038/360361a0

Chen J, Yan W, Setton LA (2004) Static compression induces zonal-specific changes in gene expression for extracellular matrix and cytoskeletal proteins in intervertebral disc cells in vitro. Matrix Biol 22:573–583. doi:10.1016/j.matbio.2003.11.008

Collod G, Babron MC, Jondeau G, Coulon M, Weissenbach J, Dubourg O, Bourdarias JP, Bonaïti-Pellié C, Junien C, Boileau C (1994) A second locus for Marfan syndrome maps to chromosome 3p24.2-p25. Nat Genet 8:264–268. doi:10.1038/ng1194-264

Dai J, Ikegawa S (2010) Recent advances in association studies of osteoarthritis susceptibility genes. J Hum Genet 55:77–80. doi:10.1038/jhg.2009.137

Dickinson ME, Kobrin MS, Silan CM, Kingsley DM, Justice MJ, Miller DA, Ceci JD, Lock LF, Lee A, Buchberg AM, Siracusa LD, Lyons KM, Derynck R, Hogan BLM, Copeland NG, Jenkins NA (1990) Chromosomal localization of seven members of the murine TGF-β superfamily suggests close linkage to several morphogenetic mutant loci. Genomics 6:505–520

Erlebacher A, Filvaroff EH, Ye JQ, Derynck R (1998) Osteobiastic responses to TGF-β during bone remodeling. Mol Biol Cell 9:1903–1918

Filvaroff E, Erlebacher A, Ye J, Gitelman SE, Lotz J, Heillman M, Derynck R (1999) Inhibition of TGF-β receptor signaling in osteoblasts leads to decreased bone remodeling and increased trabecular bone mass. Development 126:4267–4279

Fisher LW, Termine JD, Young MF (1989) Deduced protein sequence of bone small proteoglycan I (biglycan) shows homology with proteoglycan II (decorin) and several nonconnective tissue proteins in a variety of species. J Biol Chem 264:4571–4576

Ghadami M, Makita Y, Yoshida K, Nishimura G, Fukushima Y, Wakui K, Ikegawa S, Yamada K, Kondo S, Niikawa N, Tomita H (2000) Genetic mapping of the Camurati–Engelmann disease locus to chromosome 19q13.1-q13.3. Am J Hum Genet 66:143–147. doi:10.1086/302728

Gill MR, Oldberg A, Reinholt FP (2002) Fibromodulin-null murine knee joints display increased incidences of osteoarthritis and alterations in tissue biochemistry. Osteoarthritis Cartilage 10:751–757. doi:10.1053/joca.2002.0527

Goudie DR, D'Alessandro M, Merriman B, Lee H, Szeverényi I, Avery S, O'Connor BD, Nelson SF, Coats SE, Stewart A, Christie L, Pichert G, Friedel J, Hayes I, Burrows N, Whittaker S, Gerdes AM, Broesby-Olsen S, Ferguson-Smith MA, Verma C, Lunny DP, Reversade B, Lane EB (2011) Multiple self-healing squamous epithelioma is caused by a disease-specific spectrum of mutations in TGFBR1. Nat Genet 43:365–369. doi:10.1038/ng.780

Grimaud E, Heymann D, Rédini F (2002) Recent advances in TGF-β effects on chondrocyte metabolism. Potential therapeutic roles of TGF-β in cartilage disorders. Cytokine Growth Factor Rev 13:241–257

Habashi JP, Judge DP, Holm TM, Cohn RD, Loeys B, Cooper TK, Myers L, Klein EC, Liu G, Calvi C, Podowski M, Neptune ER, Halushka MK, Bedja D, Gabrielson K, Rifkin DB, Carta L, Ramirez F, Huso DL, Dietz HC (2006) Losartan, an AT1 antagonist, prevents aortic aneurysm in a mouse model of Marfan syndrome. Science 312:117–121. doi:10.1126/science.1124287

Habashi JP, Doyle JJ, Holm TM, Aziz H, Schoenhoff F, Bedja D, Chen Y, Modiri AN, Judge DP, Dietz HC (2011) Angiotensin II type 2 receptor signaling attenuates aortic aneurysm in mice through ERK antagonism. Science 332:361–365. doi:10.1126/science.1192152

Henry SP, Takanosu M, Boyd TC, Mayne PM, Eberspaecher H, Zhou W, de Crombrugghe B, Hook M, Mayne R (2001) Expression pattern and gene characterization of asporin. A newly discovered member of the leucine-rich repeat protein family. J Biol Chem 276:12212–12221. doi:10.1074/jbc.M011290200

Hocking AM, Shinomura T, McQuillan DJ (1998) Leucine-rich repeat glycoproteins of the extracellular matrix. Matrix Biol 17:1–19

Holm TM, Habashi JP, Doyle JJ, Bedja D, Chen Y, van Erp C, Lindsay ME, Kim D, Schoenhoff F, Cohn RD, Loeys BL, Thomas CJ, Patnaik S, Marugan JJ, Judge DP, Dietz HC (2011) Noncanonical TGFβ signaling contributes to aortic aneurysm progression in Marfan syndrome mice. Science 332:358–361. doi:10.1126/science.1192149

Horbelt D, Guo G, Robinson PN, Knaus P (2010) Quantitative analysis of TGFBR2 mutations in Marfan-syndrome-related disorders suggests a correlation between phenotypic severity and Smad signaling activity. J Cell Sci 123:4340–4350. doi:10.1242/jcs.074773

Horikoshi T, Maeda K, Kawaguchi Y, Chiba K, Mori K, Koshizuka Y, Hirabayashi S, Sugimori K, Matsumoto M, Kawaguchi H, Takahashi M, Inoue H, Kimura T, Matsusue Y, Inoue I, Baba H, Nakamura K, Ikegawa S (2006) A large-scale genetic association study of ossification of the posterior longitudinal ligament of the spine. Hum Genet 119:611–616. doi:10.1007/s00439-006-0170-9

Ikushima H, Miyazono K (2010) TGFβ signalling: a complex web in cancer progression. Nat Rev Cancer 10:415–424. doi:10.1038/nrc2853

Iozzo RV, Murdoch AD (1996) Proteoglycans of the extracellular environment: clues from the gene and protein side offer novel perspectives in molecular diversity and function. FASEB J 10:598–614

Iqbal J, Sun L, Zaidi M (2009) Coupling bone degradation to formation. Nat Med 15:729–731. doi:10.1038/nm0709-729

Isaka Y, Brees DK, Ikegaya K, Kneda Y, Imai E, Noble NA, Border WA (1996) Gene therapy by skeletal muscle expression of decorin prevents fibrotic disease in rat kidney. Nat Med 2:418–423

Iwata J, Hacia JG, Suzuki A, Sanchez-Lara PA, Urata M, Chai Y (2012) Modulation of noncanonical TGF-β signaling prevents cleft palate in Tgfbr2 mutant mice. J Clin Invest 122:873–885. doi:10.1172/JCI61498

Janssens K, Gershoni-Baruch R, Guañabens N, Migone N, Ralston S, Bonduelle M, Lissens W, Van Maldergem L, Vanhoenacker F, Verbruggen L, Van Hul W (2000) Mutations in the gene encoding the latency-associated peptide of TGF-β-1 cause Camurati–Engelmann disease. Nat Genet 26:273–275. doi:10.1038/81563

Janssens K, ten Dijke P, Ralston SH, Bergmann C, Van Hul W (2003) Transforming growth factor-β-1 mutations in Camurati–Engelmann disease lead to increased signaling by altering either activation or secretion of the mutant protein. J Biol Chem 278:7718–7724. doi:10.1074/jbc.M208857200

Janssens K, Vanhoenacker F, Bonduelle M, Verbruggen L, Van Maldergem L, Ralston S, Guañabens N, Migone N, Wientroub S, Divizia MT, Bergmann C, Bennett C, Simsek S, Melançon S, Cundy T, Van Hul W (2006) Camurati–Engelmann disease: review of the clinical, radiological, and molecular data of 24 families and implications for diagnosis and treatment. J Med Genet 43:1–11. doi:10.1136/jmg.2005.033522

Jiang Q, Shi D, Yi L, Ikegawa S, Wang Y, Nakamura T, Qiao D, Liu C, Dai J (2006) Replication of the association of the aspartic acid repeat polymorphism in the asporin gene with knee-osteoarthritis susceptibility in Han Chinese. J Hum Genet 51:1068–1072. doi:10.1007/s10038-006-0065-6

Kalra VB, Gilbert JW, Malhotra A (2011) Loeys–Dietz syndrome: cardiovascular, neuroradiological and musculoskeletal imaging findings. Pediatr Radiol 41:1495–1504. doi:10.1007/s00247-011-2195-z

Kawaguchi Y, Osada R, Kanamori M, Ishihara H, Ohmori K, Matsui H, Kimura T (1999) Association between an aggrecan gene polymorphism and lumbar disc degeneration. Spine 24:2456–2460

Kinoshita A, Saito T, Tomita H, Makita Y, Yoshida K, Ghadami M, Yamada K, Kondo S, Ikegawa S, Nishimura G, Fukushima Y, Nakagomi T, Saito H, Sugimoto T, Kamegaya M, Hisa K,

Murray JC, Taniguchi N, Niikawa N, Yoshiura K (2000) Domain-specific mutations in TGFB1 result in Camurati–Engelmann disease. Nat Genet 26:19–20. doi:10.1038/79128
Kinoshita A, Fukumaki Y, Shirahama S, Miyahara A, Nishimura G, Haga N, Namba A, Ueda H, Hayashi H, Ikegawa S, Seidel J, Niikawa N, Yoshiura K (2004) TGFB1 mutations in four new families with Camurati–Engelmann disease: confirmation of independently arising LAP-domain-specific mutations. Am J Med Genet A 127A:104–107. doi:10.1002/ajmg.a.20671
Kizawa H, Kou I, Iida A, Sudo A, Miyamoto Y, Fukuda A, Mabuchi A, Kotani A, Kawakami A, Yamamoto S, Uchida A, Nakamura K, Notoya K, Nakamura Y, Ikegawa S (2005) An aspartic acid repeat polymorphism in asporin inhibits chondrogenesis and increases susceptibility to osteoarthritis. Nat Genet 37:138–144. doi:10.1038/ng1496
Kou I, Nakajima M, Ikegawa S (2007) Expression and regulation of the osteoarthritis-associated protein asporin. J Biol Chem 282:32193–32199. doi:10.1074/jbc.M706262200
Kou I, Takahashi A, Urano T, Fukui N, Ito H, Ozaki K, Tanaka T, Hosoi T, Shiraki M, Inoue S, Nakamura Y, Kamatani N, Kubo M, Mori S, Ikegawa S (2011) Common variants in a novel gene, FONG on chromosome 2q33.1 confer risk of osteoporosis in Japanese. PLoS One 6:e19641
Krusius T, Ruoslahti E (1986) Primary structure of an extracellular matrix proteoglycan core protein deduced from cloned cDNA. Proc Natl Acad Sci USA 83:7683–7687
Loeys BL, Chen J, Neptune ER, Judge DP, Podowski M, Holm T, Meyers J, Leitch CC, Katsanis N, Sharifi N, Xu FL, Myers LA, Spevak PJ, Cameron DE, De Backer J, Hellemans J, Chen Y, Davis EC, Webb CL, Kress W, Coucke P, Rifkin DB, De Paepe AM, Dietz HC (2005) A syndrome of altered cardiovascular, craniofacial, neurocognitive and skeletal development caused by mutations in TGFBR1 or TGFBR2. Nat Genet 37:275–281. doi:10.1038/ng1511
Loeys BL, Schwarze U, Holm T, Callewaert BL, Thomas GH, Pannu H, De Backer JF, Oswald GL, Symoens S, Manouvrier S, Roberts AE, Faravelli F, Greco MA, Pyeritz RE, Milewicz DM, Coucke PJ, Cameron DE, Braverman AC, Byers PH, De Paepe AM, Dietz HC (2006) Aneurysm syndromes caused by mutations in the TGF-β receptor. N Engl J Med 355:788–798. doi:10.1056/NEJMoa055695
Lorenzo P, Neame P, Sommarin Y, Heinegård D (1998) Cloning and deduced amino acid sequence of a novel cartilage protein (CILP) identifies a proform including a nucleotide pyrophosphohydrolase. J Biol Chem 273:23469–23475. doi:10.1074/jbc.273.36.23469
Lorenzo P, Aspberg A, Onnerfjord P, Bayliss MT, Neame PJ, Heinegard D (2001) Identification and characterization of asporin. A novel member of the leucine-rich repeat protein family closely related to decorin and biglycan. J Biol Chem 276:12201–12211. doi:10.1074/jbc.M010932200
Lorenzo P, Bayliss MT, Heinegård D (2004) Altered patterns and synthesis of extracellular matrix macromolecules in early osteoarthritis. Matrix Biol 23:381–391. doi:10.1016/j.matbio.2004.07.007
Matsunaga S, Sakou T (2012) Ossification of the posterior longitudinal ligament of the cervical spine: etiology and natural history. Spine 37:E309–E314. doi:10.1097/BRS.0b013e318241ad33
Mátyás G, Arnold E, Carrel T, Baumgartner D, Boileau C, Berger W, Steinmann B (2006) Identification and in silico analyses of novel TGFBR1 and TGFBR2 mutations in Marfan syndrome-related disorders. Hum Mutat 27:760–769. doi:10.1002/humu.20353
Mizuguchi T, Collod-Beroud G, Akiyama T, Abifadel M, Harada N, Morisaki T, Allard D, Varret M, Claustres M, Morisaki H, Ihara M, Kinoshita A, Yoshiura K, Junien C, Kajii T, Jondeau G, Ohta T, Kishino T, Furukawa Y, Nakamura Y, Niikawa N, Boileau C, Matsumoto N (2004) Heterozygous TGFBR2 mutations in Marfan syndrome. Nat Genet 36:855–860. doi:10.1038/ng1392
Nakajima M, Kizawa H, Saito M, Kou I, Miyazono K, Ikegawa S (2007) Mechanisms for asporin function and regulation in articular cartilage. J Biol Chem 282(44):32185–32192. doi:10.1074/jbc.M700522200
Nakamura T, Shi D, Tzetis M, Rodriquez-Lopez J, Miyamoto Y, Tsezou A, Gonzalez A, Jiamg Q, Kamatani N, Loughin J, Ikegawa S (2007) Meta-analysis of association between the ASPN D-repeat and osteoarthritis. Hum Mol Genet 16:1676–1681. doi:10.1093/hmg/ddm115

Ng CM, Cheng A, Myers LA, Martinez-Murillo F, Jie C, Bedja D, Gabrielson KL, Hausladen JM, Mecham RP, Judge DP, Dietz HC (2004) TGF-β-dependent pathogenesis of mitral valve prolapse in a mouse model of Marfan syndrome. J Clin Invest 114:1586–1592. doi:10.1172/JCI22715

Paassilta P, Lohiniva J, Göring HH, Perälä M, Räinä SS, Karppinen J, Hakala M, Palm T, Kröger H, Kaitila I, Vanharanta H, Ott J, Ala-kokko L (2001) PerIdentification of a novel common genetic risk factor for lumbar disk disease. JAMA 285:1843–1849. doi:10.1001/jama.285.14.1843

Pannu H, Fadulu VT, Chang J, Lafont A, Hasham SN, Sparks E, Giampietro PF, Zaleski C, Estrera AL, Safi HJ, Shete S, Willing MC, Raman CS, Milewicz DM (2005) Mutations in transforming growth factor-β receptor type II cause familial thoracic aortic aneurysms and dissections. Circulation 112:513–520. doi:10.1161/CIRCULATIONAHA.105.537340

Pattison ST, Melrose J, Ghosh P, Taylor TK (2001) Regulation of gelatinase-A (MMP-2) production by ovine intervertebral disc nucleus pulposus cells grown in alginate bead culture by transforming growth factor-β1 and insulin like growth factor-I. Cell Biol Int 25:679–689. doi:10.1006/cbir.2000.0718

Saetia K, Cho D, Lee S, Kim DH, Kim SD (2011) Ossification of the posterior longitudinal ligament: a review. Neurosurg Focus 30:E1. doi:10.1097/BRS.0b013e318241ad33

Saito T, Kinoshita A, Ki Y, Makita Y, Wakui K, Honke K, Niikawa N, Taniguchi N (2001) Domain-specific mutations of a transforming growth factor (TGF)-β 1 latency-associated peptide cause Camurati–Engelmann disease because of the formation of a constitutively active form of TGF-β 1. J Biol Chem 276:11469–11472. doi:10.1074/jbc.C000859200

Sambrook PN, MacGregor AJ, Spector TD (1999) Genetic influences on cervical and lumbar disc degeneration: a magnetic resonance imaging study in twins. Arthritis Rheum 42:366–372. doi:10.1002/1529-0131(199902)42:2<366::AID-ANR20>3.0.CO;2-6

Scholzen T, Solursh M, Suzuki S, Reiter R, Morgan JL, Buchberg AM, Siracusa LD, Iozzo RV (1994) The murine decorin. Complete cDNA cloning, genomic organization, chromosomal assignment, and expression during organogenesis and tissue differentiation. J Biol Chem 269:28270–28281

Seki S, Kawaguchi Y, Chiba K, Mikami Y, Kizawa H, Oya T, Mio F, Mori M, Miyamoto Y, Masuda I, Tsunoda T, Kamata M, Kubo T, Toyama Y, Kimura T, Nakamura Y, Ikegawa S (2005) A functional SNP in CILP, encoding cartilage intermediate layer protein, is associated with susceptibility to lumbar disc disease. Nat Genet 37:607–612. doi:10.1038/ng1557

Serra R, Johnson M, Filvaroff EH, LaBorde J, Sheehan DM, Derynck R, Moses HL (1997) Expression of a truncated, kinase-defective TGF-β type II receptor in mouse skeletal tissue promotes terminal chondrocyte differentiation and osteoarthritis. J Cell Biol 139:541–552. doi:10.1083/jcb.139.2.541

Shukunami C, Shigeno C, Atsumi T, Ishizeki K, Suzuki F, Hiraki Y (1996) Chondrogenic differentiation of clonal mouse embryonic cell line ATDC5 in vitro: differentiation-dependent gene expression of parathyroid hormone (PTH)/PTH-related peptide receptor. J Cell Biol 133:457–468

Song YQ, Cheung KM, Ho DW, Poon SC, Chiba K, Kawauchi Y, Hirose Y, Alini M, Grad S, Yee AF, Leong JC, Luk KD, Yip SP, Kaoppinen J, Cheah KS, Sham P, Ikegawa S, Chan D (2008) Association of the asporin D14 allele with lumbar-disc degeneration in Asians. Am J Hum Genet 82:744–747. doi:10.1016/j.ajhg.2007.12.017

Specchia N, Pagnotta A, Toesca A, Greco F (2002) Cytokines and growth factors in the protruded intervertebral disc of the lumbar spine. Eur Spine J 11:145–151. doi:10.1007/s00586-001-0361-y

Stapleton CJ, Pham MH, Attenello FJ, Hsieh PC (2011) Ossification of the posterior longitudinal ligament: genetics and pathophysiology. Neurosurg Focus 30:E6

Tang Y, Wu X, Lei W, Pang L, Wan C, Shi Z, Zhao L, Nagy TR, Peng X, Hu J, Feng X, Hul WV, Wan M, Cao X (2009) TGF-β1-induced migtation of bone mesenchymal stem cells couples bone resoorption with formation. Nat Med 15:757–765. doi:10.1038/nm.1979

Thys M, Van Camp G (2009) Genetics of otosclerosis. Otol Neurotol 30:1021–1032. doi:10.1097/MAO.0b013e3181a86509

Thys M, Schrauwen I, Vanderstraeten K, Janssens K, Dieltjens N, Van Den Bogaert K, Fransen E, Chen W, Ealy M, Claustres M, Cremers CR, Dhooge I, Declau F, Claes J, Van de Heyning P, Vincent R, Somers T, Offeciers E, Smith RJ, Van Camp G (2007) The coding polymorphism T263I in TGF-β1 is associated with otosclerosis in two independent populations. Hum Mol Genet 16:2021–2030. doi:10.1093/hmg/ddm150

Tolonen J, Grönblad M, Virri J, Seitsalo S, Rytömaa T, Karaharju E (2001) Transforming growth factor β receptor induction in herniated intervertebral disc tissue: an immunohistochemical study. Eur Spine J 10:172–176

Warman ML, Cormier-Daire V, Hall C, Krakow D, Lachman R, LeMerrer M, Mortier G, Mundlos S, Nishimura G, Rimoin DL, Robertson S, Savarirayan R, Sillence D, Spranger J, Unger S, Zabel B, Superti-Furga A (2011) Nosology and classification of genetic skeletal disorders: 2010 revision. Am J Med Genet A 155A:943–968. doi:10.1002/ajmg.a.33909

Yamada Y, Miyauchi A, Goto J, Takagi Y, Okuizumi H, Kanematsu M, Hase M, Takai H, Harada A, Ikeda K (1998) Association of a polymorphism of the transforming growth factor-β1 gene with genetic susceptibility to osteoporosis in postmenopausal Japanese women. J Bone Miner Res 10:1569–1576. doi:10.1359/jbmr.1998.13.10.1569

Yang X, Chen L, Xu X, Li C, Huang C, Deng CX (2001) TGF-β/Smad3 signals repress chondrocyte hypertrophic differentiation and are required for maintaining articular cartilage. J Cell Biol 153:35–46. doi:10.1083/jcb.153.1.35

Chapter 17
TGF-β in Brain Disorders

Kerstin Krieglstein

Abstract Transforming growth factor beta (TGF-β) is known to regulate numerous cell functions in the nervous system development, adult maintenance, and degeneration. TGF-β carries roles in neurons and glia and is involved in the regulation of proliferation, differentiation, neuron survival and death, as well as orchestrating its response to lesion. In the context of brain disorders the current understanding of TGF-β action is discussed for brain tumors, neurodegenerative disease, such as Alzheimers' and Parkinson's disease, in insults such as ischemia, stroke, and vascular damage, as well as changes in neuronal activity, such as hyperactivity as seen in epilepsy, or in neuronal depression.

Keywords Cell cycle regulation • ECM • Neuroprotection • Cell death • Blood • Neurons • Astrocytes • Microglia • Myelination • Neurogenesis

Abbreviations

Aβ	Amyloid β
AD	Alzheimer's disease
Alk	Activin-receptor like kinase
ALS	Amyotrophic lateral sclerosis
APP	Amyloid precursos protein
BBB	Blood–brain barrier
BDNF	Brain-derived neurotrophic factor
Cdk	Cyclin-dependent kinase
CNS	Central nervous system

K. Krieglstein (✉)
Department of Molecular Embryology, FRIAS School of Life Science,
University of Freiburg, Albertstr 17, Freiburg, 79104, Germany
e-mail: kerstin.krieglstein@anat.uni-freiburg.de

A. Moustakas and K. Miyazawa (eds.), *TGF-β in Human Disease*,
DOI 10.1007/978-4-431-54409-8_17, © Springer Japan 2013

CNTF	Ciliary neurotrophic factor
CSF	Cerebrospinal fluid
E	Embryonic day
ECM	Extracellular matrix
EGF	Epidermal growth factor
FGF	Fibroblast growth factor
Fox	Forkhead box
GDNF	Glial cell line-derived neurotrophic factor
GFAP	Glial fibrillary acidic protein
GFRα	GDNF receptor
Id4	Inhibitors of DNA binding/differentiation
TIEG	TGF-β immediate early gene
IL	Interleukin
Kir	Inward recitfying potassium channels
MGP	Matrix GLA protein
MHC	Major histocompatobility class
MMP	Matrix metalloproteinase
MPP+	1-methyl-4-phenylpyridinium
MPTP	1-methyl-4-phenyl-1,2,3,6-tetrahydropyridine
MS	Multiples sclerosis
NGF	Nerve growth factor
PAI	Plasminogen activator inhibitor
PD	Parkinson's disease
PDGF	Platelet-derived growth factor
PNS	Peripheral nervous system
SOD	Superoxide dismutase
TGF- β	Transforming growth factor β
TβR	TGF-β receptor
TNF-α	Tumor necrosis factor α
TRAF	TNF-α receptor associated factor
t-PA	Tissue plasminogen activator
VEGF	Vascular endothelial growth factor

17.1 Introduction

The isolation and characterization of transforming growth factors-β (TGF-β) by Anita Roberts has introduced a versatile extrinsic signaling molecule affecting numerous events in the life of almost each cell (Roberts and Sporn 1990; Derynck et al. 1985). Best characterized events include the regulation of cell cycle, composition of the extracellular matrix (ECM), and thereby cell migration or differentiation, as well as regulation of cell survival and death. Imbalance of TGF-β availability is therefore likely to affect tissue development, maintenance, and homeostasis. This certainly also accounts for the nervous system. Once accepted that TGF-β is also expressed in the nervous system, the functional contribution of TGF-β is now more and more understood in numerous events in the development of the nervous system, its maintenance, and consequently also in the context of many brain diseases. This review will discuss the current knowledge of TGF-β in brain development and function and consequently in brain disorders.

Brain disorders include brain tumors, neurodegenerative disease, such as Alzheimers' disease and Parkinson's disease, insults such as ischemia and stroke, and vascular damage, as well as changes in neuronal activity, such as hyperactivity as seen in epilepsy, or neuronal depression.

17.2 TGF-β Expression in the Central Nervous System

Localisation of TGF-β isoforms in mice and rats has been performed by immunohistochemistry and in situ hybridization studies demonstrating a widespread distribution of TGF-β2 and TGF-β3 during development (Flanders et al. 1991; Pelton et al. 1991a, b; Unsicker et al. 1991). TGF-β1 is confined to meninges and choroid plexuses. During mouse development TGF-β2 and -β3 immunoreactivities become first detectable along peripheral nerves, in radial glial cells and along the central nervous system (CNS) axon tracts at embryonic age (E)12. Neuronal cell bodies become immunoreactive from E15 onwards. Most notably, TGF-β immunoreactivity is not detectable in the ventricular zone throughout the neural tube, suggesting that TGF-β may not be involved in the regulation of cell division of neural stem cells during development (Flanders et al. 1991). In contrast, on day E16, cells in the subventricular zone, subplate, and lamina I of the cortex stain positive for TGF-β. As they develop, astrocytes are also immunoreactive for TGF-β2 and -β3. In the adult nervous system both neurons and astroglia are immunoreactive for TGF-β2, -β3. Immunoreactive neuron populations include cortical layers 2, 3, and 5, hippocampus, piriform cortex, retinal ganglionic cells, hindbrain aminergic neurons, as well as spinal and hindbrain motoneurons (Unsicker et al. 1991). TGF-β1 is most prominent within the choroid plexus and meninges, it may, however, be expressed in other cells below levels of detectability. Upon lesioning, TGF-β1 may by upregulated in astrocytes as well as in neurons in vivo. TGF-β1 becomes also detectable in tissue culture, possibly mimicking a lesion-like situation. In primary neural tissue culture, treatment with all three TGF-β isoforms usually results in identical responses, suggesting that the recombinant proteins used have similar affinities for their shared receptor complex (Krieglstein and Unsicker 1994; Massague 2000).

In addition to the distribution of TGF-β within the peripheral nervous system (PNS) and CNS, its subcellular localization and mode of secretion is of importance in order to elaborate on its possible functions. Taking PC12 cells as a model to study sorting in the trans-Golgi network, Specht et al. (2003) could show that TGF-β2 may be sorted and released to a large proportion via the regulated path of secretion. Secretory vesicles provide a milieu of pH 5, which is suitable for TGF-β activation within the vesicle, enabling release of active TGF-β (Specht et al. 2003). Activity-depended release of TGF-β2 may suggest a function as a modulator for synaptic plasticity (Lacmann et al. 2007).

17.3 TGF-β in Brain Tumors

The role of TGF-β in cancer biology is complex and involves both aspects of tumor suppression (Bartholin et al. 2013; Seoane 2006) as well as tumor promotion (Wendt and Schiemann 2013; Joseph et al. 2013; Roberts and Wakefield 2003).

TGF-βs are well known for their capacity to regulate cell proliferation in a context-dependent manner. There are at least four scenarios in which regulation of cell proliferation is an important issue in nervous system development and maintenance: (a) neurogenesis, (b) proliferation of neuroblasts (neural crest cells), (c) proliferation of glial cells during development or upon lesioning, and (d) upon transformation in tumors.

Neurogenesis in the neural tube requires definite exit of the cell cycle to generate postmitotic neurons. In the past years there is increasing evidence for the role of TGF-β in developmental and adult neurogenesis (Vogel et al. 2010; Aigner and Bogdahn 2008). For some years there was indirect evidence available that neural stem cells in the neuroepithelium need to be protected from the action of TGF-β, in order to prevent premature growth retardation (Seoane et al. 2004; Hanashima et al. 2002). Seoane and coworkers have demonstrated on the basis of protein interaction analysis in human HaCaT keratinocytes that expression of the cyclin-dependent kinase (cdk) inhibitory protein 1 (p21Cip1) is regulated by TGF-β-dependent Smad complexes in combination with the Forkhead box (Fox) family member FoxO. This FoxO-Smad complex is inhibited by FoxG1, which has been shown to be essential for proliferation of telencephalic progenitor cells (Xuan et al. 1995). Indeed, FoxG1 mutants, which display reduced proliferation of telencephalic progenitor cell, premature differentiation and early depletion of the progenitor population (Xuan et al. 1995) show high levels of p21Cip1 expression in TGF-β-sensitive progenitor cells (Seoane et al. 2004). Exit from the cell cycle during terminal differentiation, as required for neurogenesis, has been described to be regulated by Ink4d and Kip1 inhibitors of cyclin-dependent kinases (Zindy et al. 1999; Cunningham and Roussel 2001). P27Kip1 has been identified as a TGF-β-dependent target gene; however, there is no evidence for a TGF-β-dependent regulation of p19Ink4d. This suggests that TGF-β may serve as an extracellular regulator to induce cell cycle arrest at the G1 phase in neural stem cells but may probably not be sufficient to regulate cell cycle exit required for terminal differentiation. In mouse hippocampal progenitor cells, TGF-β causes induction of p21Cip1 and downregulation of Cdk activators Ccnd1 and Ccnd2, leading to cell cycle exit and neuronal differentiation (Vogel et al. 2010). Along this line, TGF-β2/-β3 double knockout mice display increased cell proliferation and reduced numbers of neurons in the developing cerebral cortex and hippocampus, as the progenitor cells failed to differentiate into neurons because they did not exit cell cycle. Furthermore, TGF-β1 has been implicated as a negative modulator of adult neurogenesis (Wachs et al. 2006).

By affecting the cell cycle prior to terminal differentiation, TGF-β may, of course, regulate proliferation of neuroepithelial cells, including neuroblasts, neural crest cells, and glial progenitors (Zhang et al. 1997a; Anchan and Reh 1995).

Furthermore, TGF-β2 has been shown to regulate cell proliferation in neural crest-derived chromaffin cells (Rahhal et al. 2004) with the capacity of lifelong proliferation.

Tumors of the CNS include primitive neuroectodermal tumors, such as gliomas and medulloblastomas (Fogarty et al. 2005; Nieder et al. 2003). They derive from dividing glial cells, or neural stem and progenitor cells. Glial brain tumors are further classified using grades I–IV to express the likelihood of increased growth and malignancy. CNS tumors are characterized by rapid and infiltrative growth, angiogenesis, and immune suppression. Due to the proliferative behavior of brain cells, brain tumors show a high occurrence not only during development affecting children but also during adulthood. Particularly adult neural stem cells with the capacity to provide new neurons and glia in regions with high plasticity, following injury, or in the context of specific diseases, may escape their physiological control machinery and transform into brain cancer stem cells (reviewed in Aigner and Bogdahn 2008).

Id4 (inhibitors of DNA binding/differentiation) has been shown to serve important functions in neural stem cell differentiation and its deregulation has been implicated in glial neoplasia (Dell'Orso et al. 2010). Deregulation could occur via functional point mutations or epigenetic silencing. Martini and coworkers (2012) were able to show that epigenetic silencing of Id4 via hypermethylation resulted in reduced expression of matrix GLA protein (MGP), TGF-β1 and vascular endothelial growth factor (VEGF) which was associated with a more favorable clinical outcome. However, there are many possibilities to circumvent this effect. First, as TGF-β actions are context-dependent, the presence of certain mitogens, such as TGF-α/epidermal growth factor (EGF) or platelet-derived growth factor (PDGF) may turn TGF-β into a growth stimulating factor (Roberts et al. 1981; Leof et al. 1986; Seoane 2006). Secondly, transformed cells may become insensitive to TGF-β due to overproduction of TGF-β or due to mutations of TGF-β receptors, their signaling components, or even their target genes responsible for G1 arrest (for example Rich et al. 1999; Lyons et al. 1990; Markowitz et al. 1995; Hahn et al. 1996; Seoane et al. 2004; Rich 2003 for review).

TGF-β´s ability to regulate ECM composition puts TGF-β at high risk in the regulation of tumor invasion and metastasis. In this context TGF-β has been shown to regulate integrin expression, for example integrin $\alpha_V\beta_3$ which has been shown to play a role in glioma propagation (Uhm et al. 1999). TGF-β has also been shown to upregulate matrix metalloprotease 2 (MMP-2) and MMP-9 expression at the cell surface (Rooprai et al. 2000) that may interact with $\alpha_V\beta_3$ integrin (for review, see Platten et al. 2001).

TGF-β is a potent immunosuppressive cytokine (Wahl and Chen 2003; Roth et al. 2012; Hau et al. 2011). Brain tumors are well known for their immunosuppressive properties allowing them to escape from the host's immune surveillance. TGF-β2 and TGF-β3 are considered master molecules that upon secretion mediate this immunosupressive environment. This immunosuppressive role has been attributed to TGF-β2, which is also the preferentially expressed isoform by many glioblastomas, grade IV gliomas (Bodmer et al. 1989; Hau et al. 2011). On this basis, TGF-β2-specific antisense gene therapy strategies have been established to make tumor cells accessible to an effective anti-tumor immune response and counteract TGF-β-dependent tumor metastasis (Hau et al. 2009; Jachimczak et al. 1993; Lou 2004). Along this line, there is extensive research going on to identify TGF-β signaling inhibitors for cancer therapy (DaCosta et al. 2004; Yingling et al. 2004; Lahn et al. 2005).

17.4 Vascular Damage in the CNS

The blood–brain barrier (BBB) generates the specific milieu of the brain by building a tight boundary and thereby separating the components of the circulating blood from the brain. After injury or in neurologic diseases including trauma, ischemia/stroke, or Alzheimer's disease (AD), leakage of the BBB results in the entry of blood constituents into the brain (Abbott et al. 2006). Plasma proteins such as albumin, immunoglobulins, amyloid-β, and fibrinogen, and vascular cells such as erythrocytes and leucocytes, leaking into the brain have been associated with inflammation and restriction of repair (for review, see Beck and Schachtrup 2012). Schachtrup and coworkers (2010) identified TGF-β as a vascular-derived protein. Specifically, they could demonstrate that the plasma-derived protein fibrinogen acts as a carrier of latent TGF-β. Its activation is mediated via avβ6 and avβ8 integrins present on the surface of astrocytes.

Increased levels of TGF-β1 has been described in human brains during trauma, multiple sclerosis (MS), Parkinson's, AD, and stroke patients (Lippa et al. 1995). In patients suffering from severe head injury, high levels of TGF-β1 could be detected in the cerebrospinal fluid (CSF) within 1 day after injury (Csuka et al. 1999; Morganti-Kossmann et al. 1999). These observations strongly suggest that lesion-induced, vascular-derived TGF-β contributes to the corresponding degeneration and regeneration processes (Beck and Schachtrup 2012).

TGF-β has been profoundly investigated for its role in orchestrating the response to brain lesions (for review, see Flanders et al. 1998). With regard to astrocytes, this includes regulation of astrocytic growth, astroglial scar formation, and anti-inflammatory responses. In most contexts studied, TGF-β inhibits the growth of astrocytes (Flanders et al. 1993; Hunter et al. 1993). Most importantly, TGF-β counteracts mitogenic signals by astroglial mitogens such as fibroblast growth factor-2 (FGF-2) or PDGF. However, effects may vary depending on astrocyte culture conditions in vitro or may be brain region-dependent in vivo (Labourdette et al. 1990; Johns et al. 1992). TGF-β may also affect cell adhesion, migration, and ECM production by astrocytes, all being important steps in the cascade of shaping the reactive astrocyte phenotype. TGF-β-treated astrocytes show a slight increase in actin content, the appearance of actin stress fibers, a slight increase in the glial fibrillary acidic protein (GFAP), and an increased production of laminin and fibronectin (cf. Baghdassarian et al. 1993). Thus, treatment of cerebral wounds with anti-TGF-β2 antibodies was shown to lead to a marked reduction of glial scarring (Logan et al. 1999). Many effects of TGF-β on astroglia are anti-inflammatory and immunosuppressive, as TGF-β modulates the expression of important cytokines involved in CNS immune reactions. These include upregulation of interleukin-6 (IL-6) and nerve growth factor (NGF) (Spittau et al. 2012; Aderka et al. 1989; Lindholm et al. 1992), blocking interferon-γ mediated upregulation of major histocompatibility class (MHC) II (Dong et al. 2001), and the tumor necrosis factor α (TNF-α) and interleukin-1 β (IL-1 β)-mediated upregulation of intercellular adhesion molecule-1 (Shrikant et al. 1996).

17.5 TGF-β in Neuronal Survival and Death

TGF-β has been shown to promote survival of several neuronal populations in vitro (Krieglstein et al. 1995; Poulsen et al. 1994; Martinou et al. 1990). However, it is now well established that TGF-β may modulate the neurotrophic capacities of numerous growth factors including neurotrophins, such as NGF, brain-derived neurotrophic factor (BDNF), as well as ciliary neurotrophic factor (CNTF) (Krieglstein and Unsicker 1996) and, most importantly, glial cell line-derived neurotrophic factor (GDNF; Krieglstein et al. 1998b). GDNF was shown to crucially depend on TGF-β to exert its neurotrophic activities on peripheral as well as mesencephalic dopaminergic neurons in vitro. In vivo, GDNF's neuroprotective effect on target-deprived pre-ganglionic sympathetic neurons, as well as 1-methyl-4-phenyl-1,2,3,6-tetrahydropyridine (MPTP)-lesioned nigrostriatal dopaminergic neurons, also depends on the presence of TGF-β (Schober et al. 1999, 2007). GDNF/TGF-β cooperativity on chick ciliary ganglionic neurons has now been characterized in detail, whereby TGF-β is required for appropriate GDNF receptor (GFRα1) recruitment to the plasma membrane (Peterziel et al. 2002). Interestingly, TGF-β does not cooperate with Neurturin, a closely related factor to GDNF, and does not promote the recruitment of GFRα2 to the plasma membrane, suggesting high specificity in TGF-β/GDNF cooperativity (Peterziel et al. 2007).

Depending on the cellular context, TGF-β has also been shown to regulate ontogenetic neuron death. Upon immunoneutralization of all TGF-β isoforms in ovo (E6–E10), ontogenetic cell death of chick parasympathetic ciliary ganglionic neurons, sensory dorsal root ganglionic neurons as well as lumbar spinal motoneurons could be prevented (Krieglstein et al. 2000). Similarly, TGF-β regulates ontogenetic morphogenetic cell death in the developing retina of chick and mouse embryos (Dünker et al. 2001; Dunker and Krieglstein 2003). Another classical model for morphogenetic cell death during embryogenesis represents the removal of interdigital tissue to form individual fingers. Similarly, double deletion of TGF-β2 and -β3 in the mouse resulted in lack of cell death (Dunker et al. 2002). Furthermore, induced neuron death following embryonic limb bud ablation in chick embryos resulted in a significant neuroprotection upon immunoneutralization of TGF-β (Krieglstein et al. 2000). Together, these data suggest that TGF-β is a key regulator of ontogenetic cell death in vivo. Mechanistically, we recently identified that TGF-β-induced apoptosis in oligodendroglial progenitor cells (OLI-neu; Schuster et al. 2002) is characterized by downregulation of Bcl-xl. Furthermore, Fractin is produced as a caspase-specific cleavage product in oligodendroglial cells during TGF-β-mediated apoptosis, whereby Fractin binding to Bcl-xl induced downregulation of Bcl-xl protein levels (Schulz et al. 2009). Sorrentino and collaborators (2008) were able to show that the intracellular apoptotic cascade can be initiated via the type I receptor of TGF-β (TβR-I) and receptor-engaged TNF-α receptor associated factor 6 (TRAF6). Although TGF-β-induced apoptosis and underlying signaling pathways have been well characterized in many cells types, little is known about TGF-β-induced apoptosis in neurons (Schuster and Krieglstein 2002; Sanchez-Capelo 2005).

17.6 Cerebral Ischemia

Cerebral ischemia is caused by either a blood clot occluding a blood vessel in the brain (focal ischenia) or a more general reduction in brain blood flow (global ischemia) leading to insufficient blood flow and reduced oxygen levels in the respective brain areas, thus leading to death of brain tissue. Neuronal cell death may occur as necrosis or apoptosis. Thrombolysis is the approved treatment of stroke. TGF-β1 expressed at low levels in adult brain is rapidly upregulated following insults such as cerebral ischemia, excitatory injury, or traumatic brain injury (Klempt et al. 1992; Knuckey et al. 1996; Yamashita et al. 1999; Morganti-Kossmann et al. 1999; Zhu et al. 2000; Boche et al. 2003; Krieglstein 2006; Pál et al. 2012). TGF-β1 upregulation was observed primarily in microglial cells and in astrocytes, while TGF-β2 upregulation was seen in neurons. TGF-β3 was not upregulated; however, the levels of both TGF-β2 and TGF-β3 decreased subsequently. These data suggest a distinct spatiotemporal requirement of TGF-β isoforms action during cerebral ischemia (Pál et al. 2012). As TGF-β is a good candidate to organize the response of neurons to degeneration as well as mediating anti-inflammatory reactions, its neuroprotective potential has been widely analyzed (for review, see Flanders et al. 1998; Böttner et al. 2000; Dobolyi et al. 2012). Specifically, TGF-β1 applied either as recombinant protein or by adenoviral-based overexpression has been shown to reduce infarct size after focal cerebral ischemia and to prevent hippocampal neuronal damage after global ischemia (Gross et al. 1993; Prehn et al. 1993; Zhu et al. 2002; for review, see Buisson et al. 2003; Dhandapani and Brann 2003). Furthermore, TGF-β may also mediate tolerance of ischemic preconditioning towards subsequent ischemic insult (Boche et al. 2003). The molecular mechanism(s) by which TGF-β protects neurons from ischemic cell death relies on a signaling crosstalk between neurons and astrocytes (Prehn et al. 1994; Docagne et al. 1999) and involves the maintenance of Ca^{2+} homeostasis, modulation of the t-plasminogen activator (tPA)/plasminogen activator inhibitor (PAI-1) axis, as well as inhibition of pro-apoptotic pathways, such as Bad and caspase-3 (Zhu et al. 2001, 2002) and upregulation of anti-apoptotic proteins such as Bcl-2 (Prehn et al. 1994). An additional TGF-β-dependent anti-apoptotic pathways involving NF-κB activation has been described (Zhu et al. 2004). This pathway seems to be downstream of ALK1 (activin receptor-like kinase

1), an alternative TGF-β type I receptor first described on endothelial cells, which has been shown to be upregulated in neurons in an injury-dependent manner (Konig et al. 2005). Injury-dependent upregulation of ALK1, with signaling preference towards Smad1, may also explain numerous opposing effects of TGF-β in brain development and lesions.

17.7 Alzheimer's Disease

Alzheimer's disease (AD) is a degenerative brain syndrome characterized by a progressive decline in learning and memory, thinking, language, judgment, and other higher brain functions. Currently about 18 million people worldwide suffer from AD and it is estimated that in 2025, 34 million will be affected. The statistical risk of occurrence is 1.4 % at the age of 60 and doubles every 5 years thereafter (WHO; Huang and Mucke 2012).

AD is characterized by considerable brain shrinkage resulting in a large loss of brain weight and volume. The extent of brain volume loss suggests more general mechanisms such as shrinkage and loss of neuronal processes, including degeneration of specific neuron populations (Huang and Mucke 2012). Along this line, aberrant neuronal network activity, dysfunction, and loss of synapses may describe the cognitive decline in AD.

TGF-β has been implicated in the regulation of neurite outgrowth, transmitter synthesis as well as synapse formation (Krieglstein et al. 2011). TGF-β has been reported to cause neurite sprouting and elongation of hippocampal axons as well as promoting re-elongation of injured axons of hippocampal neurons in vitro (Ishihara et al. 1994; Abe et al. 1996). In the mouse neocortex, TGF-β can direct neuronal polarity by initiation of axon formation and neuronal migration via site-specific phosphorylation of the polarity protein Par6 (Yi et al. 2010). Extracellular signaling factors such as Wnt and TGF-βs are recognized as target-derived signals in synaptogenesis (Salinas 2005; Packard et al. 2003). In chick ciliary ganglionic neurons, developmental expression of K_{Ca} channels coincides with synaptogenesis. Dryer and coworkers have shown that target-derived TGF-β1 regulates the developmental expression of Ca^{2+}-activated K^+ currents in vitro and in vivo (Cameron et al. 1999). The acute effect of TGF-β1 relies on the translocation of K_{Ca} channels from intracellular stores to the plasma membrane involving signaling via the Ras GTPase, extracellular regulated kinases (Erk), and phosphoinositide 4′ (PI4) kinase (for review, see Dryer et al. 2003). In conclusion, TGF-β may well be suited to modulate synaptic plasticity and cognition (for review, see Krieglstein et al. 2011).

The pathogenesis of AD is focusing on amyloid β (Aβ) peptides, the main constituent in plaques, derived from amyloid precursor protein (APP) upon proteolytic cleavage (De Strooper et al. 2010; Bertram et al. 2010). There are several lines of evidence suggesting that TGF-β1 may contribute to the pathology of Alzheimer's disease, particularly through promoting Aβ precursor expression and Aβ deposition (Burton et al. 2002; Wyss-Coray et al. 1997a, b; Flanders et al. 1995;

van der Wal et al. 1993). Mice overexpressing TGF-β1 in astrocytes develop AD-like vascular and meningeal abnormalities with age (Gaertner et al. 2005). These chronic alterations could be correlated with reduced brain tissue perfusion, leading to an increased amount of fibrillar and soluble Aβ peptides. However, in brain parenchyma, astroglial TGF-β1 expression leads to a reduction of overall Aβ as well as decreased numbers of dystrophic neurites (Wyss-Coray et al. 2001). The reduced plaque burden in brain parenchyma is thought to depend on TGF-β-dependent microglial activation and microglial Aβ-clearance. Furthermore, a genetic association study of three polymorphisms of the human TGF-β1 gene with AD suggests that there is no correlation between TGF-β1 and AD on the basis of TGF-β1 gene variability (Araria-Goumidi et al. 2002).

Most recently, bioactive TGF-β has been shown to be associated with lipoproteins, specifically those containing ApoE3, but not ApoE4 (Tesseur et al. 2009). Association of TGF-β1 with lipoproteins may facilitate its diffusion and signaling and possibly also other biological functions of TGF-β1. This observation is of particular interest in this context as ApoE4 has been genetically linked with late-onset AD (Bertram et al. 2010). As TGF-β1 is a neuroprotective agent and may be beneficial in the AD condition, for example through reduction of plaque burden, the preferential binding of TGF-β1 to ApoE3 versus ApoE4 may put ApoE4 carriers on higher risk for developing AD (Tesseur et al. 2009).

Several lines of evidence suggest impairment of TGF-β-activated Smad signaling, with ectopic localization of phosphorylated Smad2/3 within amyloid plaques and neurofibrillary tangles (Lee et al. 2006; Tesseur et al. 2006; Ueberham et al. 2006; Chalmers and Love 2007a, b). Furthermore, AD patients have been shown to have reduced plasma levels of TGF-β1 (Mocali et al. 2004; Juraskova et al. 2010) as well as reduced neuronal expression of the TGF-β type II receptor (TβRII) (Tesseur et al. 2006). Lack of TβRII signaling via neuronal expression of kinase-deficient TβRII in AD transgenic mice promoted Aβ deposition and loss of dendrites (Tesseur et al. 2006), while Aβ may downregulate expression of TβRII via induction of miR-106b (Wang et al. 2010). Finally, injection of synthetic Aβ in combination with blocking TGF-β signaling via application of the TβRI kinase inhibitor SB431542 significantly increased the vulnerability of hippocampal neurons to Aβ, leading to neuronal degeneration (Caraci et al. 2008).

Together, as TGF-β1 signaling is beneficial in the AD environment, rescuing TGF-β1 levels and TGF-β signaling may represent a new strategy for neuroprotection in AD (Caraci et al. 2012).

17.8 Parkinson's Disease

Parkinson's disease (PD) is a neurodegenerative disease characterized by a progressive loss of nigrostriatal neurons and in consequence by marked reduction of striatal dopamine resulting in impaired voluntary movement (for review, see Braak et al. 2004; Dunnett and Björklund 1999). In addition to the loss of neurons, a further morphologic hallmark of PD is the presence of Lewy bodies and Lewy neurites (Forno 1996). The formation of these proteinaceous inclusions involves interaction of several proteins, including α-synuclein (Spillantini et al. 1998). The etiology of PD is likely due to combinations of environmental and genetic factors (for review, see, for example, Valente et al. 2012; Gasser et al. 2011).

TGF-β2 and TGF-β3 are expressed in adult nigral dopaminergic neurons (Unsicker et al. 1991), and TGF-β1 and -β2 were elevated in biopsies of PD patients (Nagatsu et al. 2000). TGF-βs have been shown to promote midbrain dopaminergic neuron survival in vitro and in vivo (Krieglstein and Unsicker 1994; Poulsen et al. 1994; Roussa et al. 2004), as well as protection against 1-methyl-4-phenylpyridinium (MPP+) intoxication (Krieglstein et al. 1995; Roussa et al. 2009). Most importantly, TGF-β cooperates with GDNF to promote dopaminergic neuron survival (Krieglstein et al. 1998b). GDNF is well known as a potential therapeutic agent coping with PD (for review, see Björklund and Lindvall 2000). However, also in vivo GDNF-dependent neuroprotective effects are based on the cooperativity with TGF-β, as shown in the MPTP-mouse model, an animal model of PD (Schober et al. 2007). This neuroprotective strategy has already been used incidentally by grafting extra-adrenal chromaffin cells obtained from Zuckerkandl's organ in parkinsonian rats (6-hydroxydopamine model). The behavioral improvements of parkinsonia deficts were in addition to the supply of catecholamines attributed to the release of the survival promoting proteins GDNF and TGF-β1 (Fernandez-Espejo et al. 2005).

Most recently, a new animal model for PD has been introduced. Viral transformation of rats with α-synuclein showed a slow progression of nigral dopaminergic neurons (Ulusoy et al. 2010). Most interestingly, GDNF, for long considered as the gold standard in neurotrophic-based neuroprotection of PD, is not able to rescue α-synuclein-mediated degeneration of dopaminergic neurons (Decressac et al. 2011).

17.9 Epilepsy

Epilepsy is a common neurological disorder affecting 0.5–2 % of the population worldwide. Epilepsy is characterized by seizures resulting from abnormal neuronal activity. So far, there is no cure known for the disease. The mechanisms leading to the disease are only poorly understood. However, epilepsy is often seen following brain trauma, ischemic or infectious brain injury, or drug and alcohol misuse. These conditions may be accompanied by vascular damage and leakage of the blood–brain barrier (see above).

Cacheaux et al. (2009) have identified the involvement of TGF-β signaling in epileptogenesis. The group has previously demonstrated that serum albumin causes epileptic field potentials when exposed to brain slices in vitro (Ivens et al. 2007). Serum albumin was taken up by astrocytes leading to down regulation of inward-rectifying potassium (Kir 4.1) channels, resulting in reduced buffering of extracellular potassium in neuronal hyperexcitability and epileptiform activity. As the albumin uptake was shown to occur in a TGF-β receptor-mediated mechanism, blocking of the TβR in vivo reduced the likelihood of epiletogenesis in albumin exposed brains (Ivens et al. 2007). In follow-up experiments, the group was able to show that TGF-β1-mediated signaling is sufficient to induce epileptoforming activity (Cacheaux et al. 2009). These data strongly link the TGF-β pathway with epileptogenesis and identify the TGF-β pathway as a therapeutic target for the prevention of injury-induced epilepsy (Friedman 2011).

17.10 Depression

Major depressive disorder is a mental disorder characterized by low mood, low self-esteem, and reduced interest in enjoying pleasure. There is no clinical test for depression. Patients are treated with antidepressant drugs, which improves their mental condition after several weeks of treatment. Most antidepressant medications directly or indirectly increase the levels of one or more of the monoamines, such as serotonin, noradrenaline, and dopamine, in the synaptic cleft between neurons in the brain, suggesting that depression may be the consequence of reduced synaptic activity. In addition to neurotransmitters, also neuropeptides and neurotrophic factors such as BDNF and TGF-β have been shown to be released in an activity-dependent manner (Thoenen 1995; Specht et al. 2003; Lacmann et al. 2007). Notably, KCl stimulation caused Smad translocation into the nucleus and induced TGF-β-inducible early gene (Tieg1) expression, demonstrating that activity-dependent released TGF-β may exert autocrine actions and thereby activate the TGF-β-dependent signaling pathway (Lacmann et al. 2007). These results suggest an activity-dependent release and gene transcription of TGF-β in mouse hippocampal neurons in vitro as well as subsequent autocrine functions of the released TGF-β within the hippocampal network. TGF-β is also known to have a prominent role in long-term synaptic facilitation in isolated Aplysia ganglia (Zhang et al. 1997b). Within minutes, TGF-β1 stimulated MAPK-dependent phosphorylation of synapsin, which appeared to modulate synapsin distribution, and resulted in a reduced magnitude of synaptic depression (Chin et al. 2002). Most recently, Fukushima and coworkers (2007) were able to show that TGF-β modulates synaptic efficacy and plasticity in dissociated rat hippocampal neurons. Together, increasing evidence suggests that TGF-β may be involved in synaptogenesis, modulation of synaptic transmission, and synaptic plasticity.

The delayed effects of antidepressants are thought to depend on indirect mechanisms, including the regulation of gene expression, for example antidepressant-induced upregulation of BDNF signaling, which then in turn promotes adaptive neuronal plasticity through effects on gene expression. Wibrand and coworkers (2006) identified five genes (Neuritin, Narp, Tieg1, Carp, and Arl4d) that are co-upregulated with Arc during BDNF-LTP. Tieg1 is a TGF-β-dependent immediate early gene that has also been shown to be upregulated in hippocampal neurons by TGF-β (Lacmann et al. 2007). As TGF-β has been shown to cooperate with BDNF in several scenarios, it may be quite likely that BDNF and TGF-β are the key players in antidepressant-mediated restoration of neuronal plasticity in patients suffering from major depressive disorders.

17.11 Motoneuron Disease

Amyotrophic lateral sclerosis (ALS) is a fatal disease, leading to paralysis and death. It is characterized by loss of motoneurons. Some ALS cases are due to mutation of the superoxide dismutase 1 (SOD1). TGF-β is a prominent motoneuron survival factor (McLennan and Koishi 2002; Martinou et al. 1990). Using SOD1 knockout mice as a model for ALS application of TGF-β2 caused a rapid and marked but transient improvement in the motoric performance of the mice (Day et al. 2005). In the past years, all components of the TGF-β signaling system have been localized in the presynaptic terminal of the neuromuscular junction, whereby TGF-β ligands are synthesized and localized on the postsynaptic side (McLennan and Koishi 2002; Toepfer et al. 1999). Furthermore, it has been shown that TGF-β2 alters the characteristics of the neuromuscular junction by regulating presynaptic quantal size (Fong et al. 2010).

Appropriate myelination is an important aspect of neuronal activity. Oligodendroglial cells are the myelinating cells of the CNS and Schwann cells of the PNS. Oligodendrocytes arise from a bipotential progenitor cell, the O2A progenitor. TGF-β restricts their PDGF-driven proliferation and induces oligodendroglial differentiation (McKinnon et al. 1993) but may also induce apoptosis (Schuster et al. 2002). In the PNS, TGF-β mediates developmental cell death of Schwann cells (Parkinson et al. 2001) and blocks Schwann cell myelination and expression of myelin-related proteins (Awatramani et al. 2002 and references therein). However, in adult mice TGF-β seems to stabilize compact myelin, as TGF-β1-null mice have grossly abnormal myelin (Day et al. 2003). Ski, a repressor of Smad-mediated TGF-β signaling controls Schwann cell proliferation and myelination, whereas absence of Ski abolished the formation of peripheral myelin, and myelinating Schwann cells upregulate Ski in development as well as during remyelination upon injury (Bonnon and Atanasoski 2012; Atanasoski et al. 2004).

17.12 Conclusions

TGF-β is a multifunctional and versatile molecule, effecting development, adult maintenance as well as aging of the brain. Although the prototype of a superfamily TGF-β is acting in the nervous system in a highly specific manner. In the context of brain tumors TGF-β is acting very much the way it as expected to by regulating cell cycle, adhesion, and immunosuppression. However, in all other disease scenarios TGF-β action is much more versatile reaching from orchestrating astrocyte and microglia activation to regulating growth factor responsiveness, uptake and release mechanisms, activity-dependent gene expression, and nerve myelination. TGF-β responsiveness may be mediated via alternative TβR-1 usage and may be blocked via the miRNA-dependent downregulation of TβR-II. A new and fascinating aspect is also evident from the modes of TGF-β delivery within the body through transport via lipoproteins or via fibrinogen. In conclusion, the role of TGF-β within brain disorders, either as cause or as key molecule orchestrating responses is only at the beginning of its understanding. Major open issues regard the specific action of individual TGF-β isoform, the context-specificity, the role of TGF-β activation to regulate TGF-β function and complexity of TGF-β signaling and crosstalk. This knowledge will then be also helpful to explain opposing actions of TGF-β such as promotion of survival/protection as to induction of cell death.

Acknowledgments Work from the author's laboratory is supported by grants from the *Deutsche Forschungsgemeinschaft* (including SFB780, Kr1477/10 and 11).

References

Abbott NJ, Rönnbäck L, Hansson E (2006) Astrocyte-endothelial interactions at the blood–brain barrier. Nat Rev Neurosci 7:41–53

Abe K, Chu PJ, Ishihara A, Saito H (1996) Transforming growth factor-β 1 promotes re-elongation of injured axons of cultured rat hippocampal neurons. Brain Res 723:206–209

Aderka D, Le JM, Vilcek J (1989) IL-6 inhibits lipopolysaccharide-induced tumor necrosis factor production in cultured human monocytes, U937 cells, and in mice. J Immunol 143:3517–3523

Aigner L, Bogdahn U (2008) TGF-β in neural stem cells and in tumors of the central nervous system. Cell Tissue Res 331:225–241

Anchan RM, Reh TA (1995) Transforming growth factor-β-3 is mitogenic for rat retinal progenitor cells in vitro. J Neurobiol 28:133–145

Araria-Goumidi L, Lambert JC, Mann DM, Lendon C, Frigard B, Iwatsubo T, Cottel D, Amouyel P, Chartier-Harlin MC (2002) Association study of three polymorphisms of TGF-β1 gene with Alzheimer's disease. J Neurol Neurosurg Psychiatry 73:62–64

Atanasoski S, Notterpek L, Lee HY, Castagner F, Young P, Ehrengruber MU, Meijer D, Sommer L, Stavnezer E, Colmenares C, Suter U (2004) The protooncogene Ski controls Schwann cell proliferation and myelination. Neuron 43:499–511

Awatramani R, Shumas S, Kamholz J, Scherer SS (2002) TGFβ1 modulates the phenotype of Schwann cells at the transcriptional level. Mol Cell Neurosci 19:307–319

Baghdassarian D, Toru-Delbauffe D, Gavaret JM, Pierre M (1993) Effects of transforming growth factor-β1 on the extracellular matrix and cytoskeleton of cultured astrocytes. Glia 7:193–202

Bartholin L, Vincent DF, Valcourt U (2013) TGF-β as tumor suppressor: in vitro mechanistic aspects of growth inhibition. In: Moustakas A, Miyazawa K (eds) TGF-β in human disease. Springer, Tokyo, pp 113–138.

Beck K, Schachtrup C (2012) Vascular damage in the central nervous system: a multifaceted role for vascular-derived TGF-β. Cell Tissue Res 347:187–201

Bertram L, Lill CM, Tanzi RE (2010) The genetics of Alzheimer disease: back to the future. Neuron 68:270–281

Björklund A, Lindvall O (2000) Parkinson's disease gene therapy moves towards the clinic. Nat Med 6:1207–1208

Boche D, Cunningham C, Gauldie J, Perry VH (2003) Transforming growth factor-β 1-mediated neuroprotection against excitotoxic injury in vivo. J Cereb Blood Flow Metab 23:1174–118

Bodmer S, Strommer K, Frei K, Siepl C, de Tribolet N, Heid I, Fontana A (1989) Immunosuppression and transforming growth factor-β in glioblastoma. Preferential production of transforming growth factor-β 2. J Immunol 143:3222–3229

Bonnon C, Atanasoski S (2012) c-Ski in health and disease. Cell Tissue Res 347:51–64

Böttner M, Krieglstein K, Unsicker K (2000) The transforming growth factor-βs: structure, signaling, and roles in nervous system development and functions. J Neurochem 75:2227–2240

Braak H, Ghebremedhin E, Rub U, Bratzke H, Del Tredici K (2004) Stages in the development of Parkinson's disease-related pathology. Cell Tissue Res 318:121–134

Buisson A, Lesne S, Docagne F, Ali C, Nicole O, MacKenzie ET, Vivien D (2003) Transforming growth factor-β and ischemic brain injury. Cell Mol Neurobiol 23:539–550

Burton T, Liang B, Dibrov A, Amara F (2002) Transcriptional activation and increase in expression of Alzheimer's β-amyloid precursor protein gene is mediated by TGF-β in normal human astrocytes. Biochem Biophys Res Commun 295:702–712

Cacheaux LP, Ivens S, David Y, Lakhter AJ, Bar-Klein G, Shapira M, Heinemann U, Friedman A, Kaufer D (2009) Transcriptome profiling reveals TGF-β signaling involvement in epileptogenesis. J Neurosci 29:8927–8935

Cameron JS, Dryer L, Dryer SE (1999) Regulation of neuronal K(+) currents by target-derived factors: opposing actions of two different isoforms of TGFβ. Development 126:4157–4164

Caraci F, Battaglia G, Busceti C, Biagioni F, Mastroiacovo F, Bosco P, Drago F, Nicoletti F, Sortino MA, Copani A (2008) TGF-β 1 protects against Aβ-neurotoxicity via the phosphatidylinositol-3-kinase pathway. Neurobiol Dis 30:234–242

Caraci F, Spampinato S, Sortino MA, Bosco P, Battaglia G, Bruno V, Drago F, Nicoletti F, Copani A (2012) Dysfunction of TGF-β1 signaling in Alzheimer's disease: perspectives for neuroprotection. Cell Tissue Res 347:291–301

Chalmers KA, Love S (2007a) Neurofibrillary tangles may interfere with Smad 2/3 signaling in neurons. J Neuropathol Exp Neurol 66:158–167

Chalmers KA, Love S (2007b) Phosphorylated Smad 2/3 colocalizes with phospho-tau inclusions in Pick disease, progressive supranuclear palsy, and corticobasal degeneration but not with alpha-synuclein inclusions in multiple system atrophy or dementia with lewy bodies. J Neuropathol Exp Neurol 66:1019–1026

Chin J, Angers A, Cleary LJ, Eskin A, Byrne JH (2002) Transforming growth factor β1 alters synapsin distribution and modulates synaptic depression in Aplysia. J Neurosci 22:RC220

Csuka E, Morganti-Kossmann MC, Lenzlinger PM, Joller H, Trentz O, Kossmann T (1999) IL-10 levels in cerebrospinal fluid and serum of patients with severe traumatic brain injury: relationship to IL-6, TNF-α, TGF-β1 and blood–brain barrier function. J Neuroimmunol 101:211–221

Cunningham JJ, Roussel MF (2001) Cyclin-dependent kinase inhibitors in the development of the central nervous system. Cell Growth Differ 12:387–396

DaCosta BS, Major C, Laping NJ, Roberts AB (2004) SB-505124 is a selective inhibitor of transforming growth factor-β type I receptors ALK4, ALK5, and ALK7. Mol Pharmacol 65:744–752

Day WA, Koishi K, Nukuda H, McLennan IS (2005) Transforming growth factor-β 2 causes an acute improvement in the motor performance of transgenic ALS mice. Neurobiol Dis 19:323–330

Day WA, Koishi K, McLennan IS (2003) Transforming growth factor β 1 may regulate the stability of mature myelin sheaths. Exp Neurol 184:857–864

De Strooper B, Vassar R, Golde T (2010) The secretases: enzymes with therapeutic potential in Alzheimer disease. Nat Rev Neurol 6:99–107

Decressac M, Ulusoy A, Mattsson B, Georgievska B, Romero-Ramos M, Kirik D, Björklund A (2011) GDNF fails to exert neuroprotection in a rat α-synuclein model of Parkinson's disease. Brain 134:2302–2311

Dell'Orso S, Ganci F, Strano S, Blandino G, Fontemaggi G (2010) ID4: a new player in the cancer arena. Oncotarget 1:48–58

Derynck R, Jarrett JA, Chen EY, Eaton DH, Bell JR, Assoian RK, Roberts AB, Sporn MB (1985) Human transforming growth factor-β cDNA sequence and expression in tumor cell lines. Nature 316:701–705

Dhandapani KM, Brann DW (2003) Transforming growth factor-β: a neuroprotective factor in cerebral ischemia. Cell Biochem Biophys 39:13–22

Dobolyi A, Vincze C, Pál G, Lovas G (2012) The neuroprotective functions of transforming growth factor β proteins. Int J Mol Sci 13:8219–8258
Docagne F, Nicole O, Marti HH, MacKenzie ET, Buisson A, Vivien D (1999) Transforming growth factor-β1 as a regulator of the serpins/t-PA axis in cerebral ischemia. FASEB J 13:1315–1324
Dong Y, Tang L, Letterio JJ, Benveniste EN (2001) The Smad3 protein is involved in TGF-β inhibition of class II transactivator and class II MHC expression. J Immunol 167:311–319
Dryer SE, Lhuillier L, Cameron JS, Martin-Caraballo M (2003) Expression of K(Ca) channels in identified populations of developing vertebrate neurons: role of neurotrophic factors and activity. J Physiol Paris 97:49–58
Dunker N, Krieglstein K (2003) Reduced programmed cell death in the retina and defects in lens and cornea of Tgfβ2(–/–) Tgfβ3(–/–) double-deficient mice. Cell Tissue Res 313:1–10
Dunker N, Schmitt K, Krieglstein K (2002) TGF-β is required for programmed cell death in interdigital webs of the developing mouse limb. Mech Dev 113:111–120
Dunker N, Schuster N, Krieglstein K (2001) TGF-β modulates programmed cell death in the retina of the developing chick embryo. Development 128:1933–1942
Dunnett SB, Björklund A (1999) Prospects for new restorative and neuroprotective treatments in Parkinson's disease. Nature 399:A32–A39
Fernandez-Espejo E, Armengol JA, Flores JA, Galan-Rodriguez B, Ramiro S (2005) Cells of the sympathoadrenal lineage: biological properties as donor tissue for cell-replacement therapies for Parkinson's disease. Brain Res Brain Res Rev 49:343–354
Flanders KC, Lippa CF, Smith TW, Pollen DA, Sporn MB (1995) Altered expression of transforming growth factor-β in Alzheimer's disease. Neurology 45:1561–1569
Flanders KC, Ludecke G, Engels S, Cissel DS, Roberts AB, Kondaiah P, Lafyatis R, Sporn MB, Unsicker K (1991) Localization and actions of transforming growth factor-β s in the embryonic nervous system. Development 113:183–191
Flanders KC, Lüdecke G, Renzig J, Hamm C, Cissel DS, Unsicker K (1993) Effect of TGF-βs and bFGF on astroglial cell growth and gene expression in vitro. Mol Cell Neurosci 4:406–417
Flanders KC, Ren RF, Lippa CF (1998) Transforming growth factor-βs in neurodegenerative disease. Prog Neurobiol 54:71–85
Fogarty MP, Kessler JD, Wechsler-Reya RJ (2005) Morphing into cancer: the role of developmental signaling pathways in brain tumor formation. J Neurobiol 64:458–475
Fong SW, McLennan IS, McIntyre A, Reid J, Shennan KI, Bewick GS (2010) TGF-β2 alters the characteristics of the neuromuscular junction by regulating presynaptic quantal size. Proc Natl Acad Sci USA 107:13515–13519
Forno LS (1996) Neuropathology of Parkinson's disease. J Neuropathol Exp Neurol 55:259–272
Friedman A (2011) Blood–brain barrier dysfunction, status epilepticus, seizures, and epilepsy: a puzzle of a chicken and egg? Epilepsia 52(Suppl 8):19–20
Fukushima T, Liu RY, Byrne JH (2007) Transforming growth factor-β2 modulates synaptic efficacy and plasticity and induces phosphorylation of CREB in hippocampal neurons. Hippocampus 17:5–9
Gaertner RF, Wyss-Coray T, Von Euw D, Lesne S, Vivien D, Lacombe P (2005) Reduced brain tissue perfusion in TGF-β 1 transgenic mice showing Alzheimer's disease-like cerebrovascular abnormalities. Neurobiol Dis 19:38–46
Gasser T, Hardy J, Mizuno Y (2011) Milestones in PD genetics. Mov Disord 26:1042–1048
Gross CE, Bednar MM, Howard DB, Sporn MB (1993) Transforming growth factor-β 1 reduces infarct size after experimental cerebral ischemia in a rabbit model. Stroke 24:558–562
Hahn SA, Schutte M, Hoque AT, Moskaluk CA, da Costa LT, Rozenblum E, Weinstein CL, Fischer A, Yeo CJ, Hruban RH, Kern SE (1996) DPC4, a candidate tumor suppressor gene at human chromosome 18q21.1. Science 271:350–353
Hanashima C, Shen L, Li SC, Lai E (2002) Brain factor-1 controls the proliferation and differentiation of neocortical progenitor cells through independent mechanisms. J Neurosci 22:6526–6536

Hau P, Jachimczak P, Bogdahn U (2009) Treatment of malignant gliomas with TGF-β2 antisense oligonucleotides. Expert Rev Anticancer Ther 9:1663–1674

Hau P, Jachimczak P, Schlaier J, Bogdahn U (2011) TGF-β2 signaling in high-grade gliomas. Curr Pharm Biotechnol 12:2150–2157

Huang Y, Mucke L (2012) Alzheimer mechanisms and therapeutic strategies. Cell 148:1204–1222

Hunter KE, Sporn MB, Davies AM (1993) Transforming growth factor-βs inhibit mitogen-stimulated proliferation of astrocytes. Glia 7:203–211

Ishihara A, Saito H, Abe K (1994) Transforming growth factor-β 1 and -β 2 promote neurite sprouting and elongation of cultured rat hippocampal neurons. Brain Res 639:21–25

Ivens S, Kaufer D, Flores LP, Bechmann I, Zumsteg D, Tomkins O, Seiffert E, Heinemann U, Friedman A (2007) TGF-β receptor-mediated albumin uptake into astrocytes is involved in neocortical epileptogenesis. Brain 130(Pt 2):535–547

Jachimczak P, Bogdahn U, Schneider J, Behl C, Meixensberger J, Apfel R, Dorries R, Schlingensiepen KH, Brysch W (1993) The effect of transforming growth factor-β 2-specific phosphorothioate-anti-sense oligodeoxynucleotides in reversing cellular immunosuppression in malignant glioma. J Neurosurg 78:944–951

Johns LD, Babcock G, Green D, Freedman M, Sriram S, Ransohoff RM (1992) Transforming growth factor-β 1 differentially regulates proliferation and MHC class-II antigen expression in forebrain and brainstemastrocyte primary cultures. Brain Res 585:229–236

Joseph JV, Balasubramaniyan V, Walenkamp A, Kruyt FA (2013) TGF-β as a therapeutic target in high grade gliomas—promises and challenges. Biochem Pharmacol 85:478–485

Juraskova B, Andrys C, Holmerova I, Solichova D, Hrnciarikova D, Vankova H, Vasatko T, Krejsek J (2010) Transforming growth factor β and soluble endoglin in the healthy senior and in Alzheimer's disease patients. J Nutr Health Aging 14:758–761

Klempt ND, Sirimanne E, Gunn AJ, Klempt M, Singh K, Williams C, Gluckman PD (1992) Hypoxia-ischemia induces transforming growth factor β 1 mRNA in the infant rat brain. Brain Res Mol Brain Res 13:93–101

Knuckey NW, Finch P, Palm DE, Primiano MJ, Johanson CE, Flanders KC, Thompson NL (1996) Differential neuronal and astrocytic expression of transforming growth factor β isoforms in rat hippocampus following transient forebrain ischemia. Brain Res Mol Brain Res 40:1–14

Konig HG, Kogel D, Rami A, Prehn JH (2005) TGF-{β}1 activates two distinct type I receptors in neurons: implications for neuronal NF-{κ}B signaling. J Cell Biol 168:1077–1086

Krieglstein K (2006) Transforming growth factor-βs in the brain. In: Lin R (ed) Handbook of neurochemistry and molecular neurobiology, neuroactive proteins and peptides. Springer, Heidelberg, pp 123–141

Krieglstein K, Farkas L, Unsicker K (1998a) TGF-β regulates the survival of ciliary ganglionic neurons synergistically with ciliary neurotrophic factor and neurotrophins. J Neurobiol 37:563–572

Krieglstein K, Henheik P, Farkas L, Jaszai J, Galter D, Krohn K, Unsicker K (1998b) Glial cell line-derived neurotrophic factor requires transforming growth factor-β for exerting its full neurotrophic potential on peripheral and CNS neurons. J Neurosci 18:9822–9834

Krieglstein K, Richter S, Farkas L, Schuster N, Dunker N, Oppenheim RW, Unsicker K (2000) Reduction of endogenous transforming growth factors β prevents ontogenetic neuron death. Nat Neurosci 3:1085–1090

Krieglstein K, Suter-Crazzolara C, Fischer WH, Unsicker K (1995) TGF-β superfamily members promote survival of midbrain dopaminergic neurons and protect them against MPP+ toxicity. EMBO J 14:736–742

Krieglstein K, Unsicker K (1994) Transforming growth factor-β promotes survival of midbrain dopaminergic neurons and protects them against N-methyl-4-phenylpyridinium ion toxicity. Neuroscience 63:1189–1196

Krieglstein K, Unsicker K (1996) Distinct modulatory actions of TGF-β and LIF on neurotrophin-mediated survival of developing sensory neurons. Neurochem Res 21:843–850

Krieglstein K, Zheng F, Unsicker K, Alzheimer C (2011) More than being protective: functional roles for TGF-β/activin signaling pathways at central synapses. Trends Neurosci 34:421–429
Labourdette G, Janet T, Laeng P, Perraud F, Lawrence D, Pettmann B (1990) Transforming growth factor type β 1 modulates the effects of basic fibroblast growth factor on growth and phenotypic expression of rat astroblasts in vitro. J Cell Physiol 144:473–484
Lacmann A, Hess D, Gohla G, Roussa E, Krieglstein K (2007) Activity-dependent release of transforming growth factor-β in a neuronal network in vitro. Neuroscience 150:647–657
Lahn M, Kloeker S, Berry BS (2005) TGF-β inhibitors for the treatment of cancer. Expert Opin Investig Drugs 14:629–643
Lee HG, Ueda M, Zhu X, Perry G, Smith MA (2006) Ectopic expression of phospho-Smad2 in Alzheimer's disease: uncoupling of the transforming growth factor-β pathway? J Neurosci Res 84:1856–1861
Leof EB, Proper JA, Goustin AS, Shipley GD, DiCorleto PE, Moses HL (1986) Induction of c-sis mRNA and activity similar to platelet-derived growth factor by transforming growth factor β: a proposed model for indirect mitogenesis involving autocrine activity. Proc Natl Acad Sci USA 83:2453–2457
Lindholm D, Castren D, Kiefer R, Zafra F, Thoenen H (1992) Transforming growth factor-β1 in the rat brain: increase after injury and inhibition of astrocyte proliferation. J Cell Biol 117:395–400
Lippa CF, Smith TW, Flanders KC (1995) Transforming growth factor-β: neuronal and glial expression in CNS degenerative diseases. Neurodegeneration 4:425–432
Logan A, Green J, Hunter A, Jackson R, Berry M (1999) Inhibition of glial scarring in the injured rat brain by a recombinant human monoclonal antibody to transforming growth factor-β2. Eur J Neurosci 11:2367–2374
Lou E (2004) Oncolytic viral therapy and immunotherapy of malignant brain tumors: two potential new approaches of translational research. Ann Med 36:2–8
Lyons RM, Gentry LE, Purchio AF, Moses HL (1990) Mechanism of activation of latent recombinant transforming growth factor β 1 by plasmin. J Cell Biol 110:1361–1367
Markowitz S, Wang J, Myeroff L, Parsons R, Sun L, Lutterbaugh J, Fan RS, Zborowska E, Kinzler KW, Vogelstein B, Brattain M, Willson JKV (1995) Inactivation of the type II TGF-β receptor in colon cancer cells with microsatellite instability. Science 268:1336–1338
Martini M, Cenci T, D'Alessandris GQ, Cesarini V, Cocomazzi A, Ricci-Vitiani L, De Maria R, Pallini R, Maria Larocca L (2012) Epigenetic silencing of Id4 identifies a glioblastoma subgroup with a better prognosis as a consequence of an inhibition of angiogenesis. Cancer 119:1004–1012
Martinou JC, Le Van TA, Valette A, Weber MJ (1990) Transforming growth factor β 1 is a potent survival factor for rat embryo motoneurons in culture. Brain Res Dev Brain Res 52:175–181
Massague J (2000) How cells read TGF-β signals. Nat Rev Mol Cell Biol 1:169–178
McKinnon RD, Piras G, Ida JA Jr, Dubois-Dalcq M (1993) A role for TGF-β in oligodendrocyte differentiation. J Cell Biol 121:1397–1407
McLennan IS, Koishi K (2002) The transforming growth factor-βs: multifaceted regulators of the development and maintenance of skeletal muscles, motoneurons and Schwann cells. Int J Dev Biol 46:559–567
Mocali A, Cedrola S, Della Malva N, Bontempelli M, Mitidieri VA, Bavazzano A, Comolli R, Paoletti F, La Porta CA (2004) Increased plasma levels of soluble CD40, together with the decrease of TGF β 1, as possible differential markers of Alzheimer disease. Exp Gerontol 39:1555–1561
Morganti-Kossmann MC, Hans VH, Lenzlinger PM, Dubs R, Ludwig E, Trentz O, Kossmann T (1999) TGF-β is elevated in the CSF of patients with severe traumatic brain injuries and parallels blood–brain barrier function. J Neurotrauma 16:617–628
Nagatsu T, Mogi M, Ichinose H, Togari A (2000) Changes in cytokines and neurotrophins in Parkinson's disease. J Neural Transm Suppl 60:277–290
Nieder C, Schlegel J, Andratschke N, Thamm R, Grosu AL, Molls M (2003) The role of growth factors in central nervous system tumors. Anticancer Res 23:1681–1686

Packard M, Mathew D, Budnik V (2003) Wnts and TGF β in synaptogenesis: old friends signalling at new places. Nat Rev Neurosci 4:113–120
Pál G, Vincze C, Renner E, Wappler EA, Nagy Z, Lovas G, Dobolyi A (2012) Time course, distribution and cell types of induction of transforming growth factor βs following middle cerebral artery occlusion in the rat brain. PLoS One 7:e46731
Parkinson DB, Dong Z, Bunting H, Whitfield J, Meier C, Marie H, Mirsky R, Jessen KR (2001) Transforming growth factor β (TGFβ) mediates Schwann cell death in vitro and in vivo: examination of c-Jun activation, interactions with survival signals, and the relationship of TGFβ-mediated death to Schwann cell differentiation. J Neurosci 21:8572–8585
Pelton RW, Johnson MD, Perkett EA, Gold LI, Moses HL (1991a) Expression of transforming growth factor-β 1, -β 2, and -β 3 mRNA and protein in the murine lung. Am J Respir Cell Mol Biol 5:522–530
Pelton RW, Saxena B, Jones M, Moses HL, Gold LI (1991b) Immunohistochemical localization of TGF β 1, TGF β 2, and TGF β 3 in the mouse embryo: expression patterns suggest multiple roles during embryonic development. J Cell Biol 115:1091–1105
Peterziel H, Paech T, Strelau J, Unsicker K, Krieglstein K (2007) Specificity in the crosstalk of TGFβ/GDNF family members is determined by distinct GFR α receptors. J Neurochem 103: 2491–2504
Peterziel H, Unsicker K, Krieglstein K (2002) TGFβ induces GDNF responsiveness in neurons by recruitment of GFRα1 to the plasma membrane. J Cell Biol 159:157–167
Platten M, Wick W, Weller M (2001) Malignant glioma biology: role for TGF-β in growth, motility, angiogenesis, and immune escape. Microsc Res Tech 52:401–410
Poulsen KT, Armanini MP, Klein RD, Hynes MA, Phillips HS, Rosenthal A (1994) TGF β 2 and TGF β 3 are potent survival factors for midbrain dopaminergic neurons. Neuron 13:1245–1252
Prehn JH, Backhauss C, Krieglstein J (1993) Transforming growth factor-β 1 prevents glutamate neurotoxicity in rat neocortical cultures and protects mouse neocortex from ischemic injury in vivo. J Cereb Blood Flow Metab 13:521–525
Prehn JH, Bindokas VP, Marcuccilli CJ, Krajewski S, Reed JC, Miller RJ (1994) Regulation of neuronal Bcl2 protein expression and calcium homeostasis by transforming growth factor type β confers wide-ranging protection on rat hippocampal neurons. Proc Natl Acad Sci USA 91:12599–12603
Rahhal B, Dunker N, Combs S, Krieglstein K (2004) Isoform-specific role of transforming growth factor-β2 in the regulation of proliferation and differentiation of murine adrenal chromaffin cells in vivo. J Neurosci Res 78:493–498
Rich JN (2003) The role of transforming growth factor-β in primary brain tumors. Front Biosci 8:e245–e260
Rich JN, Zhang M, Datto MB, Bigner DD, Wang XF (1999) Transforming growth factor-β-mediated p15(INK4B) induction and growth inhibition in astrocytes is SMAD3-dependent and a pathway prominently altered in human glioma cell lines. J Biol Chem 274:35053–35058
Roberts AB, Sporn MB (1990) The transforming growth factor-β. In: Sporn MB, Roberts AB (eds) Handbook of experimental pharmacology, vol 95. Springer, Heidelberg, pp 419–472
Roberts AB, Anzano MA, Lamb LC, Smith JM, Sporn MB (1981) New class of transforming growth factors potentiated by epidermal growth factor: isolation from non-neoplastic tissues. Proc Natl Acad Sci USA 78:5339–5343
Roberts AB, Wakefield LM (2003) The two faces of transforming growth factor β in carcinogenesis. Proc Natl Acad Sci USA 100:8621–8623
Rooprai HK, Rucklidge GJ, Panou C, Pilkington GJ (2000) The effects of exogenous growth factors on matrix metalloproteinase secretion by human brain tumour cells. Br J Cancer 82:52–55
Roth P, Eisele G, Weller M (2012) Immunology of brain tumors. Handb Clin Neurol 104:45–51
Roussa E, von Bohlen und Halbach O, Krieglstein K (2009) TGF-β in dopamine neuron development, maintenance and neuroprotection. Adv Exp Med Biol 651:81–90

Roussa E, Farkas LM, Krieglstein K (2004) TGF-β promotes survival on mesencephalic dopaminergic neurons in cooperation with Shh and FGF-8. Neurobiol Dis 16:300–310
Salinas PC (2005) Signaling at the vertebrate synapse: new roles for embryonic morphogens? J Neurobiol 64:435–445
Sanchez-Capelo A (2005) Dual role for TGF-β1 in apoptosis. Cytokine Growth Factor Rev 16:15–34
Schachtrup C, Ryu JK, Helmrick MJ, Vagena E, Galanakis DK, Degen JL, Margolis RU, Akassoglou K (2010) Fibrinogen triggers astrocyte scar formation by promoting the availability of active TGF-β after vascular damage. J Neurosci 30:5843–5854
Schober A, Peterziel H, von Bartheld CS, Simon H, Krieglstein K (2007) Unsicker K GDNF applied to the MPTP-lesioned nigrostriatal system requires TGF-β for its neuroprotective action. Neurobiol Dis 25:378–391
Schober A, Hertel R, Arumae U, Farkas L, Jaszai J, Krieglstein K, Saarma M, Unsicker K (1999) Glial cell line-derived neurotrophic factor rescues target-deprived sympathetic spinal cord neurons but requires transforming growth factor-β as cofactor in vivo. J Neurosci 19:2008–2015
Schulz R, Vogel T, Mashima T, Tsuruo T, Krieglstein K (2009) Involvement of Fractin in TGF-β-induced apoptosis in oligodendroglial progenitor cells. Glia 57:1619–1629
Schuster N, Bender H, Philippi A, Subramaniam S, Strelau J, Wang Z, Krieglstein K (2002) TGF-β induces cell death in the oligodendroglial cell line OLI-neu. Glia 40:95–108
Schuster N, Krieglstein K (2002) Mechanisms of TGF-β-mediated apoptosis. Cell Tissue Res 307:1–14
Seoane J (2006) Escaping from the TGFβ anti-proliferative control. Carcinogenesis 27:2148–2156
Seoane J, Le HV, Shen L, Anderson SA, Massague J (2004) Integration of Smad and forkhead pathways in the control of neuroepithelial and glioblastoma cell proliferation. Cell 117:211–223
Shrikant P, Lee SJ, Kalvakolanu I, Ransohoff RM, Benveniste EN (1996) Stimulus-specific inhibition of intracellular adhesion molecule-1 gene expression by TGF-β. J Immunol 157:892–900
Sorrentino A, Thakur N, Grimsby S, Marcusson A, von Bulow V, Schuster N, Zhang S, Heldin CH, Landström M (2008) The type I TGF-β receptor engages TRAF6 to activate TAK1 in a receptor kinase-independent manner. Nat Cell Biol 10:1199–1207
Specht H, Peterziel H, Bajohrs M, Gerdes HH, Krieglstein K, Unsicker K (2003) Transforming growth factor β2 is released from PC12 cells via the regulated pathway of secretion. Mol Cell Neurosci 22:75–86
Spillantini MG, Crowther RA, Jakes R, Hasegawa M, Goedert M (1998) Alpha-synuclein in filamentous inclusions of lewy bodies from Parkinson's disease and dementia with lewy bodies. Proc Natl Acad Sci USA 95:6469–6473
Spittau B, Zhou X, Ming M, Krieglstein K (2012) IL6 protects MN9D cells and midbrain dopaminergic neurons from MPP(+)-induced neurodegeneration. Neuromolecular Med 14:317–327
Tesseur I, Zhang H, Brecht W, Corn J, Gong JS, Yanagisawa K, Michikawa M, Weisgraber K, Huang Y, Wyss-Coray T (2009) Bioactive TGF-β can associate with lipoproteins and is enriched in those containing apolipoprotein E3. J Neurochem 110:1254–1262
Tesseur I, Zou K, Esposito L, Bard F, Berber E, Can JV, Lin AH, Crews L, Tremblay P, Mathews P, Mucke L, Masliah E, Wyss-Coray T (2006) Deficiency in neuronal TGF-β signaling promotes neurodegeneration and Alzheimer's pathology. J Clin Invest 116:3060–3069
Thoenen H (1995) Neurotrophins and neuronal plasticity. Science 270:593–598
Toepfer M, Fischer P, Abicht A, Lochmuller H, Pongratz D, Muller-Felber W (1999) Localization of transforming growth factor β in association with neuromuscular junctions in adult human muscle. Cell Mol Neurobiol 19:297–300
Ueberham U, Ueberham E, Gruschka H, Arendt T (2006) Altered subcellular location of phosphorylated Smads in Alzheimer's disease. Eur J Neurosci 24:2327–2334
Uhm JH, Gladson CL, Rao JS (1999) The role of integrins in the malignant phenotype of gliomas. Front Biosci 4:D188–D199

Ulusoy A, Decressac M, Kirik D, Björklund A (2010) Viral vector-mediated overexpression of α-synuclein as a progressive model of Parkinson's disease. Prog Brain Res 184:89–111

Unsicker K, Flanders KC, Cissel DS, Lafyatis R, Sporn MB (1991) Transforming growth factor β isoforms in the adult rat central and peripheral nervous system. Neuroscience 44:613–625

Valente EM, Arena G, Torosantucci L, Gelmetti V (2012) Molecular pathways in sporadic PD. Parkinsonism Relat Disord 18(Suppl 1):S71–S73

van der Wal EA, Gomez-Pinilla F, Cotman CW (1993) Transforming growth factor-β 1 is in plaques in Alzheimer and Down pathologies. Neuroreport 4:69–72

Vogel T, Ahrens S, Büttner N, Krieglstein K (2010) Transforming growth factor β promotes neuronal cell fate of mouse cortical and hippocampal progenitors in vitro and in vivo: identification of Nedd9 as an essential signaling component. Cereb Cortex 20:661–671

Wachs FP, Winner B, Couillard-Despres S, Schiller T, Aigner R, Winkler J, Bogdahn U, Aigner L (2006) Transforming growth factor-β1 is a negative modulator of adult neurogenesis. J Neuropathol Exp Neurol 65:358–370

Wahl SM, Chen W (2003) TGF-β: how tolerant can it be? Immunol Res 28:167–179

Wang H, Liu J, Zong Y, Xu Y, Deng W, Zhu H, Liu Y, Ma C, Huang L, Zhang L, Qin C (2010) miR-106b aberrantly expressed in a double transgenic mouse model for Alzheimer's disease targets TGF-β type II receptor. Brain Res 1357:166–174

Wendt KM, Schiemann WP (2013) The multifunctional roles of TGF-β in navigating the metastatic cascade. In: Moustakas A, Miyazawa K (eds) TGF-β in human disease. Springer, Tokyo, pp 169–188

Wibrand K, Messaoudi E, Håvik B, Steenslid V, Løvlie R, Steen VM, Bramham CR (2006) Identification of genes co-upregulated with Arc during BDNF-induced long-term potentiation in adult rat dentate gyrus in vivo. Eur J Neurosci 23:1501–1511

Wyss-Coray T, Borrow P, Brooker MJ, Mucke L (1997a) Astroglial overproduction of TGF-β 1 enhances inflammatory central nervous system disease in transgenic mice. J Neuroimmunol 77:45–50

Wyss-Coray T, Lin C, Yan F, Yu GQ, Rohde M, McConlogue L, Masliah E, Mucke L (2001) TGF-β1 promotes microglial amyloid-β clearance and reduces plaque burden in transgenic mice. Nat Med 7:612–618

Wyss-Coray T, Masliah E, Mallory M, McConlogue L, Johnson-Wood K, Lin C, Mucke L (1997b) Amyloidogenic role of cytokine TGF-β1 in transgenic mice and in Alzheimer's disease. Nature 389:603–606

Xuan S, Baptista CA, Balas G, Tao W, Soares VC, Lai E (1995) Winged helix transcription factor BF-1 is essential for the development of the cerebral hemispheres. Neuron 14:1141–1152

Yamashita K, Gerken U, Vogel P, Hossmann K, Wiessner C (1999) Biphasic expression of TGF-β1 mRNA in the rat brain following permanent occlusion of the middle cerebral artery. Brain Res 836:139–145

Yi JJ, Barnes AP, Hand R, Polleux F, Ehlers MD (2010) TGF-β signaling specifies axons during brain development. Cell 142:144–157

Yingling JM, Blanchard KL, Sawyer JS (2004) Development of TGF-β signalling inhibitors for cancer therapy. Nat Rev Drug Discov 3:1011–1022

Zhang F, Endo S, Cleary LJ, Eskin A, Byrne JH (1997a) Role of transforming growth factor-β in long-term synaptic facilitation in Aplysia. Science 275:1318–1320

Zhang JM, Hoffmann R, Sieber-Blum M (1997b) Mitogenic and anti-proliferative signals for neural crest cells and the neurogenic action of TGF-β1. Dev Dyn 208:375–386

Zhu Y, Ahlemeyer B, Bauerbach E, Krieglstein J (2001) TGF-β1 inhibits caspase-3 activation and neuronal apoptosis in rat hippocampal cultures. Neurochem Int 38:227–235

Zhu Y, Culmsee C, Klumpp S, Krieglstein J (2004) Neuroprotection by transforming growth factor-β1 involves activation of nuclear factor-κB through phosphatidylinositol-3-OH kinase/Akt and mitogen-activated protein kinase-extracellular-signal regulated kinase1,2 signaling pathways. Neuroscience 123:897–906

Zhu Y, Roth-Eichhorn S, Braun N, Culmsee C, Rami A, Krieglstein J (2000) The expression of transforming growth factor-β1 (TGF-β1) in hippocampal neurons: a temporary upregulated protein level after transient forebrain ischemia in the rat. Brain Res 866:286–298

Zhu Y, Yang GY, Ahlemeyer B, Pang L, Che XM, Culmsee C, Klumpp S, Krieglstein J (2002) Transforming growth factor-β 1 increases bad phosphorylation and protects neurons against damage. J Neurosci 22:3898–3909

Zindy F, Cunningham JJ, Sherr CJ, Jogal S, Smeyne RJ, Roussel MF (1999) Postnatal neuronal proliferation in mice lacking Ink4d and Kip1 inhibitors of cyclin-dependent kinases. Proc Natl Acad Sci USA 96:13462–13467

Chapter 18
TGF-β and Metabolic Homeostasis

Grace Bennett and Sushil G. Rane

Abstract Metabolic homeostasis is achieved via a concerted and integrative action of various organ systems. The harmonious functionality of the collaborative organs maintains tight regulation of glucose levels. This intricate balance of glucose production and utilization maintains whole body glucose homeostasis and energy balance. In disease states, such as diabetes and obesity, dysfunction in one or more organ systems disturbs the metabolic homeostasis and propagates disease pathology. Understanding the avenues that dictate metabolic homeostasis and the pathways that disrupt this harmony are thus of great medical significance. We propose that the TGF-β signaling network plays an integral role in metabolic homeostasis by virtue of its actions on several organ systems that constitute the metabolic machinery. TGF-β levels are elevated in metabolic disease, which supports the utility of therapeutics aimed at targeting the TGF-β pathway to combat these diseases. Considering the complexity of the TGF-β signaling network, a rational approach is vital to designing anti-TGF-β modalities to combat metabolic diseases.

Keywords Adipocytes • Adipose tissue • Diabetes • Homeostasis • Insulin • Metabolism • Mitochondria • Obesity • Pancreas • Smad3 • TGF-β • β-cells

18.1 Metabolic Homeostasis

For living organisms to maintain a state of metabolic homeostasis there must be a steady state of nutrient and energy intake and its utilization. Macro- (carbohydrate, lipid and protein) as well as micro-nutrients (vitamins, minerals, and trace elements) are the components of the diet which provide the energy, molecular building blocks,

G. Bennett • S.G. Rane (✉)
Diabetes, Endocrinology and Obesity Branch, National Institute of Diabetes and Digestive and Kidney Diseases, National Institutes of Health, Bethesda, MD 20892, USA
e-mail: ranes@mail.nih.gov

A. Moustakas and K. Miyazawa (eds.), *TGF-β in Human Disease*,
DOI 10.1007/978-4-431-54409-8_18, © Springer Japan 2013

co-enzymes, and co-factors needed to maintain properly functioning metabolic processes of nutrient digestion, absorption, utilization, and excretion. There are specific "metabolic tissues" that are primarily responsible for the breakdown and utilization of these nutrients. Because of their critical importance in supporting the life of an organism, the normal physiology of these tissues is well characterized, in particular the insulin responsive tissues (adipose tissue, liver, skeletal muscle) and the sole insulin producing tissue, the pancreas. However, we have gained insight into the importance of these tissues in metabolic homeostasis through their pathological contexts as well (Biddinger and Kahn 2006). In particular, the obese state, along with its metabolic comorbidities, has shed light on the integrated role of these organ systems, although this has also created the problem of determining which of these tissues is the primary site responsible for disease states such as type II diabetes (Yang and Trevillyan 2008; Olefsky and Glass 2010).

The adaptation of a stomach and adipose tissue has made it possible to eat in the meal format, where, as opposed to grazing, an organism consumes food in discrete settings until they reach a level of "fullness," stop eating while still fueling their cells to carry out their basic physiological processes due to adequate nutrient stores within the body.

After a meal, carbohydrates are broken down to glucose molecules and are either utilized by the intestine or liver (the first organs they encounter) or are sent to all other tissues where they can be utilized for the synthesis of ATP after conversion to pyruvate via glycolysis. In order for glucose to gain entry into insulin sensitive tissues, i.e., adipose tissue and skeletal muscle, glucose needs insulin, an endogenously secreted peptide hormone from the β-cells of the pancreas. Insulin binds its receptor on the cell surface of insulin responsive tissues. The receptor, being a receptor tyrosine kinase, autophosphorylates and begins a signal transduction cascade that involves insulin receptor substrates (IRS), phosphatidylinositol 3 phosphate (PI3K), and Akt. Activation of this pathway by phosphorylation releases premade vesicles containing glucose transporter 4 (GLUT4) to the cell surface thereby allowing for the entry of glucose into the cell (Chang et al. 2004). This glucose is used for the basic cellular needs, and what remains is converted to triacylglycerol through the process of de novo lipogenesis occurring primarily in the liver and adipose tissue.

Fatty acids consumed in a meal undergo a different metabolic process than carbohydrates, but, in part share the same fate being utilized for immediate energetic needs or stored in liver as glycogen or in adipose as triacylglycerol for later use. Fatty acids are typically found in a meal as triglycerides, which cannot enter the intestinal cells. In the small intestines, these triacylglycerols are bound by pancreatic lipases that break them down into component fatty acids and monoacylglycerols. However, these lipases are only able to work at the surface of these fats. Emulsification by bile salts created in the liver and released from the gall bladder creates more surface area for these lipases to work ultimately allowing for their entry in enterocytes of the small intestines to be reformed into triglycerides and packaged into chylomicrons before release into the lymphatic circulation. Unlike glucose, fatty acids do not need insulin to gain entry to adipocytes. Rather, lipoprotein lipase on the cell surface releases fatty acids into the cell where they can be stored as triacylglycerol in a lipid droplet or oxidized.

18.1.1 Adipose Tissue

In obesity, chronic over nutrition leads to the need to store excess amounts of energy, manifest as increased storage of triacylglycerols in adipose tissue, with "spillover" resulting in ectopic fat deposition in liver, skeletal muscle, and pancreas (Cusi 2010; Unger and Scherer 2010). Adipocytes respond to this need to store excess energy in a depot-specific manner: subcutaneous adipocytes (under the skin) grow by hyperplasia (i.e., increasing the amount of adipocytes able to store triacylglycerols via proliferation of progenitor cell pools) (Ibrahim 2010). However, adipocytes of the visceral cavity (intra-abdominal adipose tissue) accommodate excess lipid by physically expanding the size of each cell (i.e., hypertrophy) (Ibrahim 2010). Hypertrophic adipocyte growth is considered a cellular stress in terms of both sheer and mechanical stress as the surface area to volume ratio changes, but also with regard to disruption of the normal physiological processes occurring within the cellular machinery (Bays et al. 2008; Cusi 2010; Ghorbani et al. 1997; Nishimura et al. 2007; Khan et al. 2009; Kim et al. 2007b). For example, hypertrophic adipocytes are susceptible to ER-stress (Guo et al. 2007), decreased secretion of insulin sensitizing agents such as adiponectin (Greenberg and Obin 2006), decreased insulin sensitivity (Biddinger and Kahn 2006), increased secretion of inflammatory mediators (Gutierrez et al. 2009; Halberg et al. 2009; Jager et al. 2007; Kang et al. 2008; Karalis et al. 2009; Kim et al. 2007a; Lacasa et al. 2007; Lumeng and Saltiel 2011; McGillicuddy et al. 2011; Odegaard et al. 2007; Olefsky and Glass 2010; Perfield et al. 2010; Shi et al. 2006; Shoelson et al. 2006), increased lipolysis and increased cell death (Cinti et al. 2005; Greenberg and Obin 2008; Greenberg and Obin 2006). The latter three processes initiate a vicious cycle that exacerbates the negative impact of adipocyte hypertrophy on metabolism as cell death triggers production of inflammatory mediators and releases free fatty acids into the local environment. Moreover, inflammatory mediators promote increased lipolysis and insulin resistance by blocking insulin receptor signaling through inhibitory phosphorylation of the insulin receptor, thereby blocking the anti-lipolytic actions of insulin. Increased release of free fatty acids via lipolysis promotes the secretion of inflammatory mediators through free-fatty acid activation of Toll-like receptor 4 (TLR4), resulting in transcription of inflammatory mediators (TNFα, IL-6, etc.), thereby propagating the signal which promotes insulin resistance and leads to the recruitment of many immune cell populations (Fain 2006; Weisberg et al. 2003; Zeyda and Stulnig 2007; Suganami et al. 2007; Shaul et al. 2010; Strissel et al. 2010; Strissel et al. 2007). All of the aforementioned processes have led to the hypothesis that insult to adipose tissue is the primary lesion leading to systemic hyperglycemia, hyperinsulinemia, insulin resistance, and diabetes mellitus type II.

18.1.2 Liver

As described above, it is clear that the liver plays a vital role in metabolism. The liver is the only organ in the body that can produce glucose and therefore plays an extremely important role in metabolic homeostasis during times of fasting or between meals. In times of fasting, the drop in blood glucose concentration triggers the liver to respond by utilizing glycogen stores to produce glucose for release into circulation, to maintain the blood glucose concentration within the tight range of 80–100 mg/dL. After a meal, the liver gets the first look at nutrients comprising what was consumed. Some carbohydrates are utilized to replace any glycogen stores that may have been depleted in between meals in a process regulated by the anabolic actions of insulin, before the rest are sent to the peripheral tissues for maintenance of blood glucose concentration, metabolic use or if a surplus exists, eventual conversion to triacylglycerol for storage in adipose tissue. Pathologic features of obesity that impact the ability of the liver to function normally are hepatic steatosis as well as inflammation. Hepatic steatosis can occur endogenously by increased synthesis of triacylglycerol from excess carbohydrates as well as from ectopic lipid deposition (i.e., lipid spillover from saturated adipose tissue) (Cusi 2010; Unger et al. 2010; Unger and Scherer 2010). This creates controversy when trying to ascribe the pathogenic blame of metabolic dysregulation. For example, while it is recognized that adipose tissue becomes temporarily saturated with lipid before it is able to expand its cellular pools, spilling lipids over into the liver, the liver can also produce endogenous fatty acids from de novo lipogenesis. This makes it difficult to determine the source and kinetics of ectopic fat found in liver. Be that as it may this lipid, regardless of the source, induces local inflammation within the liver and recruitment of Kupffer cells followed by the same inflammation induced insulin resistance in liver as in adipose tissue (Saberi et al. 2009). However, unlike adipose tissue, when the liver is insulin resistant, it can no longer respond to the signal from insulin after a meal to stop producing glucose and store glycogen. Therefore, insulin resistance in the liver is a primary cause of hyperglycemia, making some argue that the liver's contribution to metabolic dysregulation is more important than that of adipose tissue.

18.1.3 β-Cells

β-cells comprise about 65–80 % of the islets of Langerhans within the pancreas, where they make and store insulin. As mentioned, the liver is a strong determinant of basal blood glucose concentration. However, when blood glucose levels increase rapidly, as they do after a meal, for example, the pancreas senses this and β-cells release two waves of insulin. The first wave releases the premade insulin rapidly, within minutes, while, the second wave occurs about 30 min post-prandially.

The β-cells also release C-peptide and Amylin, and the pancreas in general secretes many endocrine and exocrine products. Here, we will focus on the role of β-cells in insulin/glucose homeostasis. β-cell destruction by auto-antibodies leads to diabetes mellitus type I, while "β-cell fatigue" is the final step in the progression from obesity, tissue-specific insulin resistance, hyperglycemia, whole-body insulin resistance, and ultimately diabetes mellitus type II (Biddinger and Kahn 2006). The former typically occurs in adolescents and young adults, while the latter is a much more chronic process which takes years to develop. However, in the current obesigenic environment, increasing numbers of adolescents are being diagnosed as type II diabetics.

18.1.4 Skeletal Muscle

Muscle is able to aid in gluconeogenesis via transfer of amino acid substrates to the liver for conversion into carbon skeletons for glucose production; however, the large size of skeletal muscle and its responsiveness to insulin are what create the major role for this tissue in glucose homeostasis as the predominant site of glucose utilization (Thiebaud et al. 1982). As with liver and adipose tissue, inflammation and free fatty acids interfere with the function of insulin receptor signaling in skeletal muscle, leading to decreased glucose uptake and hyperglycemia (Kennedy et al. 2009; Nieto-Vazquez et al. 2008; Senn 2006; Weigert et al. 2004). It has been dogma for some time that skeletal muscle insulin resistance is the most critical component to developing whole-body insulin resistance and diabetes mellitus type II (DeFronzo and Tripathy 2009); however as mentioned above, the contributions of multiple tissue types mean that it is not difficult to see how intimately intertwined adipose tissue, muscle, β-cells, and liver are in maintaining metabolic regulation. Euglycemic/hyperinsulinemic clamp studies have made it possible to tease apart the relative contribution of liver and skeletal muscle to hyperglycemia/insulin resistance phenotype in human subjects as well as experimental settings that utilize animal models (Borai et al. 2011). However, the timing of both adipose tissue's deleterious contributions as well as failure of β-cells to compensate by increasing production of insulin remains to be elucidated.

Bariatric surgeries, in which parts of the intestinal tract are removed or bypassed, have shed light onto the importance of the intestinal-brain axis and related hormones in metabolic homeostasis, but as this is beyond the scope of this chapter, we point you to the work of Andreelli et al. for a more detailed review (Andreelli et al. 2009).

In summary, metabolic homeostasis is the ability to maintain body weight and physiological processes via mechanisms which balance energy intake, expenditure, and storage. These processes involve many tissues and organ systems, and dysregulation of one part can affect the whole organism leading to metabolic complications, weight gain, and/or disease.

18.2 Role of TGF-β Signaling in Metabolic Tissues

The ability to properly specify, differentiate, and/or proliferate in the early developmental stages of metabolic tissues is of critical importance to metabolic homeostasis. TGF-β and their family members: BMPs, activins, and GDFs have all been well characterized in their roles in determination of cellular fate, in particular, in early development and stem cell biology (James et al. 2005; Chng et al. 2011; Wu et al. 2008; Watabe and Miyazono 2009; Takenaga et al. 2007; Willems and Leyns 2008; Kitisin et al. 2007; Varga and Wrana 2005; Mishra et al. 2005). Here, we review the importance of the TGF-β signaling pathway in metabolic homeostasis based on its important actions on two organs—the adipose tissue and the pancreatic β-cell.

18.2.1 Role of TGF-β Signaling in Adipose Tissue Biology

Obesity is principally characterized by fat accumulation in the adipose tissue depots (Gesta et al. 2007; Park et al. 2008). Mammalian adipose tissue has been traditionally classified as white adipose tissue (WAT) and brown adipose tissue (BAT). WAT functions predominantly as an energy storage depot characterized by large unilocular lipid droplets and serves as a prominent endocrine organ producing hormones that regulate feeding and satiety (Rosen and Spiegelman 2006). In contrast, BAT is involved in energy dissipation and is characterized by multilocular cells that harbor densely packed mitochondria and expression of uncoupling protein-1 (UCP1) (Cannon and Nedergaard 2004, 2011). Copious amount of brown fat exists in rodents, hibernating animals, and human infants—and until recently brown fat was considered to be nonexistent in adult humans. Recent findings that metabolically active brown fat exists in humans (Nedergaard et al. 2007; Cypess et al. 2009; van Marken Lichtenbelt et al. 2009; Virtanen et al. 2009), and that levels of brown fat activity might vary with age and adiposity, have re-stimulated interest concerning the therapeutic potential of augmenting brown fat to combat metabolic diseases such as diabetes and obesity (Nedergaard and Cannon 2010; Enerback 2010). Interestingly, lineage tracing data supports the notion that brown fat shares its developmental origin with muscle, and not white fat as it was long presumed (Atit et al. 2006; Timmons et al. 2007). Further, the transcription factor PRDM16 determines the fate of $Myf5^{+}$-precursor cells towards the brown fat lineage (Seale et al. 2009; Seale et al. 2008).

18.2.2 Brown Adipocyte Induction in Traditional WAT Depots

Brown adipocytes are found interspersed within the WAT depots, in response to chemical or hormonal stimulation, cold exposure, or defined genetic manipulation

(Langin 2009; Lefterova and Lazar 2009; Frontini and Cinti 2010). Expression of the brown adipocyte marker UCP1 is increased in the multiple adipose tissue depots upon exposure to cold (Loncar 1991) or upon treatment with a β3-adrenoceptor agonist, CL 316,243 (Cousin et al. 1992). In agreement, β3-adrenoceptor knockout mice have reduced numbers of brown adipocytes in white fat upon cold exposure (Jimenez et al. 2003), whereas transgenic mice overexpressing the β1-adrenergic receptor in adipose tissue exhibit abundant appearance of brown fat cells in WAT and are resistant to obesity (Soloveva et al. 1997). Further, chronic treatment with CL 316,243, (Bloom et al. 1992) promotes thermogenesis and the appearance of multilocular brown adipocytes in WAT while protecting from high-fat diet induced obesity (Himms-Hagen et al. 1994). Also, infusion of CL 316,243 reduced abdominal fat, increased resting metabolic rate and induced abundant multilocular brown adipocytes expressing UCP1 in WAT of mice (Ghorbani et al. 1997) and Zucker fa/fa rats (Ghorbani and Himms-Hagen 1997). We have recently reviewed the various mechanisms that lead to browning of white fat (Yadav and Rane 2012).

18.2.3 TGF-β Signaling Modulates Brown Adipocyte Induction in WAT

The TGF-β superfamily, inclusive of TGF-β, activins, and BMPs, regulates the developmental programs of many, if not all, diverse cell types (Attisano and Wrana 2002; Feng and Derynck 2005; Massague et al. 2000). TGF-β transmits its signals via dual serine/threonine kinase receptors and transcription factors called Smads, with Smad3 being the principal facilitator of TGF-β isoform derived signals (Feng and Derynck 2005). We recently illustrated an important role of the TGF-β/Smad3 signaling pathway in regulating glucose and energy homeostasis (Yadav et al. 2011). *Smad3*$^{-/-}$ mice exhibit enhanced insulin sensitivity and increased glucose uptake in WAT. When challenged with high-fat diet, *Smad3*$^{-/-}$ mice gained less weight, exhibited enhanced glucose tolerance and insulin sensitivity and were protected from hepatic steatosis. Further, *Smad3*$^{-/-}$ mice displayed significantly reduced fat mass, smaller adipocyte size and reduced expression of WAT-specific genes. *Smad3*$^{-/-}$ mice also exhibited reduced levels of inflammatory cytokines and less inflammatory macrophage infiltration into WAT with a switch in macrophage spectrum from inflammatory M1 to protective M2 macrophages (Lumeng et al. 2007). Interestingly, *Smad3*$^{-/-}$ epididymal WAT possessed a dark-red color and was comprised of smaller multi-locular adipocytes interspersed within larger uni-locular white adipocyte depots (Yadav et al. 2011), a morphology that resembled brown adipocytes (Cousin et al. 1992). Further, we observed that many cells in the *Smad3*$^{-/-}$ WAT stained positive for UCP1 and expressed increased mRNA levels of BAT markers (Seale et al. 2009), i.e. PGC-1α, UCP1, and PRDM16. Physiologically, *Smad3*$^{-/-}$ mice were able to maintain significantly higher body temperature, when exposed to cold for an extended time and displayed elevated rates of fatty acid oxidation.

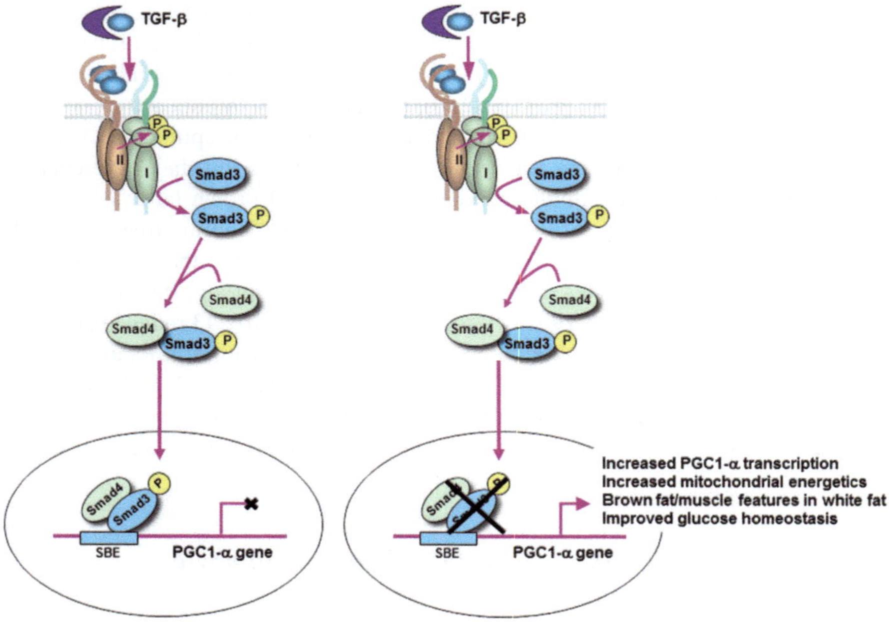

Fig. 18.1 Smad3 regulates PGC1-α promoter in the *white* adipose tissue. Binding of the TGF-β ligand to TβRII results in activation of the TβRI kinase and subsequent phosphorylation of Smad3. Phospho-Smad3 translocates to the nucleus along with Smad4. Smad3 binds to Smad-binding elements (SBE) on the PGC1-α promoter and represses PGC1-α gene transcription (*left panel*). Loss of Smad3 binding to the promoter enhances PGC1-α transcription and is associated with improved functionality of the *white* adipocytes

The molecular signature that we obtained in *Smad3*$^{-/-}$ white fat and in WAT from mice treated with the TGF-β neutralization antibody (1D11) provided interesting insight into the development of brown adipocytes within the WAT. Thus, WAT from *Smad3*$^{-/-}$ mice and mice treated with 1D11 antibodies expressed a preponderance of genes that represent brown fat, mitochondrial and skeletal muscle (Yadav et al. 2011). It is possible that TGF-β acts on a pool of cells that represents a common progenitor for white, brown, and muscle cells. Smad3 co-occupies the genome with cell-type-specific master transcription factors, including Myod1 in myotubes, PU.1 in pro-B cells and Oct4 in ESCs (Mullen et al. 2011). It is probable that in WAT, Smad3 interacts with factor/s that control the induction of cells with BAT/muscle features. Sequence analysis revealed the presence of Smad-binding elements on the promoter of PGC-1α, a transcriptional co-activator that regulates genes involved in energy metabolism and is a direct link between external physiological circuits and the regulation of mitochondrial biogenesis (Lin et al. 2005). Chromatin immunoprecipitation assays provided evidence of Smad3 binding to the PGC-1α promoter (Yadav et al. 2011), and we observed that TGF-β represses the PGC-1α promoter in a Smad3-dependent manner (Fig. 18.1). *Smad3*$^{-/-}$ WAT exhibited increased mitochondrial DNA copy number and functionality. Specifically, the *Smad3*$^{-/-}$ mitochondria displayed increased rates of basal respiration, which was consistent with the increase in the basal rate of oxygen consumption in intact *Smad3*$^{-/-}$ primary adipocytes. *Smad3*$^{-/-}$ mice also showed increased metabolic rate and exhibited lower respiratory exchange ratio in agreement with the observed increase in mitochondrial function and lipid oxidation.

We also found that elevated TGF-β levels associate with poor metabolic profile in human subjects showing a significant positive association between TGF-β and human adiposity (Yadav et al. 2011). In addition, positive correlations of circulating TGF-β were observed with fat mass, fasting insulin levels, and HOMA insulin resistance index. Importantly, administration of a TGF-β pan neutralization antibody (1D11) suppressed body weight gain, size of fat depots and fat mass, adipocyte cell size and levels of triglyceride, resistin, and leptin in mouse models of obesity and diabetes (Yadav et al. 2011). Further, treatment with 1D11 improved glucose and insulin tolerance, suppressed hyperglycemia and hyperinsulinemia, ameliorated hepatic steatosis, and increased protein levels of BAT/mitochondrial markers in the WAT (Yadav et al. 2011).

The TGF-β superfamily member, BMP7, is implicated in brown adipogenesis (Tseng et al. 2008). BMP proteins promote differentiation to either white adipocytes or brown adipocytes (Schulz et al. 2011). BMP7 triggers commitment of mesenchymal progenitor cells to a brown adipocyte lineage, and implantation of these cells into nude mice results in development of adipose tissue containing mostly brown adipocytes (Tseng et al. 2008). Subpopulation of adipogenic progenitors residing in murine brown fat, white fat, and skeletal muscle were isolated, and it was shown that muscle and white fat derived Sca-1(+) cells were able to differentiate into brown-like adipocytes upon stimulation with BMP7 (Schulz et al. 2011). Also, BMP7 knockout embryos show a marked reduction of brown fat, whereas adenoviral-mediated expression of BMP7 in mice results in a significant increase in brown fat mass and leads to an increase in energy expenditure and a reduction in weight gain (Tseng et al. 2008). However, the role of BMP7 in promoting browning of white fat remains unclear. A more recent study identified Zfp423, a BMP-Smad signaling effector, as a transcriptional regulator of both brown and white preadipocyte differentiation (Gupta et al. 2010), although whether Zfp423 plays a role in promoting brown fat features in white fat is unknown. TGF-β regulates adipocyte differentiation via a Smad3-C/EBP interaction (Choy and Derynck 2003; Choy et al. 2000). Further, Schnurri-2 interacts with Smads and C/EBP during BMP2-regulated adipogenesis (Jin et al. 2006). These studies suggest an important role for TGF-β superfamily members in WAT and BAT biology—one that is analogous to the role played by myostatin during muscle development (McPherron and Lee 2002). Further, the recently described role for TGF-β in glucose-induced cell hypertrophy (Wu and Derynck 2009) may have ramifications in conditions associated with hyperglycemia, glucose intolerance, and insulin resistance.

Circulating TGF-β levels are elevated in cardiovascular disease and hypertension (Gordon and Blobe 2008) and TGF-β and BMI are closely associated in human adipose tissue during morbid obesity (Alessi et al. 2000). Predisposition to certain cancers, atherosclerosis, myocardial infarction, hypertension, and stroke is correlated with the presence of TGF-β polymorphisms (Grainger et al. 1999). Further, polymorphisms such as the T29C polymorphism, which results in a Leu-Pro substitution at codon 10, are correlated with elevated circulating TGF-β levels. Interestingly, the T29C polymorphism is associated with increased BMI, elevated fasting insulin and glucose levels, and higher HOMA insulin resistance indices (Rosmond et al. 2003). Moreover, the *SMAD3* gene was recently identified in a type 2 diabetes genome-wide association study (Perry et al. 2009) which further supports the notion of TGF-β/Smad3 pathway as a potential target in diabetes and obesity. TGF-β antagonist approaches are being clinically evaluated to treat diseases, such as cancer, fibrosis, scarring, diabetic nephropathy, where elevated TGF-β levels are implicated. The occurrence of elevated TGF-β levels in obese individuals combined with the beneficial effect of the anti-TGF-β neutralization antibody in mouse models of obesity and diabetes offer treatment alternatives for these diseases.

18.2.4 Role of TGF-β Signaling in Pancreatic Islet β-Cell Function

The pancreatic islet β-cell, due to its unique function of insulin synthesis and glucose-stimulated insulin secretion (GSIS), is a prime target of affliction in diabetes (Weir and Bonner-Weir 2004). In addition, a majority of patients with diabetes mellitus type II develop β-cell dysfunction in concert with insulin resistance in target organs of insulin action: liver, muscle, and adipose tissue (Herman and Kahn 2006). Improved mechanistic understanding of normal β-cell function and insulin action is needed to enable early diagnosis of β-cell dysfunction and insulin resistance and to facilitate development of new rational therapeutics for diabetes.

Development of the endocrine and exocrine pancreas is controlled by factors that include members of the TGF-β superfamily (Kim and Hebrok 2001; Kim and MacDonald 2002). In addition, TGF-β signaling has been implicated in pancreatic diseases (Rane et al. 2006). BMP signaling plays an instructive role during early pancreatic development (Kim and Hebrok 2001; Kim and MacDonald 2002; Rane et al. 2006) and regulates mature β-cell function (Gannon 2007; Goulley et al. 2007), whereas activin signaling regulates islet morphogenesis and β-cell mass (Yamaoka et al. 1998; Zhang et al. 2004). TGF-β isoforms are expressed in the epithelium and mesenchyme of embryonic pancreas and in adult pancreas (Yamanaka et al. 1993). Islet cells demonstrate diffuse cytoplasmic immunostaining for TGF-β isoforms with most of the positive islet cells co-expressing insulin. TGF-β receptors (TβRI and TβRII) are present in the pancreatic epithelium and mesenchyme during early stages of development and postnatally in pancreatic islets and ducts. Further, Smad proteins are expressed in the pancreas which elucidates that components needed for activation of the canonical TGF-β signaling exist within the pancreas.

Disruption of TGF-β signaling at the receptor level using mice expressing the dominant-negative TGF-β type II receptor (DNTβRII) resulted in increased

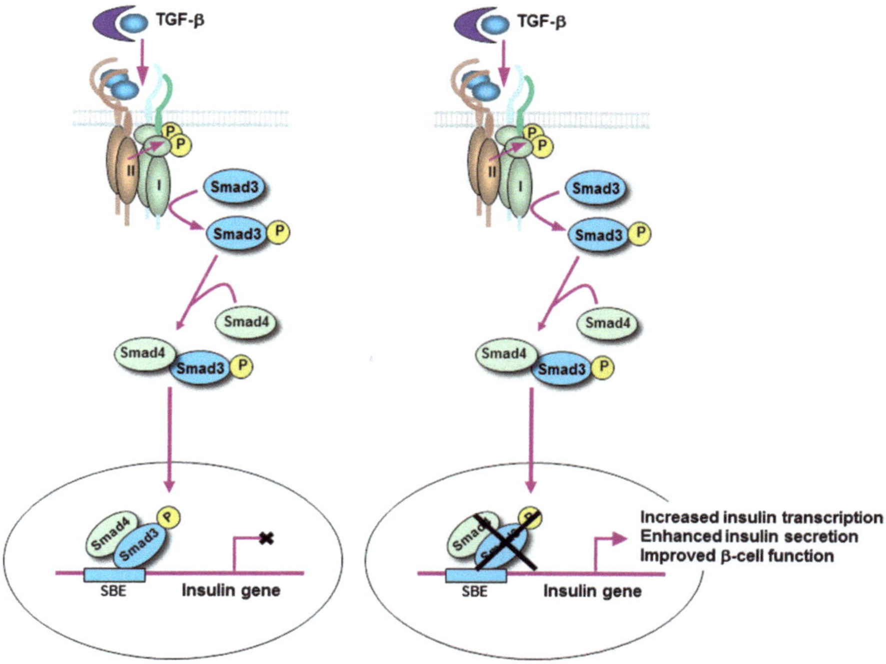

Fig. 18.2 Smad3 regulates the insulin promoter in pancreatic islet β-cells. Binding of the TGF-β ligand to TβRII results in activation of the TβRI kinase and subsequent phosphorylation of Smad3. Phospho-Smad3 translocates to the nucleus along with Smad4. Smad3 binds to Smad-binding elements (SBE) on the Insulin promoter and represses Insulin gene transcription (*left panel*). Loss of Smad3 binding to the promoter enhances Insulin transcription and is associated with improved functionality of the pancreatic islet β-cells

proliferation of pancreatic acinar cells and severely perturbed acinar differentiation (Bottinger et al. 1997). Additionally, DNTβRII mice exhibit increased endocrine precursors and proliferating endocrine cells, with an abnormal accumulation of endocrine cells around the developing ducts of mid-late stage embryonic pancreas (Tulachan et al. 2007). Transgenic mice expressing TGF-β in β-cells exhibit abnormal small islet cell clusters without formation of normal adult islets, although the overall islet cell mass is not significantly diminished (Lee et al. 1995).

We recently described an important role of the TGF-β/Smad3 pathway in regulation of insulin gene transcription and β-cell function (Lin et al. 2009). We identified insulin as a TGF-β target gene and showed that Smad3 occupies the insulin gene promoter and represses insulin gene transcription (Fig. 18.2). In contrast, Smad3 siRNAs relieve insulin transcriptional repression and enhance insulin levels. Transduction of adenoviral Smad3 into primary human and nonhuman primate islets suppresses insulin content, whereas dominant-negative Smad3 adenoviral transduction enhances insulin levels. Consistent with this, Smad3-deficient mice exhibit moderate hyperinsulinemia and mild hypoglycemia. Moreover, Smad3 deficiency results in improved glucose tolerance and enhanced GSIS in vivo. In ex vivo perfusion assays, Smad3-deficient islets exhibited improved glucose-stimulated insulin release. Interestingly, Smad3-deficient islets harbor an activated insulin-receptor signaling pathway and TGF-β signaling regulates expression of genes involved in β-cell function. These studies emphasize TGF-β/Smad3 signaling as an important regulator of insulin gene transcription and β-cell function and suggest that components of the TGF-β signaling pathway may be dysregulated in diabetes.

We find that inhibition of Smad3 signaling leads to elevated GSIS ex vivo and in vivo. It is plausible that the effects of TGF-β/Smad3 on insulin gene transcription may be independent of its role in GSIS. Levels of expression of genes involved in insulin biosynthesis, pro-insulin processing, glucose sensing, glucose metabolism, incretin signaling, insulin exocytosis, and GSIS are significantly elevated in *Smad3*$^{-/-}$ islets and, conversely, significantly repressed in cells harboring a constitutively active TβRI/Smad3 pathway. Interestingly, we observe an activated insulin receptor signaling pathway in *Smad3*$^{-/-}$ islets, which is consistent with the enhanced β-cell function in *Smad3*$^{-/-}$ mice.

TGF-β appears to exert a bimodal effect on GSIS depending on the dose, time of exposure, and concentration of coexisting glucose (Sekine et al. 1994; Sjoholm and Hellerstrom 1991). Therefore, at low glucose concentration TGF-β stimulates insulin release, whereas at a high glucose concentration the same dose of TGF-β significantly inhibits GSIS (Sjoholm and Hellerstrom 1991; Sekine et al. 1994). Interestingly, dual and opposing roles of TGF-β in cancer are believed to suppress primary tumors and paradoxically promote cancer metastases (Roberts and Wakefield 2003). Recent reports of the role of BMP4/BMPR1A and Smad7 in β-cell function further underscore the important role of TGF-β superfamily signaling in this cell type (Smart et al. 2006; Bertolino et al. 2008; Goulley et al. 2007). BMP4-BMPR1A signaling plays a key role in insulin secretion by positively regulating genes involved in glucose sensing, glucose-metabolism-coupled secretion, incretin signaling, proinsulin processing, and insulin exocytosis (Goulley et al. 2007). Our findings of TGF-β/Smad3 mediated regulation of genes associated with β-cell function are consistent with an important role of the TGF-β superfamily signaling in pancreas development, particularly, β-cell function.

The findings that amplified TGF-β/Smad3 signals repress insulin transcription and reduce insulin level and insulin secretion, whereas ablation of Smad3 leads to improved β-cell function suggesting that abnormal TGF-β signaling may promote pathogenesis of dysfunctional β-cells. We propose that TGF-β/Smad3 pathways may (Sekine et al. 1994) regulate β-cell function in settings of increased insulin demand, including insulin resistance, obesity, and during β-cell injury. Further, our results that reduced TGF-β/Smad3 signaling markedly enhance insulin content and insulin secretion suggest that pharmacological inactivation of TGF-β/Smad3 signaling might be useful for promoting β-cell differentiation and ameliorating β-cell failure during diabetes.

18.3 Perspectives and Conclusion

Together, our findings support the notion that the TGF-β signaling network plays important roles in the functionality of the pancreatic β-cell and that of the adipose tissue (Figs. 18.1 and 18.2). Further, we propose that this pathway serves an integral role in metabolic homeostasis. We believe that this role is not limited to the TGF-β isoforms and that there is evidence supporting an integrated role for various TGF-β superfamily isoforms in the process. In addition, we believe that other signaling pathways that intersect into, or, are off-shoots of, the TGF-β signaling network play a collaborative role to maintain metabolic homeostasis. As such, it is important to define the precise role of TGF-β signaling in regulating the actions of various metabolic organs—both via the canonical Smad-transcription factor loop and/or via non-canonical signaling pathways. TGF-β levels are elevated in metabolic disease, which supports the utility of therapeutics aimed at targeting the TGF-β pathway for these diseases. However, the complexity of the TGF-β signaling network warrants a cautious and rational approach to designing anti- TGF-β modalities to combat metabolic diseases.

Acknowledgment We apologize to authors whose contributions to this field of research have not been cited or have only been indirectly cited due to space limitations. Support for this work came from funds from the NIDDK, NIH intramural program.

References

Alessi MC, Bastelica D, Morange P, Berthet B, Leduc I, Verdier M, Geel O, Juhan-Vague I (2000) Plasminogen activator inhibitor 1, transforming growth factor-β1, and BMI are closely associated in human adipose tissue during morbid obesity. Diabetes 49(8):1374–1380

Andreelli F, Amouyal C, Magnan C, Mithieux G (2009) What can bariatric surgery teach us about the pathophysiology of type 2 diabetes? Diabetes Metab 35(6P2):499–507

Atit R, Sgaier SK, Mohamed OA, Taketo MM, Dufort D, Joyner AL, Niswander L, Conlon RA (2006) Beta-catenin activation is necessary and sufficient to specify the dorsal dermal fate in the mouse. Dev Biol 296(1):164–176. doi:S0012-1606(06)00732-9 [pii] 10.1016/j.ydbio.2006.04.449

Attisano L, Wrana JL (2002) Signal transduction by the TGF-β superfamily. Science 296(5573):1646–1647

Bays HE, Gonzalez-Campoy JM, Bray GA, Kitabchi AE, Bergman DA, Schorr AB, Rodbard HW, Henry RR (2008) Pathogenic potential of adipose tissue and metabolic consequences of adipocyte hypertrophy and increased visceral adiposity. Expert Rev Cardiovasc Ther 6(3):343–368. doi:10.1586/14779072.6.3.343

Bertolino P, Holmberg R, Reissmann E, Andersson O, Berggren PO, Ibanez CF (2008) Activin B receptor ALK7 is a negative regulator of pancreatic β-cell function. Proc Natl Acad Sci USA 105(20):7246–7251. doi:10.1073/pnas.0801285105

Biddinger SB, Kahn CR (2006) From mice to men: insights into the insulin resistance syndromes. Annu Rev Physiol 68(1):123–158. doi:10.1146/annurev.physiol.68.040104.124723

Bloom JD, Dutia MD, Johnson BD, Wissner A, Burns MG, Largis EE, Dolan JA, Claus TH (1992) Disodium (R, R)-5-[2-[[2-(3-chlorophenyl)-2-hydroxyethyl]-amino]

propyl]-1,3-benzodioxole-2,2-dicarboxylate (CL 316,243). A potent β-adrenergic agonist virtually specific for β 3 receptors. A promising antidiabetic and antiobesity agent. J Med Chem 35(16):3081–3084
Borai A, Livingstone C, Kaddam I, Ferns G (2011) Selection of the appropriate method for the assessment of insulin resistance. BMC Med Res Methodol 11(1):158
Bottinger EP, Jakubczak JL, Roberts IS, Mumy M, Hemmati P, Bagnall K, Merlino G, Wakefield LM (1997) Expression of a dominant-negative mutant TGF-β type II receptor in transgenic mice reveals essential roles for TGF-β in regulation of growth and differentiation in the exocrine pancreas. EMBO J 16(10):2621–2633
Cannon B, Nedergaard J (2004) Brown adipose tissue: function and physiological significance. Physiol Rev 84(1):277–359. doi:10.1152/physrev.00015.2003
Cannon B, Nedergaard J (2011) Nonshivering thermogenesis and its adequate measurement in metabolic studies. J Exp Biol 214(Pt 2):242–253
Chang L, Chiang SH, Saltiel AR (2004) Insulin signaling and the regulation of glucose transport. Mol Med 10(7–12):65–71. doi:10.2119/2005-00029.Saltiel
Chng Z, Vallier L, Pedersen R (2011) Activin/nodal signaling and pluripotency. Vitam Horm 85:39–58
Choy L, Derynck R (2003) Transforming growth factor-β inhibits adipocyte differentiation by Smad3 interacting with CCAAT/enhancer-binding protein (C/EBP) and repressing C/EBP transactivation function. J Biol Chem 278(11):9609–9619
Choy L, Skillington J, Derynck R (2000) Roles of autocrine TGF-β receptor and Smad signaling in adipocyte differentiation. J Cell Biol 149(3):667–682
Cinti S, Mitchell G, Barbatelli G, Murano I, Ceresi E, Faloia E, Wang S, Fortier M, Greenberg AS, Obin MS (2005) Adipocyte death defines macrophage localization and function in adipose tissue of obese mice and humans. J Lipid Res 46(11):2347–2355. doi:10.1194/jlr.M500294-JLR200
Cousin B, Cinti S, Morroni M, Raimbault S, Ricquier D, Penicaud L, Casteilla L (1992) Occurrence of brown adipocytes in rat white adipose tissue: molecular and morphological characterization. J Cell Sci 103(Pt 4):931–942
Cusi K (2010) The role of adipose tissue and lipotoxicity in the pathogenesis of type 2 diabetes. Curr Diab Rep 10(4):306–315. doi:10.1007/s11892-010-0122-6
Cypess AM, Lehman S, Williams G, Tal I, Rodman D, Goldfine AB, Kuo FC, Palmer EL, Tseng YH, Doria A, Kolodny GM, Kahn CR (2009) Identification and importance of brown adipose tissue in adult humans. N Engl J Med 360(15):1509–1517. doi:360/15/1509 [pii] 10.1056/NEJMoa0810780
DeFronzo RA, Tripathy D (2009) Skeletal muscle insulin resistance is the primary defect in type 2 diabetes. Diabetes Care 32(suppl 2):S157–S163. doi:10.2337/dc09-S302
Enerback S (2010) Human brown adipose tissue. Cell Metab 11(4):248–252. doi:S1550-4131(10)00078-1 [pii] 10.1016/j.cmet.2010.03.008
Fain J (2006) Release of interleukins and other inflammatory cytokines by human adipose tissue is enhanced in obesity and primarily due to the nonfat cells. Vitam Horm 74:443–477
Feng XH, Derynck R (2005) Specificity and versatility in tgf-β signaling through Smads. Annu Rev Cell Dev Biol 21:659–693
Frontini A, Cinti S (2010) Distribution and development of brown adipocytes in the murine and human adipose organ. Cell Metab 11(4):253–256
Gannon M (2007) BuMP-ing up insulin secretion by pancreatic β cells. Cell Metab 5(3):157–159
Gesta S, Tseng YH, Kahn CR (2007) Developmental origin of fat: tracking obesity to its source. Cell 131(2):242–256
Ghorbani M, Himms-Hagen J (1997) Appearance of brown adipocytes in white adipose tissue during CL 316,243-induced reversal of obesity and diabetes in Zucker fa/fa rats. Int J Obes Relat Metab Disord 21(6):465–475
Ghorbani M, Claus TH, Himms-Hagen J (1997) Hypertrophy of brown adipocytes in brown and white adipose tissues and reversal of diet-induced obesity in rats treated with a β3-adrenoceptor agonist. Biochem Pharmacol 54(1):121–131

Gordon KJ, Blobe GC (2008) Role of transforming growth factor-β superfamily signaling pathways in human disease. Biochim Biophys Acta 1782(4):197–228

Goulley J, Dahl U, Baeza N, Mishina Y, Edlund H (2007) BMP4-BMPR1A signaling in β cells is required for and augments glucose-stimulated insulin secretion. Cell Metab 5(3):207–219. doi:10.1016/j.cmet.2007.01.009

Grainger DJ, Heathcote K, Chiano M, Snieder H, Kemp PR, Metcalfe JC, Carter ND, Spector TD (1999) Genetic control of the circulating concentration of transforming growth factor type β1. Hum Mol Genet 8(1):93–97. doi:ddc008 [pii]

Greenberg AS, Obin MS (2006) Obesity and the role of adipose tissue in inflammation and metabolism. Am J Clin Nutr 83(2):461S–465S

Greenberg AS, Obin MS (2008) Many roads lead to the lipid droplet. Cell Metab 7(6):472–473

Guo W, Wong S, Xie W, Lei T, Luo Z (2007) Palmitate modulates intracellular signaling, induces endoplasmic reticulum stress, and causes apoptosis in mouse 3 T3-L1 and rat primary preadipocytes. Am J Physiol Endocrinol Metab 293(2):E576–E586. doi:10.1152/ajpendo.00523.2006

Gupta RK, Arany Z, Seale P, Mepani RJ, Ye L, Conroe HM, Roby YA, Kulaga H, Reed RR, Spiegelman BM (2010) Transcriptional control of preadipocyte determination by Zfp423. Nature 464(7288):619–623. doi:10.1038/nature08816

Gutierrez DA, Puglisi MJ, Hasty AH (2009) Impact of increased adipose tissue mass on inflammation, insulin resistance, and dyslipidemia. Curr Diab Rep 9(1):26–32

Halberg N, Khan T, Trujillo ME, Wernstedt-Asterholm I, Attie AD, Sherwani S, Wang ZV, Landskroner-Eiger S, Dineen S, Magalang UJ, Brekken RA, Scherer PE (2009) Hypoxia-inducible factor 1{α} induces fibrosis and insulin resistance in white adipose tissue. Mol Cell Biol 29(16):4467–4483. doi:10.1128/mcb.00192-09

Herman MA, Kahn BB (2006) Glucose transport and sensing in the maintenance of glucose homeostasis and metabolic harmony. J Clin Invest 116(7):1767–1775

Himms-Hagen J, Cui J, Danforth E Jr, Taatjes DJ, Lang SS, Waters BL, Claus TH (1994) Effect of CL-316,243, a thermogenic β 3-agonist, on energy balance and brown and white adipose tissues in rats. Am J Physiol 266(4 Pt 2):R1371–R1382

Ibrahim MM (2010) Subcutaneous and visceral adipose tissue: structural and functional differences. Obes Rev 11(1):11–18. doi:10.1111/j.1467-789X.2009.00623.x

Jager J, Gremeaux T, Cormont M, Le Marchand-Brustel Y, Tanti J-F (2007) Interleukin-1B-induced insulin resistance in adipocytes through down-regulation of insulin receptor substrate-1 expression. Endocrinology 148(1):241–251. doi:10.1210/en.2006-0692

James D, Levine AJ, Besser D, Hemmati-Brivanlou A (2005) TGFβ/activin/nodal signaling is necessary for the maintenance of pluripotency in human embryonic stem cells. Development 132(6):1273–1282. doi:10.1242/dev.01706

Jimenez M, Barbatelli G, Allevi R, Cinti S, Seydoux J, Giacobino JP, Muzzin P, Preitner F (2003) β 3-adrenoceptor knockout in C57BL/6 J mice depresses the occurrence of brown adipocytes in white fat. Eur J Biochem 270(4):699–705

Jin W, Takagi T, Kanesashi SN, Kurahashi T, Nomura T, Harada J, Ishii S (2006) Schnurri-2 controls BMP-dependent adipogenesis via interaction with Smad proteins. Dev Cell 10(4):461–471

Kang K, Reilly SM, Karabacak V, Gangl MR, Fitzgerald K, Hatano B, Lee CH (2008) Adipocyte-derived Th2 cytokines and myeloid PPARδ regulate macrophage polarization and insulin sensitivity. Cell Metab 7(6):485–495. doi:S1550-4131(08)00112-5 [pii] 10.1016/j.cmet.2008.04.002

Karalis KP, Giannogonas P, Kodela E, Koutmani Y, Zoumakis M, Teli T (2009) Mechanisms of obesity and related pathology: linking immune responses to metabolic stress. FEBS J 276(20): 5747–5754

Kennedy A, Martinez K, Chuang CC, LaPoint K, McIntosh M (2009) Saturated fatty acid-mediated inflammation and insulin resistance in adipose tissue: mechanisms of action and implications. J Nutr 139(1):1–4. doi:jn.108.098269 [pii] 10.3945/jn.108.098269

Khan T, Muise ES, Iyengar P, Wang ZV, Chandalia M, Abate N, Zhang BB, Bonaldo P, Chua S, Scherer PE (2009) Metabolic dysregulation and adipose tissue fibrosis: role of collagen VI. Mol Cell Biol 29(6):1575–1591. doi:10.1128/mcb.01300-08

Kim SK, Hebrok M (2001) Intercellular signals regulating pancreas development and function. Genes Dev 15(2):111–127

Kim SK, MacDonald RJ (2002) Signaling and transcriptional control of pancreatic organogenesis. Curr Opin Genet Dev 12(5):540–547

Kim F, Pham M, Luttrell I, Bannerman DD, Tupper J, Thaler J, Hawn TR, Raines EW, Schwartz MW (2007a) Toll-like receptor-4 mediates vascular inflammation and insulin resistance in diet-induced obesity. Circ Res 100(11):1589–1596. doi:10.1161/circresaha.106.142851

Kim JY, van de Wall E, Laplante M, Azzara A, Trujillo ME, Hofmann SM, Schraw T, Durand JL, Li H, Li G, Jelicks LA, Mehler MF, Hui DY, Deshaies Y, Shulman GI, Schwartz GJ, Scherer PE (2007b) Obesity-associated improvements in metabolic profile through expansion of adipose tissue. J Clin Invest 117(9):2621–2637. doi:10.1172/JCI31021

Kitisin K, Saha T, Blake T, Golestaneh N, Deng M, Kim C, Tang Y, Shetty K, Mishra B, Mishra L (2007) TGF-β signaling in development. Sci STKE 399:cm1. doi:10.1126/stke.3992007cm1

Lacasa D, Taleb S, Keophiphath M, Miranville A, Clement K (2007) Macrophage-secreted factors impair human adipogenesis: involvement of proinflammatory state in preadipocytes. Endocrinology 148(2):868–877. doi:10.1210/en.2006-0687

Langin D (2009) Recruitment of brown fat and conversion of white into brown adipocytes: strategies to fight the metabolic complications of obesity? Biochim Biophys Acta 1801(3):372–376

Lee MS, Gu D, Feng L, Curriden S, Arnush M, Krahl T, Gurushanthaiah D, Wilson C, Loskutoff DL, Fox H et al (1995) Accumulation of extracellular matrix and developmental dysregulation in the pancreas by transgenic production of transforming growth factor-β 1. Am J Pathol 147(1):42–52

Lefterova MI, Lazar MA (2009) New developments in adipogenesis. Trends Endocrinol Metab 20(3):107–114

Lin J, Handschin C, Spiegelman BM (2005) Metabolic control through the PGC-1 family of transcription coactivators. Cell Metab 1(6):361–370

Lin HM, Lee JH, Yadav H, Kamaraju AK, Liu E, Zhigang D, Vieira A, Kim SJ, Collins H, Matschinsky F, Harlan DM, Roberts AB, Rane SG (2009) Transforming growth factor-β/Smad3 signaling regulates insulin gene transcription and pancreatic islet β-cell function. J Biol Chem 284(18):12246–12257

Loncar D (1991) Convertible adipose tissue in mice. Cell Tissue Res 266(1):149–161

Lumeng CN, Saltiel AR (2011) Inflammatory links between obesity and metabolic disease. J Clin Invest 121(6):2111–2117

Lumeng CN, Bodzin JL, Saltiel AR (2007) Obesity induces a phenotypic switch in adipose tissue macrophage polarization. J Clin Invest 117(1):175–184. doi:10.1172/JCI29881

Massague J, Blain SW, Lo RS (2000) TGFβ signaling in growth control, cancer, and heritable disorders. Cell 103(2):295–309

McGillicuddy FC, Harford KA, Reynolds CM, Oliver E, Claessens M, Mills KHG, Roche HM (2011) Lack of interleukin-1 receptor I (IL-1RI) protects mice from high-fat diet-induced adipose tissue inflammation coincident with improved glucose homeostasis. Diabetes 60(6):1688–1698. doi:10.2337/db10-1278

McPherron AC, Lee SJ (2002) Suppression of body fat accumulation in myostatin-deficient mice. J Clin Invest 109(5):595–601

Mishra L, Derynck R, Mishra B (2005) Transforming growth factor-β signaling in stem cells and cancer. Science 310(5745):68–71

Mullen AC, Orlando DA, Newman JJ, Loven J, Kumar RM, Bilodeau S, Reddy J, Guenther MG, DeKoter RP, Young RA (2011) Master transcription factors determine cell-type-specific responses to TGF-β signaling. Cell 147(3):565–576. doi:S0092-8674(11)01134-2 [pii] 10.1016/j.cell.2011.08.050

Nedergaard J, Cannon B (2010) The changed metabolic world with human brown adipose tissue: therapeutic visions. Cell Metab 11(4):268–272. doi:S1550-4131(10)00077-X [pii] 10.1016/j.cmet.2010.03.007

Nedergaard J, Bengtsson T, Cannon B (2007) Unexpected evidence for active brown adipose tissue in adult humans. Am J Physiol 293(2):E444–E452

Nieto-Vazquez I, Fernandez-Veledo S, de Alvaro C, Lorenzo M (2008) Dual role of interleukin-6 in regulating insulin sensitivity in murine skeletal muscle. Diabetes 57(12):3211–3221. doi:10.2337/db07-1062

Nishimura S, Manabe I, Nagasaki M, Hosoya Y, Yamashita H, Fujita H, Ohsugi M, Tobe K, Kadowaki T, Nagai R, Sugiura S (2007) Adipogenesis in obesity requires close interplay between differentiating adipocytes, stromal cells, and blood vessels. Diabetes 56(6):1517–1526. doi:10.2337/db06-1749

Odegaard JI, Ricardo-Gonzalez RR, Goforth MH, Morel CR, Subramanian V, Mukundan L, Red Eagle A, Vats D, Brombacher F, Ferrante AW, Chawla A (2007) Macrophage-specific PPARgamma controls alternative activation and improves insulin resistance. Nature 447(7148):1116–1120. doi:nature05894 [pii] 10.1038/nature05894

Olefsky JM, Glass CK (2010) Macrophages, inflammation, and insulin resistance. Annu Rev Physiol 72(1):219–246. doi:10.1146/annurev-physiol-021909-135846

Park KW, Halperin DS, Tontonoz P (2008) Before they were fat: adipocyte progenitors. Cell Metab 8(6):454–457

Perfield JW, Lee Y, Shulman GI, Samuel VT, Jurczak MJ, Chang E, Xie C, Tsichlis PN, Obin MS, Greenberg AS (2010) Tumor progression locus 2 (TPL2) regulates obesity-associated inflammation and insulin resistance. Diabetes 60(4):1168–1176. doi:10.2337/db10-0715

Perry JR, McCarthy MI, Hattersley AT, Zeggini E, Weedon MN, Frayling TM (2009) Interrogating type 2 diabetes genome-wide association data using a biological pathway-based approach. Diabetes 58(6):1463–1467. doi:db08-1378 [pii] 10.2337/db08-1378

Rane SG, Lee JH, Lin HM (2006) Transforming growth factor-β pathway: role in pancreas development and pancreatic disease. Cytokine Growth Factor Rev 17(1–2):107–119

Roberts AB, Wakefield LM (2003) The two faces of transforming growth factor β in carcinogenesis. Proc Natl Acad Sci USA 100(15):8621–8623. doi:10.1073/pnas.1633291100

Rosen ED, Spiegelman BM (2006) Adipocytes as regulators of energy balance and glucose homeostasis. Nature 444(7121):847–853

Rosmond R, Chagnon M, Bouchard C, Bjorntorp P (2003) Increased abdominal obesity, insulin and glucose levels in nondiabetic subjects with a T29C polymorphism of the transforming growth factor-β1 gene. Horm Res 59(4):191–194. doi:10.1159/000069323 HRE59191 [pii]

Saberi M, Woods N-B, de Luca C, Schenk S, Lu JC, Bandyopadhyay G, Verma IM, Olefsky JM (2009) Hematopoietic cell-specific deletion of toll-like receptor 4 ameliorates hepatic and adipose tissue insulin resistance in high-fat-fed mice. Cell Metab 10(5):419–429

Schulz TJ, Huang TL, Tran TT, Zhang H, Townsend KL, Shadrach JL, Cerletti M, McDougall LE, Giorgadze N, Tchkonia T, Schrier D, Falb D, Kirkland JL, Wagers AJ, Tseng YH (2011) Identification of inducible brown adipocyte progenitors residing in skeletal muscle and white fat. Proc Natl Acad Sci USA 108(1):143–148

Seale P, Bjork B, Yang W, Kajimura S, Chin S, Kuang S, Scime A, Devarakonda S, Conroe HM, Erdjument-Bromage H, Tempst P, Rudnicki MA, Beier DR, Spiegelman BM (2008) PRDM16 controls a brown fat/skeletal muscle switch. Nature 454(7207):961–967

Seale P, Kajimura S, Spiegelman BM (2009) Transcriptional control of brown adipocyte development and physiological function—of mice and men. Genes Dev 23(7):788–797

Sekine N, Yamashita N, Kojima I, Miyazaki J, Ogata E (1994) Bimodal effect of transforming growth factor-β on insulin secretion in MIN6 cells. Diabetes Res Clin Pract 26(1):7–14

Senn JJ (2006) Toll-like receptor-2 is essential for the development of palmitate-induced insulin resistance in myotubes. J Biol Chem 281(37):26865–26875. doi:10.1074/jbc.M513304200

Shaul ME, Bennett G, Strissel KJ, Greenberg AS, Obin MS (2010) Dynamic, M2-like remodeling phenotypes of CD11c+ adipose tissue macrophages during high-fat diet-induced obesity in mice. Diabetes 59(5):1171–1181. doi:db09-1402 [pii] 10.2337/db09-1402

Shi H, Kokoeva MV, Inouye K, Tzameli I, Yin H, Flier JS (2006) TLR4 links innate immunity and fatty acid-induced insulin resistance. J Clin Invest 116(11):3015–3025

Shoelson SE, Lee J, Goldfine AB (2006) Inflammation and insulin resistance. J Clin Invest 116(7):1793–1801. doi:10.1172/JCI29069

Sjoholm A, Hellerstrom C (1991) TGF-β stimulates insulin secretion and blocks mitogenic response of pancreatic β-cells to glucose. Am J Physiol 260(5 Pt 1):C1046–C1051

Smart NG, Apelqvist AA, Gu X, Harmon EB, Topper JN, MacDonald RJ, Kim SK (2006) Conditional expression of Smad7 in pancreatic β cells disrupts TGF-β signaling and induces reversible diabetes mellitus. PLoS Biol 4(2):e39. doi:10.1371/journal.pbio.0040039

Soloveva V, Graves RA, Rasenick MM, Spiegelman BM, Ross SR (1997) Transgenic mice overexpressing the β 1-adrenergic receptor in adipose tissue are resistant to obesity. Mol Endocrinol 11(1):27–38

Strissel KJ, Stancheva Z, Miyoshi H, Perfield JW, DeFuria J, Jick Z, Greenberg AS, Obin MS (2007) Adipocyte death, adipose tissue remodeling, and obesity complications. Diabetes 56(12):2910–2918. doi:10.2337/db07-0767

Strissel KJ, Defuria J, Shaul ME, Bennett G, Greenberg AS, Obin MS (2010) T-cell recruitment and Th1 polarization in adipose tissue during diet-induced obesity in C57BL/6 mice. Obesity (Silver Spring) 18:1918–1925. doi:oby20101 [pii] 10.1038/oby.2010.1

Suganami T, Tanimoto-Koyama K, Nishida J, Itoh M, Yuan X, Mizuarai S, Kotani H, Yamaoka S, Miyake K, Aoe S, Kamei Y, Ogawa Y (2007) Role of the toll-like receptor 4/NF-{κ}B pathway in saturated fatty acid-induced inflammatory changes in the interaction between adipocytes and macrophages. Arterioscler Thromb Vasc Biol 27(1):84–91. doi:10.1161/01.ATV.0000251608.09329.9a

Takenaga M, Fukumoto M, Hori Y (2007) Regulated Nodal signaling promotes differentiation of the definitive endoderm and mesoderm from ES cells. J Cell Sci 120(12):2078–2090. doi:10.1242/jcs.004127

Thiebaud D, Jacot E, DeFronzo R, Maeder E, Jequier E, Felber J (1982) The effect of graded doses of insulin on total glucose uptake, glucose oxidation, and glucose storage in man. Diabetes 31:957–963

Timmons JA, Wennmalm K, Larsson O, Walden TB, Lassmann T, Petrovic N, Hamilton DL, Gimeno RE, Wahlestedt C, Baar K, Nedergaard J, Cannon B (2007) Myogenic gene expression signature establishes that brown and white adipocytes originate from distinct cell lineages. Proc Natl Acad Sci USA 104(11):4401–4406

Tseng YH, Kokkotou E, Schulz TJ, Huang TL, Winnay JN, Taniguchi CM, Tran TT, Suzuki R, Espinoza DO, Yamamoto Y, Ahrens MJ, Dudley AT, Norris AW, Kulkarni RN, Kahn CR (2008) New role of bone morphogenetic protein 7 in brown adipogenesis and energy expenditure. Nature 454(7207):1000–1004

Tulachan SS, Tei E, Hembree M, Crisera C, Prasadan K, Koizumi M, Shah S, Guo P, Bottinger E, Gittes GK (2007) TGF-β isoform signaling regulates secondary transition and mesenchymal-induced endocrine development in the embryonic mouse pancreas. Dev Biol 305(2):508–521

Unger RH, Scherer PE (2010) Gluttony, sloth and the metabolic syndrome: a roadmap to lipotoxicity. Trends Endocrinol Metab 21(6):345–352

Unger RH, Clark GO, Scherer PE, Orci L (2010) Lipid homeostasis, lipotoxicity and the metabolic syndrome. Biochim Biophys Acta 1801(3):209–214

van Marken Lichtenbelt WD, Vanhommerig JW, Smulders NM, Drossaerts JM, Kemerink GJ, Bouvy ND, Schrauwen P, Teule GJ (2009) Cold-activated brown adipose tissue in healthy men. N Engl J Med 360(15):1500–1508. doi:360/15/1500 [pii] 10.1056/NEJMoa0808718

Varga AC, Wrana JL (2005) The disparate role of BMP in stem cell biology. Oncogene 24(37):5713–5721

Virtanen KA, Lidell ME, Orava J, Heglind M, Westergren R, Niemi T, Taittonen M, Laine J, Savisto NJ, Enerback S, Nuutila P (2009) Functional brown adipose tissue in healthy adults. N Engl J Med 360(15):1518–1525. doi:360/15/1518 [pii] 10.1056/NEJMoa0808949

Watabe T, Miyazono K (2009) Roles of TGF-[β] family signaling in stem cell renewal and differentiation. Cell Res 19(1):103–115

Weigert C, Brodbeck K, Staiger H, Kausch C, Machicao F, HÃring HU, Schleicher ED (2004) Palmitate, but not unsaturated fatty acids, induces the expression of interleukin-6 in human myotubes through proteasome-dependent activation of nuclear factor-κB. J Biol Chem 279(23):23942–23952. doi:10.1074/jbc.M312692200

Weir GC, Bonner-Weir S (2004) Five stages of evolving β-cell dysfunction during progression to diabetes. Diabetes 53(Suppl 3):S16–S21

Weisberg SP, McCann D, Desai M, Rosenbaum M, Leibel RL, Ferrante AW Jr (2003) Obesity is associated with macrophage accumulation in adipose tissue. J Clin Invest 112(12):1796–1808. doi:10.1172/JCI19246 112/12/1796 [pii]

Willems E, Leyns L (2008) Patterning of mouse embryonic stem cell-derived pan-mesoderm by Activin A/Nodal and Bmp4 signaling requires fibroblast growth factor activity. Differentiation 76(7):745–759. doi:10.1111/j.1432-0436.2007.00257.x

Wu L, Derynck R (2009) Essential role of TGF-β signaling in glucose-induced cell hypertrophy. Dev Cell 17(1):35–48

Wu Z, Zhang W, Chen G, Cheng L, Liao J, Jia N, Gao Y, Dai H, Yuan J, Cheng L, Xiao L (2008) Combinatorial signals of activin/nodal and bone morphogenic protein regulate the early lineage segregation of human embryonic stem cells. J Biol Chem 283(36):24991–25002. doi:10.1074/jbc.M803893200

Yadav H, Rane SG (2012) TGF-β/Smad3 signaling regulates brown adipocyte induction in white adipose tissue. Front Endocrinol (Lausanne) 3:35

Yadav H, Quijano C, Kamaraju Anil K, Gavrilova O, Malek R, Chen W, Zerfas P, Zhigang D, Wright Elizabeth C, Stuelten C, Sun P, Lonning S, Skarulis M, Sumner Anne E, Finkel T, Rane Sushil G (2011) Protection from obesity and diabetes by blockade of TGF-β/Smad3 signaling. Cell Metab 14(1):67–79. doi:10.1016/j.cmet.2011.04.013

Yamanaka Y, Friess H, Buchler M, Beger HG, Gold LI, Korc M (1993) Synthesis and expression of transforming growth factor β-1, β-2, and β-3 in the endocrine and exocrine pancreas. Diabetes 42(5):746–756

Yamaoka T, Idehara C, Yano M, Matsushita T, Yamada T, Ii S, Moritani M, Hata J, Sugino H, Noji S, Itakura M (1998) Hypoplasia of pancreatic islets in transgenic mice expressing activin receptor mutants. J Clin Invest 102(2):294–301

Yang R, Trevillyan JM (2008) c-Jun N-terminal kinase pathways in diabetes. Int J Biochem Cell Biol 40(12):2702–2706

Zeyda M, Stulnig TM (2007) Adipose tissue macrophages. Immunol Lett 112(2):61–67

Zhang YQ, Cleary MM, Si Y, Liu G, Eto Y, Kritzik M, Dabernat S, Kayali AG, Sarvetnick N (2004) Inhibition of activin signaling induces pancreatic epithelial cell expansion and diminishes terminal differentiation of pancreatic β-cells. Diabetes 53(8):2024–2033

Part V
Disease Prognosis and Therapy

Chapter 19
Targeting Pro-Angiogenic TGF-β Signaling in the Tumor Microenvironment

Kristian Pietras and Sara I. Cunha

Abstract The recently developed targeted anti-angiogenic agents have been introduced into clinical practice over the course of the past decade. High hopes were placed on targeting the VEGF signaling pathway in endothelial cells following the preceding successful drug development in the preclinical setting. Indeed, the therapeutic efficacy observed in mouse models of cancer has in some cases been translated into clinical benefit for patients. Nevertheless, many anti-angiogenic therapies have failed to provide substantial improvement in survival in large phase III clinical trials. In the search for attractive and complementary angiogenic signaling pathways, the TGF-β family stands out as one of the most interesting. Our expanding knowledge on TGF-β signaling in the tumor vasculature has led to the development of specific inhibitors targeting TGF-β, ALK1, and endoglin. Many clinical trials exploring the concept of targeting pro-angiogenic TGF-β signaling are currently underway, and preliminary reports are encouraging. Here, we will discuss opportunities and challenges of targeting the TGF-β system for anti-angiogenic therapy of cancer.

Keywords Angiogenesis • BMP • Cancer • Targeted therapy • TGF-β

K. Pietras (✉)
Division of Vascular Biology, Department of Medical Biochemistry and Biophysics, Karolinska Institutet, Stockholm, Sweden

Department of Laboratory Medicine Malmö, Lund University Cancer Center, Lund University, Lund, Sweden
e-mail: Kristian.Pietras@med.lu.se

S.I. Cunha
Division of Vascular Biology, Department of Medical Biochemistry and Biophysics, Karolinska Institutet, Stockholm, Sweden

A. Moustakas and K. Miyazawa (eds.), *TGF-β in Human Disease*,
DOI 10.1007/978-4-431-54409-8_19, © Springer Japan 2013

Abbreviations

ALK	Activin receptor-like kinase
BMP	Bone morphogenetic proteins
EC	Endothelial cell
HHT	Hereditary hemorrhagic telangiectasia
TGF-β	Transforming growth factor-β
VEGF	Vascular endothelial growth factor

19.1 Introduction

Based on several successful preclinical studies, the oncologic clinical panorama anticipated with hope the use of anti-angiogenic therapies to combat cancer. Nevertheless, the introduction of vasculature targeting drugs led to limited clinical benefit measured in months, with almost negligible effect on overall survival and mainly in combination with cytotoxic agents. Recent studies provide evidence for the development of resistance and acquisition of a more aggressive and invasive phenotype in tumors that have been treated with vascular endothelial growth factor (VEGF) anti-angiogenic approaches. This points to the need for additional examination of the biology of tumor-nurturing blood vessels, in order to accurately pinpoint promising and novel anti-angiogenic targets. Furthermore, it is reasonable to assume that curing cancer is a complex task that requires combined strategies aiming at different tumor cellular compartments, in addition to aiming at the ample spectrum of signaling networks that are commonly deregulated in malignant cells.

Often altered in neoplasms and involved in multiple cellular functions, the large transforming growth factor-β (TGF-β) family is, as a result, an obvious candidate for aiming such efforts. More specifically, because TGF-β signaling encompasses a number of vascular restricted receptors, including activin receptor-like kinase (ALK)1 and endoglin, they represent attractive anti-angiogenic therapeutic targets for cancer. For a thorough summary of endothelial cell signaling by components of the TGF-β family, please see Chap. 14.

19.2 TGF-β Signaling as Tumor Angiogenesis Targets

19.2.1 Clinical Relevance of ALK1 in Cancer

ALK1 expression in the early developing mouse embryo coincides with sites of vasculo- and angiogenesis (Roelen et al. 1997), with prevailing expression in developing arterial endothelium, while nearly absent from small capillaries (Seki et al. 2006). During early mouse development ALK1 is strongly expressed, whilst it tends to be concealed in the adult quiescent vasculature. The expression of ALK1 is reversibly turned on during neo-angiogenesis events in wound healing or in tumors (Seki et al. 2006).

Information about pattern and intensity of ALK1 expression in human normal and cancerous tissues is unfortunately still very scarce. The public Human Protein Atlas program (Ponten et al. 2008) has characterized ALK1 immunostaining of both normal and neoplastic tissues. In this study, ALK1 exhibited frequent strong expression most notably in neuronal cells of the cerebral cortex, hippocampus, ventricle, and cerebellum, in the gall bladder, GI tract, and in tubular cells in the kidney, in line with the murine ALK1 expression profile (Panchenko et al. 1996; Wu et al. 1995). The same organization identified ALK1 prevailing expression in neoplasms in the colorectal tract, pancreas, stomach, and thyroid, as well in malignant lymphomas and melanomas. A preliminary descriptive study reported a weak but widespread pattern of expression in the vasculature of normal tissues, including positive staining in lymphatic tissues, lung, intestines, and pancreas. In a follow-up study by the same entity, ALK1 was found to be extensively present on tumor blood vessels, especially in lymphomas, and malignant tissues of the prostate, skin, thyroid, kidney, ovary, lung, pancreas, and liver (Hu-Lowe et al. 2011). Thorough studies analyzing the prognostic strength and diagnostic significance of ALK1 are highly warranted.

19.2.2 Inhibitory Drugs Targeting ALK1

Given the extensive literature describing paradoxical effects of signaling stemming from ALK1 in endothelial cells (ECs), the prediction for the net outcome for acute inhibition of ALK1 in the context of cancer is an intricate task. Furthermore, the complex ligand-receptor binding specificity and/or redundancy within the TGF-β family add multiple hurdles to the estimate of therapeutic efficacy, benefit and secondary effects for the various ALK1 inhibitors currently under development.

Small molecules comprising broad inhibitory spectrum against bone morphogenetic protein (BMP) type I receptor kinases, including ALK1, have been generated. Smad-dependent and -independent signaling, induced by BMP can be blocked by compounds, such as dorsomorphin and its structural derivative LDN-193189 (Boergermann et al. 2010; Yu et al. 2008b). These drugs can be useful and potent in inhibiting BMP type I receptor signaling in a range of diseases and familial syndromes (Yu et al. 2008a); however, their effect on BMP-induced tumor angiogenesis remains to be determined. Nonselective inhibitors also raise the concern of off-target effects, as documented for dorsomorphin (Cannon et al. 2010; Vogt et al. 2011). Further development of small molecule compounds with a more specific inhibition profile aiming primarily at BMP type I receptors, and specifically towards ALK1, should be considered.

Antibodies or soluble extracellular receptor domain traps use may alternatively provide a tighter inhibition of specific ALK1 activity. ALK1 inhibitors are currently under development for cancer treatment as anti-angiogenic drugs. Up to now, several biological inhibitors against ALK1 have been generated for in vivo use. Firstly, Pfizer is currently conducting phase I-trials of PF-3446962, a fully human monoclonal antibody against ALK1. Secondly, the use of an ALK1-Fc fusion protein (amino acids 23–119 of mouse ALK1) was reported by Genentech studies on hematogenous and lymphatic vessel development in the mouse (Niessen et al. 2010). Another ALK1 targeting agent is ACE-041 (mouse counterpart RAP-041, amino acids 22–117 of mouse ALK1), a human ALK1-Fc fusion protein, currently undergoing a phase II clinical trial, coordinated by Acceleron Pharma.

19.2.3 PF-3446692

Preclinical tumor studies using PF-3446962 were recently reported (Hu-Lowe et al. 2011). ALK1 blockade exhibited attenuation of VEGF-induced EC proliferation and tube formation in vitro. In addition, therapeutic treatment with the ALK1-neutralizing antibody delays tumor growth in vivo in MDA-MB-231 breast carcinoma and M24met/R melanoma mouse models.

The anti-hALK1 antibody from Pfizer has completed phase I clinical trials for patients with advanced solid tumors. Preliminary evidence from the trial indicates that the anti-hALK1 antibody reduced the amount of ALK1-positive circulating ECs, which were found to be present in increased levels in advanced stage cancer patients. The PF-3446962 antibody was well tolerated up to 10 mg/kg lacking concerning adverse events in all 44 patients participating in the phase I trial. The most common side effects included transient thrombocytopenia and asymptomatic elevation of pancreatic enzymes. Hence, preliminary observations evoke encouraging clinical activity where three partial responses were observed in patients who have previously received fruitless anti-angiogenic regimens.

The mechanism of action of this antibody has recently been described (van Meeteren et al. 2012). The anti-hALK1 antibody selectively recognizes human ALK1 and interferes with BMP9-induced signaling in ECs. Moreover, the anti-hALK1 antibody competitively obstructs BMP9 and TGF-β binding to ALK1 receptor and prevents BMP9-dependent recruitment of endoglin into the angiogenesis-mediating signaling complex, which may eventually hinder the BMP9/ALK1 proangiogenic effects.

19.2.4 ACE-041/RAP-041/dalantercept

The effects of RAP-041 were recently analyzed in the RIP1-Tag2 transgenic mouse model of pancreatic neuroendocrine tumorigenesis (Cunha et al. 2010). The RIP1-Tag2 tumors readily express ALK1 exclusively on ECs, mirroring the expression profile of common vascular markers during tumor progression.

RIP1-Tag2 mice treated with 2 weekly doses between 1 and 12 mg/kg RAP-041 results in a dose-dependent retardation of tumor growth and the highest dose effectively prevents further tumor expansion. Incidentally, Alk1 heterozygous mice under the RIP1-Tag2 context recapitulated the effects obtained by systemic inhibition by RAP-041. Specificity of the ALK1-targeting drug treatment was validated by decreased expression of ALK1 downstream target genes in tumors from mice treated with the RAP-041. Alternative studies with the same drug demonstrated that RAP-041 possesses growth-inhibitory response in orthotopic MCF-7 breast carcinomas (Mitchell et al. 2010).

Acceleron Pharma recently concluded a phase I trial for the human ALK1-Fc fusion protein ACE-041, which interestingly only binds and neutralizes BMP9 and BMP10, but not to any of the TGF-β isoforms (Cunha et al. 2010; Mitchell et al. 2010). Thirty seven patients with solid tumors or refractory multiple myeloma were recruited to assess safety and tolerability of ACE-041, as well as changes in tumor metabolism evaluated by ^{18}F deoxyglucose positron emission tomography. ACE-041 regimen included an administration every three weeks and was well tolerated at doses up to 1.6 mg/kg. Common collateral effects comprised peripheral edema, fatigue, nausea, headache, anorexia, and anemia. However, toxicities usually associated with VEGF inhibition, such as hypertension, proteinuria, GI tract perforation, and hemorrhaging, were not observed, possibly because ALK1 is predominately present in the actively cycling endothelium within the tumor milieu and preferentially localizes in arterial endothelium, whereas VEGFRs present a more global distribution (Seki et al. 2003).

In this clinical study, one patient with refractory head and neck squamous carcinoma exhibited a partial response (>30 weeks) and six other patients exhibited stable disease. Rapid reduction in tumor metabolic activity (>20%) was observed in ten patients, measured by FDG-PET scanning. Of note, many of the patients included in this trial had been previously inefficiently treated with other therapeutic regimens, including VEGF-targeting drugs.

After such an encouraging phase I clinical trial, the ALK1 inhibiting agent ACE-041 is undertaking a phase II clinical study with an expanded cohort on head and neck squamous carcinoma patients (http://clinicaltrials.gov). Apart from its anticancer specification, ACE-041, is further being developed for testing in age-related macular disease treatment.

19.2.5 Possible Side Effects from ALK1 Inhibition

Anti-angiogenic therapies, mainly in the form of inhibitors of VEGF signaling, have been in routine clinical use for several years for various malignancies. Side effects from inhibiting angiogenesis are in general milder than those arising from conventional chemotherapeutic treatment and include bleeding, hypertension, fatigue, and nausea. Specifically, given the causal relationship between impaired ALK1 signaling and hereditary hemorrhagic telangiectasia (HHT)-related symptoms, inhibition of ALK1 signaling in the vasculature may induce de novo arteriovenous malformations and hemorrhaging. In fact skin telangiectasis was observed in a patient treated with ACE-041, validating the on-target effect of the drug and indicating that the appearance of telangiectases may be useful as a surrogate marker for efficacy. Loss-of-function mutations in ALK1 linked with hereditary pulmonary arterial hypertension (Machado et al. 2009) highlight the risk for pulmonary circulation hemodynamics perturbations with ALK1 inhibitors. Interestingly, complete blockade of ALK1 signaling triggered by both BMP9 and BMP10 resulted in lung hemorrhaging (Ricard et al. 2012), an organ that should thus be primarily monitored. Furthermore, given the high expression levels of BMP9 in the liver, this organ should be also carefully monitored (Bidart et al. 2012). Finally, as ALK1 is reported to be expressed by, and conceivably important for lymphatic ECs, cells of the pituitary gland, hepatic stellate cells, chondrocytes, and pancreatic ductal cells, special care should be taken to record adverse events from the treatment of ALK1 inhibitors related to processes regulated by these particular tissues (Alexander et al. 1996; Finnson et al. 2008; Niessen et al. 2010; Ungefroren et al. 2007; Wiercinska et al. 2006).

19.3 Clinical Relevance of Endoglin in Cancer

Endoglin positive intratumoral microvascular density strongly correlates with poor prognosis in cancer, being associated with shorter survival and relapse-free survival rates (Bernabeu et al. 2009; Fonsatti et al. 2010). Moreover, subcutaneous tumor neovascularization and growth are impaired in endoglin heterozygous mice, reiterating the relevance of endoglin in tumor angiogenesis (Duwel et al. 2007). An enormous body of literature highlights the potential of endoglin as a tumor vasculature marker in preclinical and clinical studies (Bredow et al. 2000; Costello et al. 2004; Fonsatti et al. 2000; Fonsatti et al. 2010). In this respect, endoglin can be a more specific marker for new, immature tumor vessels, unlike other EC markers, which are expressed in both mature and immature vessels (Beresford et al. 2006).

High levels of endoglin expression have therefore been confirmed in several experimental models, such as breast, prostate, and colorectal cancer, for example (Akagi et al. 2002; Beresford et al. 2006; Wikstrom et al. 2002).

Endoglin expression has also been associated with predisposition of colorectal mucosa dysplastic tissues evolution into fully developed carcinomas (Akagi et al. 2002). Furthermore, in prostate cancer endoglin-positive microvessel density correlated with Gleason score, metastasis and tumor stage (Wikstrom et al. 2002). Paradoxically, the same study indicated that endoglin-positive vessels were more poorly covered by α-smooth muscle cells and correlated with survival. Another study in colorectal cancer patients reported that vessel density evaluated by endoglin staining significantly correlated with survival. Additionally, other reports correlate loss of endoglin with prostate cancer progression and aggressiveness

(Liu et al. 2002) and endoglin presence with decrease in prostate tumor cell motility (Craft et al. 2007). Our own work strengthens the vascular role of endoglin as protective against tumor cell metastatic seeding (Anderberg et al. 2013).

Endoglin exists in the body in two different forms: membrane-bound and circulating (Bernabeu et al. 2009). Levels of soluble endoglin have been reported in plasma of pregnant women suffering from preeclampsia (Venkatesha et al. 2006) and also in patients suffering from colorectal, breast, prostate, and leukemic cancers, correlating positively with metastatic disease (Fujita et al. 2009; Karam et al. 2008; Li et al. 2000; Takahashi et al. 2001b). However, the role of soluble endoglin in cancer is still poorly understood. Since soluble endoglin contains the binding site for different ligands of the TGF-β family, it may act as a scavenger of circulating ligands, preventing their binding to the functional receptors, hence interfering with vascular function and angiogenesis (Perez-Gomez et al. 2010). An intriguing question concerns the source of soluble endoglin detected in cancer patients. Since endoglin levels are higher in tumor vessels, soluble endoglin may conceivably derive from shedding from tumor ECs and, importantly, it may represent a surrogate marker of angiogenic activity (Fonsatti et al. 2003).

19.3.1 TRC105

The potential of endoglin-targeting monoclonal antibodies to be used as a therapeutic anti-angiogenic strategy in human cancer has received considerable support from preclinical studies.

Intravenous systemic administration of anti-endoglin monoclonal antibody TRC105 was shown to suppress angiogenesis, tumor growth, and metastasis without overt toxicity in mice (Seon et al. 1997; Tabata et al. 1999; Takahashi et al. 2001a). The combination of endoglin-targeting antibody with cyclophosphamide and doxorubicin was reported to exhibit synergistic antitumor efficacy in human skin/SCID mouse chimeras, including in metronomic regimens (Shiozaki et al. 2006; Takahashi et al. 2001a).

A total of 50 patients with advanced refractory cancer disease were included in TRC105 phase I clinical trial (Seon et al. 2010). Doses up to 1 mg/kg were administered every other week to assess efficacy, toxicity, and tolerability of TRC105. One patient with castrate-refractory prostate cancer remained in the study even after 3 years of TRC105 at 0.01 mg/kg with a complete response and bone scan normalization. In addition, an ovarian cancer patient presented with 6-month stable disease. Dose-limiting toxicities incorporate GI hemorrhage and anemia.

A phase I clinical trial has been initiated with breast metastatic cancer patients to determine maximum tolerated dose of TRC105 in combination with capecitabine, a DNA synthesis blocking agent, approved by FDA as adjuvant treatment for colon cancer, as well as, for metastatic breast and colorectal cancers (www.clinicaltrials.gov).

19.3.2 Endoglin-Fc

More recently, Acceleron Pharma characterized a soluble mouse and human endoglin extracellular domain fused to an immunoglobulin Fc domain (human endoglin amino acid sequence 26–359). This endoglin ligand trap binds specifically and with high affinity to BMP9 and BMP10 in vitro. This agent significantly impaired VEGF-induced chick chorioallantoic membrane assay in vivo. Finally, murine soluble endoglin extracellular domain acted as an anti-angiogenic factor decreasing blood vessel sprouting in VEGF/fibroblast growth factor-induced angiogenesis in in vivo angioreactors and tumor burden in the colon-26 mouse tumor model (Castonguay et al. 2011). Together these findings indicate an important role for soluble endoglin in the regulation of angiogenesis and evoke the prospective efficacy of endoglin-Fc as an anti-angiogenic therapeutic agent.

19.3.3 Possible Side Effects of Targeting Endoglin

Only two dose-limiting toxicities were reported in patients who received TRC105: one grade 4 gastric ulcer bleeding in a patient treated with 0.1 mg/kg after 4 days, which resolved spontaneously and one patient experienced grade 3 infusion reaction.

Based on a recent study, a careful follow-up of the patients treated with endoglin targeting antibodies should be in place, since the preclinical studies in mouse models, either endoglin heterozygous or endoglin-EC specifically ablated led to the emergence of worsened phenotypes with tumors acquiring a refractory behavior with increased metastatic seeding (Anderberg et al. 2013).

19.4 Clinical Relevance of TGF-β Signaling in Cancer

The role of TGF-β in cancer biology is complex and controversial, involving aspects of tumor suppression as well as tumor promotion. The ability of TGF-β to potently inhibit the proliferation of epithelial, endothelial, and hematopoietic cell lineages is central to the tumor-suppressive mechanism (Yingling et al. 2004). For example, the TGF-β receptors and their Smad signaling mediators are tumor suppressors that frequently become inactive in gastrointestinal, pancreatic, ovarian, and hepatocellular carcinomas, as well as in subsets of gliomas and lung adenocarcinomas (Bierie and Moses 2006a). As tumors evolve, they often become noncompliant to TGF-β-mediated growth inhibition and overexpress TGF-β, which in turn has a manifested autocrine impact on the biology of the malignant cells themselves. A tumor microenvironment that favors tumor growth and metastasis in a paracrine fashion is therefore created. Interestingly, in breast carcinoma, glioblastoma, melanoma, and other cancer types, specific loss of TGF-β-mediated growth inhibitory responses often accumulate through alterations downstream of Smad, leaving the rest of the TGF-β pathway operational and available to co-opt in detriment of tumor progression (Massague and Gomis 2006).

19.4.1 Inhibitory Drugs Targeting TGF-β

Improvements in understanding how TGF-β impinges upon the tumor microenvironment have led to the development of TGF-β inhibitors for cancer treatment. Three categories of TGF-β inhibitors have been characterized: soluble antisense oligonucleotides, monoclonal antibodies, and small molecule inhibitors. At first sight, this pathway presents an attractive target for the development of cancer therapeutics that simultaneously attacks the tumor and its microenvironment. One could, thus, anticipate that TGF-β targeting drugs would have a very efficacious effect on malignant progression. Nonetheless, the ubiquitous and pleiotropic nature of TGF-β signaling and its dual role in tissue homeostasis and in tumorigenesis pose on this kind of inhibitory strategies a risk that cannot be underestimated in preclinical and clinical drug development programs. Multiple clinical trials targeting the TGF-β pathway are currently in progress (see Table 19.1).

19.4.2 TGF-β Antisense Compounds

The specificity of hybridization draws antisense oligonucleotides as targeted and functional therapeutic tools to selectively modulate the expression of a variety of genes involved in the pathogenesis of malignancies and other genetic diseases (Stahel and Zangemeister-Wittke 2003; Tamm et al. 2001; Tamm and Wagner 2006).

Trabedersen (AP-12009), developed by Antisense Pharma GmbH, is a synthetic antisense oligodeoxynucleotide designed to block the production and tumorigenic effects of TGF-β2. Trabedersen is indicated for the treatment of malignant brain tumors and other TGF-β2 overexpressing solid tumors, such as those of the skin, pancreas, and colon. Preclinical studies provided evidence that trabedersen reduced the secretion of TGF-β2 in cultured tumor cells and exhibited antitumor activity ex vivo. Chronic intracerebral or systemic administration of trabedersen does not cause life-threatening collateral effects in animals, confirmed in early clinical trials in advanced cancer patients. In the initial phase I and II open-label dose-escalation study the compound was able to significantly prolong the median time to relapse compared with the published relapse times for temozolomide chemotherapy (Bierie and Moses 2006b).

Table 19.1 Summary of TGF-β family targeting clinical trials implicated in cancer

Drug target	Malignant condition	Intervention	Study type	Status	Evaluations
TGF-β2	Advanced metastatic carcinoma	Trabedersen with GM-CSF gene modified autologous tumor cell vaccine	Phase I	Completed	Primary outcome: – Safety Secondary outcomes: – Time to progression – Effect on immune stimulation
TGF-β2	Pancreatic neoplasms Melanoma Colorectal neoplasms	Trabebersen	Phase I	Completed	Primary outcome: – Maximum tolerated dose – Dose-limiting toxicity Secondary outcomes: – Safety and tolerability – Plasma pharmacokinetics – Method to assess the urine Pharmacokinetic profile – Effect on TGF-β2 plasma levels – Antitumor activity:effect on tumor size and tumor markers
TGF-β2	Glioblastoma Anaplastic astrocytoma	Trabedersen	Phase IIb	Completed	Primary outcome: – Overall response rate evaluated by brain MRI scans Secondary outcomes: – Overall survival – Response rates – Progression-free survival – Time to progression – Time to response – Survival and variation of tumor size on brain MRI – Change in quality of life – Safety and tolerability

TGF-β2	Anaplastic astrocytoma Glioblastoma	Trabedersen	Phase III	Completed	Primary outcome: – Survival rate Secondary outcomes: – Progression rate – Time to death – Overall response rate – Tumor control rate – Duration of response – Time to progression – Survival rate – Quality of life
TGF-β	Advanced malignant melanoma or renal cell carcinoma	GC1008	Phase I	Completed	Primary outcome: – Maximum tolerated dose – Dose-limiting toxicity – Safety Secondary outcomes: – Pharmacokinetic and dynamic
shFurin	Colon carcinoma with liver metastasis	FANG vaccine	Phase II	Recruiting	Primary outcome: – Overall survival Secondary outcome: – Immune response
shFurin	Advanced melanoma	FANG vaccine	Phase II	Recruiting	Primary outcome: – Immune response Secondary outcome: – Survival

(continued)

Table 19.1 (continued)

Drug target	Malignant condition	Intervention	Study type	Status	Evaluations
TGF-β1, 2, 3	Metastatic breast cancer	GC1008 + Radiation	Safety	Recruiting	Primary outcome: – Adverse events Secondary outcome: – Local response rate by PET/CT
TGF-β RI	Malignant glioma	LY2157299 with Temozolomide-based Radiochemotherapy	Phase Ib/2a	Recruiting	Primary outcome: Phase 1b – Safety and tolerability – dose Primary outcome: Phase 2a – Confirm tolerability – Pharmacodynamics
TGF-β 1, 2, 3	Relapsed malignant pleural mesothelioma	GC1008	Phase II	Ongoing	Primary outcome: – 3-month PFS rate Secondary outcomes: – Progression-free survival – Immune response parameters

ALK1	Advanced solid tumors	PF-03446962	Phase I	Recruiting	Primary outcomes: – Maximum tolerated dose Secondary outcomes: – Pharmacokinetics – Evaluation of immunogenicity – Evidence of antitumor activity – Effect on vascular function – Circulating ECs, circulating endothelial progenitors, and soluble proteins related to ALK1 signaling – Safety
ALK1	Squamous cell carcinoma of the head and neck, renal cell carcinoma, endometrial cancer	ACE-041/dalantercept	Phase II	Recruiting	Primary outcome: – Response rate Secondary outcome: – Adverse events – Serum concentration – Progression-free survival – Overall survival – Time progression – Duration of response – Disease control rate – Expression of ALK1 in tumor tissue and blood
Endoglin	Glioblastoma, renal cell carcinoma, breast carcinoma, others	TRC105	Phase I/II	Recruiting	Primary outcome: – Progression-free survival Secondary outcome: – Response rate

In a phase IIb trial, improved survival was observed in refractory high-grade gliomas patients who were administered trabedersen intratumorally, by convection-enhanced delivery (Bierie and Moses 2006b; Vallieres 2009; Yingling et al. 2004). One hundred and forty-five patients with histopathology of recurrent/refractory glioblastoma multiforme or anaplastic astrocytoma were randomly assigned to receive trabedersen or standard chemotherapy (Bogdahn et al. 2011). One patient achieved a complete response in all tumor sites after a single cycle of trabedersen, and several patients achieved tumor reductions of more than 80 % (Bogdahn et al. 2011; Yingling et al. 2004). In addition, disease stabilized in seven out of 24 patients, and two patients were in complete remission after treatment. However, this observation requires validation by an ongoing large-scale phase III randomized controlled trial (Vallieres 2009). Meanwhile, continued research on trabedersen should help determine the roles of TGF-β2 in cancer. As a result of these studies, trabedersen received orphan drug status in the European Union in 2002.

Overexpression of TGF-β2 in pancreatic malignancies is suggested to be a pivotal factor for malignant progression by inducing proliferation, immunosuppression, angiogenesis, and metastasis. Antitumor activity was reported for trabedersen in human pancreatic cancer cells and in an orthotopic xenograft mouse model of metastatic pancreatic cancer (Schlingensiepen et al. 2010), significantly reducing tumor growth, lymph node metastasis, and angiogenesis. According to these promising results, trabedersen appears attractive for the treatment of pancreatic adenocarcinoma but warrants further clinical insight.

19.4.3 TGF-β Antibodies

GC1008, the only TGF-β antibody in clinical trials, is a human IgG4 monoclonal antibody capable of neutralizing all TGF-β isoforms. Cohorts of patients with advanced malignant melanoma or renal cell carcinoma, who had failed at least one prior therapy were treated to assess effectiveness with GC1008 at doses from 0.1 to 15 mg/kg. Twenty-two patients were included (21 malignant melanoma, one renal cell carcinoma) and treated in this trial. No dose-limiting toxicities were observed and the highest dose of GC1008 was determined to be safe. Adverse events included skin rash, fatigue, headache, epistaxis, gingival bleeding, and gastrointestinal symptoms. So far, five patients reached stable disease and received extended treatment. Out of these five responders, three patients presented with metastasis shrinkage in the liver and other sites. One melanoma patient achieved a partial response with a lesion reduction of >75 %. All in all, GC1008 is well tolerated and neutralization of all TGF-β isoforms seems to hold promise in the treatment of one of the most aggressive cancer types, for which a phase II protocol is being expanded.

19.4.4 Small-Molecule Kinase Inhibitors Targeting TGF-β Receptors

There are several well-characterized small-molecule kinase inhibitors identified to target the TGF-β pathway, including A-80-01, LY364947, LY580276, LY566578, SB-505124, SD-093, SD-208, and SB-431542 (Bierie and Moses 2006a). The above ALK5 small-molecule inhibitors can, in general, also target ALK4 and ALK7 receptor, so results obtained through their application might not directly correlate with TGF-β/ALK5-specific signaling and dissecting specific effects stemming from each of the three receptors may require substantial validation (DaCosta Byfield et al. 2004; Inman et al. 2002; Peng et al. 2005). The inhibition effect on ALK4 and ALK7, in addition to the TGF-β-specific ALK5 effect on tumorigenesis, is yet to be determined. In particular ALK4 could be important, as its upregulation in the MMTV-Neu mouse model has been shown to correlate with activated Smad2 and loss of ALK5 expression (Landis et al. 2005). Furthermore, in the absence of ALK5, ALK4 mediates Smad2 phosphorylation and consequently αSMA expression during mouse yolk sac vasculogenesis (Carvalho et al. 2007).

Many of the TGF-β inhibitors were tested on breast and lung cancer xenograft tumor-bearing mice demonstrating significant tumor growth delay, providing supporting evidence for further development of these kinase inhibitors for clinical investigation (Lahn et al. 2005).

Treatment of syngeneic R3T or 4T1 tumor-bearing mice with orally supplied SD-208 inhibited primary tumor growth, angiogenesis, and number and size of metastasis. In contrast, SD-208 failed to inhibit R3T tumor growth or metastasis in athymic nude mice, suggesting that the antitumor effect is predominantly affecting immune responses (Ge et al. 2006).

SD-208 also inhibited kinase activity in SMA-560 gliomas in syngeneic mice, exhibiting a 25 % survival advantage in treated animals. No effect was detected on microvascular density. However, histological analysis of SD-208 treated tumors revealed an increase of infiltrated NK cells, CD8 T cell, and CD11b positive macrophages and neutrophils into responding tumors despite the negligible effects on overall tumor burden (Uhl et al. 2004).

Consistent with these observations, TGF-β signaling inhibition with SD-208 in the Panc-1 orthotopic pancreatic cancer model showed significant reduction of primary tumor weight and decreased incidence of metastasis. Histological evaluation revealed that SD-208 treatment reduced proliferation and induced apoptosis and vessel density in the tumor microenvironment. Additionally, an immune system contribution was observed with a greater B-cell infiltration in SD-208-treated tumors (Gaspar et al. 2007; Medicherla et al. 2007). Therapeutic benefit may be primarily driven by the host immune response against the tumor.

SD-208 also rendered beneficial effects on a model of melanoma bone metastasis (Mohammad et al. 2011). In this study 1205Lu melanoma cells were inoculated into the left cardiac ventricle of nude mice and metastasis formation and dissemination was monitored with and without administration of SD-208 prior to tumor cell injection. SD-208 treatment prevented the development of osteolytic bone metastases compared with vehicle treated mice. Moreover, in mice where bone metastases did form, the size of the osteolytic lesions was significantly reduced after 4 weeks treatment.

All in all, these results evoke the therapeutic potential of TGF-β inhibitory agents by primarily interfering with TGF-β-mediated immune suppression and, thus, generate a more potent immune response, a result anticipated taking into consideration the phenotype obtained by complete genetic ablation of TGF-β1 in mice.

19.4.5 Possible Side Effects of TGF-β Targeting Agents

All the studies with genetically deficient mice demonstrate that TGF-β signaling pathway is essential during development. Nevertheless, the potential deleterious effects of TGF-β inhibition in adult mice still remain to be determined.

TGF-β systemic inhibition affects the entire tumor microenvironment, from the malignant epithelium to all stromal players. Given the importance of TGF-β in normal tissue homeostasis, broad inhibition is predicted to affect a wide array of normal cell functions. One concern with TGF-β targeting therapies is the potential for detrimental side effects, despite that long-term treatments with TGF-β inhibitors do not seem to significantly alter animal morbidity. In particular, because of the immune-mediated disease and lethality associated with the genetic ablation or inhibition of TGF-β signaling in mice, it was unclear if inhibiting this pathway to treat cancer would be compatible with patient survival when delivered for a sustained period.

In contrast to predictions of severe toxicity stemming from TGF-β signaling inhibition, it has been shown that lifetime exposure to systemic soluble TGF-β1, TGF-β3, or pan-TGF-β neutralizing antibody 1D11 in mouse models does not result in significant adverse effects (Ruzek et al. 2003; Yang et al. 2002; Yingling et al. 2004). The soluble TGF-βRII antagonist SR2F was expressed in mice to determine its impact on tumor growth and long-term effect. No severe toxicities were observed over the lifespan of these SR2F-overexpressing mice (Yang et al. 2002); however, lifelong inhibition of TGF-β at biological levels is sufficient to block experimentally induced or implanted mammary tumors and metastatic dissemination (Lahn et al. 2005; Yang et al. 2002).

Expansion of the hypertrophic and proliferation zones of the physes in the femur and tibia was observed in rats treated with TGF-βRI inhibitors (Lahn et al. 2005). Chondrocytes and chondroid matrix were also increased. The degree of hypertrophy was dose- and time-dependent and only affected young animals, presumably due to active physeal growth. The evident role of TGF-β in remodeling of bone claims for monitoring of bone metabolism in clinical studies with TGF-β inhibitors (Lahn et al. 2005).

The effect on the immune system is more complicated to estimate. In animals deficient for TGF-βRII expression on CD4+ T cells, antigen-activated T-cell activity is enhanced and tumor growth inhibited (Chen and Wahl 2002; Lahn et al. 2005). However, increasing the activity of antigen-activated T cells by blocking TGF-β can result in autoimmune reactions.

The effect on the tumor vasculature is usually not a common readout for TGF-β inhibition therapeutic benefit. However, multiple are the preclinical reports providing evidence that blocking TGF-β signaling has a direct and indirect effect on tumor angiogenesis. Caution should be taken in VEGF-dependent tumors as type-I TGF-β receptor kinase inhibitors (SB-431542 and LY-2157299) and VEGF ligand availability were shown to synergistically promote blood-vessel formation via integrin 5 (Liu et al. 2009). Similar results were reported on SB-431542 facilitating proliferation and sheet formation of ESC-derived ECs (Watabe et al. 2003). VEGF is typically abundant in hypoxic tumors; hence, this may pose a risk for using TGF-β signaling inhibitors that may potentiate VEGF/VEGFR signaling and undesirably improve the angiogenic response.

19.5 Potential for Combinatorial Therapeutic Studies

The still scarce preclinical and clinical data currently available advocate for an anti-angiogenic and growth inhibitory effect of attenuated ALK1 signaling in cancer, hence sustaining the clinical development of drugs blocking ALK1. Furthermore, while ALK1 targeting monotherapies have been incredibly successful both in preclinical and clinical settings, so far, it is reasonable to anticipate that a combined targeted therapy can hold an agonistic effect fighting cancer disease. The multitude of cancers and heterogeneity within each malignancy combined with case-to-case special demands highly requests for incorporation of two or three different targeted drugs that have independently been beneficial. Future studies should therefore embark on such endeavurs.

Interestingly, VEGF levels are elevated in the aorta, lungs, liver, and intestine of ALK1-deficient mice (Shao et al. 2009), suggesting that double targeting VEGFR and ALK1 signaling pathways may not only be the route to a more efficacious treatment plan but may also circumvent the risk of refractoriness to anti-angiogenic drugs. Of note, bevacizumab was recently reported to attenuate VEGF-induced angiogenesis in ALK1 deletion-induced vascular malformations in the adult mouse brain (Walker et al. 2012).

In a human/mouse chimera tumor model, targeting human ALK1 decreased tumor vessel density and improved antitumor efficacy when combined with bevacizumab (Hu-Lowe et al. 2011). This study raised thus the question whether ALK1 signaling may be part of a set of adaptive mechanisms in tumors refractory to VEGF inhibition (Hu-Lowe et al. 2011). Moreover, ACE-041 used in combination with sunitinib impaired tumor growth in two xenograft models of VEGF-inhibitor resistant renal cell carcinoma, A498 and 786-O. These two models represent surrogates of renal cell carcinoma tumors that typically transiently shrink upon VEGF inhibition but quickly restore angiogenesis and resume tumorigenic program, despite continuation of treatment.

None of the ALK1 targeting studies have so far analyzed the potential adaptive effects of ALK1 targeting drugs in prolonged regimens. In order for ALK1 inhibitors to prevail and make a stand as opposed to VEGF targeting drugs, such analysis is mandatory.

Given the recent studies suggesting a regulatory crosstalk loop amongst BMP9/ALK1 and Notch signaling coordinating tip versus stalk cell specification, one may immediately anticipate possible therapeutic benefit arising from a combinatorial targeting of both pathways in tumor biology.

Neutralization of Dll4-Notch signaling in tumors results in excessive, nonproductive angiogenesis with subsequent inhibitory effects on tumor growth, due to poor perfusion-induced hypoxia (Noguera-Troise et al. 2006; Ridgway et al. 2006). Furthermore, Dll4 has been reported to mediate tumor resistance to bevacizumab in vivo as a compensatory mechanism to VEGF neutralization (Li et al. 2011). Pharmacological targeting of Dll4/Notch signaling in preclinical tumor models has been achieved by several different inhibitory strategies. Specific targeting with anti-Dll4 antibodies does not induce overt toxicity and Dll4 has, thus, emerged as an attractive target for anti-angiogenic cancer therapy (Kuhnert et al. 2011). As Dll4 inhibitors are entering clinical trials for the treatment of solid malignancies, this may pose a novel combinatorial therapeutic opportunity in breast, colon, and renal cancer, where Dll4 is selectively expressed by the endothelium of malignant tissues (Jubb et al. 2006; Jubb et al. 2009; Patel et al. 2005) and where patients may profit from combinatorial actions.

The phase I clinical trial on TRC105 evokes the combinatorial use of drugs impinging their inhibitory effect on both endoglin and VEGFRs, which have shown to generate an ameliorated therapeutic benefit both in the primary tumor burden but also on metastatic dissemination. In fact, a clinical trial analyzing the effect of TRC105 in combination with standard-dose bevacizumab in advanced solid tumors for which bevacizumab is indicated has been launched (www.clinicaltrials.gov). Interestingly, a study reported on a patient with HHT1 who had a substantiated response to bevacizumab (Bose et al. 2009). In this study, the epistaxis episodes became sparser and shorter. In different studies, a patient with HHT1 who received bevacizumab for malignant mesothelioma had a dramatic reduction in gastrointestinal bleeding from arteriovenous malformations (Flieger et al. 2006). Also, a patient with severe hepatic HHT who received six courses of bevacizumab no longer required liver transplantation and was well 6 months after completing the treatment (Mitchell et al. 2008). These studies strongly suggest a strong collaborative action of double targeting simultaneously endoglin and VEGF signaling to reacquire endothelial homeostasis. Furthermore, studies in our laboratory provide evidence for an agonistic effect by using endoglin and VEGFR targeting strategies reducing primary tumor burden and metastatic dissemination (Anderberg et al. 2013).

TGF-β has been reported to promote migration, invasion, and survival in breast cancer cells overexpressing the HER2 oncogene and to accelerate the metastasis of neu-induced mammary tumors in mice (Muraoka-Cook et al. 2005; Muraoka et al. 2003; Siegel et al. 2003). A clearer understanding of the molecular mechanisms underlying the crosstalk between TGF-β and HER2 has started to emerge.

In recent studies the synergistic effect of TGF-β and HER2 on tumor progression has been shown to likely be a combined result of two distinct features: loss of TGF-β tumor suppressive effect and gain of pro-survival and -migratory function through HER2-dependent mechanisms (Chow et al. 2011). In HER2-overexpressing breast cancer, this crosstalk results in increased cancer cell proliferation, survival and invasion, accelerated metastasis, and resistance to chemotherapy and HER2-targeted therapy (Chow et al. 2011). The transformed cellular context stemming from HER2 amplification disrupts the tumor suppressive role of TGF-β and promotes its oncogenic role. In turn, TGF-β potentiates oncogenic HER2 signaling by eliciting shedding of the ERBB ligands and clustering of HER2 with integrins (Wang et al. 2009). Blockade of TGF-β-HER2 crosstalk may suppress breast cancer progression and metastasis and enhance the efficiency of conventional therapies in patients with HER2-overexpressing breast cancer, which afflicts 25–30 % of all breast cancer patients. Moreover, targeting both TGF-β and EGFR/HER2 signaling can represent a stronger action towards the tumor neoendothelium as both have been shown to, among other functions, impair tumor angiogenesis (Izumi et al. 2002). More specifically, Herceptin, a HER2 neutralizing antibody, upregulates TGF-β target genes, such as PAI-1 and Thrombospondin-1 in vivo, rendering a decrease in vessel diameter and in tumor burden (Izumi et al. 2002).

Combinatorial studies on TGF-β inhibition with VEGFR impairment have not yet been established; however, some studies suggest that targeting both pathways may bypass the tumor adaptive actions by means of exploring alternative pathways to maintain tumor growth (Liu et al. 2009).

19.6 Perspective

The pleiotropy and intricacy of TGF-β family signaling conveys effects virtually in all cell types in the body. In cancerous disease, it is well established that TGF-β holds a bipolar role in carcinogenesis, acting as tumor suppressor during the initial stages of tumor development, whilst promoting tumor growth and metastatic spread in advanced stages (Ikushima and Miyazono 2010). As a consequence, the potential of using to our advantage the knowledge on TGF-β biology is still not fully embraced. The development and use of inhibitors of TGF-β family activity for the treatment of cancer may conduct to disparate effects depending on the stage of the disease. Furthermore, in ECs, the overall result of signaling from TGF-β family receptors is manifold and determined by a plethora of factors: ligand specificity or redundancy, engaged type I, type II receptors, and co-receptors (Cunha and Pietras 2011). In spite of the challenges of the elaborated signaling network outcomes from the TGF-β receptors in the various cell types of a tumor, the dividends for modulating the TGF-β network in therapeutic regimens may be rewarding. The current preclinical data and preliminary clinical results readily support the feasibility of using ALK1 and endoglin inhibitors as angiogenesis counteracting agents and TGF-β inhibitors affecting all the tumor cellular compartments.

References

Akagi K, Ikeda Y, Sumiyoshi Y et al (2002) Estimation of angiogenesis with anti-CD105 immunostaining in the process of colorectal cancer development. Surgery 131:S109–S113

Alexander JM, Bikkal HA, Zervas NT et al (1996) Tumor-specific expression and alternate splicing of messenger ribonucleic acid encoding activin/transforming growth factor-β receptors in human pituitary adenomas. J Clin Endocrinol Metab 81:783–790

Anderberg C, Cunha SI, Zhai Z, Cortez E, Pardali E, Johnson JR, Franco M, Páez-Ribes M, Cordiner R, Fuxe J, Johansson BR, Goumans MJ, Casanovas O, Ten Dijke P, Arthur HM, Pietras K (2013) Deficiency for endoglin in tumor vasculature weakens the endothelial barrier to metastatic dissemination. J Exp Med 210(3):563–579

Beresford MJ, Harris AL, Ah-See M et al (2006) The relationship of the neo-angiogenic marker, endoglin, with response to neoadjuvant chemotherapy in breast cancer. Br J Cancer 95: 1683–1688

Bernabeu C, Lopez-Novoa JM, Quintanilla M (2009) The emerging role of TGF-β superfamily coreceptors in cancer. Biochim Biophys Acta 1792:954–973

Bidart M, Ricard N, Levet S et al (2012) BMP9 is produced by hepatocytes and circulates mainly in an active mature form complexed to its prodomain. Cell Mol Life Sci 69:313–324

Bierie B, Moses HL (2006a) TGF-β and cancer. Cytokine Growth Factor Rev 17:29–40

Bierie B, Moses HL (2006b) Tumour microenvironment: TGFβ: the molecular Jekyll and Hyde of cancer. Nat Rev Cancer 6:506–520

Boergermann JH, Kopf J, Yu PB, Knaus P (2010) Dorsomorphin and LDN-193189 inhibit BMP-mediated Smad, p38 and Akt signalling in C2C12 cells. Int J Biochem Cell Biol 42: 1802–1807

Bogdahn U, Hau P, Stockhammer G et al (2011) Targeted therapy for high-grade glioma with the TGF-β2 inhibitor trabedersen: results of a randomized and controlled phase IIb study. Neuro Oncol 13:132–142

Bose P, Holter JL, Selby GB (2009) Bevacizumab in hereditary hemorrhagic telangiectasia. N Engl J Med 360:2143–2144

Bredow S, Lewin M, Hofmann B et al (2000) Imaging of tumour neovasculature by targeting the TGF-β binding receptor endoglin. Eur J Cancer 36:675–681

Cannon JE, Upton PD, Smith JC, Morrell NW (2010) Intersegmental vessel formation in zebrafish: requirement for VEGF but not BMP signalling revealed by selective and non-selective BMP antagonists. Br J Pharmacol 161:140–149

Carvalho RL, Itoh F, Goumans MJ et al (2007) Compensatory signalling induced in the yolk sac vasculature by deletion of TGFβ receptors in mice. J Cell Sci 120:4269–4277

Castonguay R, Werner ED, Matthews RG et al (2011) Soluble endoglin specifically binds bone morphogenetic proteins 9 and 10 via its orphan domain, inhibits blood vessel formation, and suppresses tumor growth. J Biol Chem 286:30034–30046

Chen W, Wahl SM (2002) TGF-β: receptors, signaling pathways and autoimmunity. Curr Dir Autoimmun 5:62–91

Chow A, Arteaga CL, Wang SE (2011) When tumor suppressor TGFβ meets the HER2 (ERBB2) oncogene. J Mammary Gland Biol Neoplasia 16:81–88

Costello B, Li C, Duff S et al (2004) Perfusion of 99Tcm-labeled CD105 Mab into kidneys from patients with renal carcinoma suggests that CD105 is a promising vascular target. Int J Cancer 109:436–441

Craft CS, Romero D, Vary CP, Bergan RC (2007) Endoglin inhibits prostate cancer motility via activation of the ALK2-Smad1 pathway. Oncogene 26:7240–7250

Cunha SI, Pietras K (2011) ALK1 as an emerging target for antiangiogenic therapy of cancer. Blood 117:6999–7006

Cunha SI, Pardali E, Thorikay M et al (2010) Genetic and pharmacological targeting of activin receptor-like kinase 1 impairs tumor growth and angiogenesis. J Exp Med 207:85–100

DaCosta BS, Major C, Laping NJ, Roberts AB (2004) SB-505124 is a selective inhibitor of transforming growth factor-β type I receptors ALK4, ALK5, and ALK7. Mol Pharmacol 65:744–752

Duwel A, Eleno N, Jerkic M et al (2007) Reduced tumor growth and angiogenesis in endoglin-haploinsufficient mice. Tumour Biol 28:1–8

Finnson KW, Parker WL, ten Dijke P et al (2008) ALK1 opposes ALK5/Smad3 signaling and expression of extracellular matrix components in human chondrocytes. J Bone Miner Res 23:896–906

Flieger D, Hainke S, Fischbach W (2006) Dramatic improvement in hereditary hemorrhagic telangiectasia after treatment with the vascular endothelial growth factor (VEGF) antagonist bevacizumab. Ann Hematol 85:631–632

Fonsatti E, Jekunen AP, Kairemo KJ et al (2000) Endoglin is a suitable target for efficient imaging of solid tumors: in vivo evidence in a canine mammary carcinoma model. Clin Cancer Res 6:2037–2043

Fonsatti E, Sigalotti L, Arslan P et al (2003) Emerging role of endoglin (CD105) as a marker of angiogenesis with clinical potential in human malignancies. Curr Cancer Drug Targets 3: 427–432

Fonsatti E, Nicolay HJ, Altomonte M et al (2010) Targeting cancer vasculature via endoglin/CD105: a novel antibody-based diagnostic and therapeutic strategy in solid tumours. Cardiovasc Res 86:12–19

Fujita K, Ewing CM, Chan DY et al (2009) Endoglin (CD105) as a urinary and serum marker of prostate cancer. Int J Cancer 124:664–669

Gaspar NJ, Li L, Kapoun AM et al (2007) Inhibition of transforming growth factor β signaling reduces pancreatic adenocarcinoma growth and invasiveness. Mol Pharmacol 72:152–161

Ge R, Rajeev V, Ray P et al (2006) Inhibition of growth and metastasis of mouse mammary carcinoma by selective inhibitor of transforming growth factor-β type I receptor kinase in vivo. Clin Cancer Res 12:4315–4330

Hu-Lowe DD, Chen E, Zhang L et al (2011) Targeting activin receptor-like kinase 1 inhibits angiogenesis and tumorigenesis through a mechanism of action complementary to anti-VEGF therapies. Cancer Res 71:1362–1373

Ikushima H, Miyazono K (2010) TGFβ signalling: a complex web in cancer progression. Nat Rev Cancer 10:415–424

Inman GJ, Nicolas FJ, Callahan JF et al (2002) SB-431542 is a potent and specific inhibitor of transforming growth factor-β superfamily type I activin receptor-like kinase (ALK) receptors ALK4, ALK5, and ALK7. Mol Pharmacol 62:65–74

Izumi Y, Xu L, di Tomaso E et al (2002) Tumour biology: herceptin acts as an anti-angiogenic cocktail. Nature 416:279–280

Jubb AM, Hurwitz HI, Bai W et al (2006) Impact of vascular endothelial growth factor-A expression, thrombospondin-2 expression, and microvessel density on the treatment effect of bevacizumab in metastatic colorectal cancer. J Clin Oncol 24:217–227

Jubb AM, Turley H, Moeller HC et al (2009) Expression of δ-like ligand 4 (Dll4) and markers of hypoxia in colon cancer. Br J Cancer 101:1749–1757

Karam JA, Svatek RS, Karakiewicz PI et al (2008) Use of preoperative plasma endoglin for prediction of lymph node metastasis in patients with clinically localized prostate cancer. Clin Cancer Res 14:1418–1422

Kuhnert F, Kirshner JR, Thurston G (2011) Dll4-Notch signaling as a therapeutic target in tumor angiogenesis. Vasc Cell 3:20

Lahn M, Kloeker S, Berry BS (2005) TGF-β inhibitors for the treatment of cancer. Expert Opin Investig Drugs 14:629–643

Landis MD, Seachrist DD, Montanez-Wiscovich ME et al (2005) Gene expression profiling of cancer progression reveals intrinsic regulation of transforming growth factor-β signaling in ErbB2/Neu-induced tumors from transgenic mice. Oncogene 24:5173–5190

Li C, Guo B, Wilson PB et al (2000) Plasma levels of soluble CD105 correlate with metastasis in patients with breast cancer. Int J Cancer 89:122–126
Li JL, Sainson RC, Oon CE et al (2011) DLL4-Notch signaling mediates tumor resistance to anti-VEGF therapy in vivo. Cancer Res 71:6073–6083
Liu Y, Jovanovic B, Pins M et al (2002) Over expression of endoglin in human prostate cancer suppresses cell detachment, migration and invasion. Oncogene 21:8272–8281
Liu Z, Kobayashi K, van Dinther M et al (2009) VEGF and inhibitors of TGFβ type-I receptor kinase synergistically promote blood-vessel formation by inducing α5-integrin expression. J Cell Sci 122:3294–3302
Machado RD, Eickelberg O, Elliott CG, Geraci MW, Hanaoka M, Loyd JE, Newman JH, Phillips JA 3rd, Soubrier F, Trembath RC, Chung WK (2009) Genetics and genomics of pulmonary arterial hypertension. J Am Coll Cardiol 54(1 Suppl):S32–S42
Massague J, Gomis RR (2006) The logic of TGFβ signaling. FEBS Lett 580:2811–2820
Medicherla S, Li L, Ma JY et al (2007) Antitumor activity of TGF-β inhibitor is dependent on the microenvironment. Anticancer Res 27:4149–4157
Mitchell A, Adams LA, MacQuillan G et al (2008) Bevacizumab reverses need for liver transplantation in hereditary hemorrhagic telangiectasia. Liver Transpl 14:210–213
Mitchell D, Pobre EG, Mulivor AW et al (2010) ALK1-Fc inhibits multiple mediators of angiogenesis and suppresses tumor growth. Mol Cancer Ther 9:379–388
Mohammad KS, Javelaud D, Fournier PG et al (2011) TGF-β-RI kinase inhibitor SD-208 reduces the development and progression of melanoma bone metastases. Cancer Res 71: 175–184
Muraoka RS, Koh Y, Roebuck LR et al (2003) Increased malignancy of Neu-induced mammary tumors overexpressing active transforming growth factor β1. Mol Cell Biol 23:8691–8703
Muraoka-Cook RS, Dumont N, Arteaga CL (2005) Dual role of transforming growth factor β in mammary tumorigenesis and metastatic progression. Clin Cancer Res 11:937s–943s
Niessen K, Zhang G, Ridgway JB et al (2010) ALK1 signaling regulates early postnatal lymphatic vessel development. Blood 115:1654–1661
Noguera-Troise I, Daly C, Papadopoulos NJ et al (2006) Blockade of Dll4 inhibits tumour growth by promoting non-productive angiogenesis. Nature 444:1032–1037
Panchenko MP, Williams MC, Brody JS, Yu Q (1996) Type I receptor serine-threonine kinase preferentially expressed in pulmonary blood vessels. Am J Physiol 270:L547–L558
Patel NS, Li JL, Generali D et al (2005) Up-regulation of δ-like 4 ligand in human tumor vasculature and the role of basal expression in endothelial cell function. Cancer Res 65:8690–8697
Peng SB, Yan L, Xia X et al (2005) Kinetic characterization of novel pyrazole TGF-β receptor I kinase inhibitors and their blockade of the epithelial-mesenchymal transition. Biochemistry 44:2293–2304
Perez-Gomez E, Del Castillo G, Juan Francisco S et al (2010) The role of the TGF-β coreceptor endoglin in cancer. Sci World J 10:2367–2384
Ponten F, Jirstrom K, Uhlen M (2008) The Human Protein Atlas – a tool for pathology. J Pathol 216:387–393
Ricard N, Ciais D, Levet S et al (2012) BMP9 and BMP10 are critical for postnatal retinal vascular remodeling. Blood 119(25):6162–6171
Ridgway J, Zhang G, Wu Y et al (2006) Inhibition of Dll4 signalling inhibits tumour growth by deregulating angiogenesis. Nature 444:1083–1087
Roelen BA, van Rooijen MA, Mummery CL (1997) Expression of ALK-1, a type 1 serine/threonine kinase receptor, coincides with sites of vasculogenesis and angiogenesis in early mouse development. Dev Dyn 209:418–430
Ruzek MC, Hawes M, Pratt B et al (2003) Minimal effects on immune parameters following chronic anti-TGF-β monoclonal antibody administration to normal mice. Immunopharmacol Immunotoxicol 25:235–257
Schlingensiepen KH, Jaschinski F, Lang SA et al (2010) Transforming growth factor-β 2 gene silencing with trabedersen (AP 12009) in pancreatic cancer. Cancer Sci 102:1193–1200

Seki T, Yun J, Oh SP (2003) Arterial endothelium-specific activin receptor-like kinase 1 expression suggests its role in arterialization and vascular remodeling. Circ Res 93:682–689

Seki T, Hong KH, Oh SP (2006) Nonoverlapping expression patterns of ALK1 and ALK5 reveal distinct roles of each receptor in vascular development. Lab Invest 86:116–129

Seon BK, Matsuno F, Haruta Y et al (1997) Long-lasting complete inhibition of human solid tumors in SCID mice by targeting endothelial cells of tumor vasculature with antihuman endoglin immunotoxin. Clin Cancer Res 3:1031–1044

Seon BK, Haba A, Matsuno F et al (2010) Endoglin-targeted cancer therapy. Curr Drug Deliv 8:135–143

Shao ES, Lin L, Yao Y, Bostrom KI (2009) Expression of vascular endothelial growth factor is coordinately regulated by the activin-like kinase receptors 1 and 5 in endothelial cells. Blood 114:2197–2206

Shiozaki K, Harada N, Greco WR et al (2006) Antiangiogenic chimeric anti-endoglin (CD105) antibody: pharmacokinetics and immunogenicity in nonhuman primates and effects of doxorubicin. Cancer Immunol Immunother 55:140–150

Siegel PM, Shu W, Cardiff RD et al (2003) Transforming growth factor β signaling impairs Neu-induced mammary tumorigenesis while promoting pulmonary metastasis. Proc Natl Acad Sci USA 100:8430–8435

Stahel RA, Zangemeister-Wittke U (2003) Antisense oligonucleotides for cancer therapy-an overview. Lung Cancer 41(Suppl 1):S81–S88

Tabata M, Kondo M, Haruta Y, Seon BK (1999) Antiangiogenic radioimmunotherapy of human solid tumors in SCID mice using (125)I-labeled anti-endoglin monoclonal antibodies. Int J Cancer 82:737–742

Takahashi N, Haba A, Matsuno F, Seon BK (2001a) Antiangiogenic therapy of established tumors in human skin/severe combined immunodeficiency mouse chimeras by anti-endoglin (CD105) monoclonal antibodies, and synergy between anti-endoglin antibody and cyclophosphamide. Cancer Res 61:7846–7854

Takahashi N, Kawanishi-Tabata R, Haba A et al (2001b) Association of serum endoglin with metastasis in patients with colorectal, breast, and other solid tumors, and suppressive effect of chemotherapy on the serum endoglin. Clin Cancer Res 7:524–532

Tamm I, Wagner M (2006) Antisense therapy in clinical oncology: preclinical and clinical experiences. Mol Biotechnol 33:221–238

Tamm I, Dorken B, Hartmann G (2001) Antisense therapy in oncology: new hope for an old idea? Lancet 358:489–497

Uhl M, Aulwurm S, Wischhusen J et al (2004) SD-208, a novel transforming growth factor β receptor I kinase inhibitor, inhibits growth and invasiveness and enhances immunogenicity of murine and human glioma cells in vitro and in vivo. Cancer Res 64:7954–7961

Ungefroren H, Schniewind B, Groth S et al (2007) Antitumor activity of ALK1 in pancreatic carcinoma cells. Int J Cancer 120:1641–1651

Vallieres L (2009) Trabedersen, a TGFβ2-specific antisense oligonucleotide for the treatment of malignant gliomas and other tumors overexpressing TGFβ2. IDrugs 12:445–453

van Meeteren LA, Thorikay M, Bergqvist S et al (2012) Anti-human activin receptor-like kinase 1 (ALK1) antibody attenuates bone morphogenetic protein 9 (BMP9)-induced ALK1 signaling and interferes with endothelial cell sprouting. J Biol Chem 287:18551–18561

Venkatesha S, Toporsian M, Lam C et al (2006) Soluble endoglin contributes to the pathogenesis of preeclampsia. Nat Med 12:642–649

Vogt J, Traynor R, Sapkota GP (2011) The specificities of small molecule inhibitors of the TGFβ and BMP pathways. Cell Signal 23:1831–1842

Walker EJ, Su H, Shen F et al (2012) Bevacizumab attenuates VEGF-induced angiogenesis and vascular malformations in the adult mouse brain. Stroke 43(7):1925–1930

Wang SE, Xiang B, Zent R et al (2009) Transforming growth factor β induces clustering of HER2 and integrins by activating Src-focal adhesion kinase and receptor association to the cytoskeleton. Cancer Res 69:475–482

Watabe T, Nishihara A, Mishima K et al (2003) TGF-β receptor kinase inhibitor enhances growth and integrity of embryonic stem cell-derived endothelial cells. J Cell Biol 163: 1303–1311
Wiercinska E, Wickert L, Denecke B et al (2006) Id1 is a critical mediator in TGF-β-induced transdifferentiation of rat hepatic stellate cells. Hepatology 43:1032–1041
Wikstrom P, Lissbrant IF, Stattin P et al (2002) Endoglin (CD105) is expressed on immature blood vessels and is a marker for survival in prostate cancer. Prostate 51:268–275
Wu X, Robinson CE, Fong HW et al (1995) Cloning and characterization of the murine activin receptor like kinase-1 (ALK-1) homolog. Biochem Biophys Res Commun 216:78–83
Yang YA, Dukhanina O, Tang B et al (2002) Lifetime exposure to a soluble TGF-β antagonist protects mice against metastasis without adverse side effects. J Clin Invest 109:1607–1615
Yingling JM, Blanchard KL, Sawyer JS (2004) Development of TGF-β signalling inhibitors for cancer therapy. Nat Rev Drug Discov 3:1011–1022
Yu PB, Deng DY, Lai CS et al (2008a) BMP type I receptor inhibition reduces heterotopic ossification. Nat Med 14:1363–1369
Yu PB, Hong CC, Sachidanandan C et al (2008b) Dorsomorphin inhibits BMP signals required for embryogenesis and iron metabolism. Nat Chem Biol 4:33–41

Index

A. Moustakas and K. Miyazawa (eds.), *TGF-β in Human Disease*,
DOI 10.1007/978-4-431-54409-8, © Springer Japan 2013

MIX
Papier aus verantwortungsvollen Quellen
Paper from responsible sources
FSC® C105338

If you have any concerns about our products,
you can contact us on
ProductSafety@springernature.com

In case Publisher is established outside the EU,
the EU authorized representative is:
Springer Nature Customer Service Center GmbH
Europaplatz 3, 69115 Heidelberg, Germany

Printed by Libri Plureos GmbH
in Hamburg, Germany

MIX
Papier aus verantwortungsvollen Quellen
Paper from responsible sources
FSC® C105338

If you have any concerns about our products,
you can contact us on
ProductSafety@springernature.com

In case Publisher is established outside the EU,
the EU authorized representative is:
Springer Nature Customer Service Center GmbH
Europaplatz 3, 69115 Heidelberg, Germany

Printed by Libri Plureos GmbH
in Hamburg, Germany